Soviet Naval Policy

edited by
Michael MccGwire
Ken Booth
John McDonnell

Published for the Centre
for Foreign Policy Studies,
Department of Political Science,
Dalhousie University, Halifax, N.S.

The Praeger Special Studies program—
utilizing the most modern and efficient book
production techniques and a selective
worldwide distribution network—makes
available to the academic, government, and
business communities significant, timely
research in U.S. and international economic, social, and political development.

Soviet Naval Policy
Objectives and Constraints

PRAEGER SPECIAL STUDIES IN INTERNATIONAL POLITICS AND GOVERNMENT

Praeger Publishers   New York   Washington   London

Library of Congress Cataloging in Publication Data

MccGwire, Michael, comp.
  Soviet naval policy.

  (Praeger special studies in international politics and government)
  1. Russia (1923-   U. S. S. R.). Voenno-Morskoi flot—Addresses, essays, lectures. 2. Russia—Military policy—Addresses, essays, lectures.
I. Booth, Ken, joint comp. II. McDonnell, John, joint comp. III. Title.
VA573.M32   359'.03'0947   74-11923
ISBN 0-275-09720-X

PRAEGER PUBLISHERS
111 Fourth Avenue, New York, N.Y. 10003, U.S.A.

Published in the United States of America in 1975
by Praeger Publishers, Inc.

Second printing, 1976

All rights reserved

© 1975 by Praeger Publishers, Inc.

Printed in the United States of America

PREFACE

This book derives from a seminar on Soviet naval developments held at Dalhousie University, Halifax, Nova Scotia in October 1973; this was the second in what is now an annual series, with the third being held in September 1974. The seminar's approach is interdisciplinary and cross-occupational, bringing several perspectives and analytic techniques to bear, with the emphasis always on the Soviet point of view. The purpose is to raise the level of informed analysis and debate on the subject by ensuring (1) that hard data on Soviet naval capabilities and deployments are made readily available, and (2) that analysis of Soviet naval policy is carried out within the full and proper context.

This collection is intended to complement the first volume of the series,* whose contents spanned the main boundaries of the subject and included much factual and technical data that could be used to support further analyses. The present book makes important additions to that data base. It extends the range of background coverage and places particular emphasis on the decision-making processes that produce Soviet policy. Six chapters consider various institutional aspects of foreign, military, and naval policy-making. In addition, there are six detailed case studies of Soviet military involvements in the third world. Four of these concentrate on the United Arab Republic (UAR); each of them provides a different perspective on this important test case of Soviet overseas involvement.

Admiral S. G. Gorshkov's series of articles "Navies in War and Peace" provided the backdrop to this year's discussions. These 11 articles totaling 50,000 words were published in Morskoi sbornik between February 1972 and February 1973 and evoked widespread attention; they had been analyzed in depth by three members of the group. The greater part of these articles was clearly concerned to demonstrate the continuing and increasing importance of navies as instruments of state policy in peace and war. The exact significance of the series' publication is less clear. In particular, was Gorshkov mainly advocating policies and priorities that had not been adopted by the political leadership, or was he mainly announcing newly determined policies and ordering of priorities? This central question acquired added significance after the Politburo changes following the

---

*Soviet Naval Developments: Capability and Context, Michael MccGwire, ed. (New York: Praeger Publishers, 1973).

Central Committee Plenum in April 1973, which included the elevation to the Politburo of the ministers of Defense (Andrei A. Grechko) and Foreign Affairs (Andrei Gromyko). Did the changes mean that some major decision had been reached on the future direction of Soviet defense and foreign policy and on the nature of military doctrine? If some such decision had been taken, speculation was invited on its possible implications for future naval policy.

Depending to some extent on the assessment of the balance between "advocacy" and "announcement," various inferences could be drawn from the Gorshkov series. These concerned matters such as the relative emphasis to be placed on deterrence as opposed to war-fighting capabilities; the navy's deterrent-political role in war; the extent and nature of the opposition to naval aspirations that may derive from competing institutional interests, or from economic, ideological, military-doctrinal, and political-pragmatic factors; arguments about the relative utility of naval power as an instrument of state policy in peacetime; and debate about the range and scope of naval tasks in peace and in war, the size of the fleet required to carry out such tasks, and nonnaval methods of discharging traditionally naval tasks. All these questions were considered by the seminar.

Although the contents of this book do not follow the foregoing agenda exactly, summaries of the discussion have been included where this seemed appropriate. The group did not attempt to arrive at a set of agreed conclusions, but the following points indicate the sense of the meeting on the chief topics of discussion.

Soviet Foreign Policy

Current trends toward lowering tension are likely to persist, because the objective factors underlying Soviet foreign policy are relatively stable. However, this does not rule out a good deal of activism in certain local areas, within the limits of "detente diplomacy."

Soviet Policy in the Third World

In general there has been much exaggeration of the political influence the Soviet Union has achieved in third-world countries. However, in recent years, Soviet attitudes and policy vis-a-vis the third world have become more realistic and pragmatic, so that earlier trends cannot be extrapolated with complete confidence. The Soviet Union has important (but not vital) interests in third world countries, including the denial of Western and Chinese influence and the fostering of Soviet influence. In addition, the economic factor has become of greater significance in some of the relationships. It was stressed that besides self-imposed restraints, Soviet policy in the third world will always be constrained by local regional problems and by the domestic and nationalistic priorities of the individual countries. It was recognized that we needed to develop improved tools for analyzing Soviet "influence" in the third world and that there was a requirement for more detailed case studies.

The Soviet Navy and Soviet Foreign Policy

There was no general agreement on the impact of the Soviet Navy's forward deployment on Soviet foreign policy, and to what extent the one led the other. Whereas it was difficult to identify specific instances where a Soviet naval presence had "caused" particular developments or had produced shifts in Soviet policy, it was felt that Soviet presence had tended to "neutralize" Western influence and had certainly complicated Western planning. The extent of this neutralization in the future will depend upon Western determination and policy, as well as Soviet capabilities and intentions. While Soviet naval actions had generally been restrained and calculating, naval forces have been used more actively on a number of recent occasions. It was too soon to say whether these episodes were aberrations or precedents.

Naval Policy: Continuity and Change

The original reasons underlying the size, shape, deployment, and building programs of the Soviet Navy, up until the middle 1960s, now seem to be relatively clear and widely accepted. There was general agreement on the imperatives that drew the Soviet Navy forward in strategic defense, and it was accepted that there was a good deal of continuity in Soviet naval policy, which had shown impressive consistency in mission structure from the early 1950s through to the present day. The major war-fighting missions had remained the same, although changes in the nature of the threat had induced changes in deployment patterns and ship characteristics.

There was less agreement on current operational concepts. The relative importance of the tasks of strategic retaliation and of countering the West's seaborne nuclear delivery systems was strongly disputed. The significance of withholding Soviet submarine-launched missiles from the initial exchange and the relative importance of protecting their seaborne retaliatory forces were left unresolved. There was disagreement about Soviet aspirations in the field of anti-submarine warfare (ASW), and whether the Soviets still hoped to develop a counter for Polaris. The term "sea denial" was accepted as a useful way of describing the Soviet Navy's concept of operations in certain areas. The term "balanced fleet" tended to be misleading because of its variable usage.

It was agreed that the Soviet Union would use its naval forces to further "state interests" in peacetime, but that the scope of the term could not be precisely defined at this stage. The Soviet Navy had been active in a number of recent episodes in support of foreign policy, and these episodes are significant in themselves. However, it is too early to say whether this activism represents a temporary phase, resulting from the thrust of events rather than the disposition of the Soviet leadership; or whether it represents a distinct departure point in Soviet naval policy, reflecting a major reorientation of naval missions toward priority for the peacetime uses of naval forces. It was noted that the number of combatants on sustained deployment had now leveled off in all areas, although there were still substantial buildups during crises.

It seemed likely that the Soviet Navy worked to a 20-year replacement program, with planned procurement based on a 10-year lead time, and final decisions on numbers and weapons fit being made at about the five-year mark. There did not appear to have been any substantial increase in the surface-warship-building trend. But while we might expect by now to be aware of the end products of decisions taken prior to 1966, it was still too early for the results of decisions made in 1970 to be visible.

The Gorshkov Series

The articles were important and should be taken seriously. It was the majority view that the series advocated certain policies and priorities that had not been adopted by the political leadership, either in whole or in part, but that they also served to clarify existing operational concepts. The series provided many insights into the naval, defense, and foreign policy debate within the Soviet Union. A definitive assessment of the articles would have to wait on greater hindsight and more external evidence.

The Formulation of Policy

It was recognized that our restricted knowledge of the institutional setting and the organizational processes of Soviet decision-making limited our understanding of Soviet policy and of a whole body of questions relating to matters such as strategic priorities, weapons procurement, interservice rivalries, arms races, and arms control. It appeared that an important bureaucratic struggle had been taking place between the navy on the one hand and the strategic rocket forces and the air force on the other. It would appear that the institutional significance of the navy grew in the latter 1960s, but it is not certain whether this rising trend has persisted. In addition to interservice rivalries, there is at present a good deal of introversion among the Soviet armed forces, concerned with the workings of their military system and with manpower problems.

Economic and Technological Constraints

The difficulties of making satisfactory comparisons are considerable, and in some areas (for example, comparative technology) it is difficult if not impossible to be other than impressionistic. It was thought that, in some relevant aspects, Soviet technology is not as good as the best in the West but that improvements are gradually being made. It is likely that in some sectors of the Soviet economy considerable sacrifices have had to be made in order to maintain and develop a modern navy, with all its manifold specialized and expensive requirements. There was no general agreement on whether the present time is a particularly good or a particularly bad moment to be arguing the case for the expansion of naval requirements. Both standpoints are supportable.

Methodology

The interaction of different analytical techniques again proved valuable, as did the exposure of the underdevelopment of the necessary techniques in important areas—for example, criteria for measuring "influence" and methods for assessing economic "burden." Semantic questions figured prominently in discussion. It is important that we understand exactly what the Russians mean when they use particular words or expressions. It is also necessary to avoid translating such words and expressions into Western strategic-theoretical terminology unless we are absolutely certain that __identical__ strategic concepts are being discussed.

Requirements for Research

In few areas could it be said that the amount of information and the analytical tools for handling it were satisfactory. Areas where the relative paucity of research was felt particularly keenly included the following: the politico-military decision-making processes; the military institutional framework and interservice rivalries; the occupancy pattern of higher appointments and the career pattern of senior officers within the Soviet Navy; naval aspects of defense economics and comparative technology; naval arms control and operational limitation; the measurement of overseas influence and the naval contribution; attitudes within the USSR toward overseas military involvement and toward the utility of military power in extending Soviet influence.

Reviewing the last two years, it would seem that the formation of a Soviet naval studies group has helped to provide a focus for work in this field and to identify those areas of analysis needing more detailed research and the type of data required to support such research. We are now beyond the stage of impressionistic surveys of Soviet naval policy, and in planning the third seminar for Soviet naval developments the emphasis has been on detailed case studies and the new analytic techniques. We still need to extend the range and depth of analysis of Soviet naval developments, and, by making this information more generally available, we hope to raise the level of informed debate. We are not concerned to promote any particular interpretation of Soviet naval policy but want to ensure that all properly researched interpretations are given due consideration before political judgments are made.

Michael MccGwire, Ken Booth,
John McDonnell

May 1974

ACKNOWLEDGMENTS

Without the very full support, both administrative and material, of the Maritime Warfare School, Halifax, it would not have been possible to organize the seminar on Soviet naval policy from which this book stems. The school does not, however, bear any responsibility for the contents of this publication; views expressed herein are those of the individual authors.

The cost of the seminar was borne by a grant from the Department of National Defence, Ottawa, and from funds attached to the chair of Maritime and Strategic Studies at Dalhousie University, which emanate from the Department.

CONTENTS

	Page
PREFACE	v
ACKNOWLEDGEMENTS	xi
LIST OF TABLES, FIGURES, AND MAPS	xxii
LIST OF ABBREVIATIONS	xxv

PART I: THE CONTEXT OF SOVIET NAVAL POLICY

Chapter

1   TRENDS IN SOVIET FOREIGN POLICY
Marshall D. Shulman   3

       Factors Favoring Change   5
       The Internal Debate   8
       U.S.-Soviet Relations   10
       Conclusions   15
       Summary of Discussion   16

2   THE SOVIET MILITARY'S INFLUENCE ON FOREIGN POLICY
Malcolm Mackintosh   23

       Military Access to Decision-Making   25
       The Riddle of Military Influence   28
       Case Studies   29
       Some Conclusions   35
       Notes   37

3   THE MILITARY ROLE IN SOVIET DECISION-MAKING
Matthew P. Gallagher   40

       Western Concepts and Soviet Military Policy-Making   41

Chapter		Page
	The Soviet Decision-Making Process	51
	Soviet Military Doctrine	55
	Soviet Definitions of "Military Doctrine" and Related Terms	56
	Notes	57
4	SOVIET DEFENSE POLICIES AND NAVAL INTERESTS John Erickson	59
	The Soviet Debate on "Missions"	60
	The Soviet Navy and the Soviet Military System	62
	The Perception of the "Naval Mission" and Naval Requirements	65
5	THE BUREAUCRATIC POLITICS OF WEAPONS PROCUREMENT Edward L. Warner, III	66
	Soviet Force Development	67
	Planning in the Defense Sector	69
	Soviet Weapons Acquisition	73
	Conclusion	79
	Notes	79
6	THE SOVIET DEFENSE INDUSTRY AS A PRESSURE GROUP John McDonnell	87
	Evolution of the Soviet Defense Industry	88
	The Management of the Defense Industry	91
	Defense Industry "Alumni" as Managers of the Soviet Economy	95
	Defense Industry Membership in Party Bodies	102
	Conclusions	106
	Notes	107
	Annex A: Ministries of the Defense Industry	114
	Annex B: Biographical Sketches of Defense Industry Leaders	116

Chapter		Page
7	ANALYSIS OF SOVIET DEFENSE EXPENDITURES Philip Hanson	123
	Estimates of "Effort"	124
	Estimates of "Burden"	127
	Economic Constraints and Soviet Policy	131
	The Influence of Technology	132
	Conclusions	135
	Notes	135
8	NATIONAL ECONOMIC PRIORITIES AND NAVAL DEMAND John P. Hardt	137
	Current Commitments to New Military Programs	138
	The Increased Naval Share in New Military Programs as Compared with Missiles and Ground Force Demands	140
	How Will the Soviet Leaders Answer Admiral Gorshkov's Demands?—A Speculative Assessment	142
	Notes	143
9	SUMMARY OF DISCUSSION IN PART I Ken Booth	145
	Aspects of the Policy Process	145
	Economic Factors	148

PART II: SOVIET POLICY AND THE THIRD WORLD: CASE STUDIES

10	THE SOVIET-EGYPTIAN INFLUENCE RELATIONSHIP SINCE THE JUNE 1967 WAR Alvin Z. Rubinstein	153
	Methodological Problems	153
	Moscow's Strategic Involvement in Egypt	157
	To the War of Attrition	164
	The Treaty of Friendship and Cooperation	173

Chapter		Page
	The "Lessons" of 1968-69 and 1971	175
	Notes	179
11	SOVIET DECISION-MAKING IN THE MIDDLE EAST, 1969-73 Uri Ra'anan	182
	The Bases of Analysis	183
	Military Intervention and Withdrawal, February 1970-July 1972	191
	Implementing Military Intervention	198
	The Decision to Withdraw	199
	A New Forward Policy	204
	Brezhnev's Plan	205
	Implications for the Future	210
12	THE SOVIET UNION AND SADAT'S EGYPT Robert O. Freedman	211
	Initial Soviet Policy	212
	The Ouster of Ali Sabry and the Abortive Coup in the Sudan	213
	Sadat's Fruitless Journeys to Moscow	215
	The Decision to Expel the Russians	216
	The Effect of the Munich Massacre	219
	Arab Unity and the Oil Weapon	222
	From the Summit to the War	224
	The October 1973 War and Its Aftermath	226
	Conclusions	230
	Notes	231
13	THE SOVIET UNION'S QUEST FOR ACCESS TO NAVAL FACILITIES IN EGYPT PRIOR TO THE JUNE WAR OF 1967 George S. Dragnich	237
	1925-55: Minimal Involvement in the Mediterranean	238
	1955-61: The Shaping of Egypt's Perception of Soviet Naval Power	240
	1961-65: The Growing Need for Naval Facilities in the Mediterranean	248
	1966-67: Increasing Pressure on Egypt	255

Chapter		Page
	1967: The June War and Its Aftermath	266
	Conclusion	267
	Postscript	269
	Notes	270
14	SOVIET POLICY IN THE PERSIAN GULF Oles M. Smolansky	278
	The Persian Gulf in the Middle East Context	279
	Detente and Persian Gulf Oil	281
	Constraints on Soviet Policy	283
	Avoiding Foreign Entanglements	284
	Conclusion and Prognosis	285
15	THE SOVIET NAVAL PRESENCE DURING THE IRAQ-KUWAIT BORDER DISPUTE Anne M. Kelly	287
	Details of the Dispute	287
	Soviet Actions and Possible Objectives	289
	Analysis of Soviet Actions	291
	Implications for Future Soviet Behavior	300
	Notes	301
	Annex: Chronology of Events	303
16	SOVIET POLICY IN THE INDIAN OCEAN Geoffrey Jukes	307
	Soviet Interests in the Area	309
	The Threat from Seaborne Strategic Systems	311
	Deployment into the Indian Ocean	313
	Underlying Purposes	315
	Notes	317
17	THE SOVIET PORT-CLEARING OPERATIONS IN BANGLADESH Charles C. Petersen	319
	Damage to Chittagong and Chalna Ports	321
	Arrangements for Port-clearing Work	324
	The Initial Port-clearing Operation	325
	A New Salvage Agreement	328

Chapter		Page
	The United Nations Has Its Day	329
	The Second Phase of Port-clearing Operations	330
	Political Objects and Effects of the Operation	333
	Conclusion	336
	Notes	337
18	THE SOVIET POSITION AT THE THIRD U.N. LAW OF THE SEA CONFERENCE Robert L. Friedheim and Mary E. Jehn	341
	Background	342
	Present Soviet Position in LOS Bargaining	345
	The Soviet Navy's Position	351
	The Soviet Dilemma	356
	Notes	360
19	SUMMARY OF DISCUSSION IN PART II Ken Booth	363
	Soviet Policy and Influence in the Third World	364
	Navies and Foreign Policy	367
	Examples of Soviet Naval Involvement	368
	The Wider Implications	370
	Naval Power as an Instrument of Peacetime Policy	371

PART III: SOME ANALYTICAL MATERIAL

20	SOVIET NAVAL OPERATIONS: 10 YEARS OF CHANGE Robert G. Weinland	375
	The Expanding Scale of Operations	376
	The Political Applications of Naval Power	380
	Notes	385

Chapter		Page
21	FOREIGN-PORT VISITS BY SOVIET NAVAL UNITS Michael MccGwire	387
22	SOVIET NAVAL STRENGTH AND DEPLOYMENT Robert Berman	419
23	CURRENT SOVIET WARSHIP CONSTRUCTION AND NAVAL WEAPONS DEVELOPMENT Michael MccGwire	424
	Current Submarine Construction	425
	Submarine Yard Capacity	429
	Submarine Weapon Systems	433
	Review of Submarine Programs	435
	Current Surface Ship Construction	437
	Surface Yard Capacity	441
	Surface Weapon Systems	441
	Review of Surface Programs	442
	Overview	444
	Notes	445
24	COMPARATIVE CAPABILITIES OF SOVIET AND WESTERN WEAPON SYSTEMS Nigel D. Brodeur	452
	Analytical Problems	452
	Submarine Design and Construction	454
	Anti-Air Warfare	455
	Surface-to-Surface Warfare	458
	Air-to-Surface Warfare	460
	Antisubmarine Warfare	460
	Electronic Warfare	462
	Mine Warfare	464
	Trend Significance: Soviet Ship Weapons	464
	Conclusion	468
	Notes	468

Chapter		Page

PART IV: THE SOVIET UNDERSTANDING OF DETERRENCE AND DEFENSE

25 THE SEMANTICS OF DETERRENCE AND DEFENSE
Peter H. Vigor 471

26 THE MILITARY APPROACH TO DETERRENCE AND DEFENSE
Geoffrey Jukes 479

    The Context of Military Decision-Making 479
    Soviet Military Concepts 481
    Deterrence and Defense 484
    Notes 485

27 SOVIET STRATEGIC WEAPONS POLICY, 1955-70
Michael MccGwire 486

    Two Basic Assumptions 487
    Implications: Objectives in the Event of War 489
    Operational Concepts and Priorities 490
    Conclusions 499
    Notes 501

PART V: ASPECTS OF SOVIET NAVAL POLICY

28 THE EVOLUTION OF SOVIET NAVAL POLICY: 1960-74
Michael MccGwire 505

    Background 505
    1960-1964 507
    The 1963-64 Decisions 514
    1964-1968 518
    1968-1971 522
    1971-1974 532
    Notes 541

Chapter		Page
29	ANALYSIS OF ADMIRAL GORSHKOV'S "NAVIES IN WAR AND PEACE" Robert Weinland	547
	Facts of Publication	548
	The Statement	549
	The Question of Context	557
	Summary Judgments	569
	Notes	572
30	THE SOVIET NAVAL GENERAL PURPOSE FORCES: ROLES AND MISSIONS IN WARTIME Bradford Dismukes	573
	U.S. Behavior	574
	Soviet Views	576
	The Range of Possible Soviet Responses	577
	Role of the Air-Capable Ship	579
	Implications of Current Soviet SSBN Operating Policy	581
	Conclusions	582
	Notes	583
31	THE COUNTER-POLARIS TASK Harlan Ullman	585
	Official Soviet Pronouncements	586
	The Feasibility of Counterdeterrence	592
	An Operational Concept Based on Damage Limitation	593
	Conclusions	596
	Notes	597
32	SOVIET UNDERSTANDING OF "COMMAND OF THE SEA" Peter H. Vigor	601
	The Meaning Attached to the Term	602
	Change in Attitude toward the Concept	606
	The Postwar Denigration of the "Mahan" Concept	610
	"Command of the Sea" in a New Context	613
	Gorshkov and "Command of the Sea"	617
	Notes	620

Chapter		Page
33	COMMAND OF THE SEA IN SOVIET NAVAL STRATEGY	
	Michael MccGwire	623
	Differing Usage	623
	The Imperial Russian Navy	625
	The Interwar Years	626
	The Postwar Period	629
	The Relevance to Contemporary Strategy	631
	Summary and Conclusions	633
	Notes	635
34	THE TACTICAL USES OF NAVAL ARMS CONTROL	
	Franklyn Griffiths	637
	Arms Control and an Aid to Arms Racing	637
	Gorshkov as a Dove	641
	Taking the "Diplomatic Route"	646
	Deployment Limitations	648
	Confidence-building Measures	650
	Force Limitations	651
	Maritime Peacekeeping Forces	653
	The "Diplomatic Route" Reassessed	655
	The Duality of Arms Control	656
	Notes	658
CONTRIBUTORS TO THIS VOLUME		661

LIST OF TABLES, FIGURES, AND MAPS

Table		Page
5.1	Defense Industrial Ministries and Their Military Products	72
7.1	Expenditure Trends for Principal Uses of GNP	130
10.1	Soviet-Egyptian Influence Relationship	158
17.1	Vessels Assigned to Soviets for Salvage in Chittagong, March 1972	326
18.1	Three Variables	353
18.2	Comparison of Gorshkov's Remarks with USSR and U.S. National Scores and Estimates	355
18.3	Comparison of Gorshkov and USSR Scores with Mean Bloc Preferences	357
18.4	Comparison of Developed (Major Ocean Users) and Developing (Group of 77) by Mean Bloc Preferences	358
20.1	Examples of Politically Oriented Soviet Naval Operations, 1967-72	382
22.1	Order of Battle: Submarines	421
22.2	Order of Battle: Surface Units	422
22.3	Order of Battle: Naval Aviation	423
27.1	Pattern of ICBM Deliveries	494
29.1	Publication of "Navies in War and Peace" in Morskoi sbornik	550
29.2	Distribution of Attention in "Navies in War and Peace"	554
29.3	Dates Morskoi sbornik "Signed to Typesetting" and "Signed to Press"	559

Table		Page
29.4	Context of Publication of "Navies in War and Peace"	566

Figure		
6.1	Evolution of the Soviet Defense Industry	89
6.2	Economic Planning and Science Management Agencies	94
6.3	Membership of the Central Committee	104
20.1	Annual Ship Days by Region	377
20.2	Regional Distribution of Worldwide Ship Days	377
20.3	Ship Days in the Atlantic	378
20.4	Ship Days in the Mediterranean	378
20.5	Ship Days in the Pacific	379
20.6	Ship Days in the Indian Ocean	379
24.1	Submarine Maximum Speeds and Power/Displacement Ratios	456
24.2	Surface-to-Air Missiles: Range and Weight	457
24.3	Engagement Capability: Surface to Air	458
24.4	Engagement Capability: Surface to Surface	459
24.5	Engagement Capability: Air to Surface	461
24.6	ASW Capability	463
29.1	Date Morskoi sbornik Signed to Typesetting	560
29.2	Date Morskoi sbornik Signed to Press	561

Map		Page
15.1	Iraq-Kuwait Border Dispute	288
17.1	Bangladesh	320
17.2	Wrecks in Chittagong Port	322
17.3	Minefields and Wrecks in Approaches to Chittagong	323

LIST OF ABBREVIATIONS

AA	Anti-Aircraft
AAA	Automatic Anti-Aircraft
AAW	Anti-Aircraft Warfare
ABM	Anti-Ballistic Missile
ACDA	Arms Control and Disarmament Agency (Washington)
ASU	Arab Socialist Union (Egypt)
ASW	Anti-Submarine Warfare
CBMs	Confidence-building measures
C-in-C	Commander in Chief
COMECON	Council for Mutual Economic Assistance
CPSU	Communist Party of the Soviet Union
CSCE	Conference on Security and Cooperation in Europe
CV	Aircraft carrier
CVA	Attack aircraft carrier
DOSAAF	Voluntary Society for Assisting the Army, Navy, and Air Force
FBS	Forward Based Systems
FOBS	Fractional Orbital Bombardment System
GRT	Gross Registered Tons
ICBM	Intercontinental Ballistic Missile
IISS	International Institute for Strategic Studies (London)
IRBM	Intermediate Range Ballistic Missile
LOS	Law of the Sea
MBFR	Mutual, Balanced Force Reductions, Conference on
MIRV	Multiple, independently-targeted reentry vehicle
MRBM	Medium Range Ballistic Missile
MRV	Multiple re-entry vehicle
MT	Megatons
n.m.	nautical miles
PVO (Strany)	National Air Defense Forces
p.s.i.	pounds (of pressure) per square inch
RV	Re-entry vehicle
SAF	Soviet Air Forces
SALT	Strategic Arms Limitation Talks
SAM	Surface-to-Air Missile
SA-N-1,3,4	Surface-to-Air Missile, Naval, 1, 3, 4
SIPRI	Stockholm International Peace Research Institute
SLBM	Submarine Launched Ballistic Missile
SSB	Ballistic Missile Submarine (diesel)
SSBN	Fleet Ballistic Missile Submarine (nuclear-powered)
SSGN	Cruise Missile Submarine (nuclear-powered)
SSM	Surface-to-surface Missile
SSN	Attack Submarine (nuclear-powered)

SS -7,9,11,etc.	Surface-to-surface Missile, land-based, 7, 9, 11, etc.
SS-N-3,4,5,etc.	Surface-to-surface Missile, naval, 3, 4, 5, etc.
UAA	United Arab Amirates
UAR	United Arab Republic
UNROD	United Nations Relief Operation in Dacca (Bangladesh)
VDS	Variable Depth Sonar
VTOL	Vertical take-off and landing
V/STOL	Vertical/Short take-off and landing

See also the list of Naval Type Designators, p. 420

PART

I

THE CONTEXT OF
SOVIET NAVAL POLICY

CHAPTER

1

**TRENDS IN SOVIET
FOREIGN POLICY**
Marshall D. Shulman

Are we witnessing a historic shift in Soviet foreign policy? According to Leonid Brezhnev, the answer is an emphatic yes. At Bonn in May 1973, the general secretary told the people of West Germany that the 24th Soviet Party Congress in 1971 set, and the April 1973 plenum of the Party's Central Committee reaffirmed, the foreign policy goal of implementing a "radical turn toward detente and peace on the European continent." To achieve a better life for the Soviet people, he said, the Soviet leadership had turned resolutely away from isolation and autarky and was bending its energies toward peaceful construction at home and comprehensive cooperation with the outside world.

Then, on June 22, 1973, in a talk to U.S. businessmen in Washington, Brezhnev went further. Looking back over 42 years of party and government experience, he said, "We have certainly been prisoners of those old tendencies, those old trends, and to this day we have not been able fully to break those fetters. . . ." The cold war, he said, "put the brake on the development of human relations, of normal human relations between nations, and it slowed down the progress and advance of economic and scientific times. And I ask you gentlemen, as I ask myself, was that a good period? Did it serve the interests of the peoples? And my answer to that is no, no, no and again no." Summing up, he said, "It has been and is my very firm belief that human reason and common sense and the human intellect will always be victorious over obscurantism."

This chapter formed the substance of Marshall Shulman's opening presentation to the Soviet Naval Studies Group Seminar. It is adapted by permission from Foreign Affairs, October 1973, where it appeared under the title "Towards a Western Philosophy of Coexistence." Copyright 1973 by Council on Foreign Relations, Inc.

And, again, in Washington: "I wish especially to emphasize that we are convinced that on the basis of growing, mutual confidence, we can steadily move ahead. We want the future development of our relations to become a maximally stable process, and what is more, an irreversible one."

That this is the ascendant sentiment of the Soviet leadership was underscored by the award, on May Day 1973, of the Lenin Peace Prize to the general secretary and an orchestrated wave of praise of Brezhnev in the Soviet press for his "personal contribution" to the Party's "peace program."

In the rest of the world, which has seen other "peace campaigns" come and go, Brezhnev's affirmations have been welcomed with a certain reserve. Do they represent more than a tactical turn toward a low-tension policy to gain economic help and political advances? Will the new policy last?

Undoubtedly, the present course offers tactical advantages to the Soviet Union, but there is a reason to believe that something more fundamental may be involved, that the Soviet leadership is responding to "objective factors" in the situation that require a long-term commitment to a policy of low tension abroad and consolidation in the Soviet sphere. It is essential to view the present Soviet policy in the perspective of 20 years of halting, inconsistent, incomplete, resisted efforts to shake off the Stalinist legacy in Soviet foreign policy. In a significant sense, Brezhnev's foreign policy represents the culmination of a process that Nikita Khrushchev began but was unable to carry through.

From the Geneva summit of 1955 and the landmark 20th Party Congress of 1956, Khrushchev sought to break away from the Leninist doctrine of the "fatal inevitability of war" and the Stalinist spirit of isolation and unmitigated hostility, and to establish the basis for "businesslike" relations with the West. A combination of factors prevented the consistent realization of his purpose: his own flamboyant and impulsive temperament, the strength of the political opposition, which he inflamed with a series of Party reorganizations, and the fatal effects of a series of misfortunes—the U-2 affair, the failures in agriculture, the Cuban missile episode, and the open conflict with the Chinese communists. By his injudicious efforts to exploit the first Soviet Sputnik and intercontinental missile as symbols of a "shift in the balance of power," he galvanized the U.S. missile program and further deepened the Soviet strategic inferiority. By his polemical rhetoric about "wars of national liberation," he encouraged U.S. preparations for "counterinsurgency" and the apprehensions that contributed to the U.S. involvement in Vietnam. Nevertheless, it was Khrushchev who dared to start the process of de-Stalinization, who faced the implications of the nuclear age, and who foresaw the advantageous possibilities of a long-term political strategy of "peaceful coexistence."

For its first five years, from 1964 to 1969, the Brezhnev-Kosygin group was occupied with the consolidation of a consensual leadership at home; the effects of the Vietnam war and a perceived U.S. propensity for intervention; and the accelerated effort to build strategic forces, a large modern navy, and modernized and mobile ground forces. During the 16 months between the first U.S. proposal of strategic arms limitation talks (SALT) and the first Soviet response, a debate raged over the desirability and the possibility of an agreement with the Americans to stabilize the strategic military competition. Then came Czechoslovakia, and another year went by, while new weapons were introduced into the strategic competition.

Three events in 1969 helped open possibilities for a further development in Soviet policy: in the Federal Republic, the election of Willy Brandt as chancellor, whose Ostpolitik overture offered the possibility for clearing away the obstacle of the German issue; in the United States, Richard Nixon's becoming president, with a declaration that the "era of negotiation" was a possible option; and in November, the start of the SALT process. Perhaps it was the working of the dialectic, but the following year found Soviet relations with the United States in a downward spiral as a result of events in the Middle East and the Caribbean, while SALT became bogged down in the issue of whether offensive or defensive weapons were to be limited first.

FACTORS FAVORING CHANGE

The decisive turn in Soviet policy and in U.S.-Soviet relations came in the early months of 1971. It was then that Brezhnev took personal charge of relations with the United States and the Federal Republic of Germany and of the Soviet position in the SALT negotiations. A channel of confidential communications was opened between Brezhnev and Nixon, which was to lead to the May 1971 agreement that broke the impasse at SALT. Vietnam was, in its own dialectical way, beginning to wind down. By February, internal debate in the Soviet Union on the policy to be promulgated at the 24th Party Congress in March and on the Ninth Five-Year Plan was brought to an abrupt close by the decisive commitment of Brezhnev's personal prestige to the line of "normalization" of relations with the United States. The move in this direction, which was to culminate in the Moscow summit of May 1972, was reinforced by the change in Chinese policy toward a more flexible diplomacy and the opening of contacts with the United States, which made improved Soviet relations with the United States both possible and necessary.

Among the factors responsible for this sequence of evolution in Soviet policy, the following six appear to have been of major importance:

The condition of the Soviet economy is clearly the primary determinant of present Soviet foreign policy. The five-year plan that began in 1971 projected widespread modernization of technology, improvements in productivity, and large increases in consumer goods, but the performance of the Soviet economy has fallen far short of expectations. Poor harvests have created substantial shortages of both food and feed grains, compounding the effects of low agricultural and industrial productivity and a shortage of industrial manpower. Rather than face the politically painful choice of instituting substantial economic reforms, the Soviet leadership has opted for a massive effort to overcome its shortcomings by increasing the flow of trade, advanced technology, and capital from abroad. To overcome its shortage of hard currency, the Soviet Union seeks help in developing its manufactures for Western markets and invites Western capital and technology to help exploit Soviet natural resources, such as its large Siberian reserves of natural gas, to be paid for out of the export of these resources. In his meetings with West German and U.S. businessmen, Brezhnev has projected opportunities for vast joint production ventures over periods of 20 to 30 years. The realization of these expectations manifestly requires an international climate of reduced tension.

The achievement of strategic parity with the United States has made it possible for the Soviet leadership to consider a stabilization of the strategic competition on the basis of the principle of "equal security," which it understands to mean the end of the U.S. policy of negotiating from "positions of strength." The Soviet leadership has expressed its awareness that the stark alternative to this stabilization would be a further upward spiral into increasingly complex and costly weapons systems and that this would further impede the development of Soviet industrial technology.

Soviet perceptions of the United States encourage it to believe that the president's proffered "era of negotiation" represents a serious and durable option because, according to Soviet analysts, it is a realistic and necessary response to such "objective factors" as the rise of economic and social problems in the United States and the decline of U.S. power and influence in the world. These in turn create opportunities for relative increases in Soviet political influence in a climate of reduced tension.

Soviet perceptions of Europe as an emerging economic power-center present both a potential problem and an opportunity. The Soviet leadership has responded with a determined effort to encourage Europe to develop in the direction of a neutral "independence" rather than toward a closer Atlantic association with the United States. It anticipates that in a climate of reduced tension, symbolized by the European security conference, Europe will not develop its military capabilities and will diminish its support for NATO. On the economic side, the Soviet Union no longer mounts a rearguard action against West European integration. Having accepted the reality of the European Economic Community, it now bends its efforts to keep open and further develop trading relations between it and Comecon, the East European economic organization. Similarly, Soviet perceptions of the growing economic strength of Japan led it to anticipate and encourage a competitive struggle between the "triangle" of Western industrial powers: the United States, Western Europe, and Japan. This, too, is more likely to flourish in a climate of "peaceful coexistence."

Soviet apprehensions of China are difficult to weigh as a factor in Soviet foreign policy, but it is clear that this is a matter of visceral intensity to the Russians and that it has both immediate and long-term dimensions. One aspect of the change in Chinese policy toward a policy of enlarged contacts with the West was that it relieved the Soviet Union of the inhibiting charge by the Chinese of "collusion with imperialism" against the policy of "peaceful coexistence." Moreover, the fear of a U.S. Chinese alliance, or of U.S. aid to China, has increased the Soviet incentive to accelerate the "normalization" of its relations with the United States. On at least three occasions, beginning in 1970, the Soviet Union has sought to enlist the United States in an agreement to take joint action with the Soviet Union in the event of "provocative action" by a third nuclear power—presumably China—but the proposal was converted at U.S. insistence into Article IV of the Agreement on the Prevention of Nuclear War, signed in Washington on June 22, 1973.

This article commits the two countries to enter into urgent consultations in the event of a risk of nuclear war between them, or involving other countries. According to Henry Kissinger, it was felt that this article, and the commitment in Article II of the agreement to refrain from the threat or use of force against each other or against other countries, would instead serve to reduce the danger of war between Russia and China. Meanwhile, Moscow is concerned with Chinese diplomatic efforts in Western Europe warning against the dangers of a detente with the Soviet Union, and even more with Chinese efforts to stimulate a greater degree of independence on the part of the countries of Eastern Europe. The 4,500-mile border between the Soviet Union and China is a further source of conflict that four years of negotiations have been unable to resolve, and the Soviets maintain a massive army on this front. It seems plausible that Soviet interest in quiescent relations in the West is strengthened by the necessity of avoiding a two-front engagement in the event of active hostilities with China. In July, Brezhnev's declaration to a North Vietnamese delegation in Moscow of Soviet interest in "the establishment of equal and good-neighborly cooperation among all Asian states without exception" has been regarded as an invitation to China to work toward a modus vivendi, particularly in a post-Mao situation.

The Soviet desire to consolidate its position in Eastern Europe may be dealt with more briefly, but it is by no means a negligible factor impelling the Soviet leadership to a policy of "peaceful coexistence." The persistence of nationalism and the social and political effects of advancing industrialization combine to make this area one of unrest and potential disturbances, and the Soviet problem of control is likely to be made more difficult by the increasing contacts of the West with the states of Eastern Europe. The Soviet Union seeks assurance that there will be no exacerbation of these difficulties from the West and no interference in the event of trouble. It clearly would like to avoid the embarrassment of another Czechoslovakia, although there can be no doubt that the Soviet leadership is determined to maintain the position to which it feels it is entitled in Eastern Europe as a result of World War II and that it believes it requires for reasons of security and as a symbol of its historical and ideological advance. Although the climate of detente creates complications for the Soviet Union in Eastern Europe, as at home, it is also a necessary condition, the Soviets believe, for Western acceptance of the status quo in what they regard as the Soviet sphere. The progress that has been made toward the Western acceptance of the German Democratic Republic as a separate state is regarded as an encouraging mark of the success of this policy.

THE INTERNAL DEBATE

These six factors, however, do not tell the whole story, for foreign policy in the Soviet Union, as elsewhere, is not purely an exercise in rational choices but also involves the interplay of domestic politics. As Brezhnev has indicated, his movement toward the fuller implementation of a policy of "peaceful coexistence" has not been without opposition, and occasional rumbles of dire forebodings may still be observed as reminders that some interests in the Soviet Union are watching for signs that their skepticism is justified.

As might be expected, some of the skepticism is to be found among the professional military services, which, like their opposite numbers in the United States, identify their claims upon the national budget with national security, with mistrust of the SALT process and the assumed deviousness of their adversary. The main source of opposition, however, comes from the orthodox wing of the Party and its large ideological apparatus, and from the even larger apparatus of the political police. For them, "peaceful coexistence" means trouble—a weakening of the ideological elan that is their stock in trade, an opening of the country to influences that they can only regard as "subversive," increased trouble with intellectuals and nationality groups, and an erosion of the image of the "imperialist threat" that legitimizes their power and on which their careers depend.

The burden of their argument, as it is illuminated by an occasional tracer shot fired from the military newspaper Red Star, or Kommunist, the Party's theoretical organ, or even Pravda, is that the abandonment of autarky opens the way to a fatal dependence upon the capitalist countries, that the bid for foreign trade and investment is unlikely to be productive, that the operational effects of a detente policy will weaken the Soviet system at home and in Eastern Europe, and that behind the facade of SALT, the U.S. "imperialists" are improving their lead in new weapons technology. Some remain unenthusiastic about the reconciliation with West Germany, against which residual mistrust is still strong and whose Social-Democratic leadership represents a traditional ideological enemy.

The debate is joined by spokesmen of the "peaceful coexistence" policy from different lines of defense. Some, like Georgy Arbatov, the head of the Institute on the U.S.A., writing in Kommunist in February 1973, seek to persuade the hardliners that under present circumstances, "peaceful coexistence" represents the most effective form of struggle against U.S. imperialism. Others, like Dmitry Tomashevsky, of the Institute of World Economics and International Relations, in Red Star in July of the same year argue forcefully and openly the need for Western capital and technology as the paramount considerations of the moment. An unusually broad perspective was represented in an article in Izvestia in February 1973 entitled, "The Logic of Coexistence," by Vladimir Osipov, an observer on the staff of the newspaper. Osipov wrote of "a whole new series of new factors in the life of the international community of states which now speak for all-around cooperation" and concluded that "the global nature of the interdependence of states makes anachronistic foreign policy concepts of former centuries based on the opposition of some countries to others and the knocking together of military alliances."

The net result of these conflicting domestic pressures has been that Brezhnev has won a free hand to implement his policy of "peaceful coexistence" abroad, while the apparatus of orthodoxy and control has been given a free hand to tighten the lines of ideological vigilance at home, and to prosecute the "ideological struggle" between capitalism and Soviet socialism with renewed vigor. Perhaps this too represents the dialectic at work.

At the plenum of the Central Committee in April 1973, Brezhnev was strengthened by the removal from the Politburo of Pyotr Y. Shelest, an apparent hard-liner, but at the same time the prime spokesmen for military and secret police interests respectively, Marshal Andrei A. Grechko and Yuri V. Andropov, were added.

Although some skeptics in the West believe that Brezhnev speaks publicly of his domestic opposition to encourage Western responsiveness, it seems probable that he does feel the need of some early and tangible signs that the policy with which he has identified himself is successful. Hence, the Soviet impatience for the symbolism of an East-West summit meeting before the end of 1973 to cap the proceedings of the Conference on Security and Cooperation in Europe and the urgency of his presentations to West German and U.S. businessmen. The ratification of the Moscow-Bonn Treaty and the commitment of the United States to "peaceful coexistence" in the statement of Basic Principles have been widely hailed in the Soviet press as early evidence of Brezhnev's success.

U. S. -SOVIET RELATIONS

Soviet-American relations are clearly of central importance to both countries. The barometer of tension between them has risen and fallen many times since World War II. To clarify our outlooks for the future we must attempt to characterize the present stage of their relations.

For those who live by words or phrases that sum up the entire situation at a glance, there is no simple substitute for the term "cold war." That term was once defined by the late George Lichtheim as "competitive attempts to alter the balance of power (between the Soviet Union and the United States) without overt resort to force." By this definition, the term still has a certain validity, although it does not convey the elements of collaborative action that have lately become evident; moreover, the term has acquired such emotional baggage, such connotations of absolute and intractable hostility, that it deserves to be retired. The ambiguities of the word "detente," which has come into wide usage, have led to much confusion. In its simplest meaning, "detente" suggests a relaxation of tension, but some have taken this to mean a "rapprochement," while others see it as signifying only a subjective easing in the symptoms of tension without any real change in its causes; they sometimes use the term "true detente" to distinguish a more fundamental moderation in the adversary relationship.

The Soviet preference is for the term "peaceful coexistence," which they have generally defined as a form of struggle between states with different social systems without resort to war, but specifically emphasizing the continuing ideological conflict. In earlier periods, the term suggested a temporary and tactical turn of events, but in recent Soviet usage "peaceful coexistence" has come to imply a long-term political strategy. The acceptance by the United States of the determination that in a nuclear age there is no alternative to "peaceful coexistence"—in the statement of Basic Principles of Relations between the two countries signed at the Moscow summit in 1972—is regarded by the USSR as the fundamental contractual basis for the "normalization" of the relationship. In the context of this statement of Basic Principles, the term implies a mixture of competition, restraint, and cooperation—which may be as good a working definition as any.

Leaving aside questions of nomenclature, the important point about the nature of the association is that it has become a multilevel relationship, and the movements on the various planes on which the two nations now interact are not always in the same direction. It is therefore necessary to bring to bear a more differentiated analysis of the relationship, in order to distinguish our interests in its various aspects.

Briefly, we can distinguish the following seven planes in the relationship between the Soviet Union and the United States:

The Plane of Strategic-Military Competition. Clearly this deserves to be considered first, for both sides have come to a sober recognition that their most urgent requirement is to avoid a general nuclear war. The strategic arms limitation talks have begun an important educational process, in which the Soviet Union and the United States are moving toward a more enlightened understanding of their real security interests, of the limited political advantages of their strategic arsenals, of the increased dangers and high costs of an unrestrained strategic-military competition, and of the desirability and complexity of finding an equilibrium at moderate levels. Despite SALT, however, the strategic-military competition is not yet stabilized, for both countries continue to raise the quantitative or qualitative levels of their nuclear arsenals.

The Plane of Conventional Military Competition. During the past decade, both countries have greatly increased their capabilities for conventional war and for reaching distant conflicts with modernized forces. Although each shows signs of moving toward restraint in avoiding direct involvement with the other, this remains a potential source of danger in the coming decade, for there has not yet evolved a codification of the rules of the game for the establishment of bases and the use of conventional forces in areas of strategic importance and political instability. What is more imminently dangerous is the large, competitive, and unregulated traffic in arms to the developing countries, which is likely to exacerbate local conflicts and increase the risk of involvement of the great powers.

The Plane of Political Competition. In the present fluid environment, the two great powers are engaged in the competitive politics of maneuver for relative political influence in Europe, the Middle East, Asia, Africa, and Latin America. The easing of the German problem, which had appeared to be the most intractable and decisive territorial issue between the United States and the Soviet Union until about 1968, has been a key factor in opening the way to an improvement in relations generally and also to a period of flexible maneuvering for influence in Western Europe. The Soviet Union is not a status quo power, except in Eastern Europe, and it is in a historical phase of development in which it is seeking a global presence and influence commensurate with its status as a great power. It is encouraged in this effort by its perception of the United States as having passed the zenith of its influence as a world power. Urgent aspects of the political competition from the Soviet point of view are its effort to limit the developing U.S. relationship with the People's Republic of China and to contain the widening diplomatic activities of China on the world stage, particularly in East and West Europe. But also to be noted on the political plane of the relationship are some elements of cooperation. In the Middle East, which both sides have recognized as an area of imminent danger, the political competition is accompanied by consultation and a substantial degree of restraint to reduce the danger of their direct involvement with each other. There have also been consultations and tacit cooperation in regard to Southeast Asia and Berlin, in which the Soviet Union balanced relations with its allies against larger considerations.

The Plane of Economic Competition and Cooperation. The competitive side of economic relations concerns the use of trade and economic assistance as a source of political influence, particularly in areas rich in energy resources. In Europe and Japan, where the United States is involved in trade, monetary, and investment problems, the Soviet Union is more than an interested spectator. The cooperative side of the economic relationship is reflected in the massive Soviet effort to expand its imports of grain, technology, and consumer goods, and to develop Western markets for Soviet goods to pay for these imports in the future. U.S.-Soviet trade increased from a little over $200 million in 1971 to $642 million in 1972; for 1973, trade was running at an annual rate of $1.4 billion, of which almost $800 million was in agricultural products. Currently, Soviet imports from the United States exceed its exports by more than five times. Of greater significance is the determined Soviet effort to seek long-term, large-scale Western investment in the development of Soviet natural resources in Siberia and other areas.

The Plane of Ideological Conflict. Although Soviet policy is characterized by increasing pragmatism, the Soviet leadership insists upon the continuation and the intensification of the ideological struggle, at home and abroad, against an enemy identified as "U.S. imperialism." This insistence clearly has its roots in organizational politics within the Soviet system, but it presents operational problems in foreign policy, for the continued reliance of the Soviet Union upon an external ideological adversary, as a device necessary to its system of political control, sets limits in practice on the realization of its policy of "peaceful coexistence." In the United States, once virulent expressions of anticommunist ideology have been defused by the fact that a conservative U.S. president, formerly of that persuasion, now serves as the instrument of conciliation. The pragmatic American temper is inclined to allow this plane of the relationship to be expressed in terms of the relative performance of the two systems, without benefit of an accompanying verbal barrage.

The Plane of Cultural Relations. In a period in which the technology of transport and communications has advanced rapidly, international life has been inescapably characterized by increasing interpenetration of each other's societies. This presents serious operational difficulties for the Soviet system of political control, at home and in Eastern Europe. Moreover, the widening of human contacts is understood in the West as a necessary ingredient of "peaceful coexistence," as a solvent of hostile stereotypes and a means of moderating residual adversary sentiments. This problem was dramatically illustrated at the Helsinki meeting of the Conference on Security and Cooperation in Europe, where the Western commitment to freedom of information and travel was countered by Soviet efforts to contain cultural exchanges in controllable channels. Nowhere are the asymmetries of the Soviet and Western systems more in evidence than in the inequalities to be observed in the implementation of cultural relations, less in the performing arts than in the exchange of scholars, students, and journalists.

The Plane of Functional Cooperation. In the course of the 1972 and 1973 summit meetings, the Soviet Union and the United States signed more than 10 bilateral agreements covering such areas of functional cooperation as environmental protection, medical science and public health, outer space, science and technology, agriculture, oceanography, transportation, commerce, and the peaceful uses of atomic energy. Many of these provide for joint commissions to implement the agreements. Although these agreements are of limited scope and are in fields of peripheral significance, they perform a symbolic function as a token that the two political leaderships recognize some degree of commonality of interests, and they may be of increasing practical importance as awareness grows of the urgency of environmental problems. Taken together with the agreements related to security, commerce, taxation, maritime affairs, and cultural relations, these forms of cooperation constitute the "web of interdependency" the two countries are consciously weaving.

Several general observations are needed to make this contrapuntal analysis more complete. Although the level of "atmospherics" is properly suspect as fickle and subject to manipulation, it is worth recording that the tone of the relationship has been businesslike, frank in its acknowledgment of differences, but free of the emotional inflammation of those differences that marked earlier periods.

It is also important to remind ourselves that the background against which this relationship has been developing is one of rapid transformation in international politics. Partly as a consequence of the reduction in tension between the Soviet Union and the United States, international politics is marked less by intense polarization than by fluidity and a blurring of alignments. Nonmilitary forms of power, particularly economic and technological, have become increasingly important as sources of political influence. The return of Japan and Western Europe as significant factors in world politics and the emergence of China from her diplomatic isolation have transformed the play of international politics. Against this background, it is clear that the Soviet-American relationship is less the dominant axis of international politics than heretofore and, further, that the major transforming forces of the world are less subject to the control of the two superpowers than each had taken for granted in an earlier period. The widening gap between the industrialized and the developing nations is among the most ominous of the trends pointing to the possibility of anarchic and violent ruptures in the international system.

Finally, we have become more conscious of how deeply the internal politics of each country is involved in the relations between the Soviet Union and the United States. At one level, we watch the fascinating drama of the summits between a general secretary of the Communist Party and a president who have much in common—both conservative, pragmatic realists, former hard-liners. Behind the president is a distracted society, and a shifting balance in which entrenched promilitary pressures contend with a growing impulse toward antimilitarism and a reduction in America's involvements abroad, while the stanchest champions of "peaceful coexistence" are to be found among the private interests of the business community. Behind the general secretary is a society of paradoxes: militarily strong but economically weak, tightly controlled but nervously insecure, in which the support for "peaceful coexistence" from the champions of economic modernization is ranged against military interests and the orthodox party apparatus whose vested interest in an "imperialist enemy" is combined with a fear of the effect of modernization upon the system. Clearly the future course of events depends only in part upon the chieftains, however committed they may be; it is to the inner politics and the underlying forces operating in the two societies that we must look in order to judge the prospects for continuity of the present stage of their relationship.

Whether it will take years or decades for the sense of living on the same small planet to loom larger in the consciousness of men than the rivalry of nations, no one of course can say. Change sometimes moves like a glacier, sometimes like an avalanche. We have negotiated the passage from the simplified enmities of the past to that patchwork-quilt mixture of striving with and against each other that has no simple designation. To move now from the ambiguities of coexistence to a more constructive and less dangerous state will take patience and faith in our sense of direction in the world, while we sustain an effectively functioning society at home.

CONCLUSIONS

In the light of the foregoing discussion, what can we conclude about the prospects for continuity in Soviet foreign policy? It is surely conceivable that if Soviet expectations of a substantial expansion of trade and foreign investment are unrealized, if the arms competition mounts, if another Czechoslovakia should occur in Eastern Europe, or if a conflict in the Middle East or elsewhere should threaten a Soviet-American confrontation, there would be pressures upon Brezhnev for a policy change or even the possibility of his replacement by a coalition of disaffected interests. Even without these events, given the age of the present Soviet leadership, it is always possible that younger men may soon come to power in the Soviet Union, and by no means is it clear what their propensities would be—at least, it cannot be taken for granted that they would automatically subscribe to the pragmatic inclination because they belong to another generation.

A reasonable conclusion would seem to be that, in the absence of extreme irrationality, the margins within which the present policy would change would be relatively limited in the event any of the contingencies described above should come to pass. Although it is possible, and may even be probable, that we will go through periods in which the policy of "peaceful coexistence" may be inflected to a somewhat more militant degree, the underlying conditions determining Soviet foreign policy would constrain a return to the more extreme forms of militancy and hostility of the past. It seems apparent that even to the extent such changes may stem from the workings of Soviet domestic politics, the amplitude of their effects would be substantially influenced by our own actions.

SUMMARY OF DISCUSSION

Discussion of the above material by Marshall Shulman concentrated on eight main areas: current trends, the economic factor, Eastern Europe, Soviet "militancy," the effect of SALT, current ideology, the problem of dissidents, and Politburo changes.

The Permanence of Current Trends

There was strong general agreement with Shulman's analysis of the character of current policy and his view that current trends (toward the lowering of tension) were likely to persist. This was because the major underlying factors outlined in the presentation were thought to be relatively stable. Although there had been fluctuations, detente had to be seen as a trend in Soviet policy from the mid-1950s, not only from the last few years.

However, a number of warnings were given against taking too confident and complacent a view of the apparent trend toward relaxation: (1) An analysis of Soviet foreign policy produced many ambiguities, conflicting trends, and varied long-term and short-term considerations. These made all-encompassing explanations very difficult. Different forces pushing in different directions might produce inconsistent external behavior. (2) The tactical ingredient in Soviet foreign policy should not be forgotten. At this level of analysis it was possible to discern several fluctuations in Soviet behavior. Examples of recent and more "activist" developments at this level might be thought to be in conflict with the underlying "objective" factors. (3) The issue of detente diplomacy was not closed in the Soviet Union. Discussion was still taking place, and it was possible that Brezhnev might have been extending himself on this issue against, for example, the military establishment or those worried about the implications of detente for the problem of Soviet control in eastern Europe. (4) The increase in ideological fervor might be an obstacle to the further development of any long-term trend toward a low-tension policy.

(There was some disagreement on this particular factor; see below, "The Current Ideological Position.")

The Economic Factor

There was general agreement about the importance of the economic factor in current Soviet foreign policy—that is, that detente was a major priority for the Brezhnev regime in order to expand economic collaboration with the industrialized world in order to assist the improvement of economic development in the USSR.

Shulman's view that the Soviet Union appeared to be moving away from its traditional policy of autarky was questioned. Was the existing evidence sufficient (for example, the limited number of invitations to Western firms to build plants)? Furthermore, the political value of East-West trade from a Western viewpoint was also questioned. It could be argued, for example, that "trade-offs" in other areas (such as human rights and security) would not materialize easily, if it would come at all.

In his summary Shulman underlined the importance of the economic factor in East-West relations and argued that it was an instrument worth using. As the chief aims of the Western powers were to lower the danger of war, to increase the ingredient of restraint in Soviet behavior, and in general to stabilize the relationship with the Soviet Union, he thought that it would be useful to respond affirmatively to Soviet economic initiatives, but at a moderate level. He warned against the dangers of going too far, since the balance of long-term gain was likely to accrue to the Soviet Union. He considered that the ideal response was to be modestly affirmative, possibly concentrating initially on the sale of consumer goods but with a gradually increasing technological and investment ingredient in order to maintain Soviet incentives.

The Question of Eastern Europe

The Soviet interest in detente raised the problem of the balance of Soviet interests in Eastern Europe. Some argued that it was difficult to understand how Soviet interests in CSCE (Conference on Security and Cooperation in Europe) could be serious, because detente would presumably lead to a complication of the Soviet position in Eastern Europe, the control of which has always been assumed to be among the first priorities in Soviet foreign policy.

In reply, Shulman agreed that there were conflicting factors involved for the USSR in its attitude to Eastern Europe and detente. It was true that in periods of low tension the Eastern Europeans have had more room for maneuver, and it was true that the development

of economic and cultural contacts with the West would complicate
the problem of control. However, the Soviet aim was to have their
control of Eastern Europe recognized in the context of reduced
tension. They want both good working relations with the West and
de facto recognition of Eastern Europe as a Soviet sphere. Therefore, against the complications that might result in Eastern Europe
from detente must be set the gains the Soviet leaders are seeking
from the increased flow of goods and investments and so on into both
Eastern Europe and the Soviet Union itself. On balance, in order to
achieve the prospective gains, the Soviet leadership appears to be
willing to take the risks involved in the increasing flow of people and
contacts, and in widening latitudes of the Eastern European countries
for diplomatic maneuver.

Soviet "Militancy"

Several questions arose out of the use of the word "militancy"
to describe the whole or parts of Soviet foreign policy. How should
it be defined? What would a "return to militancy" look like? What
were the best indicators?

On the question of indicators, it was argued that they might
include, for example, the rapid development of MIRVs, and increases
in the size of forces in Eastern Europe, the active pursuit of Western
support vis-a-vis China, and so forth. There was some disagreement,
however, about the validity of the modernization and development of
weapons systems as indicators of "militancy." One view was that
the continued growth of Soviet military capabilities was basically a
result of their general preoccupation with security and was essentially
a reflection of U.S. expenditures: It could not therefore be taken
as an indicator of "militancy" (why, for example, if the USSR develops
multiple independently targeted reentry vehicles (MIRVs), should it
be regarded as any more "militant" than the United States if the
latter developed the same system?). Another view was that in the
field of military hardware, the Soviet Union moved as rapidly as
technology and the country's economic constraints allowed. Particular
strategic postures were always open to diverse interpretations,
including the impact of bureaucratic politics. In general, it was
thought that the best indicators were types of foreign policy behavior
(such as risky actions or a decline in functional cooperation). Crises
in themselves might not be good indicators: What mattered was not
the fact of a crisis, but how and why it was created.

In terms of actual behavior, it was generally felt that in the
last few years there had been no signs of any "real" militancy in
Soviet policy. (In the outbreaks of violence in the eastern Mediterranean
since 1969, for example, the Soviet Union had adopted a "low profile"

despite its large military presence in Egypt until July 1972.) In the current (fourth) Middle Eastern War, it had given the Arab states diplomatic support and had supplied them with arms, but this was not "out of line" or extraordinary. Furthermore, it was noted that—to the extent that it was possible to rely on such statements—Soviet public pronouncements in recent years suggested that they would not engage in militant interventions abroad.

Summing up, Shulman agreed that behavior indicators were more revealing than hardware indicators but said that the latter should not be ignored. For example, a large Soviet MIRV deployment would call into question the fundamental balance on which SALT was based, and the Soviet leadership was aware that such a development would create anxieties in the United States. Therefore if the Soviets went ahead regardless, it would be an indication that they did not accept the basis of the current stabilization. Both indicators therefore had to be looked at.

The Effect of SALT

There was discussion of the idea current in some U.S. policy circles that the achievement of nuclear parity by the Soviet Union might increase their willingness to take risks. In fact there was no support for the assumption of the argument—that is, that Soviet parity or even superiority would have a paralyzing effect on the United States and an emboldening effect on the Soviet Union. Shulman's main point in reply was that mutual deterrence had set in many years ago and that once a level of finite deterrence had been reached, the acquisition of further capabilities had diminishing political returns. The stability of deterrence was not being undermined therefore. Furthermore, even if some "superiority" was achieved by the USSR, it would not be sufficient to remove their uncertainty about the consequences of nuclear war and therefore would not have the result of paralyzing the United States and emboldening the USSR.

Some of the "spin-off" aspects of SALT were then discussed. Note was taken of the possible role of new civilian institutes in the USSR devoted to strategic analysis. In the view of some, this development had increased the potential for the wider (if still limited) discussion of military affairs in the USSR between military and non-military bureaucrats. Such discussion was still difficult, but it was nevertheless easier than before. On the other hand, it was argued that too much importance should not be given to the new institutes; they were mainly staffed by men with military backgrounds, and the general staff was still the place where important military business was conducted. The institutes were mostly engaged in intelligence analysis and in the reporting of Western strategic doctrine.

In his summary, Shulman agreed that SALT had brought a degree of sophistication on matters of defense technology to the Soviet political leaders that they had not had at the outset. By the end of the talks, for example, there was a wider cluster of people involved on the Soviet side: Previously, Soviet participation had been very compartmentalized. Shulman agreed that too much importance should not be placed on the civilian institutes, however.

The Current Ideological Position

The question was raised whether the strengthening of ideological orthodoxy might suggest that the long-term trend toward a low-tension policy was not so strong as some tended to think. Were not the two policies contradictory?

Shulman thought that underlying the current ideological position was not so much ideological fervor per se, but a cluster of vested interests. One had to interpret the current flurry of ideological statements as reflective of organizational politics. There was certainly evidence of a great deal of hortatory rhetoric, but there was little indication of genuine ideological development. The recitations were of old formulae: No attention was being paid to current challenges to traditional ideological positions—for example, the changing structure of contemporary capitalism or the problems of multinational corporations. Part of the current effort might be explained by the need to cauterize the growing contacts with Western businessmen, to show the pragmatic relationships did not reflect ideological change.

The Problem of the Dissidents and Its Foreign
Policy Ramifications

The view that a Soviet policy of low tension abroad could be generally linked with an intensification of repression at home was challenged. Instead, it was argued that such a linkage should not be regarded as constant and at this particular time might be purely coincidental. A variety of possible linkages might be expected. Some thought that over a long time span, even if there was no protracted detente, there might still be liberalization because of the difficulties of maintaining tight controls.

Shulman agreed that too dogmatic a view should not be taken of the linkage: (1) The process was dialectical. Initially, a domestic clamp-down might be aimed at obviating adverse internal effects of external contacts, yet over the long term these contacts might still generate pressures for internal liberalization. (2) There was a spectrum of Soviet opinion on the question of the dissidents. The true dissidents are very few in number, although very important, and there is no resonance between them and the general population. At the other extreme is the vast majority of the population, which is conservative, apolitical, sometimes hostile to the dissidents, fearful of anarchy, and anxious to remain uninvolved. Between these two extremes, there is a substantial group that could be called "passive dissidents," people who work in the system, have major responsibilities, and want to change things (free themselves from bureaucratic constraints) in order to make it easier to get on with the job. This is an important group, and any changes that develop in the future will probably come from its members. There are examples of this type of actor even in the policy bureaucracy. Their motives are different and their attitudes are less extreme than the real dissidents.

The Problem of Politburo Changes

A variety of views was expressed concerning the importance and implications of the Politburo changes in April 1973. Among the points raised were the following:

1. As of March 1973 the underlying conditions were the dominant ingredient of Soviet foreign policy. However, the changes had cast doubt on this. Among other things, they seemed to suggest a possible cleavage between Brezhnev and Gromyko on foreign policy, and in recent months there had been signs of more militancy, with a concomitant slight deterioration in Brezhnev's position. This swing might be tactical rather than strategic, however. The underlying conditions remain stable.

2. The recent changes amounted to an attempt to bring the military, foreign policy, and security bureaucracies into the Politburo at a time of division, and so try to obtain a consensus. Some dissent was still evident in the press.

3. In addition, it was suggested that the current leadership was likely to be replaced by bureaucrats and technocrats; if so, a return to the conditions of the 1940s and 1950s was very unlikely.

4. On the naval implications, it was suggested that the Gorshkov series of articles represented some evidence of a major cleavage of opinion within the political leadership; the series placed Gorshkov on the opposite side to Brezhnev and Kosygin. To that extent the Politburo changes suggested that Gorshkov's views had not prevailed.

In summing up, Shulman stressed the speculative nature of the discussion of such a subject. He thought, however, that Gromyko's case ought to be distinguished from that of the military and internal security appointments. In the case of Gromyko, the appointment was probably a regularization of a long-standing decision-making process. Foreign policy had traditionally been at the forefront of Politburo concerns, and Gromyko had been closely associated with it on the technical side, had often been called in for information and comment, and by now had developed considerable competence and experience. While there was little evidence of intimacy between Brezhnev and Gromyko, rather the contrary, the latter's appointment appeared to be a natural regularization of his actual role. Unlike the Gromyko case, the other Politburo changes were probably part of an attempt to reassure the bureaucracies concerned that their interests would not be sacrificed in the conduct of the new (low-tension) foreign policy.

CHAPTER

2

THE SOVIET MILITARY'S INFLUENCE ON FOREIGN POLICY
Malcolm Mackintosh

No one would deny that the Soviet Union is at present engaged in a series of active and sophisticated foreign policy initiatives on a worldwide basis. In these circumstances, the identification of the power structures in the Soviet Union that have real potential influence over decision-making in this crucial area has rarely been more intriguing or more important. Of those power structures, the large military establishment is certainly one of the most vital. It is the purpose of this article to explore the nature and extent of the influence of the military on foreign policy decision-making. There is, of course, some evidence that the military is not necessarily united in its views on foreign and defense policy; however, this is a separate study in its own right; in the present article "the military" is intended to mean the likely majority grouping within the High Command.

Before embarking on our inquiry, mention should be made of some basic background factors and issues that are important in any analysis of this complex subject. It should be noted, for example, that there is a difference between the roles of the military in influencing defense policy on the one hand and foreign policy on the other. Defense is, of course, part of the military's business, and military influence in that sphere is, on the whole, easier to trace. Foreign policy, however, is the military's business only if the Communist Party invites the soldiers to participate in decision-making on foreign issues—and here the recent promotion of Marshal Andrei A. Grechko to full membership on the Politburo is naturally significant. We shall be looking at the role of the military in both fields, but to separate them arbitrarily into two subjects would probably not correspond to Soviet reality.

This chapter was first published in Problems of Communism 22, 5 (September-October 1973).

Secondly, the physical power, sophistication, and diversification of the Soviet armed forces' capabilities have grown steadily since the war. Apart from the postwar demobilization of the 1945-47 period and some cuts in manpower in 1955 and 1958, the trend has been for the Soviet forces to increase in size and in capabilities, especially during the 1960s and 1970s. Two new branches of the armed forces were added to the traditionally large land army and its air force and the navy: the air defense command in 1954 and the strategic rocket forces in 1959. The navy has greatly expanded and modernized its fleets, and effective marine, air transport, and airborne forces have been developed in support of the traditional concepts of Soviet military power. It is very largely because of the great buildup of the Soviet armed forces, together with Soviet economic and technological development, that the Soviet Union has been able to pull itself up into the superpower class alongside the United States, and the Soviet military establishment is entitled to claim both credit for achieving this power and an interest in how it is used in support of the country's foreign policy.

The third point, which is more of an issue than a factor, relates to the problem of military access to the decision-makers. At least since the death of Stalin, the military seems on the whole to have had reasonable access to the political leaders in the sense that its members have been heard as professional experts called in to give an authoritative view on military matters. For example, the minister of Defense, even when not a member of the Politburo, may well have been able to attend Politburo meetings on invitation, although there is no evidence to support this. It is possible that most senior members of the military establishment have been content with this usually limited role. But it also seems likely that in a system in which the political leadership relies so much on its military power to achieve its goals in foreign policy and to back up the USSR's claim to be the other superpower, there must be an urge within the armed forces to have a greater say in decision-making. In other words, there must be strong pressure within the ministry of Defense in Moscow to ensure that the degree of formalistic access the military has already gained to the party decision-makers should mean both influence over and participation in decisions on foreign and defense policy. Thus, the central issue in the problem of the role of the leaders of the armed forces in this field appears to be, Does access necessarily mean influence?

MILITARY ACCESS TO DECISION-MAKING

In order to study more closely the nature and results of the access to the political decision-makers accorded to the military, let us look first at the present institutional structure (as far as we know it) of the relationship between the party leaders and the Ministry of Defense. Clearly, the Politburo is the decision-making body with respect to both defense and foreign affairs in the Soviet Union, and traditionally the general or first secretary of the Central Committee of the Communist Party of the Soviet Union (CPSU) has assumed top responsibility for directing Soviet foreign and defense policies. This is as true of Brezhnev today as it was of Khrushchev and Stalin. Moreover, we know from Soviet sources that the Politburo can deal with the details of weaponry as well as with the broader issues of force structure and employment.[1] We may assume from what we know of the Soviet system that only the Politburo can make a decision to go to war, to send troops into another country (as in the case of Czechoslovakia in 1968), to deploy combat units abroad (as to Egypt in 1970), or to use nuclear weapons. The Politburo also probably decides on the size, nature, balance, and organization of the armed forces, as well as on their armament and deployment. In addition, it decides the type of political indoctrination the military forces receive through the army's Main Political Administration.* Apart from a few turbulent months in 1957 when Khrushchev admitted Marshal Georgii Zhukov to full membership on the Politburo, the party leadership has normally excluded professional soldiers from that body. Hence, Marshal Grechko's promotion to the Politburo in May 1973 may mark a new departure in army-party relations and in the system of checks and balances hitherto maintained by the Soviet political leadership.

Since the Politburo makes the final decisions on virtually every aspect of Soviet life, it is to be expected that some of its work on military affairs is delegated to, or handled in, a more specialized forum. This appears to be the Defense Council (<u>sovet oborony</u>), whose predecessor, the Higher Military Council (<u>vysshii voennyi sovet</u>) was mentioned in the case against Marshal Zhukov in 1957.[2] At that time the Higher Military Council, whose work Zhukov was apparently trying to disrupt, included both party and military leaders and was thus the highest-level organization in which the two leaderships could meet in formal session. It presumably had consultative rather than

---

*The Main Political Administration of the armed forces is both part of the Ministry of Defense and a department of the Party's Central Committee.

decision-making functions, but it is also possible that, as a joint
military-political body, it might have been intended to form the
nucleus of a supreme military-political leadership to direct the overall Soviet war effort in the event of hostilities with the West or with
China. At some point after 1957 (perhaps after the fall of Khrushchev
in October 1964), the Higher Military Council seems to have been
renamed the Defense Council, possibly for presentational reasons.
Western experts have recently referred to it as "the Defense Council"
or as "the Defense Committee (or the Supreme Military Council)."[3]
No information is available from Soviet sources on the responsibilities
or the membership of the Defense Council today, but until the minister
of Defense was admitted to the Politburo in 1973, the council was
probably the military leaders' nearest point of access to the party
chiefs on a constitutional basis.

Next in the hierarchy comes the Council of Ministers, of which
the minister of Defense is of course a member. In this capacity,
he is in contact with and subordinate to the council's chairman, Premier
Alexei Kosygin, and the various first deputy chairmen of the Council
of Ministers. The Council of Ministers is empowered to appoint
specialist commissions on various subjects, including militaryindustrial problems, on which the armed forces are to be represented.
But the council's main military element is the Ministry of Defense,
which has so far been headed by an active professional soldier.

(In April 1967, when Marshal Malinovsky died, there was some
press speculation in Moscow that he might be succeeded as minister
of Defense by a civilian, the experienced armaments expert D. F.
Ustinov.[4] As it turned out, Marshal Grechko was appointed to the
post April 14, 1967, but the issue could reappear.)

The Ministry of Defense is a unified ministry administering
and controlling the General Staff, the five "branches" of the Soviet
armed forces—the strategic rocket forces, the homeland air defense
command, the ground forces, the air force, and the navy—and the
logistic and support directorates which maintain the armed forces
in peacetime and prepare it for war. The minister, through his commanders in chief, can issue direct orders to the strategic rocket,
strategic air, and air defense forces and the navy; and, on the evidence
of the Czechoslovak operation of 1968, he can likewise issue orders
through the General Staff to the ground and tactical air forces. Thus,
his powers of force employment are considerable. However, authorization to exercise them can come only from the Politburo, perhaps
through the Defense Council. The Ministry of Defense, moreover,
has its own "Main Military Council," which issues directives to the
armed forces on matters such as the need to improve discipline and
party political work.[5]

Members of the armed forces also come into contact with political leaders through the CPSU Central Committee, on which 36 military officers at present serve as full or candidate members. Central Committee plenums discuss foreign (as well as domestic) affairs, and military members of the committee are known to have spoken on these occasions.[6] On a routine basis, military affairs at the Central Committee level are dealt with by two departments; the Main Political Administration (already mentioned), which is responsible for the quality and direction of the political work in the armed forces; and the Administrative Organs Department. The former is headed by Army General A. A. Yepishev (who incidentally served under Brezhnev as an army political officer in 1945), and the latter by a career political officer, N. I. Savinkin, whose war service was mostly in the Far East. To judge from Soviet military press comment, the Administrative Organs Department is primarily concerned with personnel and similar matters.[7]

From this brief survey of the formal points of contact between the military and the political decision-makers, it seems clear that the Ministry of Defense receives its directives from the Politburo, probably by way of the Defense Council, and is then responsible for the formulation and implementation of the party's military doctrine and for the preparation of the armed forces to put into practice if required. Disputed points that reach the highest military level can presumably be debated in the Defense Council or perhaps in a commission set up by the Central Committee or the Council of Ministers. Two vital responsibilities are reserved to the Politburo: the final acceptance of current military doctrine, and orders to deploy forces abroad or to go to war.

Finally, in this consideration of military access to the political leadership, we should not forget the probable influence of personal relationships. A great deal of Soviet business is probably done, as in other countries, behind the scenes and through personal and social contacts. Stalin terrified his military commanders, and they had minimum personal contact with him. Khrushchev in the 1950s built up his own "group" of army leaders who had served with him at the battle of Stalingrad and in the Ukraine; but he then antagonized them by slighting their advice and treating them in some instances with contempt. All the evidence suggests that Brezhnev has cultivated the military leaders both professionally and personally. While maintaining his authority over them, he has stabilized army-party relations and cemented a close personal relationship with Grechko that seems to have created, among other things, a new and less formal channel of communication between the minister of Defense and the general secretary of the party.

THE RIDDLE OF MILITARY INFLUENCE

When we turn to consider the effectiveness of these formal and informal channels of communication and means of access for the military to approach the party leadership, we have to admit that we have no direct evidence as to what advice may have been given by the military to the political leaders on any significant aspect of foreign policy, whether that advice was unanimous among the military leaders and whether or not it was accepted by the Politburo. Looking back over the recent past, we cannot prove that the military was in favor of, or opposed, military intervention in Czechoslovakia in 1968 or the deployment of combat units to Egypt in 1970 and their enforced withdrawal in 1972. Nor have we any direct evidence on the true attitude of the Soviet armed forces with respect to the political and military confrontation with China, although we are probably justified in assuming that the military favors a tough line against so obvious a national enemy. We also know very little about recent reaction among the military to the Strategic Arms Limitation Talks and the agreement of 1972.

It is, of course, true that the Soviet military press has discussed many of these issues, generally in terms that support the line taken by the Soviet government. While it is possible to detect changes of emphasis by one or another of the well-known military writers on foreign and defense affairs—such as Lieutenant General I. G. Zavialov and Major-Generals V. Zemskov and E. Sulimov[8]—we should remember that articles published in the central military press or in the journals of the various branches of the armed forces are prepared under the supervision of the Main Political Administration acting in its capacity as a department of the party Central Committee. In many cases, therefore, it would be misleading to assume that an article appearing in these papers or journals under the signature of a general or an admiral necessarily expresses the views of the military establishment or the author of the article. There are exceptions, of course—as, for example, when an officer of the seniority of Admiral S. G. Gorshkov, who has been commander in chief of the Soviet Navy since 1956, publicizes his concept of the role of the navy in support of Soviet foreign policy.[9] In general, however, it is difficult to rely upon published articles in the Soviet military press for the purpose of ascertaining individual military views on defense or foreign policy issues or learning what part the military establishment played in crises abroad involving the Soviet Union.

Furthermore, it is very important to distinguish between the growth and diversification of Soviet military capabilities that make it possible for the Soviet Union to intervene, effect a presence, or have a say in a certain part of the world, on the one hand, and evidence that the military leaders (who clearly welcome such heightened capabilities in general terms) do or do not advocate the policies that are then followed by the Soviet Government, on the other.

Thus, the buildup of strategic nuclear forces in the late 1960s enabled the Soviet Government to take part in the Strategic Arms Limitation Talks as the equal of the United States, but we cannot prove that military opinion advocated the kind of agreement that was reached in 1972 or would have preferred a continuation of the strategic arms race. Similarly, expansion of the Soviet air and air defense forces enabled the Soviet government to deploy combat units to Egypt in 1970, but we do not know from direct evidence whether there was any military opposition to the potential risks involved in the development of the Soviet Union's first overseas military base in Egypt, a noncommunist country. The worldwide capabilities now possessed by the Soviet Navy have allowed the Soviet Government to deploy a submarine nuclear strike force off the east coast of the United States, to maintain a "Guinea patrol" off the west coast of Africa, and to move a task force into the Bay of Bengal during the war between India and Pakistan in December 1971; but we cannot say with any certainty whether any of these actions resulted from a navy initiative or aroused any doubts within the Ministry of Defense concerning the risks involved.

CASE STUDIES

In spite of these rather discouraging reservations, let us look at a few specific cases in recent years in which Soviet foreign policy has involved the use of military forces or has had military implications. Somewhat arbitrarily, we have chosen one case involving a noncommunist country, the Arab-Israeli war of 1967; one concerned with Eastern Europe, the invasion of Czechoslovakia in 1968; and one connected with the USSR's strategic relationship with the United States, SALT, the first stage of which was completed in 1972. There are, of course, other cases of equal or greater importance—such as Soviet foreign policies with respect to China and to the Vietnam war—but an examination of the three cases selected may enable us to arrive at some general guidelines with respect to Soviet military attitudes and their influence on foreign policy decisions.

The Arab-Israeli War of 1967

Let us take the Arab-Israeli crisis of 1967 first. The Soviet Union had been arming Egypt for a number of years in the period following the Suez crisis of 1956 and had also been supplying arms to Syria and Iraq, while the United States and France had kept Israel supplied with weapons and equipment. No Soviet forces were stationed on Egyptian soil in 1967, but there were considerable numbers of Soviet advisers and technicians attached to the Egyptian armed forces, and many Arabs clearly were under the impression that the Soviet Union had committed itself to their defense in the event of another war with Israel. As the 1967 crisis moved toward its climax in May and June and President Gamal Abdel Nasser closed the Straits of Tiran and ordered the United Nations force out of the Sinai, Soviet diplomacy appeared until the last minute to encourage the Egyptians to act belligerently. There were also reports that Soviet military intelligence was spreading allegations of Israeli troop concentrations against Syria and of Israeli plans to attack Syria in a matter of days.[10]

When war erupted on June 5, 1967, the Egyptian, Syrian, and Jordanian armies suffered total defeat, and the Sinai, the west bank of the Jordan, and the Golan heights were occupied by Israeli forces. Hundreds of Soviet-made guns, tanks, aircraft, small arms, and ammunition were destroyed or captured, and Soviet policy as well as the Soviet Union's reputation as a reliable ally suffered a catastrophic setback among the Arabs. There must certainly have been serious recriminations in Moscow as a consequence of the defeat of the Soviet Union's Arab clients. Moreover, although no evidence is available on the prior attitude of the military toward the government's Middle East policy as it developed, there are grounds for believing that at least one aspect of the crisis worried the military establishment.

Two weeks after the outbreak of the war, a Central Committee Plenum was held in Moscow on June 20-21, 1967. The first item on the agenda was "the policy of the Soviet Union toward Israeli aggression in the Middle East," and the sole announced speaker was Brezhnev. We do not know how Brezhnev defended Soviet policy in the Middle East crisis, but some criticism of the Soviet course appears to have been voiced at the plenum by N. G. Yegorychev, first secretary of the Moscow City Party Committee. Yegorychev was dismissed and demoted after the plenum,[11] and he never recovered his party position.

The interesting point here is that Yegorychev, as first secretary of the Moscow City Party Committee, was also a member of the Military Councils of the Moscow Military District and the Moscow Air Defense District, the latter commanded by V. V. Okunev. (These political-military councils direct the activities of all high-level Soviet force headquarters.) In July the Moscow Air Defense District Military Council held a special meeting attended by Yegorychev's successor, V. Grishin,[12] and the district then launched a lengthy air defense exercise. All this suggests that if Yegorychev did criticize

Brezhnev's handling of the Middle East crisis at the June plenum, he may have done so in his capacity as a member of the Military Council of the Moscow Air Defense District, for he would probably have had no grounds for intervening in a debate on foreign affairs in his normal capacity as the party leader of Moscow City.

While there is no evidence that Yegorychev had any military background, one can reasonably surmise that he may have been put up to his intervention in the plenum debate by members of the military establishment. One can further surmise that what the military may have been saying, through Yegorychev, was in effect that the Soviet air defense systems were unready for war, and that if the Arab-Israeli conflict had escalated into a U.S.-Soviet confrontation, the military establishment would not have been able to guarantee the air defense of the Soviet Union. If this interpretation is correct, the implication would be that in 1967 at least some members of the military considered the policy pursued by the Soviet Government in the Middle East crisis too risky. This interpretation also suggests that members of the military establishment have tried to press their views on the Politburo through individual party leaders.

Before moving on to the other two cases, it may be useful to add some observations with regard to the probable attitude of the Soviet military toward the later evolution of Soviet policy in the Middle East conflict. We do not know for certain how the Soviet military viewed Moscow's decision to deploy combat units to Egypt after the Israeli's deep penetration air raids during 1970-71, or to the decision to use Egyptian airfields for the purpose of Soviet reconnaissance patrols against NATO forces in the eastern Mediterranean. So far as the latter is concerned, there presumably would have been a good deal of professional military support since reconnaissance from Egyptian bases would augment the Soviet armed forces' overall capabilities against NATO. With regard to the former, one can speculate that the military probably regarded the close involvement of Soviet military power in the defense of Egypt from 1970 to 1972 and the establishment of a large Soviet military mission in Cairo (interestingly enough, under the same General V. V. Okunev who had been in command of the Moscow Air Defense District in June 1967) as at least assuring the Soviet Government of the kind of political control over Egyptian actions that had been lacking in 1967. It is therefore likely that the military leaders were less worried about the risks inherent in the 1970-72 involvement than they had been about those in 1967. Such an attitude would be consistent with the feeling among observers in Moscow since the expulsion of Soviet military personnel from Egypt in July 1972 that the USSR does not consider itself bound to help Egypt militarily in the event of another Arab-Israeli war.

The Invasion of Czechoslovakia

The second case study is that of the Soviet Army's invasion of Czechoslovakia, along with forces from Poland, Hungary, Bulgaria, and the German Democratic Republic (GDR), in August 1968. Here again we have to admit that we have no direct evidence of the views of the military on the decision to invade an ally's territory and overthrow its legal government by force. All we can say definitely is that the Soviet armed forces, in virtually ideal circumstances and without the urgency of an international crisis, built up a large invasion force around Czechoslovakia's frontiers between April and August 1968 and then, upon the Politburo's order to invade, carried out a swift, overwhelming, and effective operation that successfully achieved the political objectives of the Soviet leadership.

The natural assumption in these circumstances is that the Soviet military welcomed the opportunity to display the country's military might, to prevent a member of the Warsaw Pact from defecting into neutrality, and to put the armed forces to a practical test in a realistic politicomilitary situation. One Soviet general, who commanded the Trans-Caucasus Military District, did in fact write that the invasion of Czechoslovakia provided an excellent testing ground for Soviet military training.[13] Another Soviet statement declared that "the counter-revolutionaries wanted to drive a wedge into the heart of the Warsaw Pact area, and through the Czechoslovak corridor, so to speak, West German soldiers could have marched straight to the Soviet border, and with them would be American troops."[14]

There is also no doubt that the Soviet military took the occasion very seriously as a test of their command and control procedures at the highest level. A very large forward headquarters of the Soviet General Staff was set up at Lignica in Poland in early August, and Stalin's wartime chief of operations, General S. M. Shtemenko, was recalled from an administrative post to take charge of the invasion under the overall command of the commander in chief of Soviet ground forces, General I. G. Pavlovsky.[15]

One of the most interesting military lessons of the invasion was in fact the transfer of command from the regular Warsaw Pact headquarters to the Soviet High Command at the beginning of August. The Warsaw Pact, under Marshal I. I. Yakobovsky, supervised the buildup of forces around the Czechoslovak territory during the exercise known as Sumava, which was staged in Czechoslovakia in June. As soon as the deployment was complete, however, the Warsaw Pact handed over command to the Soviet General Staff—a procedure that might well be followed in the event of war in Europe.

Perhaps the only hint of Soviet military views was contained in a report published in Le Monde early in the crisis. On May 6,

1968, the Paris newspaper reported that the head of the Soviet Army's Main Political Administration, General Yepishev, had broached the possibility that "faithful communists" in Czechoslovakia might appeal to the Soviet Union for military intervention and had intimated that, if this happened, the Soviet Army would be ready to do its duty. Although the report was denied on May 19 by the Czechoslovak News Agency (Yepishev was in Prague on this date), it may well have been basically accurate. Further, Yepishev could have been speaking either in his party or in his military capacity inasmuch as it seems likely that the advice of the military establishment would have favored the invasion of Czechoslovakia—assuming that NATO intervention could be ruled out—in order to preserve the "buffer zone" west of the Soviet frontier intact.

As for the role of the Soviet military in the final decision to invade, this is hard to assess. A study of Marshal Grechko's movements during August 1968, as reported in the press, shows that he inspected the invasion forces in East Germany and Poland with Yepishev from August 15 to 18 and was not in Moscow to greet the Finnish Defense minister, who arrived there on August 17.[16] This does not prove, however, that he did not attend the critical meeting of the CPSU Central Committee in Moscow on August 18-19, which reportedly took the final decision to occupy Czechoslovakia.[17] The most likely conclusion is that, if Grechko did attend the meeting, he reported on the readiness of his troops to move and was not called upon at this late stage to press the case for invasion on an already convinced party leadership.

SALT

This brings us to the third case study, the strategic arms limitation talks between the United States and the USSR—a matter of vital concern to the Soviet military leaders since it affects the essential element of Soviet national security and defense policy more directly than even the preservation of Soviet rule in Eastern Europe, and certainly more than support for a Middle Eastern ally. Soviet agreement to strategic arms limitation talks with the United States was initially announced on June 27, 1968, and was reiterated, after the occupation of Czechoslovakia, had taken place, on January 20, 1969. The talks began in November of the latter year and were concluded at the Moscow summit of May 1972, in the form of the agreement limiting the construction of antiballistic missile (ABM) defenses in the Soviet Union and the United States and the interim agreement on offensive strategic missiles. The second round of the talks opened in November 1972.

From the point of view of the Soviet political leadership, the talks have everything to recommend them. Through them the Soviet Union once again confirms its position as the other superpower alongside the United States; the Soviet Union is enabled to save a considerable amount of money and resources by eliminating the risk of a new round in the strategic arms race (which the Soviet Union could not win); and the interim agreement, though not without its advantages for the United States, secures the latter's assent to Soviet totals of nuclear delivery vehicles that assure to the Soviet Union the capacity to destroy vital American targets. Moreover, research and development are allowed to continue unhindered on both sides.

It would appear that the 1972 SALT agreements should also satisfy Soviet military planners on several counts. First, the agreed totals allow the Soviet Union to retain numerical superiority in land-based missiles and to continue expanding its arsenal of submarine-launched missiles up to a total yet to be decided but probably in the area that the Soviet military would have wished to reach anyway. Second, the Soviet defense industry can still produce MIRVs for the strategic nuclear forces. Third, the ABM agreement prevents the United States from moving ahead with the kind of "thick" antimissile defense that only the United States could afford, while averting the vast expenditures on both sides that ABM defense demands. Finally, the leaders of the theater forces, which are not covered by the agreement, may well hope that they can absorb some of the resources that will be saved on strategic weapons.

Published statements by Grechko and his colleagues have unequivocally supported the official line on SALT, and there have been no recent indications of military disagreement with the negotiations or the agreements concluded.[18] But this was not so when the concept of strategic arms limitation was first raised. In December 1968, a Soviet military writer subordinate to the Main Political Administration of the armed forces underlined the importance of adequate strategic nuclear weapons by saying that to overemphasize the importance of conventional weapons was a "more serious error" than the mistake of relying too completely on nuclear arms.[19] In April 1969, still another officer commented that the U.S. decision to deploy the Safeguard ABM system proved that the United States was not interested in limiting the strategic arms race.[20]

These quotations suggest that both the professional military and the political officers of the armed forces were apprehensive of the SALT negotiations in their early stages. As the talks went on, however, public expressions of military concern faded—partly, no doubt, because the issue of Soviet participation in the negotiations entered the field of the highest-level political decision-making, but partly also, perhaps, because the party was able to show the military, as the outline of the ultimate agreement emerged, that their interests would not suffer.

As for military representation at the talks themselves, the Soviet delegation included the first deputy chief of the Soviet General Staff, Colonel-General N. V. Ogarkov, representing the armed services with respect to operational and intelligence aspects of the subjects under discussion, and Colonel-General N. N. Alexeev, a General Staff officer with a technical background in missiles and electronic equipment. A third senior officer who was sometimes present, Colonel-General A. A. Gryzlov, was also a career General Staff officer. (While the talks were in progress, Alexeev was promoted and became a deputy minister of Defense, and his place was taken by another representative of the General Staff, General K. A. Trusov.) The course of the talks provided some indications that the military representatives were not overly eager to pass on military information to the civilian members of the Soviet delegation, who found themselves deferring to the generals and postponing the discussion of certain items until the Ministry of Defense was ready to contribute.

To sum up the evidence in this particular case, it seems probable that some influential Soviet military figures were initially skeptical about the strategic arms limitation concept and the desirability of negotiating with the United States but that this attitude changed as the talks developed. Clearly, the conduct of the negotiations depended a great deal on the military contribution, and the promotion of one of the senior military delegates while the talks were under way may have been significant in this context, although there is no proof that the importance of the negotiations were responsible for the move. In any case, the evidence suggests that as the talks proceeded and the outline of the eventual agreement took shape, the military realized that its requirements were going to be met, and therefore modified or abandoned its early reservations. Although the party had the ability to force the military into line by "administrative measures" (to use the Soviet phrase) or by removing opponents of its policy from their positions, it clearly proved unnecessary to resort to such measures in the case of SALT.

SOME CONCLUSIONS

The lessons to be drawn from these case studies are admittedly fragmentary and inconclusive. Nevertheless, they tend to support the view that the military establishment favors hard-line or at least firm foreign policies when issues concerning the defense of the Soviet homeland or Soviet hegemony in the "buffer zone" in Eastern Europe are at stake. On the other hand, they suggest that when a proposed foreign policy may involve the commitment of Soviet armed forces beyond just naval forces to areas far away from Soviet frontiers, the military still tends to react on the side of caution and a full evaluation of all the risks involved. As for the ability of the military to translate its attitude on foreign issues into actual influence on top-level foreign policy decision-making, the case studies suggest that when the particular policy concerned requires military support, the influence of military advisers is significant. They also demonstrate the fact that when a foreign policy adopted by the leadership coincides with military views, it is difficult to distinguish whether the policy was initiated by the party or by the military. Finally, there is nothing in any of the case studies to suggest any diminution of the party's ultimate primacy in foreign policy decision-making in cases where the views of the armed forces might differ from those of the party leaders. The current presence of one military representative on the Politburo seems unlikely to affect his situation.

This brings us back to the matter of Marshal Grechko's promotion to the Politburo, the probable reasons for it, and its significance from the standpoint of military influence on foreign policy. Many explanations have been put forward for his elevation, along with those of Gromyko and the head of the KGB, to full membership in the Politburo. These include reward for past services (perhaps especially in SALT), the need to bring the armed forces more closely into the decision-making process, and the personal preferences of Brezhnev. It seems likely that Brezhnev wanted to rally all his supporters round him in the final forum of Soviet decision-making at a time when he was facing some of his most critical decisions in foreign and domestic policy—and Grechko has certainly been one of Brezhnev's most cooperative and valued colleagues.

After a cool start in 1967, the Brezhnev-Grechko relationship warmed, both professionally and on the personal level, in the difficult years of the Czechoslovak crisis and the establishment of the Soviet naval-air base in Egypt, and Grechko found himself frequently used on diplomatic tours that might normally have been expected to be entrusted to an official of the party or of the Ministry of Foreign Affairs. Grechko was the dominant figure in the delegation whose mission to Prague in the spring of 1969 effected the dismissal of Alexander Dubcek and the appointment of Gustav Husak. He was in India before the conclusion of the Indo-Soviet Treaty; in Egypt several times during the setting up of the Soviet base; in Somalia before the appearance of facilities in that country probably designed for the use of the Soviet Navy. Thus, his role and value to Brezhnev in foreign as well as military affairs have grown in recent years. But does his political promotion mean that the influence of the military in the formulation of foreign policy has also grown commensurately?

The evidence seems to suggest that no startling change in the present position is likely. For several years, ever since Admiral Gorshkov's navy undertook the overseas support of Soviet foreign policy (as mentioned earlier), the military has had to be consulted at an earlier stage and in greater depth than ever before, and this has given representatives of the armed forces more frequent and more secure access to the party leadership than they used to enjoy. Senior officers have apparently been able to present their points of view to members of the Politburo in person at the Defense Council, and, as has already been mentioned, the minister of Defense, even when not on the Politburo, has probably been summoned to attend meetings of that body when military matters were to be decided. Grechko's own close relationship with Brezhnev in recent years has given him access to the party's top leader on the personal as well as the professional plane. His appearance among the full members of the Politburo now gives him a vote on the full range of Soviet decision-making, but it does not necessarily increase his power position in practical terms. He became, in any case, 70 years old in October 1973 and may not retain his ministerial post very much longer.

Finally, we must remember that even a rejuvenated Soviet High Command is still a force for stability and maintenance of the status quo in Soviet affairs. The military establishment is against dissent and innovation in Soviet society at home and in favor of an active—though not an "adventurist"—foreign policy abroad with military support. It probably favors a "special relationship" with the United States, bitterly opposes China, and hopes for a more advantageous position (in relative terms) in Europe as a result of Brezhnev's Westpolitik in this traditional area of Russian foreign policy. It is likely to be wary of new commitments in the Middle East, though it would like to be able to resume reconnaissance flights against NATO forces in the Mediterranean from a base in the area. Basically, the present Soviet military leaders want to serve—as well as have some influence over—a strong and nationalistic party and government with some sense of political mission. They do not want, either directly or by proxy, to have the responsibility of formulating Soviet foreign policy itself.

NOTES

1. In 1935, for example, the matter of what caliber the Red Army's standard antiaircraft gun should be was referred to the Politburo for a decision: See N. F. Kuzmin, Na strazhe mirnovo truda, 1921-1940 (On guard over peaceful labor, 1921-1940; Moscow: Voenizdat, 1941), p. 153.

2. Yu.P. Petrov, Partiinoe stroitelstvo v sovetskoi armii i flote (Party structure in the Soviet Army and Navy), 1st ed. (Moscow, 1964), p. 462.

3. John Erickson, Soviet Military Power (London: Royal United Services Institute for Defense Studies, 1971), p. 29; and Lt. C. David Miller, "Soviet Armed Forces and Political Pressure," Military Review (Fort Leavenworth, Kans.), December 1969, p. 65.

4. See Morning Star (London), April 5, 1967.

5. See Petrov, op. cit., 2d ed., 1968, p. 507, for a reference to a decree of the Main Military Council on party-political work, issued in April 1962.

6. For example, Marshal Grechko, then commander in chief of Warsaw Pact forces, spoke at a Central Committee Plenum on December 12-13, 1966, on "the international policy of the USSR and the struggle of the CPSU for the solidarity of the Communist movement."

7. See the Soviet Army's daily paper Krasnaya zvezda (Moscow), July 15, 1972, for an indication of the type of work done by this department.

8. For example, see General Zavialov's article in Krasnaya zvezda of April 19, 1973, on Soviet military and political doctrine in relation to modern war and the objectives of the armed forces.

9. He did this in an interesting series of articles published in Morskoi sbornik (Moscow), nos. 2-12, 1972, and no. 2, 1973.

10. The evidence of such Soviet activities is reviewed in A. S. Becker and A. L. Horelick, Soviet Policy in the Middle East, (Santa Monica, Cal.: Rand Corporation, September 1970), pp. 46-49. Specific mention is made of Soviet approval of the deployment of Egyptian forces into the Sinai in mid-May 1967.

11. Pravda (Moscow), June 28, 1967.

12. Petrov, op. cit., 2d ed., p. 506.

13. Colonel-General S. K. Kurkotkin, in Zarya vostoka (Tbilisi), February 23, 1969. General Kurkotkin was subsequently appointed to command the Soviet group of Forces in East Germany in September 1971.

14. Radio Moscow, September 4, 1968.

15. See the author's The Evolution of the Warsaw Pact, International Institute for Strategic Studies, Adelphi Paper No. 58, 1969, pp. 14-15.

16. Krasnaya zvezda, August 18, 1968.

17. Radio Free Europe Research Memorandum, "Czechoslovakia in 1968—A Chronology," Munich, January 22, 1969, p. 12.

18. See, for example, an article by Army General S. L. Sokolov, first deputy minister of Defense, in Izvestia (Moscow), February 23, 1973.

19. Lt.-Colonel V. Bondarenko, in <u>Kommunist vooruzhennykh sil</u> (Moscow), no. 24, 1968.
20. Lt.-Colonel T. Kondratkov, in ibid., no. 8, 1969.

CHAPTER

3

**THE MILITARY ROLE IN
SOVIET DECISION-MAKING**
Matthew P. Gallagher

The question what makes the Russians tick has been a source of disagreement and puzzlement among Western scholars and public figures since the Soviet Union first appeared on the international scene half a century ago. From the beginning, the central issue in dispute has been whether the Soviet Union could be regarded as a "normal" state in the sense that its policies could be predicted on the basis of commonly accepted principles of international behavior, or whether its peculiar doctrines and institutions added an element of unpredictability to its actions. For many, the issue tended to turn on what one thought about the ideological element in Soviet policy. Those who thought that the goal of world revolution was a paramount consideration in Soviet policy tended to believe that the Soviet Union was more unlike than like other states and that its policies were correspondingly less predictable by conventional standards. Those who discounted the importance of the ideological element tended, on the contrary, to believe that the Soviet Union was not essentially different from its rivals and that its policies, as a consequence, could be predicted with a more or less equal degree of reliability.

With the onset of the nuclear missile age, the terms of the argument have tended to change somewhat, but the substance of the issue remains pretty much the same. The issue is still whether the Soviet Union operates on principles that seem rational by Western standards. Today, however, the principles in question are not so much the criteria that apply to the determination of foreign policy

---

The first parts of this chapter are drawn from M. P. Gallagher and K. F. Spielmann, <u>Soviet Decision-Making for Defense: A Critique of U.S. Perspectives on the Arms Race</u> (New York: Praeger Publishers, 1972).

objectives but the peculiar technical standards that apply to the practice of nuclear strategy. In the viewpoint of many, an entirely new set of requirements and imperatives has been imposed on national policy-makers by the nature of the new military technology. The question of the predictability of Soviet policy today, therefore, concerns the question of whether the Soviet leaders share this perception of the strategic situation and whether they do in fact operate according to the dictates it seems to impose. These questions require a careful look at what U.S. strategic concepts imply about the Soviet decision-making process, and an even more careful look at how the Soviet Union decides military policy issues in practice.

WESTERN CONCEPTS
AND SOVIET MILITARY POLICY-MAKING

The Action-Reaction Concept

The most important of the ideas that affect U.S. thinking about the Soviet military policy-making is the interaction concept. It was formulated by then Secretary Robert S. McNamara in a speech in San Francisco on September 18, 1967. Noting that the Soviet Union and the United States mutually influence each other's strategic plans, McNamara went on to describe the mechanics of the phenomenon in the following terms:

> Whatever be their intentions, whatever be our intentions, actions—or even realistically potential actions—on either side relating to the buildup of nuclear forces ... necessarily trigger reactions on the other side. It is precisely this action-reaction phenomenon that fuels an arms race.[1]

As in so many of his pronouncements, McNamara's purpose in this speech was to argue for mutual restraint in the strategic competition. His description of the action-reaction mechanism, thus, should not be taken as a precise expression of his own understanding of how Soviet decisions are made, but rather as a generalization about the arms race that seemed appropriate to his argumentative purposes. In so doing, however, he gave expression to an image of the strategic competition that has been widely accepted in the American public dialogue on defense policy matters. It is a somewhat mechanical image, one that has often been interpreted to imply a large degree of automaticity and precision in the moves and countermoves of the contending sides.

It would be hard to exaggerate the influence of this image on American thinking about the Soviet threat. It is perhaps safe to say that almost everything of consequence that has been written about Soviet military or foreign policy over the past decade or so has used this concept as an interpretive tool.

The reasons for this wide acceptance are not hard to find. There is a broad sense in which the interaction concept is patently true. The Soviet Union and the United States are, indubitably, strategic adversaries. And strategy, the classical textbooks affirm, is the art of maneuvering forces preparatory to battle. Nowadays, instead of maneuvering soldiers over areas of ground, strategic adversaries maneuver scientific and industrial potentials over periods of time. And nowadays, instead of seeking to "offer" battle under conditions the enemy cannot accept, the adversaries seek to deter one another from going to war by demonstrating their retaliatory capabilities.[2] But, nowadays as before, the object of the game is to avoid being placed at a fatal disadvantage, and each side makes its decisions regarding weapons systems and forces with this thought in mind.

To say this, however, is a far cry from saying that each side makes such decisions with no other thought in mind—that each weapon system and force deployment decision is finely calibrated to take account of some threat or vulnerability on the opposing side. While no informed or responsible analyst would seriously contend that the action-reaction process takes place in this automatic and precise way, it is easy to slip into the habit of invoking some such image— or of arguing as though one were invoking it—in attempting to explain some particular Soviet decision.

The congressional hearings on the ABM issue in 1969 turned up many examples of such arguments. Many clearly reflected the assumption that the action-reaction phenomenon applied to the decisions affecting individual weapons systems, not simply to the psychological atmosphere in which such decisions were taken. The fact that a main issue in dispute was whether and how the Soviet Union might respond to an American ABM was proof enough that this assumption was widely shared. As for the specific application of this concept to particular decisions, one could easily gain the impression from piecing together various opinions expressed in the hearings that the sequence of Soviet and American weapon system decisions over the past decade or so had taken place with the inexorable progression of a biblical genealogy. In the beginning was the Soviet ABM . . . so one version of the genealogy might read. Then the ABM begot MIRV, MIRV begot the SS-9, the SS-9 begot Sentinel/Safeguard, and so on. Whatever one may think of the specific judgments involved in this, or any other, reading of the sequence, it is hard to avoid the impression that system-for-system interactions are implied by the way the action-reaction concept is widely used in American discourse. It is in this sense, at any rate, that its adequacy and utility as a tool of analysis stand to be proved in the context of the problem being examined here.

There are good reasons to question this concept on both counts. At the most general level, the evidence of the action-reaction phenomenon that has been manifested in the actual behavior of the two sides over the 10 years or so since the early 1960s is spotty at best. To take the initial deployments of new strategic weapons systems as an index of the phenomenon, the record shows the following: a cluster of American actions at the beginning of the 1960s, a cluster of Soviet actions half-way through, and a flurry of diplomatic activity toward the end aimed at holding both sides at mutually acceptable levels. While this may be good enough to show that some interaction takes place, it suggests that the effects manifest themselves over extended time periods and that a fairly large-grained graph is needed to plot the ups and down of the phenomenon.

The root of the analytical problem involved in making correlations of this kind is that there are various levels at which Soviet and American programs come into contact. The problem is to know where the critical conjunctions take place. Is it at the level of technology where the innovational process may start in the recognition by some scientist of a challenge to be met? Is it at the level of strategy, where the requirements for forces may emerge from the recognition by military planners of a threat to be countered? Or is it at the political level, where the issue of whether an action should be taken may wait on highly visible evidence of the need for action, or even turn on a question unrelated to the matter at issue?

The trouble with the action-reaction concept, in other words, is that it begs the questions that most need to be answered to provide an adequate explanation of Soviet decisions. Granting that the Soviet Union reacts to actions taken by the United States, this fact leaves unanswered the questions that might give substance to this insight. At what level do these reactions take place? What, indeed, constitutes a "reaction"—or an "action," for that matter? As these questions suggest, the action-reaction concept is like a mold without clay. To put substance into the concept requires a more detailed examination of the decision-making processes of the countries being analyzed.

The Dynamics of Technology

One effort to put substance into the concept is the thesis advanced by some analysts that the dynamic factor in the action-reaction phenomenon is supplied by the efforts to the scientific communities

on both sides to exploit technological opportunities. According to
this view, Soviet and U.S. scientists are stimulated to explore new
paths in weapons technology by the challenges they perceive in the
achievements in their adversaries and by the new ideas they get in
the course of meeting those challenges. In this view of things, to
quote the most prominent advocate of the theory,

> The arms race is not so much a series of political prov-
> ocations followed by hot emotional reactions as it is a
> series of technical challenges followed by cold calculated
> responses in the form of ever more costly, more com-
> plex, and more fully automated devices.[3]

This version of the action-reaction concept is a considerable
improvement over the simpler one discussed above, because it is
more concrete and because it is sufficiently ambiguous concerning
the mechanics of the interaction process to allow room for the variety
of ways in which this process may take place. To see how this version
of the process works, it may be useful to consider an illustration
provided by an actual historical case. The case in question is the
origin of the idea for putting multiple warheads on missiles, the so-
called MIRV system, and the reporter is Dr. Herbert York, former
director of the Office of Defense Research and Engineering and a
key participant in the events described.

The circumstances that gave rise to the idea, according to
York, evolved from the U.S. decision in 1956 to launch the Nike Zeus
ABM project. The studies that were begun as a result of that decision
quickly led to the recognition that the problem of developing an ABM
system included the problem of dealing with the ways in which an
enemy could overcome it. Thus, as York describes it, the MIRV
idea evolved out of the efforts of U.S. scientists to foresee all the
problems they would have to deal with in designing their own ABM
system.

To let him tell it in his own words:

> Very soon after [the Nike Zeus project was started], it
> was recognized that the defense problem might well be
> complicated by various hypothetical "penetration aids"
> available to the offense. The Office of the Secretary of
> Defense set up a committee to review the matter. In
> early 1958 that committee pointed out the feasibility of
> greatly complicating the missile defense problem by
> using decoys, chaff, tank fragments, reduced radar
> reflectivity, nuclear blackout, and—last but by no means
> least—multiple warheads.[4]

As this account suggests, York's analysis of the action-reaction phenomenon places the "action" phase of the process at a point one step removed from the actual physical actions of the opposing side. The whole process takes place within the confines of the scientific research establishment to one side only. The other side is a party to the process only to the extent that its presence—and the prudential concerns that it inspires—provides the rationale and the impetus for the whole enterprise.

York makes this feature of his analysis unmistakably clear in the following words:

> It is, I think, most important to note that these early developments of MIRV and ABM were not primarily the result . . . of anything which might be described as a "provocation" by the other side. Rather, they were largely the result of a continuously reciprocating process consisting of technological challenge put out by the designers of our own defense and accepted by the designers of our own defense, then followed by a similar challenge and response sequence in the reverse direction.[5]

York ties this process into the larger framework of the action-reaction cycle by arguing that the further development of the MIRV idea into the actual systems that have now been installed in the Poseidon and Minuteman III missiles was due to the subsequent Soviet programs for developing an ABM system. In other words, when the need arose for contending with an actual ABM threat on the Soviet side, the means were already at hand for meeting the threat. As York sees it, this is how the whole process works. He assumes that the Soviet scientific research establishment also works out new weapons systems in the same atmosphere of internal challenges and responses he attributes to the U.S.

York's account of the internal dynamics of the weapons development process suggests that the impetus for a strategic decision may be less a function of what the other side is doing than of what it may be deemed hypothetically capable of doing based on the state of the art in the technical field in question. Under these ground rules, obviously, the practice of threat assessment could become a highly subjective art form. York is generous enough to avoid dwelling on these possible abuses of the practice. Others, however, have pointed them out. Senator Albert Gore, for example, the then chairman of the Senate subcommittee that heard York testify on the points described above, suggested that the scientists engaged in the MIRV development were probably inspired more by the technical challenge involved than by the evidence that the Soviet Union was actually developing an ABM system. He mused that a scientist's response to a technical challenge could be likened to a mountain climber's response to an unscaled peak.[6]

A much more trenchant commentary on the same point was offered by Dr. Jerome Wiesner, former chairman of the President's Scientific Advisory Committee and thus one who like York had enjoyed an insider's view of the U.S. decision-making process affecting strategic weapons. In a letter to Senator Gore, included in the published account of the subcommittee's ABM hearings, Wiesner pointed out that the tendency to base such decisions on worst-case assumptions about future Soviet developments had the effect of turning the weapons development process into a self-guided and self-justifying enterprise. He wrote as follows:

> Here we have the arms race in its most virulent form. Here we are literally running an arms race with ourselves. If we can just stay the right amount ahead of the Soviet Union so that we have to be concerned with what they might develop, we can keep ourselves completely occupied by developing weapon systems and countermeasures to them without ever having seen the Russian counterpart to which we are responding.[7]

Whether these insights into the U.S. weapons development process can be applied to the Soviet side depends on whether the Soviet administrative system encourages a comparable degree of cross-fertilization and cooperation among technical specialists. This question will be touched upon in the last section of this chapter.

Military Logic

In addition to these more or less well-developed theories about how to approach the threat assessment problem, there are a number of more generalized assumptions about the nature of military policy-planning that affect Western thinking about Soviet military policy-making. The most important of them is the idea that the ends-means calculations that are used in the solution of many military problems, such as the design of weapons systems, for example, represent a method of analysis that is typical of military policy-planning generally. The task of the military planner in this view of things is to devise the means appropriate to achieve designated ends—hence, the term "requirements approach" that is applied to the analytical method in military jargon. To apply the approach to the analysis of Soviet military policy, it is only necessary to turn the process around and to look for the "requirements" appropriate

to explain whatever Soviet action stands to be analyzed. The various
efforts that were made during the ABM hearings in 1969 and later
to explain the Soviet supermissile, the SS-9, by reference to the U.S.
targets it seemed capable of attacking provide examples of how this
approach is used in practice.

The characteristic feature of this approach is the notion that
military policy-planning rests on a highly logical method of analyzing
military requirements. This picture of the military policy-planning
process derives from a variety of sources. Perhaps the most important from the standpoint of the effect it has had on the public imagination was the emergence of systems analysis as a principal tool
of defense policy-planning in the early 1960s. Users of the methodology
say that it is nothing more than an aid to decision-making—a set of
procedures designed to present military problems in a way that
permits the decision-maker to compare the costs and benefits of the
various alternative solutions available to him.[8] To the layman, it
might seem that a system of analysis that defines both the questions
to be asked and the criteria that apply in answering them, and that is
capable, moreover, of generating the kind of powerful argumentation
in favor of the preferred answers that was demonstrated in the annual
posture statements of Secretary of Defense McNamara during the
1960s, cuts rather far into the area of discretion that is formally
reserved to the decision-maker. There can be little argument, however, over the point that systems analysis has introduced a logical
rigor into the military policy-planning process that is widely accepted
as the norm that applies in this sphere of government activity.

A more substantial source of the idea that military policy-planning rests on a peculiarly rigorous application of logical analysis
is the fact that, like physics and astronomy, it deals with matters
that are largely physical in nature. Hence, the choices available
to the military planner are subject to the physical laws that apply
to the issue in question. If the issue concerns the design of a Mach-2 aircraft of a certain size and capacity, the laws of aerodynamics
will dictate that it should look like a TU-144 or a Concorde. If the
issue concerns the planning of a military operation, the laws that
govern the movement of physical bodies in time and space and that
determine the relationships of force that obtain between given masses
of troops (or given units of firepower, to update the classical category)
will dictate the appropriate timetables, deployments, and logistic
plan required to resolve the task in question. Military policy-planning,
in this view of things, comes down pretty much to calculating the
logical requirements imposed by a given set of goals and a given set
of conditions.

According to this picture, any military policy decision can be
reduced, theoretically, to a sequence of steps that can be described

briefly as follows. First, there is a definition of strategic goals. From the goals, specific military commitments are derived. Once the commitments are established, plans are drawn up concerning the contingencies that could arise within the areas covered by those commitments. And, finally, once the contingency plans are drawn up, specific force requirements are defined. Using this scheme as a paradigm of the Soviet decision-making process, one could work his way backward, it is assumed, from the characteristics of a particular weapon system to the strategic goals that started the whole process going.

This description of the "logical" approach to the interpretation of military policy is no mere abstract exercise in theory building. Anyone familiar with the American public dialogue on defense policy matters during the Johnson and Nixon administrations knows that this kind of thinking is applied to the interpretation of Soviet military policy every day. Its best-known exemplar is the columnist Joseph W. Alsop, whose stock in trade has been the ability to see large stragetic implications in small events. The technique involves the assumption that Soviet policy is a seamless garment—an assumption that is widely shared by many other analysts, both in and out of government.

Perhaps the most appropriate comment to be made about this assumption—and about the style of analysis that it supports—is to say that it is far too simplistic to account for the variety of considerations that affect real-life policy decisions. Dr. Charles Schultze, former director of the Bureau of the Budget, testifying before Congress on the ABM question in 1969, gave both eloquent testimony on this subject and some pertinent examples drawn from U.S. experience.

Regarding the logic of military decisions, Schultze stressed the importance of the judgment factor that is involved in the transition from one step on the decision-making ladder to the next. It is an illusion, he said, to think that the outlines of subordinate requirements are somehow embodied in the higher-level requirements that precede them.

> There is no inexorable logic tying one set of decisions to another. Do not think that once a decision has been made on commitments that the appropriate contingencies we must prepare against are obvious and need no outside review. Or that once we have stipulated the contingencies that the necessary force levels are automatically determined and can be left solely to the military for decision. Or that once force levels are given, decisions about appropriate weapon system can be dismissed as self-evident.

> There is a great deal of slippage and room for judgment
> and priority debate in the connection between any two
> steps in this process.[9]

To illustrate the point, Schultze cited the case of the U.S. Navy's policy of maintaining 15 capital ships to fulfill its missions. The formal rationale for the policy, according to Schultze, rests on the calculation that the requirements for a two-and-a-half-war capability figure out to this number of ships. Yet, it is curious, he points out, that the same number of ships were allotted to the navy by the Washington Naval Disarmament Treaty of 1921 and that the same number has been maintained by the navy throughout all the peacetime years since, despite all the changes in missions and contingencies that the intervening years have brought.[10]

Another case cited by Schultze concerns the Poseidon missile system. According to him, the sequence of events was as follows. The Poseidon deployment decision was made against a threat that never materialized, namely, the assumption that the Soviet Tallin system was an ABM defense. Despite the nonexistence of the threat, however, the Poseidon system was continued, presumably as a hedge against other potential threats. Then, new technology made it possible to design hard-target killing accuracy into the Poseidon, an accuracy not needed to deal with the original threat, nor any subsequent potential ones. Yet, the technology was exploited and applied, with the result that a new threat has been posed for the Soviet Union, with all the unforeseeable consequences that that may provoke.[11]

Unfortunately, there are no examples of a similar nature to illustrate how the Soviet Union may apply the principles of military logic in practice. It may be worth observing in this connection, however, that the Soviets seem to live rather comfortably with contradictions. It may be that in looking for the logic in Soviet military policy, we are looking in the wrong dialectical haystack. Beyond this, there is a serious theoretical argument to be made against the assumption that Soviet policy can be explained on purely rational grounds. According to Professor Yehezkel Dror, an Israeli political scientist who has written an exhaustive book on the subject, purely rational policy-making is impossible. To decide policies on purely rational grounds, he says, it would be necessary to construct complete weighted inventories of values and resources relevant to the issue to be decided, to identify all alternatives available to the policy-maker, and to predict the costs and benefits to be anticipated from each course of action. Except for a few problems that are peculiarly susceptible to quantitative analysis, these tasks, he says, are beyond the reach of human knowledge and capacity. Real-life policy decisions, he adds, are affected as much by "extra-rational" or "meta-rational" considerations as by the kind of logical calculations described above.[12]

Models and Analogues

To complete this survey of the ideas and thought forms that affect Western thinking about Soviet military policy-making, it is necessary to say a few words about certain efforts that have been made in recent years to relate the analysis of Soviet policy to the operational characteristics of the decision-making process. Most of these efforts have been distinguished by the attempt to apply insights drawn from U.S. organizational theory to the analysis of Soviet behavior. They have characteristically concentrated on the development of conceptual models suitable for organizing future research.

The two main variants may be described briefly as follows. The first may be called the Organizational Process Model. It attempts to describe the decision-making process in terms of the institutional structure of the decision-making organization. It directs attention to the importance of institutional interests and bureaucratic routines in the formation of policy. The second may be called the Bureaucratic Politics Model. It attempts to describe the decision-making process in terms of the internal contests for power and influence carried on by groups and individuals within the bureaucratic system. It directs attention to the importance of factional considerations and power issues in the formation of policy.[13]

These and other similar attempts to conceptualize the Soviet decision-making process have been prompted by a recognition that the purely rational decision-making model, implicit in all the interpretive devices discussed in the earlier sections of this chapter, fails to bring out the full range of factors that may affect Soviet policy in practice. They have given theoretical structure to an analytical approach that focuses on the internal, rather than the external, motivations of Soviet actions. They have contributed, thus, to broadening the image of the analytical problem posed by Soviet behavior and to enriching the stock of insights that bear on the interpretation of Soviet actions.

Yet, despite these contributions, the behavioral approach to the interpretation of Soviet policy has found few takers. It has been said of the model-builders, indeed, that they seem always to be packing their bags for a trip that never materializes. Instead of satisfying themselves with reasoned criticisms of the results of oversimplified approaches, they seem intent on trying to develop alternative approaches based on thoroughly worked-out models of the Soviet decision-making system. In so doing they risk falling victim to the malady they are trying to cure. For in the absence of better information on the military aspects of the Soviet policy-making process than is now available, it seems inevitable that these efforts must result in models that are highly imperfect or that are filled out on the basis of theories and insights derived from U.S. experience. What they may end up with, thus, is a description of the Soviet system not as it actually is, but as it ought to be, if the general principles of organizational and political behavior observed in U.S. practice apply to Soviet practice as well.

This observation can be applied to all the ideas and thought forms discussed above. All of them share to a greater or lesser extent the assumption that the experience the United States has derived from the arms race provides a norm that can be used to gauge the Soviet side of the same experience. As the above discussion suggests, this assumption can be misleading, not so much because it is necessarily invalid, but because it creates the appearance of knowledge where none in fact exists. In no field of inquiry is it more important to maintain clearly the distinction between what is known and what is not known than in the field of threat assessment. The exigencies of policy necessarily force the policy-maker to make decisions on the basis of imperfect knowledge. There is nothing wrong with this, providing he knows precisely what confidence he can place in the knowledge he has to go on.

To sum up these considerations very briefly, it may be said that many of the approaches that are commonly employed in Western analyses of Soviet military policy tend either to ignore the Soviet decision-making system or to treat it in an oversimplified way. They thus fail to provide a fully adequate framework for analyzing the full range of factors that may affect Soviet military policy decisions. For it is apparent from even a superficial consideration of the actual history of the arms race during the 1960s and up to the present that neither the Soviet Union nor the United States makes its decisions concerning strategic weapons on purely rational considerations alone. In both states these decisions are necessarily affected by subjective judgments. It is an important part of the task of analyzing Soviet military policy, thus, to consider the political and institutional conditions in which the Soviet leaders operate, for it is from the influence generated by these conditions that the subjective elements in their judgments are largely derived.

THE SOVIET DECISION-MAKING PROCESS

What then can be said about how the Soviet Union makes military policy in practice? The evidence suggests that the process is a much more exclusive one than in the United States. Whereas the U.S. constitutional system encourages the intrusion of congressional and extragovernmental influences into the decision-making process, the Soviet system works to restrict the active participants to those charged with direct responsibility for the matters at hand. For the main issues of military policy, this would include the top political and military leaders.

The Framework of Decision-Making

The regular, formal deliberations between the political and military leaders take place in a body known as the Defense Council.[14] This appears to be a functional equivalent of a body that was known in Stalin's and Khrushchev's times as the Higher, or the Supreme, Military Council. Then, as now, its purpose seemed to be to serve as a bridge between the party and government sides of the military leadership structure. Stalin used it as a forum for evaluating the lessons of the Finnish war of 1939-40, for example. Khrushchev used it as a forum for debating the implications of the strategic and force posture issues raised by the advent of nuclear weaponry. Judging by these precedents, the present council is probably concerned with broad issues of military policy, not with day-to-day managerial functions.

Some insights into the character of these meetings can be inferred from these precedents and from general Soviet administrative practice. This evidence suggests that they are probably highly regularized affairs, with agendas set in advance, lists of invitees agreed upon, and so on.

This seems to be the Soviet style, and it has probably been reinforced in recent years by the deliberate effort of the present regime to cultivate systematic procedures to distinguish itself from the Khrushchev regime. This is a style that would also be preferred by the military. It was, indeed, one of the demands implied by Marshal M. V. Zakharov's attack of Khrushchev's "hare-brained" scheming, in a famous article in 1965.

Secondly, the evidence suggests that the routine of this business is probably cyclical in nature. This would be in accordance with the importance of the plan in the Soviet system. The planning cycle extends throughout the year and calls for inputs from government agencies at certain set stages. One can sometimes get a dim reflection of this rhythm in the timing of Soviet press attention to economic matters. Judging by this evidence, it would seem that one critical turning point comes early in the summer and another early in the fall.

It is important to remember in this connection that the plan is not simply a fiscal instrument but a set of specific instructions governing production activity generally. Thus, unlike Western budgets, it does not allow discretionary authority for the use of resources within broad budgetary limitations but prescribes detailed production targets and precise schedules for fulfillment. According to economists the annual plan is more important than the five-year plan, in the sense that it calls for precise commitments, whereas the five-year plan is concerned more with setting general guidelines. This may be relevant to the problem of gauging decision points in long-term weapon programs. It would seem that the five-year plan might provide a good benchmark for dating the initiation of such programs, but perhaps not so good for dating a shift in programs that had already been funded.

Finally, to make a third point regarding the character of political-military consultations at the top level, it is probable that the great bulk of this business is concerned with incremental decisions—that is, with the management of ongoing enterprises. While this is probably true of government business everywhere, it is a point worth making in this context, since we suffer from a professional penchant for thinking of military policy in terms of large-scale programs and global strategies.

The Substance of Decision-Making

So much for the character of these consultations, how about the substance? What kinds of business are conducted between the military and political leaderships?

It is perhaps safe to say, first, that it is confined pretty much to the military field alone. Mixed politicomilitary questions in which the political component is dominant—such as the Middle East, say—would probably be decided by the political leaders themselves, with military experts invited as advisers only.

This still leaves a broad field, of course. It might extend from overall defense plans to the scheduling of missile test firings. The official history of World War II tells about discussions of issues of the first category in a series of meetings in 1940, and the Penkovskiy Papers mention that the question of additional appropriations for test firings was brought up at one council meeting. It would be a mistake to imagine, however, that issues in the first category arise very often, except in the formal sense of periodic reviews of standing plans. In the whole history of the Soviet Union, for example, there have been only two periods when the state's military doctrine has become a subject of broad debate—one during the early 1920s and the other during the early 1960s. As for naval issues, the political and military leaders probably keep a pretty close watch on operational as well as developmental matters. This would almost certainly include reviews of the standing orders governing the patrols of strategic-missile submarines, and probably even more detailed supervision of the port calls of ships of the Mediterranean squadron.

Finally, one would suppose that the political and military leaders would concern themselves with the problem of coordinating and integrating the military's various developmental programs. This is an area of military policy administration in which the Soviets themselves have confessed to problems. They have expressed admiration for the techniques of cost-efficiency analysis used in U.S. defense planning, and military journals in recent years have shown that the Soviets have acquired sophistication in these matters. Nevertheless, there are enough glaring inconsistencies in Soviet weapon programs to suggest that the standards of performance in this respect are not high. Soviet programs may be coordinated around a single strategic principle, but, if so, the principle seems to be that war may take many forms and that all bets should be hedged. This is something that may be worth considering in connection with the problem of discerning the strategic rationales of Soviet naval programs.

Informal Political-Military Relations

Thus far we've been considering the political-military relationship in the policy sphere in its more formal aspects. There is also an informal aspect that tends to come to the fore when the military fails to get its way through the normal channels or when it wishes to protest decisions that it regards as threatening to its interests. On such occasions, the military may take its case to the wider forum of party and government opinion and publicize its differences with the political leadership. The mechanism that is involved here may be described as a kind of political preemptive maneuver. By getting its position on the public record, the military apparently counts on putting the regime in a position where it can disregard military opinion only at the cost of a public display of differences. This can be a more or less powerful means of dissuasion, depending on the regime's general political position and the sensitivity of the issue.

There are many cases that illustrate this form of political action by the military. The best-known, of course, was the military's opposition to Khrushchev's force-reduction programs of the early 1960s. What the military did on this occasion was to get formal approval of the doctrine that all forces and means were necessary to achieve victory in war and then use this as a rationale for opposing Khrushchev's proposals. The result was a long, drawn-out battle that had the effect of slowing down Khrushchev's reform, if not changing it substantially, and a general cooling down of the atmosphere in which the restructuring of Soviet forces could proceed.

There are also instances in which the military has apparently attempted to use publicity to influence budgetary decisions. During the summers of 1966 and 1967, for example, when preliminary discussions concerning the plan and budget were presumably under way within the government, the military press was notably active in agitating for continued priority for heavy industry and defense. Strongly argumentative articles on this theme have appeared periodically since.

Arms control is another issue on which the military has tended to go public. Instances of such activity can be traced back to the early 1960s at least. Military opposition to the Test Ban Treaty was notably demonstrative—so much so that the Chinese alluded to it in their propaganda against Khrushchev. Opposition to SALT was also displayed in the military press, particularly in the period proceeding the opening of the talks. This included some indirect arguments against the advisability of the talks but was more typically expressed by withholding any favorable mention of the idea, including selective editing of official statements to exclude mentions of SALT.

In recent months there have been signs that the military may be agitating against detente diplomacy. There have been a number of articles that seem to have been arguing against the proposition that military force has lost its political utility in the modern world. This is a proposition that has been advanced off and on by Soviet writers since General Talenskiy first publicized it in 1965, and it seems to have been adopted as an argument for detente by some of the recent supporters of the regime's current policy. Georgy Arbatov used it, for example, in his Pravda article in July 1973. The military articles have been arguing, on the contrary, that the best way to achieve and maintain peace is through strength. The arguments have been formulated in such a way as to suggest that they have been specifically directed against the Talenskiy-Arbatov school of thought, and hence against the policy that this school of thought is currently serving.

SOVIET MILITARY DOCTRINE

This brings me to the final question I want to consider, Soviet military doctrine. (See the next section for the Soviet definition of "military doctrine" and the related terms "military science," "military art," and "military strategy.") This is a much misunderstood concept among nonspecialists, and I want to say a few words about it to suggest how it may fit in with the subject of this paper and the problems being discussed at this conference.

First, to be clear about the concept, it is important to recognize that it has nothing in common with Western concepts of doctrine. It is a highly formularized set of theses about the nature of a future war that reflect the broad guidelines adopted by the Soviet political and military leaderships for the development of the armed forces. Almost every description of the doctrine in the Soviet military literature contains some references to the joint nature of the doctrine, to the fact that it registers the views of both the political and military leaderships.

Secondly, judging by the descriptions we've been given of it in the Soviet military literature, it is a highly generalized and ambiguous set of guidelines. Each proposition it contains seems to be balanced by a contradictory one. For example, it holds at the same time,

- That war may be short, but it may also be long;
- That strategic missiles will be the main weapon, but that all forces and means will be necessary for victory;
- That war may be nuclear, but it may also be conventional.

As a set of guidelines for the construction of the armed forces, in other words, it is a very permissive document.

These two features of the doctrine—its joint politico-military authorship and its permissiveness—suggest the role it plays in the Soviet political-military relationship. It is a charter attesting the military's right to participate in policy formulation and general authorization for a broad and comprehensive development of the armed forces. To interpret it as a key to Soviet strategic policy would be both to misread the doctrine and to underestimate the dynamics of the Soviet decision-making process.

SOVIET DEFINITIONS OF "MILITARY DOCTRINE" AND RELATED TERMS

The definitions given below are from the General Staff Academy, Dictionary of Basic Military Terms (Moscow: Voenizdat, 1965).

Military doctrine is an officially state-adopted system of scientifically based views on the nature of modern wars and the use of armed forces in them, as well as on the requirements that ensue from these views for preparing the nation and armed forces for war.

Military doctrine has two aspects: the political and the military-technical. The basic tenets of military doctrine are determined by the state's political and military leadership in accordance with the sociopolitical system and the level of development in the economy, in science and technology, and in the military-technical equipping of the nation's armed forces, taking account as well of the conclusions of military science and the views of the probable opponent.

Military science is a system of knowledge concerning the nature, essence, and substance of armed conflict, the forces, means, and methods for the carrying out of combat actions by armed forces and their all-around protection.

Military science conducts research into the objective laws and law-abiding patterns of armed conflict, works out questions in the theory of military art, which forms the basic content of military science, and questions concerning the development and preparation of the armed forces and their military-technical equipment and also analyzes military-historical experience.

Soviet military science is based on Marxist-Leninist teachings and is guided by the methods of materialistic dialectics and historical materialism, in this regard taking into account and using other sciences that promote constant improvement and progress in military affairs.

<u>Military art</u> is the theory and practice of carrying out battles, operations, and armed conflict in general, with the use of all types of troops and branches of the armed forces, as well as of ensuring military actions in all respects. As a scientific theory, military art is a principal area of military science and includes tactics, operational art, and strategy, which occur in organic unity and mutual interdependence.

<u>Military strategy</u> is the highest realm of military art and consists in a system of scientific knowledge concerning the phenomena and law-abiding patterns of armed conflict.

Strategy (military)—on the basis of the tenets of military doctrine, the experience of past wars, and the analysis of political, economic, and military conditions of the present situation—conducts research and works out questions concerning the preparation and strategic wartime use of the armed forces in general and of the branches of the services, questions concerning the forms and means of carrying out armed conflict and commanding it, and questions concerning the all-around strategic protection of military actions by the armed forces.

NOTES

1. Remarks by Secretary of Defense Robert S. McNamara, September 18, 1967, quoted in <u>Bulletin of the Atomic Scientists</u>, December 1967, p. 28.
2. These comparisons between modern and classical aspects of strategy are taken from Gen. Andre Beaufre, who offered them in his book, <u>An Introduction to Strategy</u>, Maj. Gen. R. H. Barry, trans. (New York: Praeger Publishers, 1965), p. 100.

3. Herbert York, "Military Technology and National Security," Scientific American, August 1969, p. 27.

4. Herbert York, "ABM, MIRV, and the Arms Race," Science, July 17, 1970, p. 257.

5. Ibid.

6. U.S. Congress, Senate, Subcommittee on International Organization and Disarmament Affairs of the Committee on Foreign Relations, Strategic and Foreign Policy Implications of ABM Systems (Part III), 91st Cong., 1st sess., July 16, 1969, pp. 691-92.

7. Letter from Dr. Jerome B. Wiesner to Senator Albert Gore, dated June 4, 1969, quoted in ibid. (part II), May 14 and 21, 1969, pp. 590-95.

8. See Alain C. Enthoven, "Systems Analysis—Ground Rules for Constructive Debate," Air Force Magazine, January 1968, p. 33, for both a lucid explanation of what systems analysis can do and a modest disclaimer of the immodest claims that have sometimes been credited to it.

9. U.S. Congress, Subcommittee on Economy in Government of the Joint Economic Committee, The Military Budget and National Economic Priorities (part I), 91st Cong., 1st sess., June 1969, p. 53.

10. Ibid.

11. Ibid., pp. 51-52.

12. Yehezkel Dror, Public Policy Making Re-examined (San Francisco, Calif.: Chandler Publishing Co., 1968), pp. 132-33.

13. See Graham Allison, "Conceptual Models and the Cuban Missile Crisis," American Political Science Review, September 1969, for a rigorous analysis of these models and for an interesting attempt to apply them to the interpretation of an actual historical case.

14. The existence of this body was mentioned by David Mark of the State Department in testimony before Congress on the ABM issue in 1969. He described it as a "sort of limited National Security Council." See U.S. Congress Subcommittee on Economy in Government of the Joint Economic Committee, The Military Budget and National Economic Priorities (part III), 91st Cong., 1st sess., June 1969, p. 956.

CHAPTER
4

**SOVIET DEFENSE
POLICIES AND
NAVAL INTERESTS**
John Erickson

Though inevitably couched in general terms, this paper is concerned with three main themes: (1) the Soviet military system at large, with special reference to developments within the period 1970/71-73; (2) the position of the Soviet Navy within this system (and the configuration of pressure and opinion within the navy itself); and (3) the appreciation of the "naval mission." Each of these topics raises substantial questions of methodology, or, put more simply—semantics. It is clearly difficult to estimate the influence of the Soviet Navy within "the system," when our understanding of that same system is imperfect, to say the least: Equally, what is "influence" and, more particularly, naval influence? There are, of course, crude indicators (such as the military-political "pecking-order," or assumptions about resource allocation), but none of these help appreciably in elucidating the raison d'être of the Gorshkov series. This difficulty is compounded by recent changes in the Soviet military scene, and it is those changes that form the first qualification on the generalities involved in this paper: In brief, there are three features that dominate the period 1970/71-73 and that must be admitted in any appraisal of the system at large and the Soviet Navy in particular—the attainment of strategic parity and the consequences of SALT-1, the expansion of Soviet overseas presence (usually only with naval forces, though other combat forces cannot be excluded), and, finally, the random changes brought about within the senior levels of the Soviet command through death and retirement, thus reshaping the command in a limited but observable fashion—for example, the death of the Chief of the General Staff (CGS), Marshal M. V. Zakharov, the installation of General Kulikov, and the consequent realignment in "influence" through personalities and institutions at this highest level.

THE SOVIET DEBATE ON "MISSIONS"

The central question about the Gorshkov articles appears to be, Why were they written and why at this juncture? That they are unique in the annals of Soviet military writing needs no emphasis, for never in more than 50 years of military publication has a Commander in Chief captured his own professional journal for such a sustained exposition. Before tackling this issue directly, let us look at it indirectly, in terms of the general "debate" being conducted within the Soviet command. "Debate" is perhaps a misnomer, but there is no real equivalent for this process of argument and counterargument, conducted without public lobbies (as in the West). From the Soviet point of view SALT-1 has done much to transform the scene, instituting "parity" with the United States, inducing a measure of control without inhibiting the military but at the same time raising the fundamental question of the utility of military force—what kind of force—for the effective pursuit of foreign policy objectives. While it is true that there is no diminution of the notion that military weight is the best guarantor of national security, it has been impossible to avoid the issue of the utility of military power as a whole and, more important, the emergence of the idea of reliance on lower levels of force to support foreign policy goals.

Thus, in this context, Gorshkov's articles are part of the continuing dialogue about strategic power, to which the Soviet Navy is a contributor, but also must be considered as a facet of this ongoing debate about military power in general and, specifically, in relation to the pursuit of external political goals.

It would be too much say that the other branches of the Soviet armed forces have made efforts comparable to those of the Soviet Navy, at least in terms of publicity, but it would be unwise to ignore the recent revisions advanced by the ground forces (enunciating a modified posture that takes in a radical view about the conventional phase of possible operations in the European theater), or the Soviet Air Force (SAF), which under Kutakhov has discreetly pushed its claims for increased long- and medium-range lift, as well as greater flexibility.

In brief (and leaving aside much of the textual evidence), there is some substantial evidence of a significant struggle within the Soviet military establishment for "missions" (and thus for resources). Looked at from this point of view and within the context of the "debate" as a whole, Admiral Gorshkov has undoubtedly provided an impressive argument about the requirements of and for "seapower," but on the face of it it looks doubtful that the Politburo has "bought" the whole thing: In sum, he is arguing for more ships for general war tasks and also for extra ships for the execution of peacetime roles. In this sense, it could be construed quite sensibly that Gorshkov is fighting a rearguard action against the missile forces and elements such as the SAF (Soviet Air Force), though the navy has a major advantage in pressing a singular case for the maximizing of political gains by the exercise of power and presence below the strategic-nuclear level. (Additionally, Admiral Gorshkov seems also to be arguing against that proposition of the "air and submarine/missile" combination as a basic solution of naval tasks—if only because he is faced with special and technical problems of ship/class replacement and is obviously not anxious to tie his own hands.)

There is one final subjective element in the Gorshkov presentation: It is conceivable—and even admissible—that Admiral Gorshkov will retire gracefully in the none too distant future (even allowing for variations in the provisions of the 1967 Military Service Law, there are a number of "front-runners" either for a short-term holding appointment or other potential commanders in chief, with both Yegorov and L. V. Smirnov having been recently made admirals of the fleet). It could be that the Gorshkov series is a "last will and testament," or a classic and forceful exposition of the historical and current implications of a true "sea-power" philosophy for the Soviet Union and hence for the Soviet Navy. There could be something of ave atque vale in these articles, an impression reinforced by the fact that—save for some small textual exceptions—this series has been written by Gorshkov himself, with full emphasis on the implications of a full-blown sea-power philosophy: This could also account for the pronounced defensive tone in much of the writing, as well as some recherche comments on the naval implications of the present strategic situation . . . such as the assertion that what the Soviet Union faces is a coalition of maritime powers (unlike the usual presentation of the potential enemy from other quarters).

The briefest conclusion here is that the Gorshkov series, for all the singularity, cannot be isolated from the general debate on missions and force levels (politically useful force levels) within the Soviet command: Certainly, the Gorshkov series represents a total philosophy of "sea power" (though there is the strong reservation that the Soviet Navy does not intend any inroads on the preserves of the Soviet military at large), and it is more complete than any other representation from other arms or services: Finally, it could be that there is an element of the "naval testament" assigned by Gorshkov to his successors (and this would account for the full historical panoply of Russian and Soviet naval power), both an apologia pro vita sua and a prescription for the future. All of this would fit into the prevailing climate of opinion and the mood of dispute in intramilitary terms within the Soviet establishment. Perhaps the most extravagant thing that can be said in Gorshkov's favor, vis-a-vis his series, is that he "got his blow in fust," but, for all that, "sea power" still remains something of an arcane and ill-understood affair within the Politburo at large, to say nothing of the remainder of the military establishment.

THE SOVIET NAVY AND THE SOVIET MILITARY SYSTEM

This section must of necessity fall into two parts: (1) the navy and the system at large; and (2) developments within the Soviet Navy that increase its institutional effectiveness and at the same time the role of divisions within this service. Inevitably, this raises the question of how important or potentially "war-winning" are individual branches of any military establishment: This question must rest in the last resort on historical tradition or special contemporary roles— in this sense, the Soviet Navy is a newcomer in most aspects, and it is hard to derive tests to indicate the progress of the navy. For example, there is a case to argue that even now the Soviet Navy is not entirely a "full partner" in the strategic strike role: In this context, it is worth noting that Gorshkov emphasized three roles for the navy—strategic strike, counterdeterrence, and support from the seaward side of the ground forces flanks (ironic indeed, that the latter should reemerge, when the Soviet Navy spent years trying to rid itself of this "hand-maiden role" to the Red Army).

The Soviet Navy within the Military System

So, to the first question, the nature of the Soviet military system and the place of the Soviet Navy. Few, if any, would deny that this remains in toto a "soldier-dominated" system—that is, senior officers drawn from the missile and ground forces, with a lesser representation for the specialist arms (navy and air force). There is presently a naval representation at the General Staff level, with Lobov, but this is nothing new, for preceding Chabanenko—and Ivanov and Vinogradov— there is much to suggest that the navy had at least a nominal seat at this military high table. The "influence" of the navy within the high command must be a tenuous affair: While purely naval issues demand greater attention to professional naval opinion, this is not to say that the navy has made the great breakthrough. The navy has certainly made its case with singular forcefulness, the peculiar nature of the naval commitment, the striking aspect of the potential enemy (as a maritime coalition). Thus, it is hard to measure "naval influence" simply by looking at the roster of high command appointments: At least the navy is represented within this senior echelon, while the SAF is not. And even this begs the question a little, for the main weight of the Soviet system seems to lie as yet (and as before) with the buildup in strategic missile power, with new systems emerging from the momentum of the existing program, all with a marked and steady momentum dating back to the very early 1960s.

In general terms the Soviet Navy has achieved "equal status" ranking with some of its senior posts—and most recently the two senior fleet commanders (Yegorov/northern and Smirnov/Pacific) brought up to the rank of admiral of the fleet (with two air force appointments and three army appointments to full general of the army). This says little or nothing, save for a continuation of the careful balancing act. In one respect, however, the Soviet Navy is an exception in that at a much earlier stage it cleared out its World War II commanders come to the fore in the navy. There is not as yet, however, any sign of an independent naval breakthrough into the senior levels of the military command, which remains (and is likely to remain) "soldier-dominated" for sometime to come, in spite of the recent spate of naval upranking.

Not that the Soviet Navy has done all that much, however heretical this may sound. The brunt of Soviet naval activity has been fundamentally to extend Soviet fleet areas, with a forward deployment that was ill-prepared and seems to have caught Gorshkov unawares, or napping somewhat. As for other advances, these have enhanced the standing and the cause of the navy—the new mentality with its ocean orientation, the "buying into" the strategic delivery system, and even the whole mercantile effort—but they have not appreciably advanced its position within the establishment as a whole (for example, the prospect of a naval officer as CGS is remote in the extreme, even if there is an assistant chief/Navy at CGS level). This, and other things, would explain the motivation for the Gorshkov series as a serious attempt to explain the implications of true sea power to a skeptical command and an unheeding Politburo.

Configurations of Naval Opinion

There is undoubtedly a defensive tone in parts of Gorshkov's treatise, and he can be defending himself against none other than his fellow professionals. Much in the same way that the Soviet Navy is, in terms of shipbuilding and commissioning, several navies, the one overlapping the others, so the same can be said of the whole naval command, which has a fascinating profile. Oddly enough, the Soviet Navy has moved from being the most purged* (and unstable) command in the Soviet system to one enjoying the greatest stability and planned continuity, with the naval command showing considerable foresight in planning manpower and command changes in the early 1960s. Such

---

*Quantitatively and qualitatively, all the way from the mid-1930s through to the end of the 1940s (and allowing for the fact that the change of the mid-1950s was a form of "cold purge").

a pace has obviously altered the face and composition of the naval command, with rapid technological change introducing new elements into the command structure as a whole, making further divisions by age and experience. Viewed against this background (and who would know it better than Admiral Gorshkov himself?) the present series of articles can be seen as an engagement—not always hostile, by the way—with several naval "lobbies," of which five can be identified with a certain degree of precision:

The Naval Professionals. These senior officers are the guardians of the "naval mission" and the proponents of morskaya kultura—they include older and even elder officers who even in the dark days pressed for the independent naval mission and looked for an oceanic outlook— they include the professional naval/strategic thinkers, operational and Naval Academy staff and a special group of retired admirals—here the disputes are narrowly technical and professional, and it is possible that some elements do not like the kind of navy Gorshkov has built up (or not built up, as the case may be). Gorshkov is obviously defending himself in a narrowly professional sense in his own testament of naval policy and his view of sea power—this could be checked out through a careful reanalysis of the text, particularly in the choice of terminology. Thus, this is an "in-house" argument and, from Gorshkov's point of view, probably an attempt to arrive at a professional consensus about peace and war roles for navies.

Command Officers. These comprise a tight group of officers (captain-1st class) and above, concerned with specific operational aspects of naval warfare and identified more often than not by their branch. Gorshkov does not, in fact, enter into much operational detail, and it is interesting to note that subsequent articles in Morskoi sbornik have in this specific context generally supported Gorshkov's main operational arguments—for example, the vulnerability of fleets in bases to enemy attack. . . .

Engineers/Cyberneticians/Managers. This is technical expertise, almost pure and simple, and Gorshkov's writings are largely bereft of any discussion on this theme, save for generalizations about technological advance in toto.

Strategists/Popularizers/Political Staff. Numbering several polemicists, retired admirals, and popularizers (including the veterans organizations), this group is a "naval lobby" in the general sense, and it may be that the Gorshkov series is grist for their mill, not that such mills grind exceeding small (if at all, save for keeping up interest in and enthusiasm for naval service).

The Naval Air Force. Gorshkov is undoubtedly aware of the importance of naval aviation and he has ignored (or deleted) any detail about naval air operations as such, except to emphasize the principle that "naval air" operations must come under naval control—this would not preclude close collaboration between naval aviation and the SAF (which is presently the case), but the navy must retain control and now it has its own Marshal of Aviation (navy), a precedent in its own right and presumably a move to hold the door open for some up-and-coming officer (Kuznetsov?).

All of these "interest groups" are represented—and, in some cases, countered—in the Gorshkov series: From that point of view, this is certainly an "in-house" exposition and is both defensive and assertive in form—indeed, here is a whole theme that can be explored by textual and source analysis.

THE PERCEPTION OF THE "NAVAL MISSION" AND NAVAL REQUIREMENTS

In understanding the Gorshkov series, to concentrate on the purely "naval" side is perhaps myopic: Certainly, not to place these articles in the context of an ongoing Soviet "debate" is misleading. If there is any logic (which there is) in trying to see why these articles were written, then it is necessary to look at the whole Soviet scene and the post-1970/71 changes. Additionally, but perhaps peripherally, this could be Gorshkov's "testament," a justification of his past record and a projection of a sensible approach to the utility of sea power, aimed at his fellow professionals and the political leadership alike.

Gorshkov is pressing the case for both diversification and flexibility: In addition, he seems to be making a strong case for the survivability and effectiveness of appropriate surface strike forces, thus offsetting the "air/submarine" school: He proposes naval forces suitably equipped to carry out a wide range of tasks, for which reason he is prepared to play fast and loose with the term "state interests" (too rigid a definition would harm his case), since this is evidently the common parlance of a wider discussion.

His complaint about the neglect of Russian sea power tends to reinforce the view that he has put the case but by no means won it at the highest levels: Nonetheless, since he has displayed it in public, he cannot be criticized on this score.

Though possessing a growing and formidable navy, the Soviet Union is still not a "sea power" in the accepted sense. Those who wish to know how to accomplish this transformation should read Gorshkov, according to Gorshkov.

CHAPTER 5

THE BUREAUCRATIC POLITICS OF WEAPONS PROCUREMENT
Edward L. Warner III

A central element of a state's defense policy is its force posture—that is, the aggregate of military capabilities it possesses. The numbers, characteristics, and deployment dispositions of a nation's military forces rarely, if ever, reflect the implementation of a coherent master plan. Rather, a nation's force posture at any given time represents the cumulative product of past decisions on the development, procurement, and deployment of numerous individual weapons systems. In most cases, these armaments have been acquired over an extended period by various political and military leaders, who acted for differing reasons under a variety of domestic and international circumstances.

The composition of a state's force posture, including its naval capabilities, is affected by a myriad of foreign and domestic policy considerations. From the international side, weapons acquisitions and deployments are influenced by the perceptions and judgments of the political leadership regarding: (1) the nation's foreign policy objectives; (2) the utility of military power in international politics; (3) the nature of the threats and opportunities that confront the state; and (4) the desirable military relationships that should be maintained with adversaries and allies. Domestically, it reflects the economic and technological capabilities of the state, the relative priority assigned to military preparedness by the political leadership, the self-interested activities undertaken by the military services and the defense industrial producers to promote the development and acquisition of specific weapons systems, and the standardized procedures that shape the processes of defense decision-making in this area.

The views expressed in this chapter are those of the author. They do not necessarily reflect the views of the U.S. Air Force, of which he is a member, nor of the Department of Defense.

SOVIET FORCE DEVELOPMENT

The development of Soviet military capabilities since World War II has occurred within the context of an arms race with the United States—that is, Soviet force posture has developed in an environment characterized by (1) conscious antagonism between the Soviet Union and the United States; (2) a mutual structuring of forces with attention to their deterrent and combat effectiveness against one another; and (3) an ongoing competition in terms of both the quantitative size and the qualitative characteristics of their armed forces.[1] While this armaments competition has produced irregular spurts of deployment activity on both sides, it has been marked by mutually sustained efforts in weapons research and development across a wide spectrum of systems.

The pattern of Soviet-American weapons interactions has been highly complex. In light of the lengthy lead times involved in the development of modern weaponry, their competitive efforts to acquire sophisticated offensive and defensive systems have often begun simultaneously. Khrushchev noted, for example, that the Soviet ICBM and ABM development programs were both initiated at the same time.[2] Under these circumstances, it is difficult to establish who is reacting to whom.

The interactive U.S.-Soviet competition is complicated by the extreme secrecy that surrounds the major weapons development programs in both countries. Largely denied information about the early stages of the opponent's research and design efforts, the political-military leaderships on both sides are prone to fear the worst. In the name of prudence, they frequently employ worst-case analysis, which attributes to the adversary maximal weapons development efforts and optimum operational performance for a wide variety of weapons. Initial development efforts for offsetting systems are often undertaken based upon anticipated or vaguely perceived activities rather than the directly observed programs of the opponent. As a practical matter, the efforts attributed to the enemy are likely to be those that one's own weapons researchers have conceived and proposed. As a result, programs undertaken in response to such anticipated developments are often the product of an "arms race against oneself" in which one's own offense is pitted against one's defense in a manner that fortifies the claims of each. This process can easily produce an action-overreaction pattern, where, although the anticipated threat fails to materialize, the "response" nevertheless results in the procurement of a major weapons system.[3]

Arms race competition with the United States is not the sole external factor influencing the evolution of contemporary Soviet force posture. Concerns about the projection of Soviet political influence throughout the world,[4] about the political reliability of the Eastern European communist states, and increasingly in recent years, about the political and military challenge presented by the People's Republic of China are additional considerations that are likely to continue to shape the composition and deployment of the Soviet force posture.

All of these foreign policy considerations acquire significance only when perceived and acted upon by influential members of the Soviet defense policy-making community. The perceptions and actions of these participants tend to be importantly influenced by their varying political preferences, and, in many cases the institutional roles they occupy. Conservative elements, like the members of the Soviet military, are prone to emphasize the threats posed by the West and probably China as well and to counsel the pursuit of strategic superiority over these enemies. In contrast, persons viewing these countries less malevolently will be inclined to endorse small military procurements.

The contours of Soviet force posture have also been fundamentally influenced by several domestic considerations. The Soviet political leadership has consistently accorded the highest priority to the nation's military preparedness. This precedence has been manifested in the generous allocation of budgetary support, high-quality manpower, and the most advanced technological inputs—computers, high-precision machine tools, and so on—to the defense sector of the economy.[5] These priorities have strengthened the political importance of those engaged in this effort.

The maintenance of comprehensive weapons development and production programs is strongly promoted by those with a commitment to the continuous modernization of the Soviet force posture. Such a commitment to the "doctrine of quality"[6] is generally prevalent among those weapons designers and defense producers whose institutional and personal prosperity is closely tied to the level of Soviet activity in this area. These constituencies receive additional support from the services of the Soviet Armed Forces, who share their devotion to the acquisition of numerically large and qualitatively advanced weapons inventories. With each service and its attendant designers and producers seeking attention and budgetary support, the distribution of resources among these groups is bound to reflect their relative political power and respective ability to convince the political leadership of the priority of their demands. This internal dimension represents a central dynamic on both sides of the Soviet-American arms competition.

Not only the promotion of institution interests but also established traditions and standardized organizational practices in the process of weapons development and force deployment have importantly shaped the evolution of Soviet force posture. Traditions, frequently embedded in explicit military doctrine, can provide an important source of advantage for a particular military service and its associated weapons producers. Thus the Soviet doctrinal commitment to the massive, combined arms theater offensive as an integral part of modern war strengthens the claims of the ground forces and tactical aviation for the maintenance and improvement of their extensive and diversified military capabilities. Similarly traditional emphasis upon military operations in Europe may help account for the priority accorded to the procurement of the large numbers of medium and intermediate-range ballistic missiles and medium-range bombers since the early 1950s.

With regard to organizational process, weapons acquisitions and deployments can be strongly influenced by the normal operation of regularized bureaucratic routines. Thus the apparent practice of allotting sizable budgetary share to each of the five independent military services appears to stimulate across-the-board weapons developments and acquisitions.

In the area of force deployment, the Soviet movements of men and weapons into distant areas such as Cuba and Egypt appear to have been accomplished in accordance with standard operating procedures originally developed for the employment and use of these forces in the USSR and its contiguous areas. Thus the medium- and intermediate-range ballistic missiles and their associated antiaircraft surface-to-air missile (SAM) sites clandestinely moved into Cuba in 1962, were erected in the same configurations as similar systems deployed in the Western USSR—a practice that contributed importantly to their identification by U.S. intelligence.[7] In a similar manner, Soviet Ground Force units deployed to Cuba in 1962 and the Soviet equipment provided to the Egyptian Army in the mid-1960s included the complete inventory of weapons: tanks, personnel carriers, forest-clearing equipment, and so on, carried within a standard Soviet division designed to fight in central Europe, without apparent concern for local theater requirements.[8]

PLANNING IN THE DEFENSE SECTOR

The Soviet military force posture is developed, procured, and maintained in accordance with a variety of economic plans. The Ministry of Defense, like all Soviet governmental institutions, operates within guidelines established in its annual and five-year plans.[9] The plans for the Ministry of Defense are in turn components of more comprehensive plans of corresponding length, which guide the functioning of the Soviet national economy.

Given the length of time that passes between the initiation of preliminary design work and the operational deployment of a modern weapons system, a series of longer-range weapons development plans extending beyond the familiar five-year period are also likely to exist. Robert Herrick[10] and Michael MccGwire[11] have written of the existence of 10-year and 20-year Soviet naval construction plans since World War II. A recent book on planning practices within the Soviet Navy adds credence to this speculation by noting the existence of "future plans" covering "considerably prolonged segments of time on the order of 5, 10, 15, 20 and more years."[12] Because of the developmental lead times involved, plans on the order of 10 to 15 years in length are likely to guide the design and procurement of Soviet strategic offensive and defensive missiles and modern aircraft as well. The likelihood that long-term plans are employed in these cases is reinforced by the growing use of 15-year plans within the other, more "visible" sectors of the Soviet economy.[13]

Formulation of Defense Plans

Soviet military authors have published a few articles that discuss Soviet military planning during the five-year plans prior to World War II.[14] Information from these accounts has been combined with available data on the institutional framework for Soviet defense policy-making and descriptions of the routine Soviet planning practices in other areas[15] to develop a speculative depiction of the steps involved in the preparation and approval of the various plans that guide the activities of the Soviet Armed Forces.

The plans of the Ministry of Defense appear to be drafted by a section of the General Staff,[16] which probably operates with the assistance of the ministry's Central Financial Directorate.[17] These plans are evidently developed in accordance with preliminary directives provided by the minister of Defense and on the basis of requests prepared by the main staffs of the independent services and combat branches of the Soviet Armed Forces.[18] The elaboration of the ministry-wide program guidelines almost certainly requires the resolution of serious interservice conflicts. Most of these conflicts are likely to be settled within the General Staff,[19] although in some cases they may involve higher-level negotiations between the service chiefs and the minister of Defense. (Interservice negotiations at this level might take place within the collegium of the Ministry of Defense, whose membership is likely to include the minister of Defense, his three first deputies, and the nine deputy ministers of Defense.)[20]

The weapons development and production aspects of a defense plan must be closely coordinated with the corresponding plans prepared by the industrial ministries engaged in defense production (See Table 5.1). This coordination is likely to be based on the close and mutually supportive working relationships that are maintained between the services and their defense industrial producers. These ties apparently include regular contacts between weapons development directorates of the military services[21] and the design bureaus and series production plants of the industrial ministries[22] as well as direct links between the main staffs of the services and the central apparatus of the defense production ministries. Projected research and development programs and series production runs are likely to be included within the plans of both the Ministry of Defense and the appropriate defense industrial ministries.

Direct informal exchanges between the services and the industrial ministries in the parallel drafting of their individual plans is probably supplemented by a formal coordination of the Ministry of Defense's requirements with the available defense production capacity. This is likely to occur under the auspices of the interministerial Military Industrial Commission, which is apparently chaired by L. V. Smirnov.[23] This coordination is also likely to include the participation of representatives from a defense section of the State Planning Committee (Gosplan), who are responsible for integrating the defense effort within the national economy.[24]

Approval of Defense Plans

The establishment of defense spending priorities and thus the final approval of the plans of the Ministry of Defense and the defense industrial ministries almost certainly lie with the party Politburo. Supervision of the preparation and implementation of these plans prior to and following Politburo consideration probably is accomplished by party secretary and candidate member of the Politburo D. F. Ustinov, who is likely to rely in turn upon guidance provided by General Secretary Brezhnev. Ustinov probably performs this task with the assistance of his personal staff and the Defense Industries Department of the Central Committee headed by I. D. Serbin. Given his lengthy involvement in the high-level supervision of defense industrial matters, which dates from the early 1940s, and his close associations with L. V. Smirnov and the heads of the defense production ministries since that period, Ustinov is likely to enjoy enormous personal power in the direction of Soviet armaments programs.

Ustinov's relationship with the Ministry of Defense and its minister, Marshal A. A. Grechko, on military-economic planning and other defense matters is difficult to determine. Prior to Grechko's elevation to the Politburo in April 1973, Ustinov, as both a party secretary and candidate member of the Politburo, is likely to have

TABLE 5.1

Defense Industrial Ministries and
Their Military Products

Ministry	Product
Ministry of Defense Industry	Artillery, tanks, armored vehicles, small arms, fuses, primers, propellants, explosives, and possibly tactical guided missiles
Ministry of Aviation Industry	Aircraft, aircraft parts, and probably aerodynamic missiles
Ministry of Shipbuilding Industry	Naval vessels of all types
Ministry of Electronics Industry	Electronic components and parts (subassemblies not finished equipment)
Ministry of Radio Industry	Electronic systems including radio and communications equipment, radar, and computers
Ministry of General Machine Building	Strategic ballistic missiles and space vehicles
Ministry of Medium Machine Building	Nuclear devices and warheads.
Ministry of Machine Building	Possibly some portion of ballistic missiles and space vehicles

Note: Other ministries that contribute to military production include the Ministries of Instrument Manufacture, Tractor and Agricultural Machinery Building, Chemical Industry, and Automobile Industry. See Thomas W. Wolfe, "Soviet Interests in SALT: Political, Economic, Bureaucratic and Strategic Contributions and Impediments to Arms Control," P-4702 (Santa Monica, Calif.: Rand Corporation, September 1971), p. 27, footnote 48.

Source: Andrew Sheren, "Structure and Organization of Defense-Related Industries," U.S. Congress, Joint Economic Committee, Economic Performance and the Military Burden in the Soviet Union, 91st Congress, 2d Session, (Washington, D.C.: Government Printing Office, 1970), pp. 123-31.

exercised some supervisory powers regarding defense production matters over the minister of Defense, whom he outranked. Since Grechko has become a full member of the Politburo, however, Ustinov's directive role has probably decreased. Grechko is now likely to be able to represent military interests on such questions directly to Brezhnev and the other members of the Politburo. Whatever their personal relationship, in light of the symbiotic relationship between the Soviet military and its defense industrial producers, Grechko and Ustinov are both likely to lobby for generous defense allocations and extensive weapons development programs.

Prior to their consideration by the full Politburo, the plans of the Ministry of Defense are likely to be closely examined within its military policy subcommittee, the Defense Council.[25] This body apparently includes civilian Party leaders Brezhnev, Kosygin, Podgorny, and Ustinov, the leading military spokesman Grechko, and possibly Marshal I. I. Yakobovsky, the first deputy minister of Defense, and General Kulikov, the chief of the General Staff.[26] It is probably responsible both for providing initial guidelines to the appropriate agencies at the beginning of the planning process and for making recommendations to the Politburo regarding final approval of the plans developed by the Ministry of Defense and the defense-industrial ministries. The deliberations of the Defense Council are very likely to be led by Brezhnev and probably will reflect his consensus-building political style. Consequently, the plans forwarded by this committee to the Politburo for its consideration and approval are likely to reflect in advance the composite preferences of its members.

SOVIET WEAPONS ACQUISITION

The panoply of weapons held by the Soviet Armed Forces is a central element of Soviet military capability. This diversified arsenal has been designed and procured within the budgetary guidelines established in the various plans developed by the Ministry of Defense as well as the defense industrial ministries and subsequently approved by the political leadership.

The Soviets have developed a distinctive national style for weapons design and acquisition. This pattern appears to be followed for the development of a wide range of armaments ranging from small arms to ballistic missiles. It includes a standardized organizational format and distinctive set of design practices.

Memoir literature on defense matters provides abundant information about Soviet weapons development practices during the 1920s, 1930s, and 1940s. Information on these procedures has been much less plentiful since World War II. The available Western and Soviet materials that describe more recent weapons development activity, particularly with regard to aircraft production,[27] suggest that the basic pattern of institutions and procedures established in the earlier period continues to persist. Nevertheless, the description of the Soviet weapons acquisition process that follows is necessarily speculative, resting heavily on the assumption that current practices continue to resemble those of the past.

Basic Research with Military Applications

Fundamental scientific research in many areas can have military applications. Within the Soviet Union, this research is conducted in a variety of institutions. Much of it is carried out in the extensive network of research institutes supervised by the USSR Academy of Sciences.[28] Other work of this nature is conducted in the research facilities of the higher educational institutions of the Ministry of Education and other governmental ministries including the Ministry of Defense. (This research is carried out in the laboratories of the major academies run by the branches of the Soviet Armed Forces. For example, basic research on ballistics has been carried out in the Dzerzhinskii Artillery Engineering Academy, and research with aviation applications has been performed in the well-known Zhukovskii Military Air Engineering Academy controlled by the Soviet Air Forces.)[29] Finally, basic research with a very direct connection to weapons development is also conducted within scientific research institutes directly controlled by the eight defense production ministries[30] and by the Ministry of Defense.[31]

The Soviets have not written about the manner in which they manage their defense-related basic research program. Their recent efforts to improve the coordination of research and development within the civilian sector of the economy[32] suggests that a system at least as comprehensive is almost certain to exist in the high-priority defense area. Management of weapons-related research probably involves regular supervision from the highest levels of the Communist Party, most likely under the auspices of Secretary D. F. Ustinov and perhaps the Defense Industries and Science Departments of the Central Committee. Coordination of these matters in the Soviet Government also is likely to engage the efforts of the Military-Industrial Commission headed by L. V. Smirnov. The highest-priority weapons projects like the development of ballistic missiles are likely to be monitored by General Secretary Brezhnev and his Politburo colleagues on the Defense Council. These leaders probably rely upon the detailed research plans of the annual, five-year, and longer-term variety developed within the Academy of Sciences and the various government ministries and coordinated with Gosplan.

Military participation in the funding and supervision of defense-related basic research is likely to be varied. Much of the relevant research done within the Academy of Sciences or the ministerial research institutes will have its own scientific significance independent of its weapons systems applications. Other projects may be undertaken and funded specifically for their military dimensions.[33] In either case the Ministry of Defense is likely to be a persistent supporter of substantial research efforts in those areas that appear to promise eventual military utility.

Weapons Design

The creation of weapons designs and the applied engineering that must accompany their development is carried out in a smaller and more specialized group of organizations. This activity is conducted by several scientific research institutes and weapons design bureaus[34] within the various defense industrial ministries and, to a considerably lesser extent, within these organizations run by the Ministry of Defense.

The most prominent of these design entities have been traditionally headed by senior designers who enjoy enormous personal authority and considerable autonomy in their work and frequently receive the highest national honors. While senior Soviet aircraft designers such as A. I. Mikoyan, P. O. Sukhoi, A. N. Tupolev, and A. S. Yakovlev have been the most well-known within the West,[35] this pattern is also visible within other areas including the development of tanks,[36] small arms,[37] and strategic ballistic missiles.[39]

The vast majority of Soviet weapons design organizations are directly controlled by the defense industrial ministries. They must, however, remain in close touch with their customers in the Soviet military establishment. The primary channels for this interaction appear to be the weapons development directorates of the military services. These organizations have been identified in the air forces,[40] the navy,[41] and the ground forces[42] and are almost certain to exist within the national air defense forces and strategic rocket forces as well. A supervisory role may also be played by a weapons development section within the General Staff[43] and, in the strategic weapons area, by a recently formed section headed by Colonel General N. N. Alexeev.[44]

Design proposals for the creation of new weapons probably originate from three different sources: (1) from the weapons design organization; (2) from the military customer; and (3) from the civilian political leadership. Historical accounts by Soviet military personnel, weapons designers, and defense industrial executives provide many examples of designs inspired by each of these groups.

Weapons designers have a great deal at stake that prompts them to generate new weapons systems proposals. Each time a design bureau succeeds in winning approval to develop and series-produce a new weapon, it stands to gain added prestige as well as substantial financial reward.

An example of the role of entrepreneurial self-promotion within the Soviet weapons development community is found in the memoirs of a leading aircraft designer, Alexander Yakovlev. In 1951, Yakovlev was disturbed about a decision by Stalin to prohibit further design work on fighter aircraft and to concentrate instead upon modernizing the MIG-15 that had been developed by the rival design bureau headed by Artem Mikoyan. Yakovlev describes his concerns about the decision as follows:

> I was very worried about the situation developing in our design bureau. You see behind me stood 100 people, who might lose faith in me as the leader of the design collective. I understood likewise that if all experimental work was limited to the modernization of the existing production model and not to the creation of new, more advanced models, then in a very short time, this would inevitably lead us to obsolescence.[45]

Moved by these considerations, Yakovlev personally approached Stalin and succeeded in gaining permission to proceed with the development of his own aircraft. This appeal eventually led to the design and production of the YAK-25, an all-weather fighter that became a central element in the Soviet tactical aviation inventory.[46]

When the military originates a request for a new weapon, it may emanate from the weapons development experts within the General Staff or the services, who are likely to value weapons modernization as an end in itself, or from the operational planners of the main staffs and the General Staff who will probably seek a new system in order to improve the chances for the fulfillment of a particular operational mission.

V. G. Grabin, a prominent artillery designer, has written of the mix between designer initiatives and the assignment of tasks by the military in the 1930s and 1940s.

> Our design bureau always carried out two parallel assignments: that of meeting the gross production goals and that of creating new types of artillery systems. As a rule, our plant received its tactical-technical require-

ments* for the development of new guns from the Main Artillery Directorate. But several guns were developed on our own initiative. There was a special section in our design bureau which worked on long range developments. The Z1S-3 76 mm-division gun was an example of such development. And when during the war the need arose for a new and better gun we were already able to present a completed model to the State Defense Committee.[47]

Grabin's account also casts light on another important dimension of the Soviet weapons development effort. He clearly illustrates the future orientation of the successful designer and the importance attached to closely monitoring the activities and achievements of potential adversaries with these comments:

Our design bureau constantly followed the achievements of science and technology, including those of foreign countries. It was necessary for us to foresee many things— the possible velocity of moving targets, the maximum weight limits of bridges, the conditions of roads and the long range perspectives for development of materials. The artillery designer was also obliged to know the industrial potential of the enemy. And if we created a gun designed to destroy only the enemy's existing means of attack and defense, then we had not fulfilled our task of always looking ahead.[48]

Other historical accounts reiterate Grabin's point that design bureaus frequently develop weapons in response to the requests of their military customers.[49] This is likely to remain the prevalent mode of weapons development initiation today. The design specifications for these systems, the "technical-tactical characteristics" noted above, are probably established by the weapons development directorates of the services in cooperation with their main staffs and in some cases, in concert with the weapons development section of the General Staff.

---

*This phrase, "taktiko-tekhnicheskii trebovanie" in Russian, is used by the Soviets to describe the performance characteristics sought for a new system in Soviet weapons development efforts ranging from tanks to ballistic missiles.

Members of the Soviet political leadership have often played an active role in weapons development. This involvement included direct personal assignments of design responsibilities by Stalin between 1937 and 1953, during the period of his highly personalized rule.[50] There is evidence that Khrushchev was also personally involved in these matters.[51] The businesslike demeanor of the present regime suggests that while Brezhnev and Kosygin are certain to be kept well-informed about major weapons development projects, they are unlikely to be the direct inspirers of these efforts.

Whatever the source of a design proposal, a standardized format appears to be generally followed for the full-scale development and eventual production of a new weapons system. In many cases, this process includes a direct competition between design bureaus for the right to add their creation to the Soviet military inventory. These competitions may involve several design bureaus[52] and can be held at different stages of the development process. They may involve the comparison of detailed plans, full-scale mock-ups, or working prototypes. In the latter case, extensive operational tests are frequently conducted to assist in determining the winner.

Whether a competition is being held or a single design is being developed, a new weapons system is carefully monitored and evaluated during its development. This task is performed by a specially appointed scientific-technical commission. This body, composed of specialists drawn from the appropriate defense-industrial ministry and the customer military service, regularly reviews the progress of a weapons development effort.[53]

The supervisory and evaluative activities of these commissions are likely to be routinely scrutinized by the staff of the Military Industrial Commission on the government side and by the Defense Industries Department of the Central Committee acting for the party. High-priority and particularly expensive systems like a new ICBM, nuclear submarine, or aircraft carrier are likely to merit the direct attention of Party Secretary D. F. Ustinov and perhaps even the Defense Council. The Soviets may also form special commissions whose members include the highest political figures to supervise special military development efforts.[54]

Weapon Production

A successfully developed weapon is eventually certified for series production by the appropriate scientific-technical or "state commission."[55] The transition to series production must then be made. The complexity of modern weaponry will almost always require the coordination of production efforts in several different plants and often between ministries. This responsibility for this interministerial coordination is likely to rest with either with the Military Industrial Commission or with the ministry most responsible for the production of the particular type of weapon.

Frequently the original design organization is not directly affiliated with the plant selected to mass-produce the weapon. In these cases, representatives are commonly sent from the design bureau to assist in setting up series production.

During the production phase the military customer continues to be directly involved in the process. A small team of officers, probably drawn from the service's weapons development directorate, is stationed at the plant to monitor its output for conformity with the technical specifications agreed upon by the scientific technical commission. These representatives are empowered to refuse to accept delivery and withhold payment for systems that fail to meet these requirements.[56] Their acceptance of a production run, on the other hand, marks the final procurement step for a new system entering the Soviet weapons inventory.

CONCLUSION

The planning and weapons development processes discussed above play central roles in shaping the development of the Soviet force posture. They provide the organizational context within which Soviet military, industrial, and political leaders interact on a continuing basis in the design, development, and production of new weapons systems. The process is fundamentally political. It is permeated with pulling and hauling over roles, missions, budgetary priorities, and many other institutional considerations. It is also bound to be influenced by the differing perceptions of its key participants regarding a variety of domestic and international political matters. Within this framework, the multitude of program decisions are made that determine the aggregate size and composition of the Soviet military arsenal.

NOTES

1. Colin S. Gray, "The Arms Race Phenomenon," World Politics, October 1971, p. 41.
2. Arthur Sulzberger, Interview with Khrushchev, New York Times, September 8, 1961.
3. George W. Rathjens, "The Dynamics of the Arms Race," Robert J. Art and Kenneth N. Waltz, eds., The Use of Force: International Politics and Foreign Policy (Boston: Little, Brown and Company, 1971), pp. 488-91.

4. This concern appears most evident with regard to the dramatic increase in Soviet distant area naval deployments since the mid-1960s. See Franklyn Griffiths, "Forward Deployment and Foreign Policy" and Robert Weinland, "The Changing Mission Structure of the Soviet Navy" in Michael MccGwire, ed., Soviet Naval Developments: Capability and Context (New York: Praeger Publishers, 1973), pp. 9-15, 292-305.

5. For comments on these priorities, see Vernon V. Aspaturian, "The Soviet Military-Industrial Complex—Does It Exist?" Journal of International Affairs 26, 1 (1972): 18-19; Richard Armstrong, "Military-Industrial Complex—Russian Style," Fortune, August 1, 1969, pp. 124-26.

6. For a discussion of this concern and its impact within the U.S. defense community, see Richard G. Head, "Doctrinal Innovation and the A-7 Attack Aircraft Program" in Richard G. Head and Ervin J. Rokke, eds., American Defense Policy, 3d ed. (Baltimore: Johns Hopkins Press, 1973), pp. 432-35.

7. Graham T. Allison, Essence of Decision: Explaining the Cuban Missile Crisis (Boston: Little, Brown and Co., 1971), pp. 102-13.

8. Ibid., p. 105; Lecture by Colonel Eliyahu Ze'ira, Chief of Operations, Israeli Army General Staff, U.S. Air Force Academy, Fort Collins, Col., November 1967.

9. For references to the existence of annual and five-year plans within the Ministry of Defense, see Major General A. Baranenkov, "Financial Support to the Troops under Annual Planning Conditions," Tyl i snabzhenie Sovietskikh Vooruzhennykh Sil (Rear services and supply of the Soviet Armed Forces), no. 10, 1972, pp. 57-61; Colonel General V. Dutov, "Improving Economic Operations in the Army and Navy," Kommunist vooruzhennykh sil, no. 2, January 1972, p. 34; and V. D. Sokolovskii, ed., Voennaia strategiia (Military strategy), 3d ed. (Moscow: Voenizdat, 1968), p. 378.

10. Herrick quotes a "former Soviet naval officer" regarding the existence of a 10-year Soviet naval shipbuilding plan approved around 1950. Robert W. Herrick, Soviet Naval Strategy: Fifty Years of Theory and Practice (Annapolis, Md.: United States Naval Institute, 1968), pp. 63-64.

11. Michael MccGwire, "Soviet Naval Procurement" in The Soviet Union in Europe and the Near East: Her Capabilities and Intentions (London: Royal United Services Institution, August 1970), pp. 76-77. MccGwire has written in considerable detail about the classes of ships included within this postwar 20-year naval rebuilding program and their scheduled delivery dates. This plan, however, is MccGwire's own analytical construct, developed on the basis of observed Soviet ship construction patterns and a single quote from

Admiral Weakley, U.S. Navy, on Soviet submarine construction plans in the 1950s, (New York Times, February 4, 1959). Michael MccGwire, "The Soviet Navy in the Seventies," unpublished manuscript, Chapter 12, "Naval Shipbuilding Practices."

12. V. D. Skugarev and L. V. Kudin, Setovoye Planirovaniya no flote (Critical path planning method in the navy) (Moscow: Voenizdat, 1973), p. 2.

13. See Gertrude E. Schroeder, "Recent Developments in Soviet Planning and Incentives" in U.S. Congress, Joint Economic Committee, Soviet Economic Prospects for the Seventies, 93d Cong., 1st sess. (Washington, D.C.: Government Printing Office, 1973), pp. 13-18; and Theodore Shabad, "Soviet Economists Split on Flexibility Planning," New York Times, October 9, 1973.

14. Marshal M. V. Zakharov, "On the Eve of World War II: May 1938-September 1939," Novaia i noveyshaia istoriia (New and newest history), no. 5, September-October 1970, pp. 3-27; and "The Communist Party and the Technological Rearmament of the Army and Navy in the Years of the Pre-war Five Year Plans," Voenno-istoricheskii zhurnal, no. 2, February 1971, pp. 3-12; Major General Ye Nikitin and Lt. Colonel V. Tret'Yakov, "Historical Experience of the Party Leadership of Soviet Military Construction," Voenno-istoricheskii zhurnal, no. 8, August 1973, pp. 3-10.

15. Barry Richman, Soviet Management: With Significant American Comparisons (Englewood Cliffs, N.J.: Prentice-Hall, 1965), pp. 94-107; Herbert S. Levine, "Economics," in George Fischer, ed., Science and Ideology in Soviet Society (New York: Atherton Press, 1967), pp. 107-38.

16. Zakharov, "On the Eve of World War II," op. cit., p. 6.

17. For an account of the planning activities of the Central Financial Directorate, see Col. General V. Dutov, "Leninist Principles of Financing the Soviet Armed Forces," Tyl i snabzhenie Sovetskikh Vooruzhennykh Sil (Rear services and supply of the soviet armed forces), no. 3, March 1970, pp. 8-13.

18. Nikitin and Tret'Yakov, p. 7.

19. Interview, V. M. Kulish, Colonel, Soviet Army (retired) and deputy director of the Institute of World Economics and International Affairs (IMEMO), Princeton, N.J., May 1970.

20. For a reference to this collegium, see "In the Ministry of Defense of the USSR," Krasnaia Zvezda, April 7, 1972.

21. See pp. 75-78 and notes 40-42 in this chapter for the identification of these organizations and a discussion of their role in the weapons acquisition process.

22. Konstantin K. Krylov, "Soviet Military-Economic Complex," Military Review, November 1971, p. 96.

23. Smirnov, who played a major role in the strategic arms negotiations at the Moscow summit in May 1972, was identified as chairman of the Military-Industrial Commission in John Newhouse, Cold Dawn: The Story of SALT (New York: Holt, Rinehart and Winston, 1973), pp. 251-52. See also David Holloway, Technology Management and the Soviet Military Establishment, Adelphi Paper no. 76 (London: Institute of Strategic Studies, 1971), p. 38.

24. This section, probably headed by the first deputy chairman of Gosplan, V. M. Ryabikov, a veteran defense industrial manager, apparently includes a number of "alumni" from the defense industrial ministries. See John A. McDonnell, "The Soviet Defense Industry as a Pressure Group," Chapter 6 of this volume, note 61.

25. For a recent discussion of this body, its role, and composition, see Malcolm Mackintosh, "The Soviet Military's Influence on Foreign Policy," Chapter 2 of this volume.

26. Kulish interview, May 1970.

27. Arthur J. Alexander's "R & D in Soviet Aviation," R-589-PR (Santa Monica, Calif.: RAND Corporation, November 1970) and "Weapons Acquisition in the Soviet Union, United States and France" F. B. Horton, A. Rogerson, and E. L. Warner, III, eds., Comparative Defense Policy (Baltimore: Johns Hopkins Press, 1974), pp. 526-44.

28. E. Zaleski et al., Science Policy in the USSR (Paris: Organization for Economic Cooperation and Development, 1969), pp. 216-92; Alexander G. Korol, Soviet Research and Development: Its Organization, Personnel and Funds (Cambridge: The MIT Press 1965), passim.

29. Major General M. Serebryakov, "Scholar-Artilleryist N. F. Drozhdov," Voenno-istoricheskii zhurnal, no. 7, July 1968, pp. 116-19; Robert A. Kilmarx, A History of Soviet Airpower (New York: Praeger Publishers, 1962), pp. 68, 116.

30. For example, Central Aerohydrodynamics Institute (TsAGI), Central Institute of Aviation Motors (TsIAM), and All-Union Institute of Aviation Materials (VIAM) within the Ministry of Aviation Industry; Alexander, "R & D in Soviet Aviation," op. cit., p. 5.

31. For example, the Scientific Research Institute of the Red Army Signals Command (NIIS-KA) played an important role in Soviet radar development. John Erickson, "Radio Location and the Air Defense Problem: The Design and "Development of Soviet Radar, 1934-1940," Science Studies, no. 2, 1972, pp. 255-59.

32. This effort has included the formation of the State Committee for Science and Technology in 1965 and measures to identify priority areas for research and to improve the ties between research organizations and production facilities. Robert Adamson, "Mobilizing Soviet Science" Scientific Research 3, 2 (January 22, 1968): 25-34.

33. For example, the Main Artillery Directorate (GAU) funded the research work of both the Leningrad Electro-Physics Institute (LEFI) and the Central Radio Laboratory (TsRL) in the early development of Soviet radar. Erickson, p. 247.

34. A design organization that possesses only limited prototype construction facilities like those in the aircraft industry is called an opytno-konstruktorskoe byuro (OKB) experimental design bureau. In contrast, a design group colocated with a major armaments production plant, as in the tank and artillery areas, is simply designated a konstruktorskoe byuro (KB) design bureau. Zaleski, et al., p. 541.

35. Their fame has been aided by the Soviet practice of designating aircraft with the initials of the chief designer. Thus, Mikoyan and his partner M. Gurevich were responsible for the MIG series, Sukhoi for the SU's, Tupolev for the TU's Yakovlev for the YAK's, and so on.

36. The leading tank designers have included M. I. Koshkin, A. A. Morozov, I. L. Kukhov, Zh, Ya Kotin, and L. S. Troyanov.

37. The most prominent Soviet small arms designers, who like the aircraft designers have their products designated with their initials, included F. V. Tokarev, B. Shpital'nyy, V. A. Degtyarev, M. T. Kalashnikov, and S. T. Simonov.

38. Although naval designers are not known by name, MccGwire has written, "The Ministry of Shipbuilding Industry has its own specialist design offices and submarine design teams which have stayed together from the earliest postwar diesel programs to the latest nuclear boats and this presumably applies to surface types as well." Michael MccGwire, "Soviet Naval Procurement," op. cit., p. 74, which draws on the statement in U.S. Congress, House, Seapower Subcommittee, Committee on the Armed Services, Status on Naval Ships (Washington, D.C.: Government Printing Office, 19 March 1969), p. 419.

39. Leading designers in Soviet missile development have included S. P. Korolev, L. A. Voskresenskii, V. P. Glushko, A. M. Isayev, V. N. Chelomei, and M. K. Yangel. Nicholas Daniloff, The Kremlin and the Cosmos (New York: Alfred A. Knopf, 1972), pp. 67-88.

40. The Air Forces' Chief Administration for Aviation Engineering Service is identified in Kilmarx, op. cit., p. 113, and in Raymond L. Garthoff, "Soviet Air Power: Organization and Staff Work" in Asher Lee, ed., The Soviet Air and Rocket Forces (New York Praeger Publishers, 1959), p. 181.

41. The naval high command includes the post of the deputy commander in chief for Shipbuilding and Armaments, suggesting the existence of a weapons development directorate of that title. Directory of USSR Ministry of Defense and Armed Forces Officials (Washington, D.C.: Government Printing Office, April 1973), p. 10.

42. The ground forces apparently possess two weapons-development directorates, the Chief Armor Directorate and an updated version of the prestigious Main Artillery Directorate (GAU). In recent years GAU has probably been expanded to include the management of operational-tactical missile development in addition to its traditional direction of artillery and small arms production. John Milsom, Russian Tanks, 1900-1970 (Harrisburg, Pa.: Stackpole Books, 1971), p. 80.

43. The General Staff has played an important role in weapons development in the past. During the late 1930s, it assumed the broad responsibilities in this area that had been exercised by the chief of Armaments and his staff. The role of the General Staff with regard to these matters in recent years, however, is unknown. Zakharov, "Communist Party and Technological Rearmament," op. cit., p. 4, footnote 4.

44. Speculation about the creation of a new high-level organization to coordinate strategic weapons development within the Ministry of Defense was sparked by the appointment in October 1970 of Colonel General N. N. Alexeev to the post of deputy minister of Defense without a publicly identified area of responsibility. Alexeev had been one of the six chief Soviet delegates to SALT and is reported to have an extensive technical background in missiles and electronic equipment. John Erickson, Soviet Military Power (London: Royal United Services Institution, 1971), p. 27; Mackintosh Chapter 2 of this volume, p. 29. Twice in the past, special high-level organizations were formed within the Ministry of Defense to manage weapons development. The first, the chief of Armaments and his staff, functioned during the late 1920s and early 1930s at the time of the initial five-year plans. The second, headed by the deputy minister of Defense for New Weapons operated during the late 1940s and 1950s and apparently played a central role in the management of the Soviet strategic missile program. General Alexeev may head an organization created to supervise the latest cycle of strategic weapons programs in the complex arms limitation environment. Zakharov, "Communist Party and the Technological Rearmament," op. cit., p. 4; Malcolm Mackintosh, "The Role of Institutional Factors in Soviet Decisions on Weapons Development," unpublished manuscript, 1967, p. 10; Oleg Penkovskiy, The Penkovskiy Papers, translated by Peter Deriabin (New York: Avon Books, 1966), p. 309.

45. Aleksander S. Yakovlev, Tsel'zhizni: zapiski aviakonstruktora (The goal of life: Notes of an aviation designer) 2d ed. (Moscow: Izdatel'stvo Politicheskoi Literatury, 1968), p. 491.

46. Numerous cases of designer promotion of their own weapons are described within Soviet memoir literature. Noteworthy examples include the creation of the T-34 tank by designers

M. Koshkin and A. Morozov—Col. V. Mostovenko, "Steps in Tank Construction," Tekhnika i vooruzhenie (Technology and armaments), no. 9, 1966, p. 14; S. Ilyushin's initiation of the design of the IL-2 "Stormovik" ground attack fighter-bomber—S. Ilyushin, "A Front Line Weapon" Tekhnika i vooruzhenie, no. 5, 1970, p. 20; Tupolev's development of the TU-4 medium bomber—A. N. Tupolev, "TU—The Man and the Aircraft," Znamya, no. 9, 1973, p. 41; Petrov's development of the 122 mm. howitzer—Lt. General F. Petrov, "Search for Design Perfection," Teknika i vooruzhenie no. 11, November 1968, pp. 2-4; and the self-initiated work on the 85 mm. gun for the T-34 tank by Zh. Kotin—P. Murav'yev, "Guns for Tanks," Tekhnika i vooruzhenie no. 5, 1970, p. 12.

    47. V. Grabin "Contribution to Victory," Tekhnika i vooruzhenie, no. 5, 1970, pp. 7-8.

    48. Ibid, p. 8

    49. B. L. Vannikov, "From the Notes of the People's Commissar of Armaments," Voenno-istoricheskii zhurnal no. 2, February 1962, p. 80; Murav'yev, op. cit., p. 12; Marshal N. N. Voronov, Na sluzhbe voennoi (In wartime service) (Moscow: Voenizdat, 1963), p. 235.

    50. Numerous memoirs testify to Stalin's pervasive involvement in the weapons development process. Compare Yakovlev, op. cit., on aviation, passim; and Admiral N. G. Kuznetsov on naval shipbuilding, "Before the War," Oktyabr (October) no. 11, November 1965, pp. 141-44.

    51. Khrushchev's directing role in the Soviet missile program is described in Leonid Vladimirov, The Russian Space Bluff (London: Tom Stacey, 1971), passim. Khrushchev's frequent interventions in other aspects of military policy suggest that he played a leading part in weapons acquisition as well. Penkovskiy, op. cit., pp. 230-41.

    52. Yakovlev writes of a fighter design competition in 1939, which included 11 competitors, A. S. Yakovlev, "The Aim of a Lifetime," International Affairs, no. 2, February 1973, p. 94. Soviet memoirs provide accounts of numerous design competitions throughout armaments effort.

    53. Alexander, "Weapons Acquisition in the Soviet Union, United States and France," op. cit., p. 430.

    54. This approach was frequently applied in the 1930s and 1940s, when special commissions were often formed to troubleshoot specific weapons development programs. Compare B. L. Vannikov, "The Defense Industry of the USSR on the Eve of the War," Voprosy istorii, no. 1, January 1969, pp. 122-23; "Submachine Guns," Tekhnika i vooruzhenie, no. 6, June 1971, p. 12.

    55. References to the routine approval activity of these commissions are found in A. Nikitin, "The History of the Creation of an Anti-tank Aviation Bomb," Voenno-istoricheskii zhurnal, no. 9,

September 1969, pp. 72-73; Vannikov, "Defense Industry of USSR on Eve of War," op. cit., p. 120; and Mostovenko, op. cit., p. 13.

56. Krylov, op. cit., p. 96; Vannikov, "From the Notes of the People's Commissar of Armaments," op. cit., pp. 79, 86; Major General N. E. Novovskiy, "Our Arsenal of Armaments," Voprosy istorii, no. 11, November 1970, pp. 126-27; Andrew Sheren, "Structure and Organization of Defense-related Industries," U.S. Congress, Joint Economic Committee, Economic Performance and the Military Burden in the Soviet Union, 91st Cong., 2d sess. (Washington, D.C.: Government Printing Office, 1970), p. 126.

CHAPTER

6

THE SOVIET DEFENSE INDUSTRY AS A PRESSURE GROUP
John McDonnell

This chapter has a twofold purpose: (1) to describe the evolution and organization of the Soviet defense industry and (2) to examine its potential role as a pressure group in the Soviet political system. While the former will be discussed primarily in terms of institutional structures established for the production of military equipment and supplies, the latter aspect involves a consideration of political conflict occurring within the Soviet governmental and party leadership. Since the early 1950s, important posts in the fields of economic planning and management, and the direction of scientific research, have often been occupied by individuals with extensive backgrounds as defense industry executives. These defense industry "alumni" may have significant influence in the ongoing debate over the share of resources to be allocated to defense rather than other sectors of the society. Thus the career patterns of those who have "graduated" from the defense industry provide a key to its ability to act as a pressure group.

The defense industry is a distinct sector of the Soviet economy, technologically more advanced than the civilian-oriented sectors, with better pay rates and facilities, and a priority claim on high-quality manufactured goods, materials, skilled manpower, and so on.[1] Since 1968, this sector has consisted of eight ministries, which are identifiable by their omission from the lists of "plan fulfillment indices" published quarterly and annually for the rest of the Soviet economy.[2] These eight ministries are described in detail in Annex A; nearly all produce goods for civilian use as well, and Brezhnev has stated that 42 percent of the total output of the defense industry goes to the civilian economy.[3] Much of this can be explained by the output of ministries (such as Aviation Production and Shipbuilding) producing similar goods (planes, ships) for both the civilian and military sectors. But a substantial part would stem from the manufacture of consumer goods in defense plants that might also or otherwise be engaged in related (or even nonrelated) military production.[4] This suggests that the defense industry has a deliberate measure of spare capacity over and above its present level of output. The classical example of such spare capacity is the convertible assembly line, which can turn out either tractors or tanks.

EVOLUTION OF THE SOVIET DEFENSE INDUSTRY

Figure 6.1 presents a schematic "organizational history" of the Soviet defense industry. In 1936 a People's Commissariat of Defense Production was first established. In January 1939, this was divided into separate commissariats for aviation, shipbuilding, munitions, and armaments.[5] These four continued to exist during the war years, and additional commissariats were created in September 1941, for tank production, and in November 1941, for mortars.[6] At the end of the war, the People's Commissariats were redesignated ministries; at the same time, a reorganization of the defense industry took place. The Ministries of Aviation and Shipbuilding Production were retained as separate entities, but other elements of defense production were consolidated into the Ministry of Armament. The specialized wartime commissariats were transformed into ministries serving the civilian sector of the economy.[7]

Soviet research into the possibilities of developing atomic weapons had begun in 1942-43, once the imminent danger of defeat had been overcome. In June 1946, a "semiministry," called the First Main Administration of the USSR Council of Ministers, was created to manage and direct the atomic weapons program.[8] In July 1953, this was reorganized as the new Ministry of Medium Machine-building, and several veteran defense industry executives were added to its management.[9] Since this occurred immediately following L. P. Beria's arrest, the secret police may previously have had overall jurisdiction in this area.

Following Stalin's death in March 1953, the Soviet Government was reorganized as part of the process of concentrating political control within the new collective leadership. Three new "superministries" were created, composed in each instance of several of the old industrial ministries; one of these, Transport and Heavy Machine-building, absorbed Shipbuilding Production.[10] In a number of other cases, pairs of related ministries were combined—for instance, the Ministries of Aviation Production and Armament were merged into a new Ministry of Defense Production. In September, however, Aviation Production was reestablished as a separate ministry, with Defense Production retaining its new name.[11]

FIGURE 6.1

Evolution of the Soviet Defense Industry

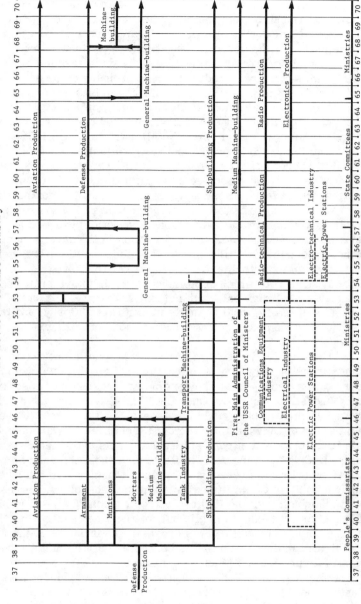

Source: Adapted from Andrew Sheren in U.S. Congress, Joint Economic Committee, *Economic Performance and the Military Burden in the Soviet Union* (Washington, D.C.: Government Printing Office, 1970), p. 127.

The governmental consolidation of March 1953 was reversed in April 1954, and most of the old industrial ministries reemerged in much the same form. At this time four ministries constituted the defense sector of the economy (Aviation Production, Defense Production, Shipbuilding Production, and Medium Machine-building), while a fifth (Radio-technical Production) was probably doing some work in this field.[12] In April 1955, an additional defense-oriented ministry was created—General Machine-building—and while its tasks were never publicized, they appear to have involved traditional weaponry.[13] In May 1957, it was merged into the Ministry of Defense Production, from which it had likely been separated originally.[14]

Khrushchev's 1957 economic reform, whereby the central industrial ministries were dissolved and regional economic councils created to act as the major administrative links, affected the defense industry in two stages. At the time of the major decentralization in May, these ministries (including Radio-technical Production) were retained in truncated form to provide central direction in the development of advanced technology.[15] But like the abolished civilian-oriented ministries, their factories were turned over to the regional economic councils, several of which established administrative branches in these fields.[16] Individuals with defense industry backgrounds emerged as leaders of the regional councils in Leningrad, Gorky, and Sverdlovsk, all of which have heavy concentrations of military-oriented plants.

The retention in Moscow of the ministerial structure for the defense industry may have been related to Khrushchev's need to secure military support in the developing leadership crisis, which finally broke in June 1957. Once the "Anti-Party Group" had been defeated,[17] this support could be dispensed with. Marshal Zhukov, the minister of Defense, was removed from his party and governmental posts in October, and strong political controls over the army were restored. Shortly thereafter, in December 1957, the defense industry ministries (with the exception of Medium Machine-building) were replaced by State Committees for Aviation Technology, Defense Technology, Shipbuilding, and Radio-electronic Technology.[18] In March 1961, the last of these was split in two, and a new State Committee for Electronics Technology was created. In March 1963, the Ministry of Medium Machine-building was converted into the State Production Committee for Medium Machine-building.

Following Khrushchev's ouster in October 1964, the defense industry was the first to be recentralized, just as it had been the last to be decentralized in 1957. On March 2, 1965, the six state committees concerned with military technology and production were reinstated as full ministries.[19] At the same time a seventh ministry, General Machine-building, was added, with responsibilities in the field of ballistic missiles and space exploration equipment. In February 1968, the eighth and last ministry of this sector was created. While little has been revealed about this new Ministry of Machine-building, its tasks appear to have been drawn from either the Ministry of Defense Production or of General Machine-building or both.

To recapitulate: For the first decade after the war, the Soviet defense industry comprised four ministries (including the First Main Administration of the USSR Council of Ministers from its inception in 1946). From the mid-1950s, there were successive additions through 1968, so that in 1973 there were eight ministries forming a distinct sector within the industrial side of the economy. Undoubtedly this expansion reflects the growing technological demands of modern weapons systems. But this doubling of the number of ministries that make up the defense industry can hardly have failed to increase the influence of its senior executives as a group.

THE MANAGEMENT OF THE DEFENSE INDUSTRY

From the mid-1930s until the mid-1950s, the defense industry was characterized by the cross-posting of senior executives among the various ministries. Aside from Stalin's desire to encourage feelings of uncertainty among his subordinates, this suggests a shortage of high-quality industrial managers, who had to be moved around to cover the most critical areas. The careers of such men as Malyshev, Khrunichev, Vannikov, and Zavenyagin, detailed in Annex B, illustrate this pattern, and it is perhaps not surprising that several died suddenly while occupying important posts.

By contrast, most of the ministers now serving have spent nearly their entire careers within the same subsection of the defense industry. Their average age is 65, with only two in their fifties and one over 75. Four have directed their ministries from the early-mid 1950s, while three others have been in charge since their ministries were created. Defense Production is the only ministry to have had several heads in the 15 years 1959-73, and with one exception these changes were necessitated by promotion of the incumbent. Thus the management picture is now one of rather remarkable stability, with the average tenure of the present group of ministers exceeding 12 years.

As we move from the individual ministries of the defense industry to higher levels of coordination and control, specific arrangements are less certain, due to the reticence of Soviet sources. We can, however, state with some confidence that the eight ministries are directly answerable to a deputy chairman of the USSR Council of Ministers,* who is responsible for supervising this sector of the economy. Since the mid-1950s, such a position has been filled by men with extensive backgrounds in the defense industry, including the direction of a ministry. In March 1955, Deputy Chairman of the Council of Ministers V. A. Malyshev was relieved of his ministerial assignment (Medium Machine-building), since he was "entrusted in the USSR Council of Ministers with the responsibility of leading a group of ministries concerned with machine-building," a frequent euphemism for the defense industry.[20] Two additional defense industry executives were appointed deputy chairmen of the Council at the same time, and when Malyshev was assigned to direct science policy in May 1955, one of these, M. V. Khrunichev, may have taken over responsibility for administering the defense industry as a whole. This post appears to have been vacant for a year beginning in December 1956, but since then there has continually been a deputy chairman of the council having the requisite defense industry background. From December 1957 until March 1963, this was D. F. Ustinov, who had been head of the armament industry since 1941. When he was promoted, the chairman of the State Committee for Defense Technology, L. V. Smirnov, succeeded him as a deputy chairman of the council. Several weeks later Khrushchev confirmed that Ustinov had had responsibility for the defense industry and that Smirnov had inherited his duties.[21]

As a deputy chairman of the council, Malyshev, Khrunichev, Ustinov, and Smirnov would have each in turn been members of a governmental executive committee called the Presidium of the USSR Council of Ministers.† On the basis of previous Soviet arrangements,

---

*Deputy chairmen of the USSR Council of Ministers constitute the third level of the government hierarchy, below the chairman, or prime minister, and the first deputy chairmen, but above the ordinary ministers. While the number of first deputy chairmen has gone as high as five, more usually there are two or three, or as now, only one. Similarly, the number of deputy chairmen has varied over the post-Stalin period from 1 to 10, with higher figures in the more recent years. At present there are nine deputy chairmen, of whom five head key agencies such as Gosplan; the rest are thought to supervise groups of ministries, such as the eight composing the defense industry.

†The Presidium of the USSR Council of Ministers is made up of its chairman, first deputy chairmen, and deputy chairmen. As described earlier, the number of members can vary enormously, from as few as 6 in 1953, to as many as 14 two years later. Since Khrushchev's ouster, membership has been stable, in the 10-12 range.

this committee would have the tasks of allocating resources among competing sectors of the economy, and coordinating policy between governmental departments, including the relationship between military policy and economic priorities. Thus the defense industry would be one of several claimants on resources available, and its relative importance would depend on prevailing domestic and international circumstances.

An executive committee with these responsibilities is known to have existed at least until the end of World War II. It was first created in November 1918, during the period of civil war and foreign intervention, and originally was called the Council of Workers' and Peasants' Defense.[22] In March 1920, the name was changed to the Council of Labor and Defense, to reflect the increased importance of peacetime economic reconstruction.[23] In April 1937, this body was replaced by an Economic Council, responsible for industrial allocations, and a Defense Committee to oversee military policy.[24] But the defense industry appears to have been placed under the latter, as in January 1938 a "Military-Industrial Commission" was established answerable to the Defense Committee.[25] Immediately after the German invasion in June 1941, these were superseded by the State Defense Committee, composed of Stalin and his most trusted lieutenants.[26] It was especially concerned with mobilizing resources for the war effort, and certain of its members had particular responsibilities with respect to specific defense industries.[27] The State Defense Committee was abolished on September 4, 1945, and there appear to be no references to the creation of a successor. Stalin may have operated through more informal arrangements, but following his death, former institutional practices seem to have been restored, as described earlier.

Below this executive committee there appears to be a second body directly related to the defense industry that is responsible for coordinating and supervising military-oriented research, development, and production. Official U.S. sources state that its membership is drawn from the Ministry of Defense, the various production ministries, the planning agencies, and the party apparatus.[28] Western analysts have sometimes referred to this body as the "Defense industries committee,"[29] but it appears to be identical to the "Military-Industrial Commission" recently identified in connection with SALT.[30] As the deputy chairman of the Council of Ministers responsible for the defense industry, Smirnov is likely the head of this commission, but he presumably remains answerable to Ustinov as the party secretary with jurisdiction in this field.

FIGURE 6.2

Economic Planning and Science Management Agencies

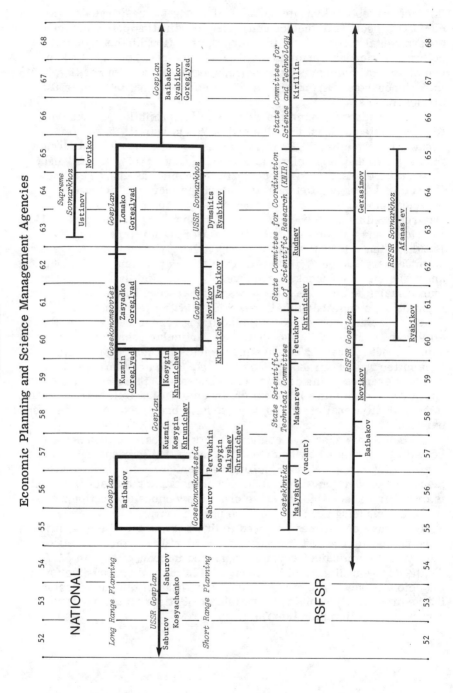

DEFENSE INDUSTRY "ALUMNI" AS MANAGERS OF THE SOVIET ECONOMY

In addition to the practice already described of appointing a senior executive to supervise the defense industry, individuals with extensive backgrounds in this sector also serve in other top managerial posts. Since the early 1950s, these defense industry "alumni" have been particularly prominent in the agencies charged with economic planning and the direction of scientific research. The post-Stalin evolution of these agencies is shown in Figure 6.2: Economic planning and coordinating agencies are included both for the Soviet Union as a whole (that is, USSR level), and for the Russian Soviet Federated Socialist Republic (RSFSR level), the largest of the 15 republics, and the most important economically, since it contains an overwhelming majority of Soviet industry, including the defense industry. The names of those chairmen (and in some cases, first deputy chairmen) of these bodies who are defense industry "alumni," are underlined.

Such appointments from the ranks of defense industry leaders did not begin until the mid-1950s. Prior to this, however, V. A. Malyshev had been a deputy chairman of the Council of Ministers (and hence a member of its presidium, or executive committee) from 1940 to 1944 and from 1947 to 1953, while also heading the tank industry during the war and serving as minister of Shipbuilding, 1950-53. Immediately following Stalin's death, membership in the council's presidium was restricted to those in the party Politburo.* Thus while Malyshev headed the "superministry" into which shipbuilding had been absorbed (Transport and Heavy Machine-building), he lost his deputy chairmanship of the Council of Ministers. After having been transferred to head the new Ministry of Medium Machine-building, thereby taking over direction of the nuclear weapons program, he regained the rank of deputy chairman in December 1953.[31] Fifteen months later he was relieved of his ministerial assignment, to become the council deputy chairman in charge of the defense industry. At the same time, two other defense industry executives were also promoted to deputy chairmanships: A. P. Zavenyagin, the new minister of Medium Machine-building, and M. V. Khrunichev, whose new responsibilities (he had been first deputy to Malyshev in Medium

---

*The Politburo was renamed the Presidium of the Central Committee at the 19th Party Congress in 1952, and the name was changed to Politburo once more at the 23d Party Congress in 1966. To avoid confusion with the Presidium of the Council of Ministers, the term Politburo will also be used for the period 1952-66.

Machine-building) were not revealed. These appointments (and two related ones, promoting industrial executives from the civilian sector) were marked by a curious departure from Soviet practice, in that published decree stipulated they had been made "at the suggestion of USSR Council of Ministers' Chairman N. A. Bulganin."[32] Three weeks earlier Bulganin had replaced Malenkov as head of the Soviet Government, after a dispute in the collective leadership concerned in part with the priority of defense versus consumer goods. These promotions could be viewed as rewards for loyalty to the winning side and/or as indications of increased emphasis on developing advanced weapons systems.

Over the course of the next decade, 1955-65, the structure of economic planning and management was constantly reorganized, partly in an effort to improve economic performance, but also as a result of domestic political conflicts that were never fully resolved. While the names of such agencies were changed frequently, their functions altered less and were related to two basic points of contention: (1) whether there should be separate bodies for long- and short-term planning; or a single, unified planning agency; and (2) whether the day-to-day managerial functions should be centralized in Moscow or decentralized on a regional basis. Beginning in this period, it became standard practice to appoint the heads of the national planning agencies to the government executive committee, as deputy chairmen of the USSR Council of Ministers; a similar policy was observed at the Russian Republic (RSFSR) level.

In late May 1955, the national economic planning agency was divided in two. One body, which retained the title USSR Gosplan, was restricted to long-range planning. The other and more important body was given the name Gosekonomkomissia (State Economic Commission) and took over responsibility for formulating the annual plan, which specifies what quantity of which goods each plant should produce, and to whom it should supply them.[33] A few days later, a third agency was established—Gostekhnika, the State Committee on New Technology—to direct scientific research and coordinate the introduction of technological advances into the economy.[34] Malyshev, who had headed a similar committee in the late 1940s, was named chairman of Gostekhnika, and Khrunichev may have thereby inherited his role as the deputy chairman of the council responsible for the defense industry. While there were no defense industry "alumni" in prominent planning posts at this time, this group did have significant access to the key economic decision-makers through its three representatives (out of a total membership of 14) in the Presidium of the USSR Council of Ministers. Thus at the midpoint of the 1950s, the defense industry had considerable potential as a pressure group.

At the end of 1956, the political struggle within the collective leadership of the party sharpened once more, and the posts of the senior economic managers, including the defense industry "alumni," were directly affected. At a Central Committee plenum held that December, "Gosekonomkomissia" (the short-range planning agency) was strengthened at the expense of "Gosplan." Economic leadership changes were also made, but these were not announced until mid-February 1957.[35] Defense industry "alumni" Malyshev and Khrunichev were among the new leaders of "Gosekonomkomissia," the former as one of two first deputy chairmen, the latter as one of three deputy chairmen. These two along with four others simultaneously lost their positions as deputy chairmen of the Council of Ministers, but such demotions may have been more titular than real.[36]

Zavenyagin, the minister of Medium Machine-building, had died at the very end of 1956, after the Central Committee plenum but before the personnel changes had been announced. While his obituary listed him as still a deputy chairman of the Council of Ministers, it is possible that he too was slated to lose the higher rank.[37] In the latter part of February Malyshev also died, and his obituary indicated he had retained the chairmanship of Gostekhnika through the shakeup that had just taken place.[38] Thus two important posts were suddenly vacant, and it was a reflection of the turmoil then enveloping the collective leadership that neither was filled quickly. Only at the beginning of May 1957 was M. G. Pervukhin appointed the new minister of Medium Machine-building. While he retained his seat on the Politburo and his post as first deputy chairman of the Council of Ministers, this was clearly a personal demotion rather than an upgrading of his new assignment.[39]

Khrushchev had by this time achieved a total reversal of policy regarding the structure of economic management. On May 10, the Supreme Soviet enacted into law his plan for economic decentralization, abolishing most of the central industrial ministries as well as Gosekonomkomissia, which had been strengthened but six months earlier.[40] The same law changed the name of "Gostekhnika" to the State Scientific-Technical Committee, and the "new" agency had substantially weakened powers.[41] It was not until July (that is, after Khrushchev's defeat of his opponents in the "anti-party group") that Yu. Ye. Maksarev was named chairman of this agency. Maksarev had been associated in the past with Malyshev and in fact had been his first deputy chairman in the old "Gostekhnika," but aside from perhaps having directed a tank factory during the war, he was not closely linked with the defense industry. More significantly, his new appointment was not coupled with a deputy chairmanship of the Council of Ministers. Thus by late 1957, through the vicissitudes of death and politics, the defense industry had lost nearly all of its direct access to the key economic decision-makers. Khrunichev had been reduced to but one of several deputy chairmen of USSR Gosplan, the unified planning agency, while there were no defense industry "alumni" in the Presidium of the Council of Ministers.

The continual shakeup in economic management had, however, opened up new posts to men with backgrounds in this field. A second generation of defense industry "alumni" now began to form, one that was much more clearly tied together in a "personal association" than had been true of the preceding group, with its intermittent connections with Malyshev, and one that still dominates the defense industry today. This new generation may be called the "Ustinov group," since all its members had served in the Ministry of Armament (later, Defense Production) headed by D. F. Ustinov from 1941. In most cases, the association with Ustinov went back to the early days of the war and in two instances extends even further back, into the 1930s.

The fortunes of the Ustinov group members would depend, as had their predecessors, on those of its protectors and supporters in the party leadership. In the late 1950s and early 1960s, the "patron" of the Ustinov group was Frol R. Kozlov, at the time generally considered Khrushchev's "heir apparent," but more accurately a pretender to and challenger of his power.[42] While the following narrative focuses on the promotions and demotions of defense industry "alumni," it should be remembered that these were largely a reflection of a more important power struggle occurring at a higher level. This reflected image portrays the successes and reversals of a conservative coalition struggling against Khrushchev on such matters as defense policy and priorities in resource allocation, and thereby indexes the defense industry's potential as a pressure group.

The Ustinov group began its rise with the creation of the regional economic councils in May 1957. The Leningrad and Gorky councils were headed by two of Ustinov's wartime deputies, V. N. Novikov and K. M. Gerasimov. Given the concentration of defense plants in each area, these appointments were relatively unsurprising; but they enabled these men to gain experience outside their own fields and paved the way for later promotions. In December 1957, Ustinov himself was promoted from minister of Defense Production to deputy chairman of the USSR Council of Ministers, filling the vacant post of overseeing the defense industry. In May 1958, Novikov was named chairman of RSFSR Gosplan, the planning agency for the Russian Republic, and simultaneously became a deputy chairman of the RSFSR Council of Ministers. His successor as chairman of the Leningrad economic council was S. A. Afanas'ev, who had also had a career in the Ministry of Armament/Defense Production.

The linkage between Soviet foreign and defense policy, on the one hand, and the role of defense industry "alumni" on the other, is suggested by the promotion of several of these figures in the spring of 1960. Khrushchev had predicated his policies of detente on the assumption that Western leaders were "reasonable men" who would refrain from aggressive actions directed at the USSR. The U-2 affair revealed not only that U.S. spy planes had been systematically violating Soviet airspace for several years but had done so on the orders of that most reasonable of men (in Khrushchev's opinion), President Dwight D. Eisenhower. A number of party and government personnel changes ensued, weakening Khrushchev's dominance; among these was Novikov's appointment as chairman of USSR Gosplan and as a deputy chairman of the Council of Ministers. Gerasimov succeeded him as chairman of RSFSR Gosplan, while another wartime leader of the armament industry, V. M. Ryabikov, was named chairman of the newly created Russian Republic Economic Council ("RSFSR Sovnarkhoz") in June.[43] The planning agencies had again been divided into short- and long-range bodies in April, with defense industry "alumni" occupying the post of first deputy chairman in each.[44]

A second wave of promotions for the "alumni" occurred a year later and may be related to the Soviet response to increases in U.S. military expenditures following John F. Kennedy's inauguration. On April 8, 1961, the State Scientific-Technical Committee was strengthened and renamed the State Committee for the Coordination of Scientific Research (KNIR).[45] Khrunichev left USSR Gosplan to become its chairman and was also appointed a deputy chairman of the Council of Ministers, a move symbolic of the increased importance of this agency. The replacement process resulting from his promotion strengthened the hold of the Ustinov group on key managerial bodies, as Ryabikov succeeded Khrunichev as first deputy chairman of USSR Gosplan, and Afanas'ev took over the chairmanship of the RSFSR Sovnarkhoz. Khrunichev died at the beginning of June 1961, but in contrast to similar circumstances in 1957, his successor was named almost immediately. The new chairman of KNIR was K. N. Rudnev, whose ties to Ustinov also went back to the 1940s and whose most recent position had been as chairman of the State Committee for Defense Technology, the functional equivalent of minister of Defense Production.

Since Rudnev was also appointed a deputy chairman of the USSR Council of Ministers, the Ustinov group now had three seats out of a total membership of eight in the council's presidium. Two of these positions, moreover, were coupled with the direction of economic planning and science management agencies, and defense industry "alumni" occupied the "number-two" posts in both the long- and short-term planning bodies. "Alumni" also headed the economic organs for the Russian Republic, where most of the Soviet Union's industry is located. It would appear, then, that by the middle of 1961, partisans of military priorities had virtually total control over the Soviet economy.

This was in fact the period (1960-61) when Khrushchev was in a most serious dispute with the military over defense priorities, especially his attempt to deemphasize conventional weapons and the role of the ground forces, which he had announced in January 1960. The influential positions held by defense industry "alumni" are likely to have been a factor in the suspension of the demobilization program in July 1961 and in the major modifications to Khrushchevian military doctrine introduced at the 22d Party Congress the following October.[46] Thus there appears to have been a partial victory for the military and its allies in mid-1961.

But while Khrushchev had been forced to make concessions to his conservative opponents, he continued to press for a redirection of Soviet economic priorities. He railed against what he called "steel eaters," traditionalists who envisioned economic progress only in terms of increasing the output of heavy industry, and attacked "hidebound thinking" in the planning agencies.[47] V. N. Novikov, the defense industry "alumnus" heading the RSFSR Gosplan, appears to have been a particular target of these remarks,[48] and he was replaced in June 1962. He was further demoted (along with several "nonalumni" industrial leaders) in November, losing his seat in the Presidium of the Council of Ministers as well. Immediately after the Cuban missile crisis, Khrushchev pushed through yet another reorganization of the central management agencies, one more cosmetic than fundamental.*
But his prestige had been weakened by the Cuban fiasco, and soon his control diminished. His speeches in early 1963 were pessimistic about alleviating the burden of high military expenditures.[49] In March his latest economic reforms were superseded by the creation of a new central agency to coordinate and supervise the entire economy, the Supreme Council of the National Economy, or Supreme Sovnarkhoz. D. F. Ustinov was named chairman of this powerful body and was promoted to first deputy chairman of the Council of Ministers. Five of the eight deputy chairmen of the council were subordinated to him, including L. V. Smirnov, his replacement as chief of the defense industry.[50]

Ustinov's promotion was linked with a sharpened challenge to Khrushchev on the part of Frol Kozlov. But Kozlov suffered a heart

---

*The long-range planning agency (formerly "Gosekonomsoviet") was now named "Gosudarstvennyi planovyi komitet SSSR"—that is, "USSR Gosplan." The short-range planning body, which had been called Gosplan, was renamed "Sovet narodnogo khoziaistva SSSR" (USSR Council of the National Economy) or "USSR Sovnarkhoz."

attack in mid-April 1963, and as a consequence many of those associated with his more conservative views underwent political reversals, among them Ustinov. Although Khrushchev did not abolish the month-old Supreme Sovnarkhoz, he criticized the defense industry for inefficiencies and boasted of his ability to "shakeup" both Ustinov and the younger Smirnov.[51] Rather than being promoted to at least candidate membership in the Politburo, the party rank associated with his governmental post of first deputy chairman of the Council of Ministers, Ustinov had difficulty in preserving some distinction between himself and the "ordinary" deputy chairmen, who were nominally his subordinates.[52] Khrushchev proceeded to ignore Ustinov's Supreme Sovnarkhoz, preferring joint meetings of the Politburo and the Council of Ministers, and it would not be influential for the remainder of Khrushchev's tenure.[53]

The defense industry, like the military, apparently played only minor roles in the ouster of Khrushchev in October 1964.[54] Some improvement in the status of the defense industry and its "alumni" occurred in March 1965, when the state committees for military technology were reinstated as centralized ministries. Toward the end of that month, Ustinov was appointed a Central Committee secretary and finally promoted to candidate member of the party Politburo. Novikov returned from his semidisgrace to fill the post of chairman of the Supreme Sovnarkhoz, also regaining a deputy chairmanship of the Council of Ministers.

This agency was, however, abolished in yet another economic reform in September 1965, one that returned the managerial structure roughly to that existing before Khrushchev had begun tampering with the economy in 1957. Although Novikov thereby lost the position that had given him official status, he remained a deputy chairman of the council, and hence a member of its Presidium. But since Smirnov is clearly in charge of the defense industry, it is possible that Novikov now has responsibilities outside this field. The September 1965 reorganization also resulted in the demotion of K. N. Rudnev, who lost both his post as chief science research coordinator and his rank of deputy chairman of the council. He was now made head of the Ministry of Instrument-building, Automation Devices, and Control Systems, which is related to but outside the defense industry. He was replaced in the Presidium by V. A. Kirillin, chairman of the newly renamed State Committee on Science and Technology, a man with no past connections to the defense industry or military-related research. Thus by the mid-1960s the defense industry "alumni" occupied only two seats out of a total of 12 in the Presidium of the Council of Ministers, and this reduced role has since remained unchanged.

DEFENSE INDUSTRY MEMBERSHIP IN PARTY BODIES

The discussion in the previous section focused on the part played by defense industry "alumni" in managing the Soviet economy and directing scientific research. Representation in the Presidium of the USSR Council of Ministers has been used as one measure of the defense industry's ability to influence policy. But this is an imperfect index at best; the presidium as a governmental executive committee is not the highest policy-making body in the Soviet political system. That prerogative is formally reserved to the Central Committee of the Communist Party but is in fact exercised by the Politburo, composed of the highest party and government leaders.

No individual with an extensive background in the defense industry has been a full member of the Politburo in the post-Stalin period. (Membership in party bodies such as the Politburo and the Central Committee is divided into two levels, full and candidate. In the post-Stalin period, total Politburo membership has varied from a low of 11 to a high of 24, with one-fourth to one-third usually candidates. As of April 1974, there are 16 full and 7 candidate members.) V. A. Malyshev had been made a full member in 1952, but under unusual circumstances, and he lost this position at the time of Stalin's death.55 There were, however, several indications in the spring and summer of 1953 that Malyshev was in line for at least candidate status in this body. As minister of Transport and Heavy Machine-building, he was the only one of the three men heading "superministries" not to have a seat in the Politburo. More intriguingly, he was the only non-Politburo figure to be demonstratively associated with its members at the time of Beria's arrest and removal from the collective leadership.56 Shortly thereafter he was selected to take charge of the nuclear weapons program, as minister of Medium Machine-building. But these "signals" of elevation to the Politburo were not followed by actual promotion, possibly due to a too close association with one of the contending factions in the power struggle for Stalin's succession.57

Ustinov was finally promoted to candidate membership in the Politburo in March 1965 and was also appointed one of the 10 secretaries of the Central Committee. This was clearly an increase in his own political stature and by reflection raised the prestige of the defense industry, since leaders of the party monitoring agencies are not usually recruited directly from the institutions they are to monitor.*

---

*In addition to Ustinov's post as party secretary "for the defense industry," there is a department of the Central Committee

Ustinov's precise duties as a party secretary are not known but presumably include oversight of the defense industry and may extend to the military as well. He has not, however, subsequently been raised to full membership in the Politburo but has been passed over several times, most recently in late April 1973, when Minister of Defense Grechko and Minister of Foreign Affairs Andrei Gromyko, neither of whom had been candidate members, were both "jumped" over him to full membership status. Thus he continues to have only a "consultative voice" in Politburo deliberations, rather than one of the determining votes. (Precise "decision-making rules" in the Politburo are not known. In 1957, Khrushchev stated that a consensus is usually reached through discussion, but when this is not possible, decisions are made through a simple majority vote.[58] More recently, Brezhnev assured Western journalists that a consensus is achieved 99.99 percent of the time, but when disagreement persists, a subcommittee is appointed to resolve the question. It was also observed that the conference table in the Politburo meeting room had 16 chairs around it, precisely the number of full members of the Politburo at the time of the interview, while "dozens" of chairs lined the walls for lower-ranking officials attending to report on specific matters.)[59]

(In addition to Ustinov, one full and two candidate members of the Politburo began their careers in the Defense Industry. A. P. Kirilenko, a full member and party secretary, graduated from the Rybinsk Aviation Institute in 1936, and worked briefly as a designer in an aircraft plant. In 1942-43, he was a special representative of the State Defense Committee at an aircraft plant in Moscow. G. V. Romanov, first secretary of the Leningrad Regional Party Committee, was a designer in the shipbuilding industry, and graduated from the Leningrad Shipbuilding Institute in 1953, a year before transferring to full-time party work. M. S. Solomentsev, chairman of the RSFSR Council of Ministers, was an executive in an armaments plant in Chelyabinsk during the war.)

The Central Committee, to which the Politburo and the party secretaries are nominally answerable, is a much larger body, and the distribution of its membership among key functional groups making up the Soviet elite provides another approach to assessing the relative

---

apparatus for defense industry affairs, one of 22 such sections. Its tasks would include checking the credentials of party members in the defense industry, and directing the work of party organizations in this field. Since 1958, this department has been headed by Ivan D. Serbin, who has worked in the Central Committee apparatus since 1942. Serbin has been a candidate member of the Central Committee since 1961.

FIGURE 6.3

Membership of the Central Committee

THE WHOLE

	1952	1956	1961	1966	1971
Full Members	128	133	175	195	242
Candidate	108	122	155	165	156
Total Membership ...	236	255	330	360	398

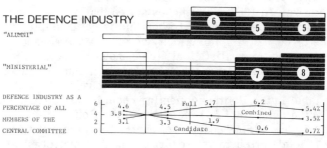

THE DEFENCE INDUSTRY
"ALUMNI"

"MINISTERIAL"

DEFENCE INDUSTRY AS A PERCENTAGE OF ALL MEMBERS OF THE CENTRAL COMMITTEE

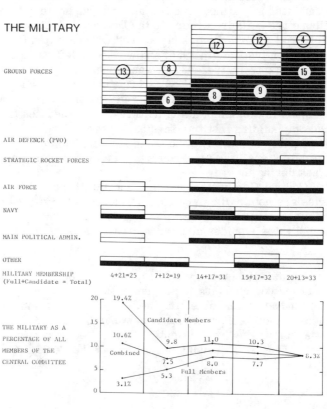

THE MILITARY

GROUND FORCES

AIR DEFENCE (PVO)

STRATEGIC ROCKET FORCES

AIR FORCE

NAVY

MAIN POLITICAL ADMIN.

OTHER

MILITARY MEMBERSHIP (Full+Candidate = Total)   4+21=25   7+12=19   14+17=31   15+17=32   20+13=33

THE MILITARY AS A PERCENTAGE OF ALL MEMBERS OF THE CENTRAL COMMITTEE

importance of such groups.⁶⁰ The top part of Figure 6.3 shows the
number of full and candidate members associated with the defense
industry, from the last Central Committee "elected" under Stalin in
1952 to the most recent, formed in 1971. The growth in the number
of defense industry executives and "alumni" in the Central Committee
has no more than kept pace with the expansion of the committee's
size, as the percentage diagram indicates. There has, however, been
a shift from an even "mix" of full and candidate members, to a near
total concentration in the more prestigious, and presumably more
important, full memberships. Military representation, which is shown
for comparison, is considerably larger, but the shift toward full membership here has been less pronounced. Even taken together, as the
two percentage diagrams show, the military and defense industry
contingents form but a small minority of the Central Committee,
indicating their need for political allies, especially in the party apparatus, to secure the adoption of policies they would prefer. In recent
years a small number of scientists who appear to be involved in
weapons research have also been elected to the Central Committee,*
and there may be more than coincidence in the fact that several regional party leaders from areas with defense industry concentrations
have had technical training in this field.

    These individuals are L. I. Grekov, 2d secretary, Moscow City
Party Committee; B. F. Korotkov, 1st secretary, Perm Regional
Party Committee; V. P. Lomakin, 1st secretary, Primorsk Territorial
Party Committee; N. I. Maslennikov, 1st secretary, Gorky Regional
Party Committee; and V. I. Vorotnikov, 1st secretary, Voronezh
Regional Party Committee. All but Maslennikov attended aviation
institutes in the late 1940s to early 1950s—Maslennikov went to the
Zhdanov Industrial Institute in Gorky—and worked in factories, as
foremen and/or engineers, before going into party work. Korotkov
has since retired due to ill health.

    (Izvestia, on May 7, 1974, announced the retirement of K. M.
Gerasimov, the defense industry "alumnus" who had been chairman
of RSFSR Gosplan since 1960. Named as his successor was one N. I.
Maslennikov who, although not further identified in the announcement,

---

    *The most prominent of these is M. V. Keldysh, the president
of the USSR Academy of Sciences, a full member since 1961, who
has worked on a number of weapons-related programs, including jet
aircraft and guided missiles. Others are P. D. Grushin (full member,
1966 and 1971), V. P. Makeev (candidate member, 1971), and M. K.
Yangel' (candidate member, 1966 and 1971). (See the biographical
sketches of these individuals in Yezhegodnik Bol'shoi Sovetskoi
Entsiklopedii, 1971, pp. 590, 600, 610, and 643.)

is almost certainly the individual described above as party 1st secretary in Gorky. Maslennikov was born in 1921, joined the party in 1951—the same year he finished his higher technical training—and was elected a full member of the Central Committee in 1971. For over a decade he worked at the "Krasnaia Etna" factory in Gorky before transferring to party work in 1961. All of his party posts have been in Gorky, where Gerasimov headed the regional economic council in the late 1950s.)

CONCLUSIONS

The defense industry is a coherent sector of the Soviet economy, one that is administratively separate in that there is a deputy chairman of the Council of Ministers who supervises the eight ministries in this field. These eight ministries receive special treatment, most obviously in the secrecy surrounding their operations, but more importantly in their higher standards and benefits. There may even be a special section in the planning agencies dealing with the defense industry.[61] Career patterns in the defense industry over the past several decades indicate a trend away from cross-postings with civilian industry and toward specialization within single ministries or closely linked subgroupings.

Defense industry "alumni" continue to be present in the Presidium of the Council of Ministers, although with somewhat less prominent roles since the mid-1960s. The defense industry is perhaps disproportionately represented in the Central Committee, for it is the only economic sector whose ministers are uniformly accorded full membership. Aside from agriculture, the defense industry is unique among the economic sectors in having a party secretary with Politburo status. The importance of the defense industry in formulating military and foreign policy is suggested by the key role played by Smirnov in the final negotiations of the first SALT agreements.[62] While it is probable that he was drawn into the negotiations for his technical expertise rather than to advance a political viewpoint, these two elements are intertwined.

Decisions on whether to proceed with advanced weapons systems have long-term implications for the allocation of resources. The increased complexity of such systems obliges the political leadership to rely in some measure on the professional judgments of experts from the military-scientific-defense industry sector. Others have noted that this technological trend has been accompanied by a tendency toward greater professional autonomy for the Soviet officer corps.[63] While a similar process is not as readily observable in the less publicized defense industry, increased specialization of personnel would encourage it, and it is doubtful if such an evolution could be entirely prevented.

What can certainly be seen is the rise to positions of authority of a group of defense industry "alumni" in the early 1960s. For a short time the Ustinov group occupied many of the key posts in the economy, and their views must have had an impact on policies concerned with the allocation of resources. Given their backgrounds, it is not surprising that these men were part of a conservative coalition, which also included many military and some party leaders. This coalition opposed Khrushchev's reforms and innovations in domestic, military, and foreign policy, and it eventually forced his retirement. The coalition subsequently dissolved, and the collective position of the defense industry appears to have deteriorated somewhat. Nonetheless, the defense industry, like the military, remains an important interest that the political leadership must continue to acknowledge.

NOTES

1. The following items, and especially the first, are the most useful writings on Soviet defense industry: Andrew Sheren, "Structure and Organization of Defense-Related Industries," in U.S. Congress, Joint Economic Committee, Economic Performance and the Military Burden in the Soviet Union (Washington, D.C.: Government Printing Office, 1970), pp. 123-32; Vernon V. Aspaturian, "The Soviet Military-Industrial Complex—Does It Exist?" Journal of International Affairs 26, 1 (1972): 1-28; Joseph J. Baritz, "The Organization and Administration of the Soviet Armament Industry," Bulletin of the Institute for the Study of the USSR 4, 11 (November 1957): 12-21; Richard Armstrong, "Military-Industrial Complex—Russian Style," Fortune, 1 August 1969, pp. 84-87 and 122-26.

2. See, for example, Izvestia, 26 January 1974, pp. 1-2.

3. "The Report of the CPSU Central Committee to the 24th Congress of the Communist Party of the Soviet Union," Pravda, 31 March 1971; translation in Current Digest of the Soviet Press (hereafter, CDSP) 23, 13 (27 April 1971): 7.

4. See S. Zverev, Minister of Defense Production, "Vozmozhnosti Ostrasli" (The potentialities of the sector), Izvestia, 7 July 1971, p. 1; and P. Dement'ev, Minister of Aviation Production, "More Than Just Airplanes," Izvestia, 22 May 1971, p. 1, translation in CDSP 23, 21 (22 June 1971): 25.

5. Pravda, 12 January 1939, p. 1.

6. M. V. Zakharov, chairman of the Editorial Commission, 50 let vooruzhennykh sil SSSR (Moscow: Voenizdat, 1968), p. 264. A People's Commissariat of Medium Machine-building was also

created about this time, but it should not be confused with the later Ministry of Medium Machine-building. Its tasks appear to have been no more sinister than manufacturing motor vehicles for the Soviet Army.

7. See Sheren, op. cit., p. 127. The following transformations occurred at this time: Munitions to Agricultural Machine-building; Mortars to Machine and Instrument-building; Medium Machine-building to Motor Vehicle and Tractor Production; Tank Industry to Transport Machine-building (that is, railroad equipment).

8. Arnold Kramish, Atomic Energy in the Soviet Union (Stanford, Cal.: Stanford University Press, 1957), p. 117.

9. Pravda, 17 July 1953, p. 2. The ministry's leadership included: Minister V. A. Malyshev (previously in charge of the tank and shipbuilding industries); First Deputy Ministers M. V. Khrunichev (postwar minister of Aviation Production) and B. L. Vannikov (a veteran defense industry executive who had headed the First Main Administration, 1946-53); Deputy Ministers V. M. Ryabikov (formerly first deputy minister of Armament), Ye. P. Slavskii (deputy chief of the First Main Administration), and Z. P. Zavenyagin (a former deputy minister of Internal Affairs). A biography of Soviet nuclear scientist I. V. Kurchatov identifies Malyshev, Vannikov, Slavskii, and Zavenyagin as leaders in the atomic weapons program from the mid-1940s. See "The Kurchatov Story" by Igor Golovin, Sovetskaia Rossia, August 3 to 16, 1966; excerpts in CDSP 18, no. 33 (7 September 1966): 3-6, and no. 34 (14 September 1966): 6-9, especially p. 7 of the latter.

10. Pravda, 7 March 1953, p. 1. The "superministries" were created by consolidating old ministries as follows: Ministry of Machine-building—Motor Vehicle and Tractor Industry, Machine and Instrument-building, Agricultural Machine-building, and Machine-tool Construction. The new minister was M. Z. Saburov. Ministry of Electric Power Stations and Electrical Industry—Electric Power Stations, Electrical Industry, and Communications Equipment Industry. The new minister was M. G. Pervukhin. Ministry of Transport and Heavy Machine-building—Transport Machine-building, Heavy Machine-building, Shipbuilding, and Construction Equipment Industry. The new minister was V. A. Malyshev.

11. Pravda, 16 March and 10 September 1953.

12. While plan fulfillment percentages were not published during this period for the ministries clearly engaged in defense production, they were for the Ministry of Radio-technical Production. See, for example, Pravda, 2 August 1956, as well as an article by Minister of Radio-technical Production V. D. Kalmykov, mentioning only the civilian goods manufactured by his ministry, Pravda, 6 May 1955.

13. See Sheren, op. cit., p. 130, and Grey Hodnett's biographical sketch of V. N. Novikov, who was first deputy minister of General Machine-building, in George W. Simmonds, ed., Soviet Leaders (New York: Crowell, 1967), p. 203. No biographical information appears to be available on the minister of General Machine-building, Petr N. Goremykin, but he is likely the same individual identified in B. L. Vannikov's memoirs as People's Commissar of Munitions at the outbreak of the war. See "Iz zapisok narkoma vorruzheniia," Voenno-Istoricheskii Zhurnal, 1962, no. 2, pp. 78-86, translated as "In the People's Commissariat of Armaments" in Seweryn Bialer, ed., Stalin and His Generals (New York: Pegasus, 1969), pp. 153-59.

14. See Article 10 of "Zakon o dal'neishem sovershenstvovanii organizatsii upravleniia promyshlennosti i stroitel'stvom" (Law on further improving the organization of administration of industry and construction; hereafter cited as "Zakon..."), Pravda, 11 May 1957, pp. 1-2.

15. See Article 13 of "Zakon...."

16. The diagram of the organization of the Moscow City Economic Council, in Pravda, 5 April 1957, p. 4, indicated administrations for general machine-building, aviation industry, and radio-technical industry; the Leningrad Economic Council had administrations for, among others, shipbuilding, radio-technical industry, aviation industry, and general machine-building—see S. Mitrofanov, "Leningradskii ekonomicheskii raion i voprosy upravleniia ego promyshlennosti," Pravda, 8 April 1957, p. 3, and "Upravlenie promyshlennost'iu—na uroven' vozrosshikh zadach," Pravda, 13 April 1957, p. 2; an article by the Secretary of the Gorky Party Committee, N. Ignatov, "Organizovanno provesti perestroiku upravleniia promyshlennost'iu," Pravda, 6 May 1957, pp. 3-4, mentions the creation of administrations for shipbuilding, radio-technical industry, and defense industry, among others. See also Baritz, op. cit., p. 19.

17. As originally announced, the "antiparty group" consisted of G. M. Malenkov, V. M. Molotov, and L. M. Kaganovich, all full members of the Politburo, and D. T. Shepilov, a candidate member. The first three were long-time aides of Stalin, the last a Khrushchev protégé who turned against him. All four were expelled from their government and party posts, including membership in the Central Committee. At the same time, M. G. Pervukhin was demoted from full to candidate member of the Politburo, and full member M. Z. Saburov was removed from the Politburo completely. These two senior economic managers (who had no substantial ties to the defense industry) were only later branded as members of the "antiparty group." The same was true of N. A. Bulganin and K. E. Voroshilov, who remained full members for the time being. Thus of 11 full members of the Politburo, seven were eventually accused of having plotted

against Khrushchev. See Pravda, 4 July 1957, and footnote, page 95 of this volume.

18. Pravda, 15 December 1957.
19. Pravda, 4 March 1965.
20. Pravda, 1 March 1955.
21. See Khrushchev's address to the Conference of Russian Republic Industrial and Construction Workers, Pravda, 26 April 1963; translation in CDSP 15, 17 (22 May 1963): 3-14, especially 8.
22. Zakharov, op. cit., p. 56; V. D. Sokolovskii, Voennaia strategiia, 3d ed. (Moscow: Voenizdat, 1968), p. 420; Leonard Schapiro, The Communist Party of the Soviet Union (New York: Vintage, 1971), p. 243; John Erickson, The Soviet High Command (London: Macmillan, 1962), p. 38; William J. Spahr, "The Soviet Military Decision-Making Process," paper delivered at the Fifth National Convention of the American Association for the Advancement of Slavic Studies, Dallas, Texas, 15 March 1972, p. 8.
23. Zakharov, op. cit., p. 111; Sokolovskii, op. cit., p. 421; Erickson, op. cit., Soviet High Command, pp. 305-6; Spahr, op. cit., pp. 8-9.
24. Zakharov, op. cit., pp. 199, 234; KPSS o vooruzhennykh silakh Sovetskogo soiuza: Dokumenty, 1917-1968 (Moscow: Voenizdat, 1968), p. 277; Erickson, Soviet High Command, op. cit., p. 477; Spahr, op. cit., pp. 10, 13; Alec Nove, The Soviet Economy: An Introduction, 3d ed. (London: George Allen and Unwin, 1968), pp. 70-71.
25. KPSS . . . , p. 278. The decree specified that this body would replace "the existing commission on the mobilization of industry. . . ."
26. Zakharov, op. cit., p. 264; Sokolovskii, op. cit., pp. 426-27; Erickson, The Soviet High Command, op. cit., p. 598; KPSS . . . , p. 302.
27. Molotov for tanks, Beria for armaments and munitions, and Malenkov for aircraft. See Schapiro, op. cit., pp. 498-99.
28. Committee on Government Operations, U.S. Senate, National Policy Machinery in the Soviet Union (Washington, D.C.: Government Printing Office [GPO], 1960), p. 399; U.S. Senate, Committee on Government Operations, Staffing Procedures and Problems in the Soviet Union (Washington, D.C.: GPO, 1963), p. 46. I am indebted to Edward Warner for these two sources.
29. A "defense industries committee" is stated to exist by David Holloway, Technology, Management and the Soviet Military Establishment, Adelphi Paper #76 (London: International Institute for Strategic Studies, April 1971), pp. 6, 23; and by John Erickson, Soviet Military Power (Washington, D.C.: United States Strategic Institute 1973), p. 21. Speculating on the possibility of such a committee are: Sheren, op. cit., p. 124; Spahr, op. cit., p. 29; and

Matthew P. Gallagher and Karl F. Spielmann, Jr., The Politics of Power: Soviet Decision-making for Defense (Arlington, Va.: Institute for Defense Analysis, October 1971), p. 21.

30. John Newhouse, Cold Dawn: The Story of SALT (New York: Holt, Rinehart and Winston, 1973), p. 251. The name "Military-Industrial Commission" is the same as that used in the January 1938 decree cited in note 25 above.

31. Pravda, 22 December 1953, p. 2.

32. Pravda, 1 March 1955.

33. Pravda, 26 May 1955, p. 2. Gosplan's full name was changed from "Gosudarstvennyi planovyi komitet Soveta Ministrov SSSR" to "Gosdarstvennaia komissiia Soveta Ministrov SSSR po perspektivonomu planirovaniiu narodnogo khoziaistva." Gosekonomkomissia's full name was "Gosudarstvennaia ekonomicheskaia komissia Soveta Ministrov SSSR po tekushchemu planirovaniiu narodnogo Khoziaistva."

34. Pravda, 30 May 1955, p. 2. Gostekhnika's full name was "Gosudarstvennyi komitet Soveta Ministrov SSSR po novoi tekhnike."

35. Pravda, 12 February 1957, p. 2.

36. Robert Conquest, Power and Policy in the USSR (New York: Harper and Row, 1967), pp. 294-95, points out that the net effect of these demotions and related events was to end Malenkov's anomalous position as the only one of nine deputy chairmen of the council with a seat on the Politburo. Now he was simply the only deputy chairman.

37. Pravda, 2 January 1957; the obituary was noteworthy for its description of one phase of Zavenyagin's career: During and after the war, as deputy minister of Internal Affairs (that is, the police apparatus), "he was a leader in the construction of a series of large-scale industrial and hydroelectrical structures and enterprises of the mining industry. . . ." In other words, he was in charge of prison camp labor projects.

38. Pravda, 21 February 1957, p. 3.

39. Pervukhin would be demoted again for his role in the "antiparty group" affair. In July 1957, he slipped from full to candidate member of the Politburo, lost his first deputy chairmanship of the Council of Ministers and eventually was removed from the Ministry of Medium Machine-building as well. See Pravda, 4, 6, and 25 July 1957. The fact that his punishment was less severe than that accorded others, and that his demotions came in stages, indicates he still had protectors in high places.

40. See Article 18 of "Zakon. . . ."

41. See Article 19 of "Zakon. . . ."

42. See Carl A. Linden, Khrushchev and the Soviet Leadership, 1957-1964 (Baltimore: Johns Hopkins Press, 1966); and Michel Tatu, Power in the Kremlin: From Khrushchev to Kosygin (New York: Viking, 1970), passim. Tatu notes, p. 137, that Ustinov has been

consistently elected a deputy to the Supreme Soviet from Izhevsk, where Kozlov had served during the war as an organizer of military production.

43. See Tatu, op. cit., pp. 114-22, for an excellent discussion of this series of promotions and their relationship to the power struggle within the Soviet leadership. Ryabikov had been a deputy chairman (without portfolio) of the RSFSR Council of Ministers since 1958.

44. "Gosplan" had lost its long-range planning functions to "Gosekonomsoviet" (Gosudarstvennyi nauchno-ekonomicheskii sovet Soveta Ministrov SSSR—the State Scientific-Economic Council), which had been created, with less important tasks, in February 1959. While its new chairman, A. F. Zasyadko, had no direct connections with the defense industry (his career had been in the coal industry; later events would show he was no friend of Khrushchev), his first deputy, A. A. Goreglyad, had been a leader of the tank industry during the war, minister of Shipbuilding immediately after it, and then was demoted to director of the Zhdanov Shipyard in Leningrad.

45. Pravda, 9 April 1961. Its full name was "Gosudarstvennyi komitet Soveta Ministrov SSSR po koordinatsii nauchnoissledovatel'skikh rabot."

46. Khrushchev announced the suspension of the demobilization program in his address to graduates of the military academies, Pravda, 9 July 1961, partial translation in CDSP 13, 27 (2 August 1961): 3-6. At the same time he announced a one-third increase in explicit Soviet defense expenditures, but this may have been no more than a "surfacing" of previously hidden military spending, in reply to an earlier U.S. increase of similar proportions. Defense Minister Malinovskii, in addressing the 22d Congress in October, returned Soviet military doctrine to the principle of "balanced forces," in particular reemphasizing the importance of a large ground army. See Pravda, 25 October 1961, partial translation in CDSP 14, 1 (31 January 1962): 19-22.

47. See Tatu, op. cit., p. 288.

48. Novikov was not allowed to address the 22d Party Congress (indeed, none of the prominent defense industry "alumni" spoke), a privilege given rather to one of his deputies, V. E. Dymshits, who succeeded him in 1962. Ibid., p. 175; and Robert M. Slusser, The Berlin Crisis of 1961 (Baltimore: Johns Hopkins Press, 1973), pp. 302-3.

49. See especially Khrushchev's address to the voters of the Kalinin District of Moscow, Pravda, 28 February 1963, translation in CDSP 15, 9 (27 March 1963): 6-12.

50. See Pravda, 14 March 1963, translation in CDSP 15, 9 (27 March 1963): 3-5. The full name of the "Supreme Sovnarkhoz" was "Vysshii Sovet narodnogo khoziaistva SSSR."

51. See the Khrushchev speech cited in note 21 above.
52. See Tatu, op. cit., pp. 344-45, for several examples of instances of Ustinov being listed as a "mere" deputy chairman rather than one of the three first deputy chairmen.
53. See Linden, op. cit., p. 198.
54. See Tatu, op. cit., pp. 410-20.
55. According to Conquest, op. cit., pp. 396-97, the total number of full and candidate members was raised from 12 to 34 in 1952, and Malyshev was one of 15 new full members added at this time. It would appear, however, that a subcommittee composed largely of the preexpansion members actually functioned in place of the larger group. Immediately after Stalin's death, nearly all the "expansion" members were excluded, and the Politburo was reduced to 10 full and 4 candidate members.
56. Pravda, on 28 June 1953, listed the attendance of the entire Politburo at a Bolshoi Theater performance the previous evening, with the notable exceptions of Beria and his secret police colleagues Mel'nikov and Bagirov, but with the addition of Malyshev.
57. Conquest, op. cit., p. 231, suggests he had close ties with Malenkov. However, in March 1955 he was publicly associated with Bulganin and may indeed have switched over to the winning side.
58. See his interview with Turner Catledge, editor in chief, New York Times, in Pravda, 14 May 1957.
59. New York Times, 15 June 1973, p. 3.
60. See Michael P. Gehlen and Michael McBride, "The Soviet Central Committee: An Elite Analysis," American Political Science Review 62, 4 (December 1968): 1232-41.
61. In the speech cited in note 21, Khrushchev complained that "certain undisciplined people" in the defense industry were abusing the secrecy associated with this field to cover up poorly utilized reserves. Since one of the tasks of the planning agencies is to ensure that no capacity lies idle, this implies that Gosplan and others were unable to monitor the defense industry properly. But the continued practice, since the mid-1950s, of appointing first deputy chairmen of the planning agencies with defense industry backgrounds, suggests that such individuals may head special sections closed to the regular Gosplan officials and have used this arrangement to protect their colleagues in the ministries and factories. It should be emphasized that this is speculation, and the author has no direct evidence that a "special section" actually exists.
62. Newhouse, op. cit., pp. 252-53. Two recent additions to the Soviet armed forces command structure were members of the regular delegation to SALT: Colonel-General N. V. Ogarkov, named first deputy chief of the General Staff in May 1968, and Colonel-General N. N. Alexeev, appointed a deputy minister of defense in

October 1970, reputedly with responsibilities in the field of weapons development. The latter may be the chief military representative on the "defense industries committee" described earlier. See Erickson, Soviet Military Power, op. cit., pp. 15, 18, 21. Another Soviet delegate was P. S. Pleshakov, a deputy minister of Radio Production.

63. See for example, Roman Kolkowicz, The Soviet Military and the Communist Party (Princeton, N.J.: Princeton University Press, 1967), pp. 309-21.

ANNEX A: MINISTRIES OF THE DEFENSE INDUSTRY

Ministry of Defense Production (Ministerstvo Oboronnoi (Promyshlennosti): Known prior to 1953 as the Ministry of Armament (Vooruzhenniya), which absorbed the military production of the wartime commissariats of munitions (boepripasov), tank industry, mortars, and medium machine-building when these were converted to civilian production in 1946.

Military products: "classical" armaments, such as tanks, artillery, small arms, and ammunition; and optical goods.

Civilian products: sportsmen's rifles, police weapons, explosives for construction, optical goods, petrochemicals, railroad cars, tractors, motorcycles, cameras, and so on.

Ministers: B. L. Vannikov (1939-41), D. F. Ustinov (1941-57), A. D. Domrachev (1957-58), K. N. Rudnev (1958-61), L. V. Smirnov (1961-63), S. A. Zverev (1963-   ).

People's Commissars of Munitions: I. P. Sergeev (1939-??), P. N. Goremykin (??-1942), B. L. Vannikov (1942-46).

People's Commissar of Tank Industry: V. A. Malyshev (1941-46).

People's Commissar of Mortars: unknown.

People's Commissar of Medium Machine-building: S. A. Akopov (1941-46).

Ministry of Aviation Production (Ministerstvo Aviatsionnoi Promyshlennosti): Created in 1939, merged briefly in 1953 into Defense Production.

Military products: all types of aircraft, including helicopters.

Civilian products: civilian aircraft, engines for agricultural machinery, consumer durables (refrigerators, washing machines, vacuum cleaners), aluminum kitchenware, medical equipment, and so on.

Ministers: M. M. Kaganovich (1939-40), A. I. Shakhurin (1940-46), M. V. Khrunichev (1946-53), P. V. Dement'ev (1953-   ).

Ministry of Shipbuilding Production (Ministerstvo Sudostroitel'noi Promyshlennosti): Created in 1939, merged into Transport and Heavy Machine-building, 1953-54.

Military products: warships of all types.

Civilian products: merchant ships, fishing vessels, storage tanks, pipe, chain, boilers, and so on.

Ministers: I. T. Tevosyan (1939-40), I. I. Nosenko (1940-46), A. A. Goreglyad (1946-50), V. A. Malyshev (1950-53), I. I. Nosenko (1953-56), A. M. Red'kin (1956-57), B. Ye. Butoma (1957-   ).

Ministry of Medium Machine-building (Ministerstvo Srednogo Machinostroenniia): Formally created in 1953, although the First Main Administration of the USSR Council of Ministers was its functional equivalent, 1946-53. Military products: nuclear weapons and reactor propulsion units. Civilian products: uranium mining and refining, industrial reactors, radio-isotopes. Ministers: B. L. Vannikov (1946-53), V. A. Malyshev (1953-55), A. P. Zavenyagin (1955-56), M. G. Pervukhin (April-July 1957), Ye. P. Slavskii (1957-   ).

Ministry of Radio Production (Ministerstvo Radio Promyshlennosti): Known as the Ministry of Radio-technical Production, 1954-57, and as the Ministry of Communications Equipment Industry, 1946-53. Neither of these, however, appear to have been primarily devoted to military goods.

Military products: communications equipment, radars, computers, and so on.

Civilian products: television sets, radios, tape recorders, and so on.

Ministers: I. G. Zubovich (1946-47), G. V. Aleksenko (1947-53), V. D. Kalmykov (1954-74), P. S. Pleshakov (1974-   ).

Ministry of Electronics Production (Ministerstvo Elektronnoi Promyshlennosti): Created in 1961.

Products: electronic components, parts, subassemblies, which are supplied to the Ministry of Radio Production and several civilian-oriented ministries.

Minister: A. I. Shokin (1961-   ).

Ministry of General Machine-building (Ministerstvo Obshchego Machinostroenniia): A ministry with this name existed, 1955-57, producing traditional weaponry, and was merged into Defense Production. Re-created with different tasks in 1965.

Military products: ballistic and cruise missiles.

Civilian products: space exploration equipment (?).

Ministers: P. N. Goremykin (1955-57), S. A. Afanas'ev (1965-   ).

Ministry of Machine-building (Ministerstvo Machinostroenniia): Created in 1968.

Products: unknown; possibilities include space exploration equipment, antiballistic missile systems, surface-to-air missiles, and so on.

Minister: V. V. Bakhirev (1968-   ).

ANNEX B: BIOGRAPHICAL SKETCHES OF DEFENSE INDUSTRY LEADERS

Afanas'ev, Sergei Alexandrovich. Born 1918. Joined party 1943. Central Committee 1961, 1966, 1971. Bauman Higher Technical School, Moscow, 1941. Career: 1941-46—from foreman to deputy chief mechanical engineer at armaments plants in Moscow and Perm; 1946-53—"executive posts" in Ministry of Armament; 1953-57—deputy chief, then chief of Technical Administration of Ministry of Defense Production; 1957-58—deputy chairman, then first deputy chairman of Leningrad Regional Economic Council; 1958-61—chairman of Leningrad Regional Economic Council; 1961-65—chairman of Russian Republic Economic Council (RSFSR Sovnarkhoz) and deputy chairman of RSFSR Council of Ministers; 1965 on—minister of General Machine-building.

Bakhirev, Viacheslav Vasil'evich. Born 1916. Joined party 1951. Central Committee 1971. Moscow State University, 1941. Career: 1941-65—from engineer to factory director in defense industry; 1965-68—first deputy minister of Defense Production; 1968 on—minister of Machine-building.

Butoma, Boris Yevstafevich. Born 1907. Joined party 1928. Central Committee 1961 (cand.), 1966, 1971. Leningrad Shipbuilding Institute, 1936. Career: 1937-48—from foreman to director, Vladivostok Shipyard; 1948-52—chief of a Main Administration, Ministry of Shipbuilding; 1952-53—deputy minister of Shipbuilding Production; 1953-54—member of collegium, Ministry of Transport and Heavy Machine-building; 1954-57—deputy minister of Shipbuilding Production; 1957-65—chairman of State Committee for Shipbuilding; 1965 on—minister of Shipbuilding Production.

Dement'ev, Petr Vasil'evich. Born 1907. Joined party 1938. Central Committee 1952 (cand.), 1956, 1961, 1966, 1971. Zhukovsky Military Aviation Academy, Moscow, 1931. Career: 1931-33—worked in research institute; 1933-41—from section head to director of an aircraft plant; 1941-46—first deputy people's commissar of Aviation Production; 1947-53—deputy minister of Aviation Production; 1953-57—minister of Aviation Production; 1957-65—chairman of State Committee for Aviation Technology; 1965 on—minister of Aviation Production.

Gerasimov, Konstantin Mikhailovich. Born 1910. Joined party 1939. Central Committee (cand. only) 1961, 1966, 1971. Bauman Higher Technical School, Moscow, 1935. Career: 1935-37—from designer to director of a factory design bureau; 1937-38—in Red Army; 1939-41—chief engineer in factory; 1941-49—chief of Main Administration in Ministry of Armament; 1949-51—deputy minister of Armament; 1951-54—director of scientific research institute; 1954-57—chief of Main Administration and deputy minister of Defense Production; 1958-60—first deputy chairman, then chairman of Gorky Regional Economic Council; 1960-74—Chairman of RSFSR Gosplan, and deputy chairman of RSFSR Council of Ministers. 1974—retired.

Goreglyad, Aleksei Adamovich. Born 1905. Joined party 1930. Central Auditing Commission 1966, 1971. Bauman Higher Technical School, Moscow, 1936. Career: 1931-35—trade union executive in motor tractor and aviation industry; 1935-38—member of party control commission of Central Committee; 1938-41—chief of main administrations in unspecified industries; 1941-46—deputy, then first deputy people's commissar of Tank Industry; 1946-50—minister of Shipbuilding Production; 1950-54—director of Zhdanov Shipyard, Leningrad (building destroyers); Minister of River and Merchant Fleet, then of Merchant Fleet; 1955-59—first deputy chairman, State Committee on Labor and Wages; 1959-62—first deputy chairman, Gosekonomsoviet; 1963 on—first deputy chairman, USSR Gosplan

Kalmykov, Valerii Dmitrievich. Born 1908. Died 22 March 1974. Joined party 1942. Central Committee 1956 (cand.), 1961, 1966, 1971. Power Engineering Institute, Moscow, 1934. Career: 1934-39—designer, engineer, director of research institute in Moscow; 1949-51—chief of Main Administration of Shipbuilding Production; 1951-54—deputy chief of an Administration under the USSR Council of Ministers (First Main Administration?); 1954-57—minister of Radio-technical Production; 1957-65—chairman, State Committee for Radio Technology; 1965-1974—minister of Radio Production.

Khrunichev, Milhail Vasil'evich. Born 1901. Died 2 June 1961. Joined party 1961. Central Committee 1952, 1956. Ukrainian Industrial Academy, All-Union Institute of Administration (part-time). Career: 1932-37—deputy director, then director of defense industry factory; 1937—chief of Main Administration, people's commissariat of Defense Production; 1938—deputy people's commissar of Defense Production; 1939-42—deputy people's commissar of Aviation Production; 1942-46—first deputy people's commissar of Munitions; 1946-53—minister of Aviation Production; 1953-55—first deputy minister of Medium Machine-building; 1955-56—deputy chairman of USSR Council of Ministers; 1956-57—deputy chairman of Gosekonomkomissia; 1957-61—deputy chairman, then first deputy chairman of USSR Gosplan; 1961—chairman of State Committee for Coordination of Scientific Research and deputy chairman of USSR Council of Ministers.

Malyshev, Vyacheslav Aleksandrovich. Born 1902. Died 20 February 1957. Joined party 1926. Central Committee 1939, 1952, 1956. Bauman Higher Technical School, Moscow, 1934. Career: 1934-39—from designer to factory director; 1939-41—people's commissar of Heavy Machine-building; 1940-44—deputy chairman of Council of People's Commissars; 1941-45—people's commissar of Tank Industry; 1946-47—minister of Transport Machine-building; 1948-50—chairman of State Committee for Introducing Advanced Technology into the Economy, and deputy chairman of USSR Council of Ministers; 1950-53—minister of Shipbuilding Production and Deputy Chairman of USSR Council of Ministers; 1953—(March-July) minister of Transport and Heavy Machine-building; 1953-55—minister of Medium Machine-building and deputy chairman of USSR Council of Ministers; 1955-56—chairman of Gostekhnika and deputy chairman of USSR Council of Ministers; 1956-57—first deputy chairman of Gosekonomkomissia, and chairman of Gostekhnika.

Nosenko, Ivan Isidorovich. Born 1902. Died 2 August 1956. Joined party 1925. Central Committee (cand. only) 1952, 1956. Leningrad Shipbuilding Institute 1928. Career: 1928-38—worked at Nikolaev Shipyard; 1938-39—director of Ordzhonikidze-Baltic Shipyard; 1939-40—deputy people's commissar of Shipbuilding Production; 1940-46—people's commissar of Shipbuilding Production; 1941-45—simultaneously, first deputy people's commissar of Tank Industry; 1947-51—minister of Transport Machine-building; 1952-53—first deputy minister of Shipbuilding Production; 1953—first deputy minister of Transport and Heavy Machine-building; 1953-54—minister of Transport and Heavy Machine-building; 1954-56—minister of Shipbuilding Production.

Novikov, Vladimir Nikolaevich. Born 1907. Joined party 1936. Central Committee 1961, 1966, 1971. Military-Mechanical Institute, Leningrad, 1934. Career: 1928-41—from designer to factory director in defense industry; 1941-45—deputy people's commissar to Armament; 1945-48—deputy minister of Armament; 1948-54—"executive posts in defense industry"; 1954-55—deputy minister of Defense Production; 1955-57—first deputy minister of General Machine-building; 1957-58—chairman of Leningrad Regional Economic Council; 1958-60—chairman of RSFSR Gosplan and deputy chairman of RSFSR Council of Ministers; 1960-62—chairman of USSR Gosplan and deputy chairman of USSR Council of Ministers; 1962—permanent representative of the USSR Council of Ministers to Comecon and deputy chairman of the USSR Council of Ministers; 1962-65—chairman of the Commission of the USSR Council of Ministers on Foreign Economic Questions and USSR Minister; 1965—chairman of Supreme Sovnarkhoz and deputy chairman of USSR Council of Ministers; 1965 on—deputy chairman of the USSR Council of Ministers.

Rudnev, Konstantin Nikolaevich. Born 1911. Joined party 1941. Central Committee 1961, 1966, 1971. Tula Mechanical Institute, 1935. Career: 1935-40—from designer to deputy chief of factory design bureau; 1940-47—chief engineer, then director of defense industry factory; 1947-48—director of research institute; 1948-52—chief of Main Administration of Ministry of Armament; 1952-53—deputy minister of Armament; 1953-57—deputy minister of Defense Production; 1957-58—deputy chairman of State Committee on Defense Technology; 1958-61—chairman of State Committee on Defense Technology; 1961-65—chairman of State Committee for Coordination of Scientific Research and deputy chairman of USSR Council of Ministers; 1965 on—minister of Instrument-building, Means of Automation, and Control Systems.

Ryabikov, Vasilii Milhailovich. Born 1907. Died 19 July 1974. Joined party 1925. Central Committee 1952, 1956 (cand.), 1961, 1966, 1971. Leningrad Naval Academy, 1937. Career: 1937-39—designer-engineer and party organizer at "Bol'shevik" factory in Leningrad; 1939-40—deputy people's commissar of Armament; 1940-46—first deputy people's commissar of Armament; 1946-51—first deputy minister of Armament; 1951-53—"executive posts in defense industry"; 1953-55—deputy minister of Medium Machine-building; 1955-57—"committee chairman in USSR Council of Ministers"; 1958-61—deputy chairman of RSFSR Council of Ministers; 1960-61—chairman of Russian Republic Economic Council (RSFSR Sovnarkhoz); 1961-62—first deputy chairman of USSR Gosplan; 1962-65—first deputy chairman of USSR Sovnarkhoz; 1965-1974—first deputy chairman of USSR Gosplan.

Shokin, Aleksandr Ivanovich. Born 1909. Joined party 1936. Central Committee 1961 (cand.), 1966, 1971. Bauman Higher Technical School, Moscow, 1934. Career: 1934-38—chief of design bureau at shipyard; 1938-39—chief engineer in shipyard; 1939-43—chief engineer of an administration in shipbuilding industry; 1943-49—"executive government posts"; 1949-53—deputy minister of Communications Equipment Industry; 1954-55—deputy minister of Radio-technical Production; 1955-57—first deputy minister of Radio-technical Production; 1958-61—first deputy chairman of State Committee on Radio Technology; 1961-65—chairman of State Committee on Electronics Technology; 1965 on—minister of Electronics Production.

Slavskii, Yefim Pavlovich. Born 1898. Joined party 1918. Central Committee 1961, 1966, 1971. Moscow Institute of Nonferrous Metals and Gold, 1933. Career: 1918-28—in Red Army; 1933-40—from engineer to director of factory; 1940-41—director of Dnieper Aluminum Plant; 1941-45—director of Urals Aluminum Plant; 1945-46—deputy people's commissar of Nonferrous Metallurgy; 1946-53—deputy chief of first Main Administration of USSR Council of Ministers; 1953-56—deputy minister, then first deputy minister of Medium

Machine-building; 1956-57—chief of the Main Administration of the USSR Council of Ministers for the Utilization of Atomic Energy; 1957-63—minister of Medium Machine-building; 1963-65—chairman of State Production Committee for Medium Machine-building; 1965 on—minister of Medium Machine-building.

Smirnov, Leonid Vasil'evich. Born 1916. Joined party 1943. Central Committee 1961, 1966, 1971. Novocherkaask Industrial Institute, 1939. Career: 1939-48—engineer, then shop foreman in armament industry; 1949-60—director of research institute, chief of an administration, then factory director in defense industry; 1961—deputy chairman of State Committee on Defense Technology; 1961-63—chairman of State Committee on Defense Technology; 1963 on—deputy chairman of USSR Council of Ministers.

Stepanov, Sergei Aleksandrovich. Born 1903. Joined party 1928. Central Committee 1952, 1956 (cand.), 1961 (full member). Bauman Higher Technical School, Moscow, 1931, Leningrad Naval Academy, 1933. Career: 1931-38—from design engineer to chief engineer in a machine-building plant; 1938-41—deputy people's commissar of Heavy Machine-building; 1941-45—deputy people's commissar of Tank Industry; 1946-51—first deputy minister of Transport Machine-building; 1951-53—minister of Agricultural Machine-building; 1953-54—deputy minister, then first deputy minister of Machine-building; 1954-57—minister of Transport Machine-building; 1957-62—chairman of Sverdlovsk Regional Economic Council; 1962-63—chairman of Central Urals Regional Economic Council; 1963- ?—deputy chairman of USSR Gosplan.

Ustinov, Dmitri Fedorovich. Born 1908. Joined party 1927. Central Committee 1952, 1956, 1961, 1966, 1971. Leningrad Military-Mechanical Institute, 1934. Career: 1934-36—engineer at research institute in Leningrad; 1937—from design engineer to assistant chief designer, "Bol'shevik" plant in Leningrad; 1938-41—director of "Bol'shevik" plant in Leningrad; 1941-45—People's Commissar of Armament; 1946-53—minister of Armament; 1953-57—minister of Defense Production; 1957-63—deputy chairman of USSR Council of Ministers; 1963-65—chairman of Supreme Sovnarkhoz and first deputy chairman of USSR Council of Ministers; 1965 on—secretary of Central Committee and candidate member of Politburo.

Vannikov, Boris Lvovich. Born 1897. Died 22 February 1962. Joined party 1919. Central Committee 1939, 52, 56. Bauman Higher Technical School, Moscow, 1926. Career: 1927-37—various industrial posts, including factory director of machine-building plants in Tula and Perm; 1937-39—deputy people's commissar of Defense Production; 1939-41—people's commissar of Armament; 1941—briefly under arrest as result of dispute with Stalin regarding artillery production plans; 1941-42—deputy people's commissar of Armament;

1942-46—people's commissar of Munitions; 1946—minister of Agricultural Machine-building; 1946-53—chief of First Main Administration of USSR Council of Ministers (atomic weapons program); 1953-58—first deputy minister of Medium Machine-building; 1958—retired due to ill health.

Zavenyagin, Avraamii Pavlovich. Born 1901. Died 31 December 1956. Joined party 1917. Central Committee 1934, 1952 (cand.), 1956 (full). Gornaia Akademiia, 1929. Career: 1920s—party work; 1930-33—engineering work in heavy industry; 1933-37—director of Magnitogorsk Metallurgical Combine; 1937-38—first deputy people's commissar of Heavy Industry; 1938-40—chief of construction at Noril'sk metallurgical combine; 1940-46—deputy people's commissar of Internal Affairs; 1946-50—deputy minister of Internal Affairs; 1950-53—"work in the USSR Council of Ministers"; 1953-55—deputy minister of Medium Machine-building; 1955-56—minister of Medium Machine-building and deputy chairman of USSR Council of Ministers.

Zverev, Sergei Alekseevich. Born 1912. Joined party 1942. Central Committee 1966, 1971. Leningrad Institute of Precision Mechanics and Optics, 1936. Career: 1936-47—from engineer-designer to deputy director of factory; 1947-57—executive posts in Ministry of Armament/Defense Production; 1958-63—deputy chairman of State Committee on Defense Technology; 1963-65—chairman of State Committee on Defense Technology; 1965 on—Minister of Defense Production.

A Note on Biographical Sources

Information on Soviet officials in English can be obtained from Prominent Personalities in the USSR (Metuchen, N.J.: Scarecrow Press, 1968), assembled by the Institute for the Study of the USSR, Munich, Germany. Earlier editions were published under the title Who's Who in the USSR, 1961/62 and 1965/66. An updated revision is overdue. For those who read Russian, the most convenient source is the Yezhegodnik Bol'shoi Sovetskoi Entsiklopedii, 1971, at the back of which are published short biographies of all members of the Central Committee (CC) and the Central Auditing Commission (CAC) elected at the 24th Party Congress, April 1971. Similar biographies were published for CC and CAC members elected at the 22d and 23d Party Congresses, in the Yezhegodniki 1962 and 1966, respectively. Biographies of all Supreme Soviet deputies are published in Deputaty verkhovnogo soveta SSSR, which is published shortly after each convocation is elected, as one has been in June 1974. Where military research and production are involved, Soviet biographies can be a little vague on some periods. In the sketches above, material in quotations follows the phrasing of Yezhegodnik 1971. After an individual dies, Soviet sources tend to be a little more forthcoming, which is why obituaries are quite valuable. A good example of this is the case of M. K. Yangel', elected a candidate member of the CC in 1966 and 1971. His biography in Yezhegodnik 1966 was studiously uninformative; between the 24th Congress and the publication of Yezhegodnik 1971, Yangel' died, and his biography was amended to state that he had been a "scientist and designer in the field of rocket space technology."

CHAPTER

7

**ANALYSIS OF SOVIET
DEFENSE EXPENDITURES**
Philip Hanson

Available resources of labor, capital, and technical know-how set certain limits to any country's military policies in the short term. It is natural, therefore, for anyone analyzing Soviet military policy, whether in respect of naval or other forces, to turn to economists for help. A considerable effort has been made, mainly in the United States, to provide such help, and Western estimates of Soviet military spending are now numerous and widely quoted. These estimates have all been made in the face of large information gaps; they have wide margins of error and have been the subject of a good deal of controversy. Inevitably, with such a subject, there are suspicions on all sides (usually not publicly expressed) that the other chap's arithmetic is in some degree political.

Given the state of the art of analyzing Soviet defense expenditure and given the present conjuncture of world affairs, what actual use are these economic analyses to Western policy-makers? In particular, what use are they when it comes to assessing likely future developments? I will try to answer this question in two ways. Firstly, by reviewing the main limitations and pitfalls of our assessments of Soviet military spending, I will try to show what these assessments can and cannot do. There have recently been a number of extended published discussions of problems in estimating Soviet defense spending.[1] What I have to say on these will be mainly a matter of summary and comment. Secondly, I want to draw attention to the vital question of comparative Soviet-Western technical levels and to some of the evidence that is beginning to be accumulated on them. This evidence should at least make us reconsider carefully some of our assessments of current Soviet military potential.

Everything I say will refer to published, unclassified assessments of Soviet military expenditure and potential. I hope, but cannot be sure, that it also has some relevance to classified assessments.

There are in practice, at present, two kinds of global assessment of Soviet defense spending: in U.S. dollars and in Soviet rubles. They are not merely denominated in different current units; they have different meanings and are meant to answer different questions. Dollar assessments are concerned with estimates of "comparative effort"; ruble assessments are concerned with estimates of the "defense burden" on the Soviet economy.

ESTIMATES OF "EFFORT"

The end product of Western analyses that is most widely bandied about is of the form "In such-and-such a year the Soviet Union spent x billion dollars on defense." Thus the International Institute for Strategic Studies (IISS) publication, The Military Balance 1972-73, gives (page 9) a figure of $84 billion for Soviet defense spending, including reserves, in 1971, and goes on to say, "in 1971 there was virtual parity in defense spending between the two super-powers."

Thus U.S. dollar figures are estimates of "effort," as opposed to "burden," and are intended to show what it would cost the U.S. economy, in U.S. dollar prices of the current or some specified earlier year, to mount the Soviet defense program that is being measured.*
In other words, they are meant to provide a comparison of the Soviet military "effort" with that of the United States. In recent years the IISS has stressed the problematic nature of such dollar estimates. Their use in debates on NATO military spending seems irresistible; their usefulness in understanding the world balance of military forces is, in my opinion, nil.

There are three reasons for asserting this. The first is that the Russians do not, in the first place, publish much information on their defense spending, nor are they so helpful as to give us full definitions of those figures they do publish. It appears that the overt "defense" allocation in the Soviet budget does not include expenditure on military R and D, military space and nuclear programs, military

---

*The defense program in this context, sometimes referred to as "national security expenditure," is usually so defined as to be the equivalent of expenditure in the United States by the Department of Defense, Atomic Energy Commission, and National Aeronautics and Space Agency. Other national expenditure related to defense, but less directly related, is not covered. These figures therefore do not measure the full extent to which either U.S. or Soviet economic activity is geared to defense purposes.

stockpiling, or civil defense. Ruble costs of these programs are guesstimated by Western analysts with wide margins of error.

Secondly, putting a U.S. price tag on the overall Soviet defense effort means making more guesses and assumptions about still more unknown quantities. Defense expenditure covers a variety of activities: the pay and maintenance of servicemen; operations and equipment maintenance; procurement of military hardware; R and D. Some Soviet prices for some of these things are known; many (notably military hardware prices) are not, or at least are not in the public domain. Since the USSR has a relatively low-wage economy, Soviet military manpower is relatively cheap and each ruble of the (estimated) expenditure on manpower has a relatively high dollar value—though just how high depends in part on a judgment of the relative "productivity" (military effectiveness) of one U.S. and one Soviet soldier, sailor, and so on. Similarly with expenditure on personnel in military R and D, except that here relative productivity of Soviet and U.S. scientists and technologists is perhaps to a greater extent a matter of judgment—that is, it is not, effectively known. Estimates of the dollar value of Soviet military hardware procurement are still more precarious. The usual method is to assume that relative U.S.-Soviet prices of "analogous" civilian machinery will do as a proxy for relative prices of military hardware, thus providing us with a dollar "purchasing power parity" for each ruble (of a highly uncertain number of rubles) spent on hardware procurements. This point will be taken up later.

For these reasons the scope for disagreement over the dollar value of Soviet military, or military and space, spending is very large. Thus for 1971 the official Soviet view would presumably be that the value in current U.S. dollars of their defense effort would be $19.7 billion (overt budget defense appropriation at the official exchange rate). SIPRI (Stockholm) put it at $42.6 billion, IISS (London) put it at $84 billion and ACDA (Washington) at $64 billion. Estimates for earlier years by Boretsky (Washington) and the Stanford Research Institute would suggest figures for 1971 possibly higher than those of the IISS. Anything within the range $50-$90 billion could be fairly readily defended. This range extends from well below to well above the equivalent U.S. figures for 1971 and is large enough to account for the total estimated military spending (in dollars) of the lesser members of the NATO and Warsaw Pacts in that year.

Thirdly and finally, dollar figures such as these would be of doubtful utility even if values could be far more closely agreed. They add nothing of operational use (except for public relations exercises) to direct and detailed military assessments of Soviet capabilities. If the aim is to assess the Soviet military "threat" or to compare the Soviet military "effort" to that of NATO or of the United

States alone, it is better to do this directly by looking at what Soviet expenditure actually produces in the form of troops, equipment, and so on "on the ground."

(In principle, a "dollar" estimate of the Soviet defense effort could be obtained, not by converting ruble expenditure figures into dollars but by compiling an extensive list of Soviet defense activities—personnel, operations, additions to stocks of different kinds of equipment, and so on—in physical terms and then going through the list establishing what their activities would cost in the United States and thus reaching an overall dollar valuation of them. If practicable, this building block method would avoid many of the problems mentioned here in dollar valuations. It has been asserted that only the U.S. intelligence community has the information required to do it.)[2]

Two qualifications might be made to this conclusion, though neither of them is very strong. One is that comparing U.S. and Soviet dollar totals over a period of years provides a particularly clear view of trends over time—for example, in the late 1960s or early 1970s, according to several calculations, the overall Soviet defense "effort" in "real" terms was catching up and beginning to overtake that of the United States. Once again, though, it is surely more informative to do this directly, physically, and in detail (as the IISS Military Balance publications do). In any short period of time, it is always possible for one country to be spending a great deal more than the other on, say, naval building programs, in order to bring its fleet up from a relatively weak position to one of (in some sense) parity. The annual expenditure figures for hardware procurement are flows, but it is the resulting stocks that constitute military potential. These are best assessed directly.

The other qualification concerns military R and D. This is not (so far as a mere academic can judge) as visible to the outside world as tanks and submarines. Moreover, its output is something that takes military effect after a substantial time lag; the potential military power it represents cannot, generally, be determined at the time when the R and D work is being undertaken. Is it not therefore useful to measure at least the input into Soviet R and D in a way that is comparable with U.S. R and D expenditure (that is, in U.S. dollar prices) so as to provide some guide to its likely future output? This is a reasonable argument, and the main trouble about it is the sheer practical difficulty of the comparison. Ruble totals for Soviet military and space R and D are particularly uncertain[3] and the relative productivity of Soviet and U.S. R and D personnel a particularly vexed question. The best way to judge the productivity of Soviet military R and D may indeed be by its results, which takes us back to actual weapons systems and equipment.

ESTIMATES OF "BURDEN"

More useful than dollar-value estimates of the Soviet military "effort" are ruble-value estimates of the Soviet military "burden." These are free of the data and conceptual problems of ruble-dollar conversion rates; they are however still subject, of course, to the original uncertainties surrounding "true" Soviet defense expenditure in rubles. They should nonetheless help us to understand better the pressures on Soviet policy-makers. They provide, not a comparison with other countries, but a comparison with other uses of resources in the Soviet economy.

Despite the guesstimating problems mentioned, there is some incomplete measure of agreement among Western analysts that the share of the Soviet gross national product (GNP) allocated to defense (in the sense of direct expenditure comparable to that of the U.S. department of Defense, Atomic Energy Commission, and NASA) has fluctuated around 10 percent since the early 1950s, falling from above to below that level in the late 1950s, rising sharply between 1960 and 1963, falling again up to 1965 and then rising slightly to around or above 10 percent in the late 1960s[4] and early 1970s.

U.S. national security expenditure in 1970 and 1971 was close to 8 percent of U.S. GNP. Soviet GNP was between a half and two-thirds of U.S. GNP. It may therefore seem odd that a not very different share of a much smaller national output can support a Soviet defense effort that is widely believed to "match" (however roughly) that of the United States. The explanation lies in the difference between the two sets of prices: U.S. and Soviet. Soviet manpower and (by assumption) military hardware are relatively cheap or, to put it another way, most Soviet nonmilitary goods are relatively costly, compared to U.S. prices. A ruble of military spending is worth more dollars than a ruble of civilian spending, on average. In other words, if Soviet government industrial and household purchasers were faced with U.S. prices of all final goods and services and still bought the same collection of military and civilian goods and services that they have recently been buying, Soviet military spending would come to much more than 10 percent of Soviet total final expenditure, and the apparent paradox of the relatively low Soviet defense burden would disappear.

What does a 10 percent or, for that matter, a 15 percent burden mean, though? If military spending is reduced by 10 million rubles, does this mean that resources of labor and capital are released to produce 10 million rubles worth of civilian output, or more, or less? This is not a question that requires merely a measuring of past expenditures; it requires an analysis of cause and effect relationships in the Soviet economy, which in turn requires further information.

If, for example, men are released from the services but remain unemployed or underemployed, the gain to civilian output may be relatively small. If these men have technical skills that are in short supply in the civilian economy and they are promptly employed, the gains may, on the contrary, be considerable. It may well be, in fact, that the opportunity cost in civilian production forgone through conscription is substantially understated by the pay of Soviet military personnel.[5] In this case, the "real" cost of Soviet military spending to the Soviet economy is understated by our usual "burden" figures.

This may be true of other kinds of Soviet defense spending, though the evidence is inconclusive. For example, a great many of the most capable Soviet scientists and technologists are absorbed in military-related work—probably an even larger share than their counterparts in the United States. (The U.S. Department of Defense has attributed as much as 75-80 percent of Soviet R and D expenditure to the military and space efforts, though this appears to rely on a good deal of guesswork.) The release of some of them to civilian work might, after a time lag, have a disproportionate effect on the growth of Soviet civilian output. On the other hand, it can be argued that R and D in the Soviet military sector is so much more free of administrative hindrances and disincentives than it is in the Soviet civilian economy, that a switch of resources from the former to the latter would produce disappointing results. We are short of hard evidence on these matters; and so, almost certainly, is the Soviet Politburo. When the economic effect of defense spending in Western countries remains a matter of controversy, we should not expect the Soviet leadership to be provided with agreed and reliable advice on the quantitative effect on their civilian economy of various possible changes in the level of their defense expenditure.

An important matter on which more information is available to the Soviet leadership than to Western analysts is the proportion of Soviet industrial capacity and labor force that is working indirectly—as well as directly—for military uses. For instance, how much of the Soviet industrial labor force and capital stock is engaged not only in building naval vessels but in making the components, equipment, steel, and so on that goes into these vessels? Available Soviet data on the structure and intersector deliveries of Soviet industry fudge the answers to precisely this sort of question.[6] A fortiori, it is impossible to assess with any confidence the effects on the civilian economy of changes in this proportion.

In short, studies of the economic "burden" of Soviet military programs are potentially useful, but the statement that direct Soviet military expenditure absorbs about 10 percent, say, of GNP, is of itself no guide to Western policies. The question we want to ask is: Does this burden enforce a cutback or deceleration of military spending, or does it on the contrary allow military programs to be maintained or expanded—and perhaps even to be expanded faster than Soviet national product is likely to grow? Is it such, for instance, that Soviet interest in further progress at SALT must be construed as genuine and urgent?

One has only to put the question in this way to see that it is a question about future choices by the Soviet leadership. The assessment of likely future Soviet policies is therefore, in the last resort, inherently problematic. It is not so very different from trying to predict what some particular household is going to spend on consumer durables next year. We may know a great deal about their past spending, their likely income, their other wants, the prices they face, and so on. We may even know that they have a new washing machine on order at the beginning of the year, an old fridge that keeps breaking down, and, on the other hand, a new, splendid, and economical car. We may know, further, that some members of the household are agitating for more kitchen gadgets while others are all for more holidays and ice creams instead. Who will, nonetheless, predict with any great confidence the ultimate, compromise outcome? So with Soviet military spending. The more we try and second-guess future decisions, the more we must go back to the political and military considerations. The economic pressures are only part of the picture and will not necessarily be decisive. The further ahead we want to look, the more unpredictable will be the environment of foreign policies and military technologies in which these economic pressures (themselves perhaps relatively stable over time) will work.

We may still get somewhere, however, by studying the apparent past effects on the Soviet civilian economy of variations in military programs. This has been attempted most notably, perhaps, by S. H. Cohn.[7]

One exercise that Cohn has carried out is a comparison of growth rates of Soviet defense spending in five different periods of time between 1950 and 1969 with growth rates of Soviet capital investment, private and public consumption, and administration, the five periods being selected to distinguish major shifts in the overall priority given to defense expenditure. His figures are in Table 7.1 (Defense A and Defense B are for alternative estimates of Soviet defense spending in rubles, and all the figures are from Western recalculations rather than unprocessed Soviet official data).

The slowdown in Soviet defense spending from the early 1950s to 1960 (associated especially with Khrushchev's large-scale demobilizations of troops) coincides with rapid growth of investment and GNP. The sharp upturn of military spending in the early 1960s, associated with the Berlin and Cuban crises, seems to have had its immediate impact, as Cohn observes, on investment rather than on consumption (private plus public). With the brief succeeding slowdown

TABLE 7.1

Expenditure Trends for Principal Uses of GNP
(percent per year)

	1950-52	1952-60	1960-63	1963-65	1965-69
Private consumption	7.0	6.7	3.5	4.8	6.2
Public consumption	5.1	5.1	6.8	6.2	5.3
Capital investment	12.5	12.7	4.8	8.6	6.8
Defense A	19.2	2.5	15.7	1.2	10.1
B	20.6	0.7	15.7	-0.5	7.3
Administration	6.8	2.7	2.5	3.5	6.1
GNP	5.2	6.3	4.0	7.0	4.9

Source: Stanley H. Cohn, "Economic Burden of Defense Expenditures," in U S. Congress, Joint Economic Committee Soviet Economic Prospects for the Seventies (Washington, D.C., 1973), p. 151.

in defense spending, conversely, the benefit seems to have been felt primarily by investment. The figures for 1965-69 might possibly be interpreted as suggesting that the competition between investment and defense spending has become less acute in the latterly more highly developed Soviet economy.

The interpretation of these figures is however highly problematic.* Cohn's conclusion from a more detailed analysis that defense procurement is particularly at the immediate expense of producer durables—for example, production of new machinery and equipment—illustrates the strain of Soviet defense production on the Soviet engineering industries. This should affect current output in later periods, however, and the extent to which it may have done so is uncertain. So far as the overall impact of defense spending on Soviet economic growth is concerned, the picture is quite similar. The disastrous harvest of 1963 is enough by itself to depress the 1960-63

---

*Cohn also tries to estimate statistical relationships between defense spending and other categories of final expenditure from these time-series data. The results however are inconclusive as evidence of relationships, because the number of observations is generally very small and most of the variables in most time periods are subject to dominant upward trends over time that would tend to produce spurious correlations.

growth rate of GNP considerably, and in the 1960s more generally a slowing growth of labor supply has probably had a critical effect on overall economic growth, though the relative importance of this and other factors are a subject of controversy. We are still a very long way from being able to isolate with any precision the effects on the Soviet civilian economy of variations in different categories of Soviet defense programs. So, no doubt, is the Kremlin.

ECONOMIC CONSTRAINTS AND SOVIET POLICY

What is the usefulness of all this for assessing Soviet policies in the mid-1970s? The "burden" estimates can at least help us to put some of the quantities in perspective. Let us provisionally agree that Soviet direct defense spending has lately been of the order of 10 percent of Soviet GNP. Let us further agree for the moment that a generous estimate of the current trend rate of growth of Soviet GNP in real terms would be 5 percent a year. A number of conclusions then follow. First, that defense spending could be pushed up at an average annual rate of 5 percent without reducing other (civilian) end-use shares of GNP (at any rate, as measured in constant, initial-year, prices). Secondly, that, as a matter of arithmetic, quite large year-to-year increases or decreases in Soviet defense spending could <u>prima facie</u> be accommodated without impinging in a very dramatic way on the growth of civilian consumption and investment. For example, defense spending could be increased by 15 percent between one year and the next (the approximate average rate of increase in 1960-63, according to Cohn) while still allowing civilian production to rise by 3.9 percent and the civilian share of GNP to fall by marginally less than 1 percent. Conversely, decreases of the same order of size, which could be very significant for military programs, do not involve spectacular immediate gains for the civilian economy. This arithmetic illustrates the point that with continued respectable growth rates in the Soviet economy, the economic constraints on defense spending are far from narrow.

We can also be fairly sure, however, that military hardware procurement competes strongly with civilian investment. Military research, development, and testing, moreover, keeps particularly productive resources away from civilian R and D programs (to an extent probably not compensated by civilian spin-offs from military R and D). This means that defense procurement and R and D have longer-run repercussions on the growth of the civilian economy, <u>via</u> their effects on the growth of civilian capital stock and technical progress, respectively. Neither Western analysts nor Politburo advisers can be very sure how large these opportunity costs are, but it is likely that they are quite large. It is also likely that the prevailing view among the Soviet leadership is that they are large and should be worried about. There has been a great emphasis in Soviet specialist writings and leadership speeches of the 1960s and early 1970s on the key role of technical change in Soviet economic growth and on the need to overcome weaknesses in Soviet civilian R and D and introduction and diffusion of new products and processes. This suggests that the trade-off between military and civilian technology programs is something that the Soviet leadership is really worried about.

Economic pressures against a sustained, rapid (say, around or above 10 percent a year) growth in overall Soviet defense spending are probably now quite considerable. The costs of such increases to the Soviet Union's economic growth probably loom larger now in Soviet decision-making than they did in the early 1960s and, perhaps, than the corresponding U.S. costs do in current U.S. decision-making, since (1) in 1960-62 the recent past growth rates of the two economies encouraged Khrushchev to think that the USSR was comfortably on the way to overtaking the United States, a point on which no Soviet leader can currently feel very confident; (2) the Soviet leaders—or some of them, at least—now apparently feel that problems of civilian technological progress are urgent, a view that was not influential in the USSR a decade or more ago and that is hardly fashionable now in the United States; (3) the USSR is probably still short of really well-trained, competent scientists and technologists, compared to the United States; (4) the USSR probably has less spin-off from military to civilian technology because of administrative barriers and greater Soviet secretiveness about military-related work; (5) Soviet R and D personnel released from military-related work will normally be promptly employed (though probably less efficiently employed) in civilian work whereas in the United States they may be unemployed for considerable periods of time.

THE INFLUENCE OF TECHNOLOGY

Soviet technological performance has been referred to frequently in the discussion. I want finally, and briefly, to draw attention to some questions about Soviet technology. Much that has been written by Western economists about Soviet production potential in general, and military production potential in particular, has had to be based on assumptions, rather than strong evidence, about the comparative levels of Soviet and Western technology. A number of Western specialists, including a group based on the Birmingham University Center for Russian and East European studies, are now turning their attention to detailed assessments of Soviet technology in particular fields. This work, admittedly, is not directed to answering questions about Soviet defense capacities. On the other hand, the interlinkages of an advanced economy are such that a wide range of technology that is not primarily military is nonetheless relevant to a country's military effort. A better understanding of the strengths and weaknesses of Soviet technology in various areas should improve, and might change, our assessment of Soviet policies. Some tentative preliminary results of work on comparative Soviet-Western technical levels in different branches will illustrate this.

Firstly the Western studies that try to compare Soviet and U.S. defense "efforts" in dollars rest in part, as was noted above, on the assumption that the relative prices in rubles and dollars of Soviet and U.S. defense hardware are similar to the relative prices of Soviet and U.S. civilian machinery. This is a plausible assumption, but Western academics are in no position to check it. There is also, however, a further assumption here: That we know with sufficient accuracy the relative prices of Soviet and U.S. civilian machinery in the first place. The major source for these prices is a careful and extensive study by A. S. Becker.[8] This however refers to 1955 types and prices of machinery, and subsequent "updating" has attempted only by strong assumptions and rough estimates to cope with the massive turnover of new and improved machinery types. Furthermore, Becker explained that his comparisons of prices were for machines with closely similar performance specifications. The resulting overall price ratios do not and cannot allow for the existence of sophisticated U.S. machines for which there was no Soviet machine of similar specifications, nor, of course, is a similarity in performance specifications necessarily equivalent to an overall similarity in effectiveness in use, since design and production faults may seriously affect the reliability, operating costs, and service life of a machine.

Despite these limitations, which were pointed out by the author, Becker's study has been very widely used to support calculations of Soviet engineering production capacity relatively to that of the United States. If the net affect of these limitations is to understate relative Soviet costs and prices, it follows that the ensuing "dollar value" calculations of Soviet machinery production are exaggerated relatively to U.S. levels. Engineering evaluations of (1) some standard Soviet machine tools in use and (2) Soviet machine-tool standards and standardization policy strongly suggest that basic design and production weaknesses in Soviet machine-tools (compared to U.K. tools of similar specifications) may be widespread.[9] It is likely that these effects are not restricted to machine tools; the circumstantial evidence that Soviet machinery exports to sophisticated markets have been small, supports this view. It is probable that conventional Western assessments have been understating Soviet machinery costs and prices and overstating production capacity, relative to that of Western countries.[10]

It may not follow, of course, that Western analyses of Soviet military hardware production have been misleading. We may be overstating their efficiency and capacity in civilian machinery production, but what is an overstatement for civilian machinery may nonetheless be about right for military hardware. Detailed comparative evaluations of Soviet and NATO military hardware in at least some sophisticated areas such as naval weapon systems[11] show no obvious Soviet technical inferiority.

One comes back to the vexed question of the extent to which Soviet military-related technology (and production quality) develops independently of Soviet civilian technology: the extent to which the military-related sector is administratively "hived off," subject to more effective organization and supplied with markedly better-quality inputs of scientific and technical skills.

Whatever the full answer to this question may be, a totally independent military-related production sector, independent from the rest of the economy for all its materials and components and for its inputs of skill and know-how, is scarcely conceivable. Technological studies of branches whose output is an input to both military and civilian branches are useful here. For example, the evidence that the Soviet Union has been lagging behind the United States in the production of alloy and quality steels is more relevant to an assessment of comparative military potential, I suspect, than the evidence that total Soviet iron and steel production has recently exceeded that of the United States. Thus, alloy steels account for 11.6 percent of all Soviet smelted steel production in 1970, but the equivalent U.S. share was already at about that level in 1965. Despite the isolated success of the Soviet electroslag refining process, a number of other processes important for making special steels for aerospace and assessments production in the United States are apparently more highly developed and diffused there than in the USSR: for example vacuumarc, electron-beam, and vacuum-induction refining, and vacuum degassing.[12] Similarly, if numerically controlled (n.c.) machine tools are an important technological input into current aerospace programs, the well-known Soviet lag behind the United States in stocks of n.c. machine tools in use would be important.

One suspects that systematic accumulation of information on quality and technical level in Soviet industry will tend on the whole to support the more conservative economic assessments of the "threat" of Soviet military capacity and to emphasize, on the other hand, the burden that the Soviet military effort places on the Soviet economy.[13] The fact that the more alarmist "threat" assessments tend to come from the general direction of Washington reinforces this suspicion, though one's feelings on this point are no doubt too vulgarly unquantifiable to count as evidence.

CONCLUSIONS

What economic analyses of the Soviet military effort can do is improve our understanding of Soviet policy-making. They cannot, unless we adopt a doctrine of economic determinism more extreme than anything in Marx, tell us how the Soviet leadership must allocate resources next year or the year after. Political and strategic considerations are not necessarily less precise or less compelling in the decision-making process than economic considerations, but our best guesses must rest on an assessment of all three.

NOTES

1. Campbell, R. W. et al., "Methodological Problems Comparing the U.S. and USSR Economics," and Herbert Block, "Value and Burden of Soviet Defense," both in U.S. Congress, Joint Economic Committee, Soviet Economic Prospects for the Seventies (Washington, D.C., 1973), pp. 122-47 and 175-206, respectively; Stockholm International Peace Research Institute, "Estimating Military Expenditure in the Soviet Union" (draft), 1973.
2. See Block, "Value and Burden of Soviet Defense," p. 184.
3. See Comptroller General of the United States (GAO), "Comparison of Military Research and Development Expenditures of the United States and the Soviet Union," Congressional Record, p. E 8607m (1971).
4. S. H. Cohn, "Economic Burden of Defense Expenditures," in U.S. Congress, Joint Economic Committee, Soviet Economic Prospects for the Seventies (Washington, D.C., 1973), p. 151. Several other studies agree broadly with Cohn's, though M. L. Boretsky obtains a substantially higher "burden" for 1968.
5. E. R. Brubaker, "The Opportunity Costs of Soviet Military Conscripts," in U.S. Congress, Joint Economic Committee, op. cit. (1973), pp. 163-75.
6. See V. G. Treml's work on the published Soviet input-output data, in, for example, Treml and K. W. Kruger, The Structure of the Soviet Economy (New York, 1972).
7. Cohn, "Economic Burden of Defense Expenditures," and "The Economic Burden of Soviet Defence Outlags" in U.S. Congress,

Joint Economic Committee, Economic Performance and the Military Burden in the Soviet Union (Washington, D.C.: Government Printing Office, 1970).

8. Prices of Producers Durables in the United States and the USSR in 1955, RAND RM2432, August 1959.

9. M. R. Hill, "Experience in the use of Soviet-built General Purpose Machine Tools: Initial Study," draft, 1973, idem, "Standardisation Policy and Practice in the Soviet Machine Tool Industry," Univ. of Birmingham, Ph.D. Thesis, 1970.

10. As Alec Nove argued in his criticism of Boretsky's work, Survival, October 1971.

11. See Chapter 24 of this volume, N. D. Brodeur, "Comparative Capabilities of Soviet and Western Weapon Systems."

12. J. M. Cooper, "The Technical Level of the Soviet Iron and Steel Industry, A Comparative Study," University of Birmingham, draft, 1974, pp. 16-18.

13. For an illustration of the wide range of products and technologies required, for example, in building and equipping of a modern warship, see Raymond Hutchings, "The Economic Burden of the Soviet Navy" in M. MccGwire, ed., Soviet Naval Developments: Capability and Context (New York: Praeger Publishers, 1973).

CHAPTER 8

NATIONAL ECONOMIC PRIORITIES AND NAVAL DEMAND
John P. Hardt

Admiral Gorshkov, in the special series of articles in <u>Morskoi sbornik</u> in 1972-73, argued for a larger and more powerful Soviet fleet in the years ahead. According to his concepts, this fleet would have expanded capability not only for defensive roles around the periphery of the Soviet Union but also for an offensive capability appropriate for making global political impact and challenging the United States presence in many areas not contiguous to the Soviet Union. The surface naval combat and support ships required by this new Soviet naval strategy imply a significant additional naval demand on scarce resources in the Ninth Five-Year Plan (1971-75) and the years ahead. At a time when poor economic performance and rising civilian demands for resources to modernize their economy and better serve consumption needs have been acknowledged by General Secretary Leonid Brezhnev and his colleagues in the leadership, a new, substantial set of military claims on resources would seem especially unwelcome and untimely. At the same time, the appearance of the Gorshkov series after the 24th Party Congress and the formulation of the Ninth Five-Year Plan strongly suggests that the issue is not settled and that even though Brezhnev and possibly Marshal Grechko may oppose the adoption of the new naval strategy, it does have support among top leaders. As the issue of resource claims of the new naval demand is not explicitly or publicly discussed by Admiral Gorshkov in the <u>Morskoi sbornik</u> series or elsewhere to our knowledge, we must largely speculate on the pros and cons in the resource allocation aspects of this debate. What follows is an attempt to raise questions presumably central to such a debate and, by suggesting putative answers, to provide insights on how Soviet leaders may view the issues raised by Admiral Gorshkov. The conjectural purpose of this exercise reduces the utility of detailed documentation. The references provided are illustrative of information available that may clarify if not support one or another position. The format employed will be that of a "pro" and "con" exposition. Some subjective observations on the likely Soviet leadership view will end this short discourse.

The following is to provide pro and con arguments on two questions faced by the Soviet leadership that are central to the debate on new naval demands raised by Admiral Gorshkov:

1. Is the military share of resources to continue to expand at a faster rate than economic growth as it did in the last decade?

2. Whether or not the defense share of resources continues as in the past, will the Soviet Navy fare better than missile or ground forces in the division of the defense budget?

CURRENT COMMITMENTS TO NEW MILITARY PROGRAMS

Con

The Soviet military establishment has received more than its proportionate share of economic growth during the 1960s. The decision to expand offensive missile production, apparently made in 1962, was expensive but led to the attainment of strategic parity. This military buildup used resources that might have been devoted to civilian investment programs to modernize the Soviets' material supply system, improve transportation, and raise the level of real income. Likewise, the military forces were expanded to man the China border watch and meet the Czech invasion requirements. Restrictions in the quantity and quality of the resources available for civilian programs retarded economic growth.[1]

The Ninth Five-Year Plan appeared to be formulated on the assumption that programs in civilian investment and consumption, deferred in the 1960s, would be met in the 1970s.[2] Although the plan, as approved by the 24th CPSU Congress in April 1971, may have been overcommitted, and the necessary choices between civilian and military programs may not have been made, the economic disaster of 1972 may have forced decisions against new military programs. The specific shortfalls in Soviet energy, metal production, civilian machine-building industries, and food production may have precluded the funding, at this time, of new military programs or the expansion of the military support industries. With the long lead time from decisions to develop sophisticated weapon systems to their coming on line, recent intelligence on military growth may be misleading. Many reports of expanded and improved military programs may be products of programs set in motion as much as a decade ago.

Now, with negotiations on strategic arms limitations, parity with the United States in strategic weapons appears to be assured, and future negotiations promise further reduction of the likelihood that the strategic inferiority that led to the Cuban missile crisis will return. Moreover, the manpower requirements of the China border have been met, and mutual (and balanced) force reductions in Europe may release Soviet ground forces.

If parity is assured, though, is superiority a meaningful opportunity? By most quantitative measures the Soviet order of battle exceeds that of the United States. Even with allowance for different missions—such as securing the Chinese border, providing for internal security in Eastern Europe—the quantitative edge will probably increasingly favor the USSR. With any predicted level of quantitative advantage, would the Soviet Union be able to translate its military margin into political dominance of Eurasia or the world? As long as the United States maintains a reasonably invulnerable second strike capability that could destroy the majority of Soviet political and military centers, the deterrence of the U.S. umbrella would seem likely to continue in force. The validity of deterrence has always rested on assumptions as to leadership credibility and rationality. The kind of quantitative margins postulated seems unlikely to change the past logic of mutual deterrence.

Ultimately, the question of choice may be between incremental resources for improved economic performance or additional resources for expanded military programs. The Ninth Five-Year Plan may not be fulfilled without a shift from military to civilian priorities. Brezhnev's tenure in power may be closely tied to planned economic performance. On the other hand, military claims on the order of those of the last decade may be the key factor in slow economic growth.[3] And tangible political returns from seeking military superiority may be unattainable.

Moreover, the growth stimulation required from commercial relations with the West may only be attainable if more resources are shifted from high-priority military support production. The Western "bail-out" of the Soviet economy may only work effectively if infrastructure investment is included in the new Soviet plans.[4]

Pro

It may not be possible efficiently to shift resources and technological capability from the Soviet defense establishment to the Soviet civilian sectors. The enterprises of the Ministry of Defense Industries have become increasingly specialized. Under the traditional conventional weapons programs shifts of shipbuilding from naval to merchant ships, production lines from armored tracked vehicles to agricultural equipment were technologically possible in a short-time period. However, the "tank to tractor" option is less true for the missile age systems. The unique capacity for continued military expansion may be already available or on line. Options may thus already be foreclosed.

Likewise the military may well argue that depriving them may show meager results in the civilian economy. The level of technology in civilian sectors may be one third that of the Soviet military-industrial complex.[5]

Moreover, the newly expanded commercial relations with the United States and other industrially advanced nations may provide the technology and goods necessary for meeting the shortfalls in the current Ninth Five-Year Plan and better assist the Soviet planners in meeting their future needs in energy, metal, machine-building, and food output. Expanded commercial relations may thus provide both an opportunity to avoid making the painful decisions to downgrade the military priority and to meet the economy's need for technology and products the military industries might not, in any event, be able to supply, especially in the near term.

If commercial relations with the industrial nations do fill the bottlenecks and technological needs, a revived economic growth rate may provide an ever expanding annual resource increment for military programs. With a 2 percent GNP growth in 1972, the increment was only about $10 billion, rather than an expected increment of close to $25 billion implicit in the plan.[6] With a larger growth in GNP in 1973, a continued military demand may not be as untimely as it seemed in 1971-72.

THE INCREASED NAVAL SHARE IN NEW MILITARY PROGRAMS AS COMPARED WITH MISSILES AND GROUND FORCE DEMANDS

Con

If any military programs continue to increase their future share of economic resources, it may be either missile or ground force programs rather than new naval demands. The Strategic Arms Limitation Interim agreement restricts the number of missiles for each side and limits defensive missile (ABM) development. The Soviet superiority in numbers was presumed to be balanced by superior quality of U.S. offensive missile capability. With an apparent Soviet breakthrough in MIRV technology, the best payoff to the Soviet military might be a further round of offensive missile development to exploit the potential opportunity to gain strategic weapon superiority. Soviet ground force development has been focused on the China border watch. If the Sino-Soviet antagonism continues or exacerbates, the Soviet military may wish to increase its border capability. As it is difficult for Soviet forces to outnumber the Chinese, they may choose to emphasize fire power, mobility, communications, and other aspects of ground force capability. The prospects for a substantial payoff from further development of missile and ground force capabilities may, thus, outweigh the advantages of the new naval demand in priority for future resources.

Pro

In terms of "usable military power"—a concept of Henry Kissinger's—it may be more beneficial for the Soviet leaders to opt for naval expansion than increased ground or missile forces.[7] Military power is usable in this sense in political and strategic terms, rather than in war-fighting or active military terms. If it can be assumed that Soviet strategic nuclear superiority is not possible and that large-scale military actions on the ground, including a Sino-Soviet conflict, are not likely, the political and strategic role of an expanded Red Navy may provide the greatest opportunity for political and strategic benefits from military expansion. In a historical sense, it may now be the navy's turn to be the chief political force in the Soviet military arsenal. It might be argued that the political role of the Red Army was paramount when Stalin was consolidating the East European system and threatening Western Europe in the late 1940s. Stalin's wartime question on the number of divisions the Pope had and his reported statement that Soviet power did not reach the Italian border because the Red Army stopped short are illustrative of this political role of ground forces.

Their missile buildup of the 1960s had the political benefits that accrued from U.S. acceptance of the USSR as an equal nuclear power. Indeed missile rattling began in the Suez crisis in 1956 and reached a plateau in the Cuban missile crisis of 1962.

Expanding Soviet influence in the Middle East, South Asia, the Far East—indeed the world—may now accrue from the widening Soviet naval presence. The Soviet deployment in the Pacific Ocean of both surface and submarine capability, not only in strategic waterways such as the Taiwan, Tsushima Straits, but throughout East Asia, may carry considerable weight in Sino-Soviet and Soviet-Japanese relations. Expanded Soviet naval capability in the East Mediterranean has at least offset the dominance of the U.S. Sixth Fleet in those waters. With the opening of the Suez Canal the Soviet naval arm may extend to the Indian Ocean and, at minimum, further the neutralization of the subcontinent. Thus, the political and strategic benefits that

might result to the Soviet Union from meeting Admiral Gorshkov's
new naval demands may be viewed as both substantial and not too
risky in terms of actual combat.

HOW WILL THE SOVIET LEADERS ANSWER ADMIRAL GORSHKOV'S DEMANDS? — A SPECULATIVE ASSESSMENT

<u>This is not the time for continued massive commitments to new military programs.</u>

The explicit awareness of Leonid Brezhnev since the December 1969 CPSU Plenum of serious economic problems was not ameliorated by the economic disaster of 1972—the worst year in Soviet experience. The implied official sanction for continual discussion of new military programs in the release of the Gorshkov series may be explained as a Brezhnev safety valve for dealing with those in the Politburo still committed to their same military vote or burden in the economy as in the 1960s. Moreover, Leonid Brezhnev's technique appears to be one of avoiding hard and fast decisions on resource allocations but allowing them to emerge en route. Thus, the new commitments appeared to be made on the Ninth Five-Year Plan civilian investment programs, such as those for developing the oil and gas of West Siberia in 1970. It may be that new military programs competitive with that sort of civilian development may have been delayed or strung out. Barring a leadership assessment that the chance for strategic superiority is good or that conflict is imminent (for example, on the China border), economic performance seems likely to be the dominant issue in Soviet affairs. Thus the uncontrollable allocations to programs under way designed to improve economic performance are likely to preclude the controllable commitment to new programs in the military areas.

<u>Naval demands may well be preferred over new missile and ground force claims in a relatively stable defense budget.</u>

With the U.S. withdrawal from forward deployment throughout Eurasia and Africa and the increasing independence from Western influence in those areas, the prospects for political and strategic gain from Soviet extension of naval power seems enhanced. Whether or not Soviet naval power extensions are effective in the Mediterranean, Indian Ocean, the Pacific Ocean, and elsewhere may depend on the effective use of their military force and its perceived credibility. However, Soviet leaders may be expected to assume effectiveness in their application of a new military factor in these regional power balances. At the same time, "missile rattling" and use of ground-based forces in places such as Cuba and the Arab Egyptian Republic may remind them that naval intrusions have, in recent years, been safer and more productive.

Moreover, the naval expansion may be considered a part of a unified maritime concept. Global interest in the resources of the seabed, offshore drilling, fishing, and expanded merchant marine correlate with expanded naval military capability. Specific applications of Soviet global naval capability in some forms of "gunboat diplomacy" seem less likely than the general purpose of a balanced economic and military force for extending Soviet influence throughout the world.[8]

NOTES

1. U.S. Congress, Joint Economic Committee, Soviet Economic Prospects for the Seventies (Washington, D.C.: Government Printing Office, 1973), passim; U.S. Congress, Joint Economic Committee, Soviet Economic Outlook, Hearings (Washington, D.C.: Government Printing Office, July 1973).

2. N. K. Baibakov, Gosudarstvenniy piatiletnii plan razvitiia narodnogo khoziaistva SSSR na 1971-1975 gody (State five-year plan for development of the U.S.S.R. national economy for the period 1971-75) (Moscow: Politizdat, April 1972); Norton T. Dodge, ed., Analysis of the USSR's 24th Party Congress and 9th Five-Year Plan (Mechanicsville, Md.: Cremona Press, 1971); Abram Bergson, "Toward a New Growth Model," Problems of Communism, March-April 1973, pp. 1-9.

3. Close to an inverse relationship between the change of defense allocations and the rate of economic growth has been demonstrated for the period since the Korean War using econometric methods. Stanley H. Cohn, "Economic Burden of Defense Expenditures," Soviet Economic Prospects for the Seventies (Washington, D.C.: Government Printing Office, June 1972), pp. 147-62; John P. Hardt, "West Siberia: The Quest for Energy," Problems of Communism, May-June 1973, pp. 25-36.

4. John P. Hardt, "Soviet Commercial Relations and Political Change," prepared for presentation at the Conference on "Economics, Technology and Their Impact on Foreign Policy," at Kenyon College, April 25-28, 1974. For further discussion of these problems, see John P. Hardt and George D. Holliday, U.S.-Soviet Commercial Relations: The Interplay of Economics, Technology Transfer, and Diplomacy, U.S. Congress, House of Representatives, Committee on

Foreign Affairs (Washington, D.C.: Government Printing Office, June 1973): Robert Kovach, James Noren, and John P. Hardt in Norton T. Dodge, Summit a Year Later (Mechanicsville, Md.: Cremona Foundation Symposium of the Washington Chapter, AAASS, held May 1973).

    5. Peter G. Peterson, U.S.-Soviet Commercial Relations in a New Era (Washington, D.C.: U.S. Department of Commerce, August 1972), p. 34.

    6. John P. Hardt, "Summary," Soviet Economic Prospects for the Seventies (Washington, D.C.: Government Printing Office, June 1972), pp. IX and XVI.

    7. Henry A. Kissinger, "Congressional Briefing: United States-Soviet Relations in the 1970's," in Remarks of Senator Mike Mansfield, Congressional Record, vol. 118, June 19, 1972, pp. S9599-S9608.

    8. Nicholas G. Shadrin, "The Soviet Merchant Marine, A Late Developing Economic Growth" in U.S. Congress, Joint Economic Committee, Soviet Economic Prospects for the Seventies (Washington, D.C.: Government Printing Office, June 1972), pp. 719-765.

CHAPTER

9

SUMMARY
OF DISCUSSION
IN PART I
Ken Booth

ASPECTS OF THE POLICY PROCESS

Discussion centered on the nature of the military input into the Soviet policy-making at the highest level, the character and implications of the institutional structures and processes through which weapons were procured, the particular position of the Soviet Navy, its influence within the whole defense establishment, and its influence on Soviet foreign policy. Despite some substantial contributions, the elusiveness of much of the raw material meant that a number of important questions remained unresolved. While the discussion in this part frequently drifted toward the implications of the Gorshkov series, for the sake of clarity, this aspect of the discussion has been left out of this summary.

The Institutional Setting

It was generally agreed that in order to understand Soviet military policy, one must appreciate certain aspects of its institutional setting. The old idea of simply assuming a unitary actor is increasingly unsatisfactory. While decision-making is centralized, it responds in some measure to the initiatives of lower-level interests and is limited by the standard operating procedures of the various bureaucracies.

Decisions on major issues, including approval of 10-to-15-year development programs, are probably made by the Defense Council, a subcommittee of the Politburo. The former comprises the relevant members (particularly Brezhnev, Kosygin, and Grechko) of the Politburo, which undoubtedly ratifies its decisions.

Within the Ministry of Defense, the General Staff and its various administrations act as a "central nervous system." While little is known of its procedures, it appears to function as a locus of reconciling divergent plans from the various services, so that it may direct actual military operations. A number of agencies were suggested as having influence on the longer-range aspects of defense policy-making, such as the defense industry ministries, scientific research institutes, and weapons design bureaus. These provide an internal "push" to the procurement process (on the basis of technological innovation), which supplements the "pull" of specific service requirements.

Crisis Decision-making

It was suggested that the present regime had developed a good sense of regularity in its decision-making compared to the previous one. It was thought likely that there would be a tendency to bring in the whole of the Politburo (as in the Czechoslovak case), so that no one could use decisions taken in his absence for political gain.

At the same time it was recognized that in a very intense phase of a crisis, a small standing committee such as the Defense Council might act in place of the Politburo as a whole. It was also suggested that Brezhnev's increasing primacy might permit him to act alone when time did not allow for consultations.

Presumably there exist some contingency plans for dealing with the initial phase of a crisis, as well as some latitude for the naval high command to order ships to sea in anticipation of political instructions. These factors might help explain the apparent rapidity of Soviet response in such cases as the Indo-Pakistani War. They would also suggest an explanation of Soviet behavior in response to the mining of Haiphong, when Pacific fleet units were immediately deployed to the general area, but then did nothing. Presumably in this instance the political instructions were explicitly not to take positive action.

The Growing Institutional Significance of the Soviet Navy

It was agreed that in institutional penetration and ranking the Soviet Navy has increased in importance from the early 1960s to the early 1970s. In addition, from the late 1960s onward there had been more decisions of a foreign policy nature in which the navy had to be called in for advisory purposes. The improvement in the navy's position was thought to be informal rather than formal. It was open to discussion whether the increased recognition of the navy in the General Staff was the result of the increased importance of the navy qua navy, or because of the bureaucratic need for a counterbalance to another group.

Interservice Wrangling

It was agreed that it was likely that a good deal of interservice rivalry was now current in the Soviet Union. The problem was, "Who is fighting whom for what?" One view was that the Soviet Navy was seeking to establish itself as a primary force for the strategic strike role and that consequently the strategic rocket forces will be sensitive, on the defensive, and bureaucratically hostile. Against this, it was argued that it was not self-evident that the navy wanted the strategic strike role, because such a task was not a "proper" naval role. Another view was that the "real" argument at present within the Soviet military establishment was between the navy and the air force.

In view of the alleged wranglings within the Soviet military, the meeting considered the evidence in the periodical military literature. There was some disagreement about what it actually showed and its possible significance. Some agreed that the Gorshkov series should essentially be seen in the context of the whole military debate. It was argued that the journals in which the positions of the other services had been declared were important. Particular note was made of the effort of Aviation Marshal Kutakhov, who from a "low profile" had emerged to take up a much more assertive position on the importance of the role being performed by the airforce for the Soviet Union. An alternative view suggested that, while accepting the vested interest of the other services in asserting their particular usefulness, none of the nonnaval presentations or the journals in which they appeared, could compare in substance or importance with the Gorshkov series in <u>Morskoi sbornik</u>.

Lesser Matters

<u>Groups within the Soviet Navy.</u> While it was accepted that the Soviet Navy was made up of various institutional and/or functional groupings, it was argued that it was very difficult to delineate them accurately, although the exercise would be extremely worthwhile. While nobody felt able to offer a better profile of the groups within the Soviet Navy to the one suggested by Erickson, some difficulties were pointed out to this particular categorization.

<u>The Role of Political Officers.</u> It was generally agreed that their main purpose was now to contribute to the combat readiness of troops (especially the efficiency of small units) rather than simply to work at ideological indoctrination. They were also much involved (and usefully) in welfare questions. The political officers themselves were raising new questions about their role.

Manpower Problems. Note was made of the fact that the Soviet forces were having difficulties in this area. Like their Western counterparts they were having to compete with the demands and temptations of the consumer sector.

Introversion. Because of the manpower problem and other difficulties, it was felt that the Soviet military establishment was currently "obsessed" with finding out how their military system worked. They were looking for ways to change it to their advantage at a time when the system, the society, and the international environment were facing change. There were signs that the military preoccupation was focusing on questions of manpower effectiveness rather than on weapons.

ECONOMIC FACTORS

It was recognized that there was a serious shortage of information about the economic aspects of Soviet defense efforts and especially about the naval component of it. Furthermore, the economic specialists were at pains to explain the difficulty of making satisfactory calculations and of finding sound bases for comparison. With the admitted weaknesses of analysis, nonspecialists (especially in the public domain) are very constrained in discussing Soviet naval expenditures, comparative technology, and the naval aspect of Soviet economic priorities.

Fluctuations

In his presentation, Hutchings drew attention to a tendency for budget spending on the economy and on defense to displace each other in a rhythm patterned on the timetable of the Soviet five-year plans. He suggested that naval planners would seek to use the high midplan peak of defense allocations as a springboard from which to launch plan-bridging programs, in order to secure an unassailable base for naval expenditures, whatever general budgetary economics might be demanded and in particular to tide the navy over the leaner times to be expected toward the start and end of economic plans.

Against this, it was noted that there was little cyclical fluctuation in the delivery pattern of warships in the USSR; there seemed to be an attempt to attain a steady through-put. The change in recent years in the Soviet Navy had been in deployment rather than procurement. The delivery pattern would probably be essentially "more of the same," although this would create problems later through block obsolescence.

Comparative Technology

A variety of impressions were offered. (1) On ship technology there was not a clear picture, but it was suggested that the Soviet Union might lag in the mass production of goods with fine tolerances (for example, a study of Soviet machine tools had concluded that although they were initially cheap to purchase, they were poor in cost-effectiveness). However, it was pointed out that there was a large and growing body of Soviet writings on such matters as shipyard maintenance, design, weapons forecasting, and so on. Whatever the technological level at present, it might be expected to improve. In relation to this, the reminder was given that it was always difficult in Soviet writings, technical as well as theoretical, to distinguish between what is, what will be, and what should be. (2) On innovation ability, it was pointed out that the West had been repeatedly surprised by Soviet technical innovation in the past and that due respect ought therefore to be given to it. It was suggested that while the absolute total might not be large, some of the brightest technologists do go into the military and space fields in the USSR. (3) On performance, it was argued by some that in matters such as steaming capability, refit and maintenance cycles, and operational availability, the Soviet Navy compared favorably with Western navies. Comparing combat readiness, however, would be guesswork.

The Impact of the Navy on the Economy

The meeting spent some time discussing whether there was anything in the character of Soviet naval procurement that meant that some sector of the economy was having to suffer. The general view was that there were some such sectors. One view was that the main impact of naval procurement was in engineering and electronics. Another view was that it impinged mainly on investment, which basically meant that research and development in the creation of new capacity in the civilian sector were hit. The impact was therefore long rather than short term, with the navy having to compete with all the other demands on the economy, including the other military services. It was suggested that the naval requirement also took a significant slice of highly trained and technical manpower out of the economy. The naval requirement also had a limited impact on the merchant fleet. The navy and the merchant fleet were not competitive for space in new yards, since these were not easily interchangeable, but they were competitive for older shipyard facilities. Merchant ship production would use naval yards if a naval program were canceled, but not vice versa.

The Timing of Present Naval Demands

In opposition to the view that the present moment was not a good time for new military expenditures, it was suggested that too much attention ought not to be placed on the admittedly important, but possibly short-run, problems of the Soviet economy in 1972. More positively, it was argued that the present time was possibly the best time to submit requests for allocations. (When compared to the end or start of the period of a five-year plan). Because of the time taken to debate resource allocation issues, and because of production lead times, it was argued that naval requests should be made about now if important deadlines and targets were to be achieved. This might have explained Gorshkov's articles in 1972.

PART II
SOVIET POLICY AND THE THIRD WORLD: CASE STUDIES

CHAPTER

10

THE SOVIET-EGYPTIAN
INFLUENCE RELATIONSHIP
SINCE THE JUNE 1967 WAR
Alvin Z. Rubinstein

The purpose of this chapter is to present some preliminary findings—methodological and substantive—on the nature and extent of Soviet influence in Egypt since the June 1967 Arab-Israeli war. The problem of assessing A's influence in B is at the heart of foreign policy analysis, yet we lack any uniform method of determining influence. We need some accepted criteria for assessing what the Soviet Union is accomplishing as a consequence of its diplomatic, military, economic, and cultural inputs into Egypt. How much actual influence has the Soviet Union exercised there? on which issues? and in which areas? Indeed, we need to ask, Has influence been available when desired by the USSR?

Thus there are important insights for foreign policy analysis if it turns out, for example, that Soviet influence does not directly or closely correlate with inputs of military and economic assistance; that the Soviet-Egyptian influence relationship is not a zero-sum situation—the gains to the Soviet Union do not involve heavy costs for Egypt; or that influence-building in the current international system is a highly uncertain, transient, and variable process, quite different from its prenuclear antecedents, to which it is only remotely related. Close examination suggests that answers to these questions will shed light not only on the Soviet-Egyptian relationship, important in its own right, but on other dyadic Soviet-third-world relationships, and on the limits of the superpower rivalry in areas such as the Middle East and Southern Asia.

METHODOLOGICAL PROBLEMS

Differentiating Influence from Policy

Our aim is to identify and understand influence, not to predict it:[1] We want to know when it exists and how it can be assessed. As a working definition we may use the following: Influence is manifested when A affects, through nonmilitary means, directly or indirectly, the behavior of B so that it redounds to the policy advantage of A. Definition immediately raises a semantic problem because the phenomenon of influence is both a process and a product. As defined above, influence is a process; on the other hand, what is in fact observed and assessed is the net result or outcome of influence—that is, the product. No wording can completely free us of this problem. However, our use of either of the two meanings should be sufficiently precise to make our intent clear throughout.

Like electricity, influence is more readily identified than explained: It can be detected and measured (in relative terms), and its increase or decrease can be observed. It may be considered to have a number of characteristics: It is a relational concept; it is asymmetrical and shortlived and requires continual reinforcement to be potent and operative; and it is manifested on concrete issues, being segmental and specific, not systemic or pervasive.

After Sadat's expulsion of the Soviet military personnel in July 1972, and his withdrawal of concessions that had given the Soviet military an unprecedented foothold on Egyptian soil, many Western analysts wondered what the Soviets had gained from their massive courtship of Egypt: The Soviet military advantages, which included reconnoitering of the U.S. Sixth Fleet, servicing Soviet naval vessels, and exchanging intelligence data, were sharply curtailed; the imports of Egyptian cotton—which the USSR does not need—are unlikely ever to repay the more than $2 billion in economic assistance, much less the upwards of $5 billion in military assistance, that the USSR has provided since 1967;[2] the Egyptian market has marginal commercial appeal—its problems overshadow its potential; Egypt's diplomatic significance for Soviet foreign policy has diminished somewhat as a consequence of the Soviet quest for detente with the United States and Western Europe, the shift of Soviet interest to the Persian Gulf and the Indian Ocean areas, and Moscow's awareness of the marginal role that Egypt can play in advancing Soviet interests in Black Africa. Nonetheless, the Soviet Union continues to sustain a steady level of significant economic and military support.

In light of this situation, the presumption is reasonable that a consensus exists in the Kremlin concerning the Soviet Union's Egyptian connection. Many Western analysts ask, Has the game been worth the candle? The fact of the matter is that we have no way of knowing how Soviet leaders really feel about their Egyptian policy; they aren't saying. But it may not be too important that we know the reasons for continued Soviet support of Egypt—which could well be a function more of Soviet bureaucratic politics than of rational strategic calculations—if we can agree upon a method and criteria for establishing the relative success or failure of Soviet policy and the extent to which massive Soviet inputs have, in fact, brought tangible returns of influence.

Assessing Influence

To determine influence, we need some operational criteria to help identify possible instances of it. First, manifestations of influence must be sought in the specific "issue areas"[3] that arise in the normal course of Soviet-Egyptian diplomacy and that reveal shifts in Egyptian domestic and foreign policies congenial to the USSR. From the character, degree, frequency, and implications of such domestic and/or foreign policy changes, inferences can be made concerning Soviet influence. However, often what appears to be influence turns out instead to be joint interests of the two parties. In practice, there are only a few "issue areas" deemed important by both the Soviet Union and Egypt and in which Egypt may adjust its preferences to the USSR's (or conversely, in which the USSR may adjust to Egypt's preferences). In such situations, it may be said the Soviet Union exercises significant influence. Its influence is of lesser significance when Egypt adapts to the USSR's preferences on issues that are of low salience to Egypt, as for example, Egypt's adoption of a sympathetic position toward the Soviet invasion of Czechoslovakia in August 1968, its studied avoidance of any comment on the "Brezhnev doctrine" for several years, and its adherence to the Soviet position in the U.N. Disarmament Committee.

Even such minimal adaptations are, of course, linked to the overall influence relationship; they are the "payoffs" for services rendered or requested and are readily given since the cost to Egypt in these instances is minimal. The problem of distinguishing between important and less important issues can be knotty, but in the case of Soviet-Egyptian relations, it is not: The critical instances become apparent to the close observer, and agreement on them should not be hard to reach. What is not always readily apparent in such cases, however, is who influenced whom.

Second, influence may also exist when Moscow's ability to carry out transactions in Cairo on terms favorable to the USSR noticeably improves. However, it is difficult to determine when quantitative increments connote qualitative changes in the relationship. Several examples may be cited: Sergei Vinogradov, the overbearing Soviet Ambassador in Cairo from September 1967 to early 1970, frequently bragged of his ability to see Nasser any time he wanted; assuming he did see Nasser far more often than any other ambassador, how significant was the content of the discussions? Soviet advisers are active in many important areas of the Egyptian economy, but how influential are they in fashioning Egypt's planning and development strategy or in lobbying for certain kinds of reforms? Soviet military personnel were assigned to all levels of Egyptian military units during the 1967-72 period and they manned the air defense and communication systems, but what was the extent of their control over Egyptian strategy and tactics? In the absence of authoritative information about the actual relationship in a situation where the Soviet Union cannot project its military power directly, this criterion needs to be viewed critically and used with caution.[4]

Third, a major increase in the security commitment of the Soviet Union to Egypt suggests a change in Soviet influence, one that may result in a greater Soviet ability to restrain (though not compel) the behavior of the client and consolidate a more stable dependent relationship. Yet, unless the Soviet Union is in a position to control the levers of power in Egyptian domestic politics, Egypt's accession to additional Soviet preferences must not be automatically assumed on the basis of the changed security relationship; signs of Soviet influence must still be sought, as we mentioned earlier, in specific issue areas.

Fourth, and very important, influence must be viewed within a strategic context, and not merely in terms of the tangible, short-term advantages discernible within the Soviet-Egyptian relationship itself. The success or failure of Soviet influence-building may be evaluated in terms of consequences that are observable only in the broader context of the USSR's desire to have Egypt opt for policy outcomes that Egypt prefers but that are made possible only by the USSR, which believes they will redound to its long-term regional or global advantage.

A donor may have a number of objectives in mind: The desire for immediate return may be present but not pressing. By way of illustration, the Soviet Union has given extensive aid to Egypt since the June war. In the short run, it acquired munificent strategic dividends: naval facilities at Alexandria, the use of air fields for reconnoitering the U.S. Sixth Fleet, and an expanded presence in Egyptian life. These capital gains were virtually wiped out after July 17, 1972, when President Anwar Sadat ousted most of the Soviet military personnel from Egypt. One might therefore be tempted to argue that the Soviet Union made a poor investment. If the argument relates primarily to the immediate payoff from the aid and support, the case is quite convincing. But this would overlook, or minimize, what may well have been the most important Soviet objective in providing Egypt with massive and immediate assistance, military and economic: to keep Egypt from negotiating a settlement of the Arab-Israeli conflict lest this eliminate from the Middle East a festering problem that has helped the Soviet Union to intrude itself into the politics of the

Further, to have denied Egypt aid in 1967 would have called into question the value of Soviet "friendship" for a third-world country in time of critical need. Given Soviet strategic interests in the region, the Kremlin preferred to deal with an Egypt that was dependent on Soviet support. To gauge the Soviet-Egyptian influence relationship in terms of the broader consequences for Moscow in rendering or not rendering support clearly leaves us open to the criticism of being imprecise and highly judgmental and of complicating rather than clarifying the determination of influence. Yet it does serve an explanatory function.

MOSCOW'S STRATEGIC INVOLVEMENT IN EGYPT

Egypt (with India) has been the centerpiece of Soviet policy in the third world. Since 1955 the Soviet Union has persistently supported Egypt, for a changing combination of strategic, diplomatic, ideological, and domestic political reasons. However, this support has varied in intensity and scope according to the state of Moscow's more pressing concerns with the United States, China, and West Germany. Through all the fluctuations, inevitable in international politics, the Soviet Union has nonetheless remained Egypt's mainstay among the great powers since the late 1950s; and since 1967 it has proved absolutely indispensable not only to Egypt's foreign policy interests but to the stability of the Sadat regime itself.

While relying on Soviet support for its security, Egypt has had to maintain a delicate balance between preserving independence and freedom of action in areas of permanent concern to the Egyptian leadership, on the one hand, and satisfying the minimal demands of a powerful patron and creditor, on the other. Unable to assure its security by its own efforts, Egypt sought direct, sustained, and dependable commitments from one of the great powers.* Nasser's strategy after June 1967, deliberate or intuitive, was to enmesh the Soviet Union increasingly in the defense and promotion of Egyptian interests without surrendering authority or sovereignty. Sadat, too, has followed such a policy, and with success, notwithstanding the precipitate and publicly humiliating expulsion of Soviet military personnel in July 1972. Sadat's rebuff to Moscow came at a felicitous moment for the USSR, which seized the opportunity to reduce its involvement,

---

*In similar situations, other small countries have tried the very difficult business of playing off one Great Power against another—witness Afghanistan, Nepal, Cambodia (prior to April 1970), and North Korea.

TABLE 10.1

Soviet-Egyptian Influence Relationship

Key Event/Issue Area	Date	Egypt's Behavior or Position	USSR Behavior or Position	Policy Initiator	Who Benefits?	Apparent Influencer
Egypt's effort to recover from the June war and avoid being pressured into a settlement with Israel.	post-June 10, 1967	Defeated by Israel. Extreme vulnerability.	Massive and rapid rearmament of Egypt. Convenes emergency session of U.N. General Assembly in an attempt to pressure a repeat of 1956 Israeli withdrawal.	Soviet Union	Egypt recovery starts. USSR expands its presence and commitments; its support enables Egypt to refuse a settlement with Israel	Soviet Union
A Soviet-American initiative at the United Nations to persuade both parties to the Middle East conflict to agree to a compromise settlement.	July 10-20, 1967	Nasser meets with Soviet deputy foreign minister in Cairo but refuses to accept the Soviet case for a political settlement.	Works with the United States to devise a draft resolution acceptable to Egypt and Israel. Tries to convince Nasser to agree to the compromise superpower proposal.	Soviet Union (and the United States)	A settlement of the Middle East conflict seemed within reach; certainly, it had U.S. and Soviet support. Egypt was the stumbling block.	Egypt
U.N. General Assembly Resolution 242	November 22, 1967	Demands a return to status quo ante.	Persuades Egypt to accept Resolution 242, which is somewhat less	Soviet Union and Egypt both played an active	USSR obtains Egypt's agreement to a political settlement;	Soviet Union

		Presses for U.N. action against Israel.	than a call for a total Israeli withdrawal. Presses for a political settlement.	role in United Nations.	on the other hand, Egypt has a resolution that can be interpreted to suit its needs and aims.
				Egypt	Egypt
Start of the War of Attrition	March–June 1969	Breaks the truce and resumes heavy fighting along the canal. Its action draws the superpowers into an ominous confrontation.	Calls for a political settlement. Criticizes Arab groups that contend Arab forces are strong enough to achieve a military victory. Consults with U.S. about the basis for a settlement.	Egypt	Nasser offsets sagging domestic image and the effect of student unrest. Nasser preempts role of militant leader from Palestine Liberation Organization and Fedayeen guerrillas. Nasser carries USSR along with his policy.

(continued)

TABLE 10.1 (continued)

Key Event/Issue Area	Date	Egypt's Behavior or Position	USSR Behavior or Position	Policy Initiator	Who Benefits?	Apparent Influencer
Egypt's inability to cope with Israeli air power	January 1970	Nasser flies secretly to Moscow to ask for help.	USSR agrees to send missile crews, pilots, and air defense teams to protect Egypt's heartland.	Egypt	Egypt gains greater commitment from the USSR, though its dependence brings an expanded Soviet military presence. USSR, more than ever, established as Egypt's protector; its relations with U.S. deteriorate.	Egypt
Ceasefire agreement along the Suez Canal	August 8, 1970	Agrees to a three-month ceasefire, thus implicitly acknowledging the failure of the War of Attrition. Nasser cracks down on	Favors the ceasefire and a political settlement. Pressure on Egypt to agree to the ceasefire forestalls the possibility of the USSR being further enmeshed in	Soviet Union (in support of U.S. ceasefire plan)	Egypt gains a respite from severe pounding. USSR enabled to defuse a dangerous situation that threatened its detente with U.S. and West Germany.	Soviet Union

			Soviet Union	Egypt
		Palestinian organizations, terminating their broadcasting privileges.		the actual fighting.
Soviet-Egyptian Treaty of Friendship and Cooperation	May 27, 1971	Sadat purges his domestic opponents, some of whom are known to favor closer ties with the Soviet Union, and he has the Rogers Plan for an interim settlement and reopening of the canal under consideration.	Developments in Egypt cause uneasiness in USSR. Soviet President Podgorny arrives on May 25, with a top-level government, party, and military delegation, to discuss Soviet-Egyptian relations in the light of Sadat's purge. Podgorny brings the draft of the treaty.	Sadat obtains formalization and expansion of Soviet military commitments and Soviet assurance of non-interference in Egypt's domestic policies. The USSR seeks to institutionalize its presence in Egypt. It lessens the likelihood of a resurgence of U.S. activity in Egypt.

(continued)

TABLE 10.1 (continued)

Key Event/ Issue Area	Date	Egypt's Behavior or Position	USSR Behavior or Position	Policy Initiator	Who Benefits?	Apparent Influencer
Indo-Pakistani War	December 1971	Sadat defers resumption of hostilities against Israel.	USSR withdraws missile crews and planes from the defense of Aswan Dam, supposedly to reinforce India, and cautions Sadat against starting a war.	Soviet Union	The USSR assures itself against possible military involvement at a time when it was involved in the crisis on the Indian subcontinent	Soviet Union
Expulsion of Soviet military personnel	July 17, 1972	Sadat orders the 15,000 Soviet military advisers and troops to leave Egypt.	The Soviet Union withdraws its military personnel, thus sharply diminishing its presence in Egypt.	Egypt	Sadat retains support of the army, by mollifying its pique at the USSR, and gains popularity among Egyptians: His position is stronger than ever.	Egypt

The Fourth Arab-Israeli War.	October 6, 1973	Egypt launches an attack across the Suez Canal to retake Sinai (while Syria strikes in the Golan Heights). Egypt's military initiative is reinforced by an Arab oil boycott organized by Egypt and Saudi Arabia.	The USSR airlifts arms and supplies to Egypt (and Syria) within 24 hours of the outbreak of hostilities. Prevents the U.N. Security Council from acting during early stages of the fighting, when Egypt (and Syria) were gaining ground.	Egypt
			Egypt regains the east bank of the Suez Canal. Sadat forces the United States to take an active role in seeking a Middle East settlement. Sadat's position internally and in the Arab world is enormously enhanced.	Egypt

at a time when such involvement might have jeopardized the Soviet "opening to the West." However, from early 1973 on, Egypt, with Saudi Arabian financial backing, purchased all the weaponry it needed from the Soviet Union: What Moscow was reluctant to provide gratis, it happily sold for cash.

In evaluating Soviet influence in Egypt since 1967 there is a temptation to assume that it was greatest during the period of maximum direct involvement, namely, from June 1967 to July 1972. After the June 1967 war, Egypt's dependence was total: Soviet arms and advisers flowed in; in late 1969 to early 1970 the Soviet Government committed military personnel to operational roles on a scale hitherto unapproached in Soviet maneuverings outside the communist bloc; economically, Moscow proved more munificent than Egypt's oil-rich brethren—a commentary on the relative strengths of blood and politics in international affairs; and local communists were again bobbing about in Egypt's media and economic life, presumably as a gesture of goodwill toward Moscow. Yet, even in this period of maximum and mounting commitment and involvement, Soviet influence was, according to the criteria mentioned earlier, by no means unconstrained.

As Table 10.1 shows, the critical issue areas in the Soviet-Egyptian influence relationship during the period of maximum Soviet involvement are limited in number and thus lend themselves to systematic investigation and comparison.[5] For considerations of space, we have chosen 2 out of the 10 key events: (1) the start of the "War of Attrition" and (2) the Soviet-Egyptian Treaty of Friendship and Cooperation of May 27, 1971. In these two cases, when, where, and how was the influence manifested? Examination of these two issue areas will provide some insights into the nature of influence, its parameters and possibilities, and its evanescent character. It may also illustrate, in preliminary fashion, the limits of Soviet influence-building elsewhere in the third world.

TO THE WAR OF ATTRITION

It is easier to identify the end of the phase of Egyptian-Israeli relations commonly known as "the war of attrition" than to pinpoint the date of its beginning: The "war of attrition" ended with the cease-fire agreement of August 8, 1970; it probably began on March 8, 1969, though there were several times when serious fighting erupted earlier only to peter out; but by mid-June 1969 the fighting finally assumed the character of a new war.

During the late summer of 1968, there was minor action along the Suez Canal, such as sniper fire and the laying of mines on the Israeli-held east bank. Egyptian reports of resistance against Israel dealt primarily with the activities of the Palestinian organizations. However, on September 8, 1968, Egyptian forces laid down the most extensive artillery barrage since the June war.[6] Three days later President Nasser, recently returned from a period of medical treatment in the Soviet Union, met with his cabinet and discussed the condition of the Egyptian Army and the military situation, especially "on the fighting front" (the Suez Canal area). Cairo Radio devoted increasing coverage to the "tense" situation in the Middle East, and the newspapers warned that "the situation along the front is expected to explode."[7] On September 14, at the opening session of the Arab Socialist Union (ASU) Congress, Nasser reaffirmed the commitment to liberate Egypt's land and reviewed the military and economic progress made since June 1967.

In what has become a key aim of Egyptian diplomacy, Cairo sought to generate pressure in the United Nations for an Israeli withdrawal. Through a combination of diplomacy, commentaries in the media, and military actions, it acted to produce a crisis atmosphere which would require great-power intervention. The artillery barrage of September 8 was timed to impress the Security Council, which was to meet on the following day. On September 25, on the eve of the forthcoming session of the U.N. General Assembly, the Cairo press called for "positive action" to ensure the success of Gunnar Jarring's mission. A statement issued by the Soviet Foreign Ministry on September 25, sharply condemning Israel for its intransigence and warning that it would be held responsible for any provocative actions against Egypt, Jordan, or Syria, was publicized in the Egyptian press. Two days later, Hasanein Heikal, confidante of Nasser and editor of semiofficial Al-Ahram struck an ominous note:

> The Middle East crisis is about to make a decisive and perhaps final appearance in this U.N. international arena. After this, the race between a political settlement and a military solution—and the political settlement so far has lagged behind in the race—will have ended with the final elimination of the political solution from the race.[8]

By the fall of 1968, Egypt's armed forces had attained and in certain areas even exceeded their June 1967 levels: The Soviet supply effort had largely restored Egypt's tanks, artillery, and aircraft; all that remained materially in short supply was trained pilots. Egypt's growing sense of military self-assertiveness, epitomized by the periodic artillery duels and air clashes in the Suez Canal area throughout September and October, was intended to be communicated to the representatives of the great powers in New York.

However, despite a Soviet proposal to implement the U.N. Security Council resolution 242 of November 22, 1967, and Egypt's efforts to exacerbate the mood of impending disaster about the Middle East, international developments were unkind to Cairo: The Soviet invasion of Czechoslovakia on August 21, the Soviet naval buildup in the Mediterranean, the anxiety in Yugoslavia and the Balkans that was fed by the enunciation of the "Brezhnev doctrine" on September 26, and the presidential election campaign in the United States, all overshadowed the immediate Egyptian-Israeli problem. Despite several positive elements in the proposal, the Soviet initiative was generally dismissed in the West as a ploy to distract attention from Soviet policy in Eastern Europe and from the implications of the Brezhnev doctrine.[9] The tension created by Egyptian actions along the canal failed to produce any sustained response from the great powers. Egyptian diplomacy was stymied. By late October the threat of serious fighting had passed. On November 6, Nasser briefed the ASU Central Committee on the military situation. Two days later Foreign Minister Mahmoud Riad returned early from the U.N. General Assembly session, saying that the Jarring mission was deadlocked and no further purpose was served by his remaining there.

Domestic Determinants

Egyptian domestic developments took a hand in moving Nasser closer to a new confrontation with Israel. On November 21, serious student unrest broke out, this time in Mansurah and Alexandria. For the second time in nine months, riots and disorder spread in the major cities. Triggered by the students, they were perceived by the leadership as antigovernment. The universities were suspended, and an emergency session of the ASU National Congress was convened. On December 2, Nasser addressed himself to the roots of the unrest.[10] He spoke of Egypt's recovery from the disaster of June 1967, of the democratization of the Arab Socialist Union being undertaken in accordance with the March 30 Reform Program, and of the easing of censorship; he counseled patience, observing that all these changes could not be accomplished quickly. But above all he tried to answer the question why the war to recover Egyptian territories and liberate Arab lands was not being pressed against Israel. He likened Egypt's situation to that of England after Dunkirk with the difference that "We are neither in a state of war nor of peace—hence the sad atmosphere." He pleaded for time to rebuild the armed forces but said that "at the same time we are working for an honorable political settlement," which does not mean surrender, there being "a difference between peaceful settlement and surrender." This was the last time before the outbreak of "the war of attrition" that Nasser spoke of a "political settlement" in so conciliatory a fashion. As was evident in his speech of January 20, 1969, and interviews in February with Newsweek and The New York Times, domestic pressures hardened his attitude and impelled him to "the war of attrition."

In early December, firing along the canal predictably intensified on the eve of visits by Gunnar Jarring, the U.N. envoy, and William Scranton, President-elect Nixon's personal envoy touring the Middle East on a fact-finding mission. Internally, Nasser's prestige was lower than at any time since the June debacle: Not only had the student disorders tarnished his image at home, but his role of leadership in the Arab world was under implicit challenge from the Palestinian guerrilla organizations, whose border raids against Israel from Syria, Lebanon, and Jordan, and whose terrorist exploits abroad and in Israel itself were capturing the popular imagination and presenting Israel with a new major security problem. Headlines in Egyptian newspapers spoke of resistance to Israel, but primarily by Palestinian groups.

Soviet Efforts to Restrain Nasser

The arrival in Cairo on December 21 of Soviet Foreign Minister Andrei Gromyko came at a critical juncture for Nasser. The U.N. session had ended without action being taken. Domestically, Nasser needed proof of some progress. For its part, Moscow wanted to avert a new outbreak of hostilities. On December 22, Cairo Radio announced that Gromyko had brought "an important message from the Soviet leaders." On the same day, a Soviet peace plan was submitted to Jarring in Moscow, then formally submitted on December 30 to Britain, France, and the United States. It called, inter alia, for acceptance of the Security Council resolution of November 22, 1967, by Israel and "those neighboring Arab states," that is Egypt, Jordan, and Lebanon, "willing to participate in implementation" of a plan of action, for a timetable and procedure for a phased withdrawal "under UN supervision" of Israeli forces from all the territories occupied in 1967; for "secure and recognized boundaries," "freedom of navigation in the region's international waterways" (including reopening of the Suez Canal), and "a just solution of the refugee problem."[11] (Note: Washington was not too interested at this time in reopening the canal, preferring to keep it closed in order to complicate the Soviet problem of supplying North Vietnam.)

The tone of the Joint Soviet-Egyptian communique that was issued on December 24, 1968, at the end of Gromyko's visit suggests that Cairo accommodated to Moscow, that it was giving Soviet diplomacy an opportunity to produce results. While none of the specific proposals contained in the Soviet peace plan were mentioned, the joint communique stressed the need for, and Egypt's acceptance of the principle of, a peaceful settlement based on the implementation of resolution 242. It avoided any mention of the refugee problem, thus implying that the Palestinian issue would not be permitted to block any move toward a settlement of the "aggression of 5 June 1967"; and it did not mention "imperialism," thus soft-pedaling U.S. support of Israel.

That Moscow understood there were serious differences between the Soviet and Egyptian positions may be inferred from the cursory treatment the Gromyko visit received in the Soviet press. At the United Nations, the Soviet ambassador urged an early meeting of the four powers to forestall a further deterioration of the Middle East situation. Egyptian newspapers wrote of the Soviet plan, at the same time emphasizing the mounting tension. Sniping incidents along the canal were minor, but intense fighting was reported along the Jordanian and Syrian borders.

The Egyptian leadership, skeptical of any positive results from the Soviet initiative and impatient, became increasingly truculent. In his weekly column, on January 3, 1969, Heikal wrote:

> A peaceful solution to the Middle East problem is not now feasible . . . even a political solution cannot be achieved or made possible except through some kind of military action. This action alone is capable of changing the present condition of the land and this action alone can give expression to the terms which eminent diplomats and veteran lawyers are laboring to formulate.[12]

On January 10, 1969, Heikal summarized some of the main points in the Soviet proposal. However, it is significant that he chose to ignore several others—namely, the provisions calling for an Israeli withdrawal in stages, for declarations by the Arab states recognizing each nation's sovereignty and right to live in peace within secure frontiers, and for a reopening of the Suez Canal after Israel withdrew a certain distance into Sinai, all important elements of the Soviet proposal.[13] The discrepancy between the Soviet Government's proposal and Heikal's reporting of it indicated that Cairo was not willing to agree to the minimal concessions implicit in the Soviet plan, that it was openly defining the uncompromising character of its position, in contrast to the more equivocal position it had agreed to in the joint Soviet-Egyptian communique of December 24, 1968. Heikal's analysis sharply differed from that of Soviet commentators who, though they deplored Israeli intransigence and blamed it for the tense situation in the Middle East, nonetheless reiterated the need for a peaceful settlement.

Heikal's pessimism accurately reflected Nasser's mood. In a speech before the new National Assembly on January 20, 1969—the first time he had spoken there since November 23, 1967—Nasser talked about the strengthening of the armed forces that had been accomplished in the previous 18 months and "the true and sincere cooperation" of the USSR in this enterprise. In contrast to the tenor of the Soviet-Egyptian communique of December 24, 1968, and Soviet diplomatic efforts, Nasser's speech bristled with belligerence. First, he saw no prospect of a peaceful solution: "We must realize that the enemy will not retreat unless we force him to by fighting. As a matter of fact, there is no hope of a political solution unless the enemy realizes that we can force him to withdraw by fighting. In other words, no progress can be made by military or political action unless the military front is the starting point for such progress."[14] Second, he spotlighted the significance of the Palestinian resistance and insisted that no decisions affecting them would be made without their approval. (The Palestinian issue has become a political barometer of Egypt's attitude toward dealing with Israel: When deemphasized it suggests a readiness to negotiate the 1967 war; when emphasized it signifies an intransigence on any accommodation short of returning to prepartition guidelines.) Several days later, in secret session, the minister of War, Mahmoud Fawzi, briefed the Assembly on the military situation.

The Soviet leadership took advantage of a visit to Cairo by a member of the CPSU Politburo, Aleksandr N. Shelepin, to send a message from Brezhnev to Nasser. As chairman of the All-Union Central Council of Trade Unions, Shelepin headed the Soviet delegation to the Fourth Congress of the International Confederation of Arab Trade Unions. At the Congress, he reaffirmed full Soviet support for "legitimate Arab rights," declaring that an "Israeli withdrawal behind the June 5, 1967, lines is a necessary condition for any solution." Prior to his departure on February 3, he held "a very, very important meeting" with Nasser lasting more than three hours.[15] According to Al-Ahram there was "complete agreement that Israel's continued occupation will probably create an explosion in the Middle East at any moment." That Moscow and Cairo were in actuality far apart is clear from the following: <u>Pravda</u> did not mention that Shelepin brought a message to Nasser from Brezhnev, that he met with Nasser for three hours for an important talk on the Middle East, or that he brought back a message from Nasser to Brezhnev. In a word, <u>Pravda</u> conveyed the impression that Shelepin's only function in Cairo <u>was</u> to represent Soviet trade unions, completely ignoring the Middle East aspects of the visit.

In an interview published in Newsweek on February 4, Nasser said that at the time of Gromyko's visit in December he had told the Soviets that he was not optimistic about their plan. His comments bespoke a hard line: He said that Israel's adherence to the Security

Council resolution would be adequate "for removing the effects of the aggression—in other words as far as the 1967 crisis is concerned—but it is not adequate for the Palestine problem, which is the main problem" (emphasis added). Three weeks later, in an interview with C. L. Sulzberger of The New York Times, Nasser alluded to Egypt's need to resort to force to achieve its aims.

Throughout February the number of incidents along the Suez Canal increased. They may have been designed politically, to lend urgency to Soviet and French efforts to promote the four-power talks, to induce the United States to pressure Israel to commit itself to a full withdrawal, and to satisfy impatient elements in Egypt. On February 24, a state of emergency was declared in all the governates of Egypt. The feeling of impending major military action conveyed by Egyptian newspapers was epitomized by a headline in Al-Ahram at the time of Nasser's interview with Sulzberger: "Fourth Round of Arab-Israeli War Is Inevitable."[16]

Egypt Moves to War

On March 8, Egypt escalated the fighting along the Suez Canal: Sniper fire, machine gun exchanges, and mine-laying gave way to massive artillery barrages. The fighting was the longest and severest since the June war. The ceasefires arranged by U.N. observers proved short-lived.

But in March 1969, as in September 1968, international developments elsewhere again militated against Egypt's attempt to compel great-power attention and action. On the day Al-Ahram featured Nasser's interview with Sulzberger, it carried on the same page under a Moscow dateline a small item reporting a military clash between Soviet and Chinese troops. For the next six months tensions along the Ussuri River threatened open war between the USSR and China and kept Moscow absorbed with its China problem. For the new administration in Washington, Vietnam was the most important problem, and remained so for most of the next four years. Notwithstanding these critical problems and their global rivalry, the two superpowers inched their way toward negotiations on the limitation of strategic arms.

On March 27, Nasser said that the fighting represented a new phase in the struggle against Israel, and its intensity and scope would depend on the outcome of the four power talks, scheduled to begin at the United Nations on April 3. The intense fighting along the canal was sustained to impress upon the great powers the gravity of the situation. Nasser also spoke glowingly of Moscow's generosity in supplying arms and advisers to help train Egyptian troops.* His detailed description at that time of Soviet aid may have been designed to minimize the possibility that Moscow might restrict its flow of weaponry in an effort to induce Egypt to curtail the fighting along the canal.

There is no evidence that Moscow then sought to exercise leverage by withholding weapons. However, it was concerned that the situation might get out of hand. It castigated Israel for allegedly triggering the fighting, but also criticized the "extremists," those in the Arab camp pressing for a military solution, who "have become particularly active now when there have been signs of progress in the search for paths to a political settlement of the Near East crisis, when consultations are being held between the representatives of the USSR, the United States, France, and Britain on this question."[17] Officially, Moscow blamed Israel for the "provocations," but behind the scenes it (and Washington) urged Nasser to ease up on the artillery barrages and commando attacks, which were upsetting the Israelis and making escalation inevitable. On April 21, Secretary General U Thant informed the Security Council that the action along the Suez Canal had reached "a virtual state of active war."

The fighting continued, including a heavy air battle over the canal on May 22, and Nasser's speeches were uncompromising. While Pravda and Izvestia attributed the aggravation of the Near East crisis primarily to Israel,[18] a Soviet note to U Thant on May 8 called for strict adherence to the ceasefire—an implied admonition to Egypt.[19] Western diplomats reported that in private talks their Soviet counterparts indicated "little sympathy with the Arab position."[20] Pravda's

---

*Nasser gave a number of new details about the Soviet-Egyptian arms relationship. "The Soviet Union is supplying us with the arms we need without exerting pressure on our current financial resources, which are bearing the heavy burden of the war. It is enough to tell you that we have not yet paid a single penny. The first consignment of arms we received from the Soviet Union was free. After that, all other arms consignments were paid for with long-term loans. . . . We also asked the Russians to assist us in training, in grasping arms, and in modernizing the various commands—from the supreme to the subordinate commands. I insisted on requesting Soviet experts for deployment with the armed forces because of my conviction that to confront the Israeli enemy we needed the full assistance of Soviet arms and also of those who could instruct us on the use of the arms and who could help us in command training. . . . We have benefited a great deal in the recent months from the Soviet experts and advisers who are with our units." BBC Reports, ME/3037/A5-A6.

Middle East specialist, Igor Belyayev, endorsed the principle of direct talks—with qualifications—in a domestic radio broadcast, but not in the international version, a sign of Moscow's sensitivity to Arab attitudes. Izvestia praised Nasser's "positive attitude" toward the four-power talks but cautioned the Arabs against any "adventurist" course of action.21 On June 4, Al-Ahram quoted Izvestia as having denied a U.S. report that the Soviet position was becoming more flexible on the issue of an Israeli withdrawal, calling it a ploy aimed at sowing dissension between Moscow and Cairo.22

On June 10 Cairo Radio briefly announced the arrival of Soviet Foreign Minister Gromyko. During the next four days he held intensive discussions with Egyptian leaders, including three meetings with Nasser. The joint communique of June 13 represented a victory for Cairo's position. It held that "to find a peaceful settlement in the Middle East requires the implementation of all parts and provisions of the 22 November 1967 resolution and the withdrawal of Israeli forces from all Arab areas occupied by Israel as a result of the 5 June 1967 aggression" (emphasis added). In contrast to the December 24, 1968 communique, it clearly linked a solution of the Palestinian problem to any peaceful settlement, thus congealing Cairo's views.

Moscow's grudging support of Cairo was necessary to reassure Nasser that the USSR was not going to pressure him to move closer to the U.S. position for an interim settlement. Al-Ahram's commentary the day after Gromyko's departure said that the communique dispelled the uneasiness aroused by the U.S. "psychological warfare of the past two weeks," which had insinuated that there were differences between Moscow and Cairo; indeed, "some press reports even went so far as to say that there was an agreement between the United States and the Soviet Union [on the Middle East] and that Gromyko's visit was intended to ask Cairo's support."23 We do know that Gromyko came to consult about a 13-point U.S. plan and met with Cairo's rejection. Thus, there was probably truth in the report that Al-Ahram took pains to deny.

Moscow's failure to forestall "the war of attrition" after considerable diplomatic effort showed the limits of its influence in Egypt: Despite massive Soviet inputs of military and economic assistance and Egypt's continued dependence on continued Soviet support, the USSR lacked influence when it sought to restrain Nasser from moving to war; judging by Egypt's lavish expenditure of artillery shells, for which it was completely dependent on the Soviet Union, Moscow did not seriously try to restrict supplies, for fear such blatant pressure would undermine its presence (the massive use of artillery barrages, a Soviet trademark in World War II, was a technique the Egyptians had learned from Soviet advisers, and did well with on their own, even though we must allow that Soviet advisers may have offered suggestions once the die had been cast for war). The Soviet Union, unable to dissuade Nasser from a new confrontation, went along reluctantly with Egypt's course, the road to "the war of attrition."

THE TREATY OF FRIENDSHIP AND COOPERATION

The origins of the treaty of May 27, 1971, are to be found in the political upheaval that took place in Egypt's power structure in early May and in the apparent movement toward a Middle East solution that was stirring in the spring of 1971. The domestic components are particularly important for an understanding of the Soviet-Egyptian influence relationship in this situation.

On April 17, 1971, Egyptian President Anwar Sadat returned from Tripoli with the announcement that Egypt, Libya, and Syria planned to establish a Federation of Arab Republics—a new crystallization of the modern Arab quest for unity. At a series of meetings of the Central Committee of the Arab Socialist Union, held between April 25 and 27, Vice President Ali Sabri challenged Sadat, ostensibly on the issue that Sadat had made too many concessions to Libya and Syria, but actually to mobilize support against Sadat's growing power internally. On May 3, one day before the scheduled arrival of Secretary of State William Rogers to discuss a possible interim settlement involving a phased Israeli pullback in Sinai and the reopening of the Suez Canal, Sabri was dismissed. Ten days later, Sadat launched a preemptive purge against other top leaders who, he said, were plotting a coup d'etat: The purged read like a Who's Who from the Nasser period. Some of the key figures, such as Ali Sabri and War Minister General Mahmoud Fawzi, were generally regarded as advocates of closer ties with the Soviet Union. Moscow's response was virtual silence, indicative of the uncertainty in the Kremlin about the implications of Sadat's purges and the course to be followed. The Soviet media said little about the deposed or the developments in Egypt.

Late in the evening of May 23, Cairo Radio announced that Soviet President Nikolai V. Podgorny would arrive in Egypt on May 25 for consultations. Two days after his arrival the Treaty of Friendship and Cooperation was signed. The treaty attracted widespread attention and was generally interpreted in the West as signifying an increase and institutionalization of Soviet influence in Egypt. First, this was the first treaty with security commitments that Moscow had concluded with a noncommunist, third-world country, and it went far beyond those signed with contiguous countries such as Afghanistan, Iran, and Finland. Second, Articles 7 and 8 call for consultations "regularly at various levels with regard to all important issues which concern the interests of both countries" and the promotion of "cooperation in the military sphere on the basis of suitable agreements between them. This cooperation will particularly include aid in training the personnel of the UAR Armed Forces, and in their assimilation of arms and equipment supplied to the UAR for strengthening its capability to remove the traces of the aggression and for strengthening its capability to confront aggression in general." These implied a deepened Soviet commitment to arm and defend Egypt. Third, it is to be operative for 15 years, signifying a determined Soviet effort to pursue its courtship of Egypt systematically and with long-term purposes in mind. By its all-encompassing provisions, the treaty sought to safeguard Moscow's massive stake in Egypt and preclude any significant westward leaning by Egypt's leaders. Clearly, Moscow wanted to perpetuate Egypt's dependence on the Soviet Union, and it was willing to pay a considerable price to achieve this aim. One must keep in mind that the Soviets view signed treaties as serious obligations, entered into for mutual benefit. That this seriousness regarding treaties is not characteristic of the Middle East is worth mentioning. According to one Yugoslav official whom I spoke with recently, this Soviet attitude toward treaties may well be a basic shortcoming of the USSR in its approach to the Middle East: "We," he said, "cautioned the Soviets that Sadat was agreeing to the Treaty too quickly for it to have any lasting value."

That it was the Soviet Union that desired the treaty is apparent from the following: (1) the Soviet Government initiated the Podgorny visit; (2) the treaty's provisions are remarkably similar in many respects to those concluded among Warsaw Pact nations, suggesting that Moscow came with it in hand; and (3) the composition of the Soviet delegation accompanying Podgorny was unusual: It was broadly representative of the party, government, and military—the pillars of the Soviet establishment. At the suggestion of the Soviet delegation, three working groups were formed promptly after the first Podgorny-Sadat meeting on May 26: on dealing with political affairs, headed on the Soviet side by Foreign Minister Andrei Gromyko; a second treating military matters, represented by Soviet Deputy Defense Minister General Ivan Pavlovsky; and a third, the party-to-party group, headed by Boris Ponomarev, a member of the Secretariat of the CPSU Central Committee and long responsible for Communist Party relations with foreign communist and "progressive" parties. The treaty was ratified by Egypt on June 14 and by the Soviet Union on June 18. In terms of media coverage Moscow made more of a fuss over the treaty than Cairo.

The despatch and vigor with which Moscow pressed for the treaty signified that it wanted to go far beyond its already well-established relationship with Cairo to safeguard its position in Egypt and stave off what it perceived as a U.S. effort to regain a major role in Arab affairs. Despite its already significant contribution to Egypt's stability and strength, Moscow felt impelled to up the ante dramatically to assure itself of a special role in Egypt. (To emphasize Egypt's dependence and counter Sadat's diplomatic soundings for a settlement in February 1971, Moscow had stepped up deliveries of aircraft and missiles in March and April 1971.) Moscow feared that the Rogers Plan for an interim settlement and reopening of the canal just might materialize, in which event Washington would succeed in delivering by diplomacy what Moscow had been unable to do through weapons. If it could not speak for Cairo in discussions with the United States, what function, other than that of quartermaster, was it to play in the labyrinthine Middle East affairs? The fallen pro-Moscow Egyptians were to be left to their fate, if Sadat agreed to the reinsurance treaty that Moscow proffered.

Sadat's alacrity in signing is not hard to understand. In the wake of the purges his immediate task was to entrench his power. The treaty with the Russians simultaneously secured their promise of noninterference in Egyptian domestic affairs and acceptance of Sadat's legitimacy, and satisfied the military, upon whose loyalty he depended, through the Soviet promise of new acquisitions of sophisticated weaponry and the advisers necessary to train the Egyptians in their use. The additional commitments of Soviet support could only strengthen his hand in bargaining with the United States.

The unquestioned initiators of the treaty, the Soviets expanded their commitments in order to retain the position they already held. They did not improve their situation in any of the sectors of Egyptian society, though an expanded presence in the military was immediately involved. They may have contributed to the squelching of the Rogers Plan, but the need to consolidate power probably played the more important role in Sadat's shift from an interest in exploring the possibility of an interim settlement to an emphasis on a decisive confrontation with Israel in 1971. They accepted the suppression of "progressive" elements with seeming equanimity and watched Sadat shift steadily to the reembourgeoisization of society. Their position in this strategic outpost would in little more than a year be shown to have been built on quicksand.

THE "LESSONS" OF 1968-69 AND 1971

A few general observations on the Soviet-Egyptian influence relationship during the foregoing periods may be made. First, Soviet influence over Egyptian foreign policy, even during the period of maximum Egyptian vulnerability, was limited. The Egyptians knew the nature of their total dependence in 1967-68 and acquiesced to Soviet diplomatic efforts designed to effect a political settlement agreeable to them. However, the massive Soviet military and economic inputs did not result in the Soviets acquiring any significant influence over its disposition: As far as I can determine the Egyptians decided when and for how long to mount military actions along the canal. Cairo accepted Soviet advice on how best to use the new weaponry, but it did not look to Moscow to develop a foreign policy for its new strategic situation; nor, despite much fanfare, did Nasser or Sadat really shake up Egyptian economic ministries, practices, or priorities, in line with Soviet suggestions.

Second, Soviet policy seems to have made adjustments to Egyptian domestic politics more often than Egyptian policy-makers yielded to Soviet preferences. On the basis of the two events discussed above, admittedly a limited sample, there is no evidence that when it came to matters of concern to the Egyptian leadership Moscow was able to mobilize or strengthen the position of Egyptian officials or groups that were disposed toward accommodating to suggested Soviet preferences. Egypt followed Moscow on Czechoslovakia, disarmament, and the German question, but only because these issues were of no consequence to Cairo. Moscow's leverage on issues of importance to the Egyptian leadership was marginal, at best, once Cairo had resolved upon a course of action. This suggests that Cairo's gratitude for Soviet support does not convert to any willingness to tolerate Soviet interference in Egyptian decision-making on key issues. It is as easy to overestimate the extent to which Soviet inputs into third-world countries bring influence as it is to underestimate the profound constraints on Soviet influence inherent in the institutions, practices, and political climate of third-world countries.

Third, the Soviet-Egyptian influence relationship is asymmetrical both as to aims and accomplishments. Moscow's ability to exercise influence derived from its enabling Egypt to resist pressures for a settlement with Israel and in thus perpetuating this regional conflict. Cairo's influence over Moscow was most apparent in the matter of obtaining vitally needed supplies. While Moscow may not have given all that Egyptian leaders wanted and while it may have manipulated the flow and level of weaponry to retain a residual leverage on issues whose outcomes might have jeopardized Soviet strategic interests— one of which was to avoid a confrontation with the United States—it did function as a major, though occasionally reluctant, supplier of material vitally needed by Egypt. There is truth in Nasser's oft-repeated comment that the donor is circumscribed by its commitment to a small nation: having once committed itself to giving, any move by Moscow to cut the flow of weaponry significantly would immediately have undercut its entire, expensively acquired, position in Egypt. The USSR must continue to give as long as its strategic considerations in the region remain important, or until the overall political context within which the relationship is pursued changes for reasons beyond Moscow's control. There is no indication that Moscow was able to establish a relationship with Egypt comparable to that which it has with the East European members of the Warsaw Pact, a relationship once characterized as follows: "The Soviet Union wants its client states to be strong enough to stand on their own, but weak enough to take orders." During the 1968-71 period we had the anomaly of Egypt not having been strong enough to stand on its own, but having been strong enough not to take orders. Developments since then have dramatically confirmed that the dependent member of a relationship has the ability to undertake independent initiatives, to the detriment of the donor's interests.

Soviet Exodus: July 1972

The Treaty of May 1971 enabled Moscow and Cairo to formalize their relationship in ways congenial to each. However, serious strains appeared, stemming from incompatible objectives and the dynamics of Egyptian domestic politics. Moscow reequipped Egypt's army, but Cairo complained that it was not being provided with the offensive weaponry needed to retake the Israeli-occupied Arab territories; Moscow upheld Egypt's position that "a just and lasting peace in the Middle East can be established only if all the provisions of the November 22, 1967, resolution of the Security Council are fulfilled" but shied away from supporting a full resumption of fighting.[24]

President Sadat was beset by mounting pressures from those who wanted to put an end to the interregnum of "no war, no peace," who urged either war or a political settlement; and from those who were uneasy over the growing Soviet presence and influence. In response to the student riots in December 1971, on behalf of liberalization internally and a decisive policy toward Israel, Sadat flew to Moscow in early February 1972 to request offensive weaponry. He went again in late April to press his case and to persuade Soviet leaders to obtain concessions on the Middle East from President Richard Nixon (the Moscow summit conference was held from May 22 to 29, 1972). The Soviet-Egyptian communique, issued on April 29, 1972, at the end of Sadat's visit, did state that the Arab states "have every justification for using other means ... to regain the Arab territories captured by Israel," thus implying that Moscow accepted the use of force as legitimate, a position it had heretofore been reluctant to condone.

On July 18, 1972, Sadat, in a dramatic demonstration of Egyptian sovereignty and displeasure with the USSR, ordered the removal of Soviet military advisers and personnel from the country. Frustrated by Moscow's unresponsiveness to his repeated requests for offensive weaponry, pressured by important elements in the Egyptian military who were fed up with the overbearing attitude of Soviet staff officers, and desirous of boosting Egyptian pride (and his own prestige) and patriotism on the eve of the 20th anniversary of the Egyptian Revolution, Sadat acted forcefully. He demonstrated that Egypt was not a Soviet puppet.

For a time relations between the Soviet Union and Egypt were seriously strained: Moscow lost the exclusive use of six air bases and the section of the port of Alexandria that it had been given by Nasser; no longer did it possess extraterritorial enclaves on Egyptian soil. However, Sadat did not expel all Soviet military personnel, several thousand remaining to man the air defense and communication systems. Also, the expulsion ordered did not affect Soviet economic advisers and personnel. What the incident showed was that Sadat was determined not to allow the Arab-Israeli stalemate to become permanent, Moscow's preferences to the contrary notwithstanding.

The Fourth Arab-Israeli War

On October 6, 1973, Egyptian forces launched a massive attack across the Suez Canal and established footholds on the eastern bank (at the same time, Syrian forces struck in the Golan Heights and penetrated almost to the June 5, 1967, border with Israel, before being pushed back in fierce fighting). The Soviet Union began to resupply Egypt and Syria with weapons and ammunition within 48 hours after the fighting had started; its airlift assumed massive proportions on October 10. The Soviet involvement, which threatened to turn the outcome of the fighting decisively against Israel, prompted a U.S. response. A week after the Egyptians had triggered the fourth Arab-Israeli war, the budding superpower detente was in danger of being squashed. For the first time using its naval power provocatively, the Soviet Union moved toward a collision course with the United States.

On October 17, Soviet Premier Kosygin flew to Cairo for secret talks with President Sadat. Presumably, he was concerned over the growing escalation of superpower involvement and the danger of a confrontation and urged Sadat to settle for limited tactical gains, with a promise of full support for a political settlement agreeable to Cairo. On October 20, at the urgent request of the Soviet Government, Secretary of State Henry A. Kissinger flew to Moscow. Two days later, the Soviet and U.S. governments asked for an immediate meeting of the U.N. Security Council and cosponsored a resolution calling for an end to hostilities. The fighting continued for several more days, as Israeli troops jockeyed for a better position before the ceasefire finally went into effect on October 25, 1973. With U.S. and Soviet encouragement and under the aegis of the United Nations, Egyptian and Israeli negotiators hammered out an interim agreement, which produced a disengagement of forces along the eastern bank of the Suez Canal.

Based on the available evidence, admittedly fragmentary, it would appear that the decision to start the war was made in Cairo.[25] From its intelligence on the scene, Moscow knew a war was coming at least several days before the actual outbreak of hostilities. By removing families and dependents of Soviet advisers in Egypt and Syria on October 3, 4, and 5, it may well have been trying to "signal" Washington, though this speculation requires further research. But it is hard to believe that Moscow would have undertaken the evacuation at that crucial time if Soviet leaders were in collusion with the Egyptians and if they approved of or pressed for the attack. Once the fighting began, Moscow acted to support Egypt, to demonstrate its credibility as an ally. It went along with U.S. efforts to arrange a ceasefire only after Egypt had exhausted her military momentum and her forces were in danger again of being severely defeated by Israel.

True, Egypt remains heavily dependent on the Soviet Union: All her weapons are Soviet; more important, the USSR is Egypt's protector, her ultimate shield from defeat. However, the Soviet Union is also tied to Egypt. It requires a decent relationship with Egypt to reassure doubting Arabs of Moscow's reliability and value and to help retain a major Soviet presence in the Arab world, more because of considerations inhering in its global rivalry with the United States than because of any expectation of spreading communism in the Middle East. Now that Egypt has a wealthy and willing banker in Saudi Arabia, the Soviet Union obtains hard cash for its arms; but for political and strategic, as well as economic, reasons of its own, Moscow is unlikely to cut the supply under any circumstances.

What emerges from an examination of the Soviet-Egyptian relationship is the conviction that the Soviet presence in Egypt has given Egypt the opportunity to influence the Soviet Union; and that the task of improving the criteria and methods by which we assess the influence relationship between the USSR and third-world countries is only in its infancy.

NOTES

1. The methodological and political dimensions of the Soviet-Egyptian relationship will be treated more fully by the author in a study currently in progress.

2. New York Times, March 31, 1972.

3. For a discussion of the idea of "issue areas" see James N. Rosenau, "Pre-Theories and Theories of Foreign Policy," in R. Barry Farrell, ed., Approaches to Comparative and International Politics (Evanston, Ill.: Northwestern University Press, 1966), pp. 60-92; a formal definition of the term appears on p. 81.

4. For an assessment by U.S. foreign policy specialists of the relative usefulness of different quantitative measures, see Alvin Z. Rubinstein, "U.S. Specialists' Perceptions of Soviet Policy toward the Third World," Canadian-American Slavic Studies 4, (Spring 1972): 93-107.

5. There are a number of other issue areas that can be identified, but not all of them, unfortunately, can be researched, because of a paucity of data—for example, the persistent Soviet effort to persuade Egypt to buy Soviet Ilyushin-62 and Tupelov-134 passenger aircraft, thus far without success. See Cairo Radio, October 13, 1968; and Egyptian Gazette, March 27, 1971.

6. Al-Ahram, September 9, 1968.

7. Al-Ahram, September 11, 1968.

8. Al-Ahram, September 27, 1968.

9. Egyptian newspapers avoided any mention of the Brezhnev doctrine.

10. BBC Report, ME/2942/A1-A10. In subsequent days Nasser shifted his emphasis somewhat, putting the principal blame for the unrest on "Israeli agents" and other "subversive elements." Egyptian Gazette, December 4 and December 5, 1968.

11. The text of what was purported to be the Soviet peace plan was published in the Lebanese newspaper Al-Anwar on January 10, 1969; in Al-Ahram on January 19, together with the U.S. reply of January 17; and in Pravda, which only outlined the main proposals, on January 25.

12. Al-Ahram, January 13, 1969.

13. Pravda, January 25, 1969.

14. Al-Ahram, January 21, 1969.

15. Al-Ahram, February 3, 1969.

16. Al-Ahram, March 3, 1969. Interestingly, the headline in the New York Times, March 2, 1969, had read, "Nasser Foresees 4th War Unless Israelis Withdraw."

17. Izvestia, April 3, 1969.

18. Pravda, April 30, 1969.

19. New York Times, May 9, 1969.

20. New York Times, May 23, 1969.

21. Izvestia, May 25, 1969.

22. Al-Ahram, June 4, 1969.

23. Al-Ahram, June 15, 1969. "There has been no question of the Soviet Union in any way adopting a less firm attitude than it has

done ever since the June aggression. Not that the Soviet attitude was doubted here, but the doubters, or at least the wishful thinkers, abroad . . . needed to be told exactly where the Soviet Union stood." Egyptian Gazette, June 15, 1969.

24. Alvin Z. Rubinstein, "The Soviet Union in the Middle East," Current History 63, 374 (October 1972): 168.

25. For a contrary interpretation holding that Brezhnev orchestrated the entire scenario, see Uri Ra'anan, "The USSR and the Middle East: Some Reflections on the Soviet Decision-Making Process," Orbis 17, 3 (Fall 1973): 946-77.

CHAPTER

11

SOVIET DECISION-MAKING IN THE MIDDLE EAST, 1969-73

Uri Ra'anan

Most observers will agree that the origins, course, and outcome of the 1973 "October war" in the Middle East cannot be elucidated without a simultaneous analysis of the Soviet Union's policies and decision-making process. Some will emphasize that the very key to contemporary developments in the region must be sought in the recesses of the Kremlin; few will deny that there is, at any rate, a link between two decades of conflict in the area and Soviet attempts to establish a powerful presence along the southeastern rim of the Mediterranean.

Of course, it may be argued that a new phenomenon, generally described as detente, has appeared and that, consequently, Soviet actions and motivations must be judged now by criteria that differ entirely from those applied during the "cold war." When challenged to produce evidence that the spirit of detente is really present in the Eastern Mediterranean, the protagonists of this view usually will refer to the Brezhnev-Kissinger "Little Summit" of October 1973 and the piece of paper it produced for endorsement by the UN Security Council. A little reflection will show, however, that this retort does not begin to answer the three crucial questions:

(1) What was the Soviet role with regard to the outbreak of the October war, and why did Moscow not only refuse insouciantly to assist early efforts to terminate it but actually escalate the level of conflict by initiating massive airlifts of weapons to Egypt and Syria?

(2) What caused the subsequent shift in the Kremlin's performance on the international stage and why was it suddenly in the Soviet interest to bring military operations to a halt, at least for a while?

This chapter was first presented at the Soviet Naval Studies Group seminar. It was subsequently published in <u>Orbis</u> 17, 3 (Fall 1973) and is reprinted with permission.

(3) Why, immediately after succeeding in extracting significant concessions from Dr. Kissinger in Moscow, did the USSR crudely threaten to resort to unilateral military action in the Middle East, in breach of all previous understandings with Washington?

THE BASES OF ANALYSIS

To tackle such questions, especially at a time when the political processes associated with the October war are still in high gear, is a complex task. Perhaps the simplest approach would be both to analyze Soviet decision-making patterns in general and to examine a previous Soviet policy episode toward the Middle East in particular, a period for which, to use election jargon, "all the returns were in" sometime prior to the October war. In this way, it may be possible to reconstruct enough of the mold within which Soviet policy has been, and continues to be, shaped to make current Soviet behavior in the Middle East more readily comprehensible. However, this is easier said than done, for the analysis of the Soviet decision-making process is beset by obstacles that at times seem to be well nigh insuperable.

The observer of the Soviet political scene is confronted by an old and well-known dilemma: Either he can confine himself to a mere uncritical and unanalytical narrative, rehashing in chronological order various quotations from articles in the Soviet press or statements by Soviet leaders, thus playing it safe. Or he can undertake, within the limits of his ability and understanding of the subject, a logical reconstruction of the course of events, in which causes and effects are linked coherently and motivations are explained lucidly, and in which he is aided: (1) by a knowledge of events or scraps of significant information that have become accessible to some Western analysts; (2) by a comparison with analogous situations in the past or in similar communist countries; (3) by "leakages" emanating from defectors and dissidents or by "revelations" made in the course of a purge or during an attack on the policies of a deposed leader: (4) by content analysis in depth of spoken and written words originating in the USSR; (5) by some understanding of ideological terms as a Soviet operational code; and (6) by a (hopefully) sophisticated instinct for deciphering and comprehending political processes in the Soviet Union as a special category of general bureaucratic models. However, if the Western observer chooses this second approach (as the present analysis has done), he leaves himself wide open to charges of having taken a "speculative" rather than "documentary" path and, especially in circles not inured to the difficulties and pecularities of Soviet and communist studies, he might well become suspect of resorting to methodologies that do not enjoy full academic respectability.

These are some of the considerations that ought to be borne in mind when evaluating this study. They are given not by way of apology, but to explain why any attempt at reconstructing the origins and development of Soviet politicomilitary actions in the Eastern Mediterranean from 1969 to the present cannot avoid a certain speculative flavor.

The Conventional Interpretation of Soviet-Egyptian Relations

To start, it may be useful to recall the generally accepted interpretation of these developments, which runs more or less along the following lines:

In the spring of 1969 President Nasser opened a war of attrition against Israeli forces positioned along the Suez Canal ceasefire lines. The Israelis responded by establishing heavy fortifications along the canal and launching air strikes to destroy the massed Egyptian artillary formations, which were bombarding Israeli positions; however, this did not persuade Nasser to desist. By December 1969 the Israelis were flying deep penetration raids to the vicinity of Cairo, so that the facts of the military situation would become visible to the Egyptian public and generate political pressure on Nasser to restore the ceasefire. Instead, Nasser suddenly and secretly flew to Moscow in January 1970 to ask for Soviet personnel to man Soviet surface-to-air missiles and interceptor aircraft that were to be dispatched to Egypt. The Soviet leadership immediately agreed to this request and combat personnel appeared in the delta area by February 1970; however, these Soviet units were to be used only in a purely "defensive" role and were not to be drawn into "offensive operations." Resentment against this restriction mounted after Nasser's death and was exacerbated by the new Egyptian President Sadat's suspicion that the Soviet leaders were supporting his rivals, such as Ali Sabri, and by his concern over the temporarily successful communist-supported left-wing coup in the Sudan. The Soviet leadership tried to remedy this situation by means of the Soviet Egyptian Friendship Treaty of 1971, but Cairo misread the treaty as an earnest of Soviet intentions to support directly an offensive across the canal, and as an implicit promise not to negotiate with the United States behind Egypt's back. By the spring of 1972 Sadat had come to realize his mistake, and his disillusionment was accentuated by the Kremlin's refusal to deliver certain unspecified "offensive" weapons. After a warning and an ultimatum, in the summer 1972, Sadat ordered the withdrawal of Soviet military personnel from Egypt, with certain limited exceptions, and Moscow meekly complied.

Fallacies of the Conventional Interpretation

The fatal flaw in this widely accepted interpretation, is that it does not allow for what we know of the behavior pattern and decision-making process of the Soviet leadership, and its attitude toward smaller countries in general, and client states in particular; nor is it consonant either with various data that have become available in the meantime or with certain basic requirements of technology and logistics.

Patterns of Soviet Behavior

It is not at all typical of Soviet behavior, either under the "collective" headed by Brezhnev or its Khrushchevian predecessor, to comply meekly and even abjectly with the peremptory demands of small and weak states, least of all clients dependent on Soviet protection and hardware, especially concerning such "minor" issues as whether and when Soviet military personnel with their accoutrements is to come in or leave. Questions of security, particularly military security, are zealously safeguarded as the particular prerogative of the Soviet leadership. Soviet forces come into or evacuate foreign countries in line with decisions made by the Soviet leadership and at its initiative, in accordance with its evaluation of the Soviet interest, shaped by the contemporary situation as viewed from Moscow, and influenced considerably by the internal balance of power between competing factions in the Kremlin.

Czechoslovakia. For instance, the Red Army entered Czechoslovakia largely because the "interventionists" in Moscow, although apparently originally a minority within the Politburo, were able to take the wind out of the sails of their opponents (reputedly including such surprising names as Suslov), who expressed their reservations concerning the proposed venture not so much because they were "doves" but because they hoped to inflict a serious setback upon their personal rivals among the "interventionists" by exposing them as irresponsible adventurists. The supporters of bold offensive action reportedly were able to overcome such objections by producing convincing evidence that the leaders of the Czechoslovak forces had decided not to resist, that Washington had no inclination whatever to make any reactive move, near Czechoslovakia or elsewhere, and that no other Western state would take action. The proposed move, therefore, was quite safe and not at all adventuristic; that was the ultimate reason for the Kremlin's decision, and it merely remained then to manufacture some pretext, preferably plausible but, if necessary, as barefaced as an alleged Czech "invitation."

Hungary and Cuba. It was a somewhat similar constellation in 1956 that, after some factional jousting in the Kremlin, had led to the final Soviet decision to resort to military intervention in Hungary. One of the critical factors in 1956 was not so much any unwillingness of the Hungarian forces to resist as the preparation of what the Red Army regarded as a practically foolproof plan—which, indeed, succeeded— to decapitate the Hungarian resistance by entrapping its naive military leadership into Soviet captivity. Again, in the Cuban case, 1962, as in the subsequent Czechoslovak instance, it is believed that a factional "debate" in the Kremlin, centering on Soviet global considerations rather than anything said or done by Castro, led to the inception of the missile crisis and its outcome. Some elements in the Soviet leadership, reportedly including old Kuusinen, asked whether the emplacement of Soviet military personnel and nuclear-tipped missiles in an island off the U.S. coast could not be in the category of adventuristic acts, thus implying that the more offensive-minded faction, supported by Khrushchev himself, ought to have its wings clipped. These objections and misgivings were overruled apparently on the basis of (erroneous) Soviet intelligence that Washington was wilfully shutting its eyes to Soviet actions in Cuba, partly because of demoralization following the Bay of Pigs, and would therefore take no serious counteraction. In this case, the Kremlin's "interventionist" group turned out to be mistaken, a factor that played no minor part in Moscow's rapid reversal of direction during the crisis and in Khrushchev's subsequent, if somewhat delayed, ouster.

The important point here is that the Kremlin's actions in dispatching and withdrawing military elements in disregard of clients' wishes applied not merely in states immediately adjacent to the USSR and considered part of its continental empire, but also to overseas clients, such as Castro's Cuba. Indeed, in the Cuban instance, Soviet withdrawal of missiles and their crews took place against Castro's express demands and over his official protests. What really mattered, in all these instances, were Soviet interests, defined by Moscow's view of the general situation, as established by the monetary factional balance within the Kremlin.

While there are individual and peculiar factors to be found in each instance, it is significant that on each occasion the initiative whether to send in Soviet military forces, and whether they should stay or leave, emanated from the Kremlin; the wishes of the client either constituted a fraudulently manufactured pretext, such as the Hungarian and Czech "invitations," or were blatantly ignored, as were the protests of Nagy, Dubcek, and Castro. Of course, where the desires of a client coincided conveniently with Soviet plans and decisions, as with Castro's initial enthusiasm for the clandestine importation of Soviet personnel and missiles, the Kremlin was happy to exploit this fact. The Egyptian case, if the generally current version is accepted, would be practically sui generis since the initiative supposedly always emanated from Cairo, with Moscow merely following suit. On these grounds alone it would warrant skeptical scrutiny. Nor would it be a valid objection that the other instances mentioned are simply nonanalogous precedents, since at least Cuba involved a somewhat difficult, semiautonomous, overseas client, much like Moscow's friends in the Middle East, rather than a member state of the Warsaw Pact, along the very borders of the USSR.*

Patterns of Soviet Decision-Making

The modus operandi of the Soviet leadership in the decision-making process, as evidenced in the various instances cited in this article, seems to follow certain patterns.†

Although factions and factionalism are formally prohibited, political processes, of course, cannot be outlawed by fiat or ukase, even in the USSR, and there are kaleidoscopic formations and reformations of alliances and coalitions within the leadership each time a specific decision has to be reached. These factions are not necessarily stable, since they are not specifically issue-oriented—at least not at the very top (groups with a vested professional bias are to be found slightly lower down the ladder, including, of course, the military services); rather, the leadership reveals a strangely "feudal" pattern, with each top personality "investing" subordinates wherever possible

---

*It is true, of course, that even in Eastern Europe there have been instances in which, despite Soviet desires, the dispatch of Soviet military elements has had to be canceled or delayed—for example, Yugoslavia 1948 and, probably, 1968 and Romania 1968. However, the factor that made Moscow hesitate in these cases was not the absence of an "invitation" from the governments of the countries to be entered but rather the <u>deterrent</u> effect created by the prospect of protracted local resistance to the Red Army, consequent possible foreign reaction, and the potential damage to Soviet interests that might be caused by both.

† The following two paragraphs are taken from the author's paper, "Some Political Perspectives Concerning the U.S.-Soviet Strategic Balance," presented to the May 1973 Conference on the U.S.-Soviet Strategic Doctrine and Nuclear Multipolarity, held under the auspices of the International Security Studies Program of the Fletcher School of Law and Diplomacy. The proceedings of the conference, including the paper, will be published by Heath Press.

with power and position in return for personal "fealty" and allegiance, and these subordinates similarly distribute whatever privileges they command among their various personal collaborators in return for political support and loyalty, until a veritable "feudal" pyramid of "enfeoffments" is created from one of the top leaders all the way down the ladder to minor local functionaries who may not be in direct touch with him but belong to his "party" through a chain of personal links of allegiance. Sometimes, these links are even reinforced by familial ties, including political marriage to create bonds of blood, in truly "feudal" style. This system strongly influences the modus operandi at the very apex of the Soviet leadership. Given a delicate balance of "collectivity" at the top, each one of the main leaders, conscious of the particular power structure he personally commands and attempts to preserve and enhance, will attempt temporary coalitions with his colleagues and competitors, above all to prevent the strongest and, therefore, most dangerous among them at any one moment from gaining the degree of overwhelming power, momentum, and control that would lead to the total subordination and possible elimination of the weaker members. The actual issues over which these conflicts are fought out are not necessarily of overriding importance to the various competitors nor are they inevitably tied to a consistent line on specific issues. The real battle concerns personal power and influence rather than a particular political question, but the latter can and does serve a useful tactical purpose in gathering recruits and allies for one group and alienating supporters from another.

Moreover, the outcome of a conflict over a specific issue can be of great symbolic and psychological impact in demonstrating which factional alliance is gaining and which is declining. Thus it is almost axiomatic that the advocacy of a proposal by one group will lead to some measure of opposition by various of its rivals and adversaries. There is reason to believe, however, that such opposition rarely takes the form of outright confrontation or rebuff, but rather is expressed by way of carefully worded reservations and caveats, for the record.*
On at least three occasions (the Cuban Missile Crisis, the Soviet invasion of Czechoslovakia, and, as hopefully will be demonstrated, the dispatch of Soviet pilots and ground crews to Egypt), and probably on several others, it seems that a proposal supported by the number-one

---

*There is evidence to the effect that this is the traditional way in which communist leaders express their opposition to proposals advanced by rival factions; a prominent member of Imre Nagy's entourage, who was able to reach the West, has described such scenes, and similar evidence has been received from leading émigrés from other communist countries, as well as from various leaks and revelations.

leader in the "collective" favoring military or semimilitary action of a potentially risky nature, has led to the expression of such reservations by other members of the leadership, nor necessarily because they were more "dovish" but because they felt that they had been given an opportunity for clipping the wings of the "primus inter pares" or even for removing him altogether, should the venture misfire and he could then be accused of "adventurism"—one of the most damning of pejoratives in the Bolshevik dictionary. In such circumstances, the various opposing elements have always been delighted to use as ammunition and to maximize the specter both of potential local resistance to the direct involvement of Soviet forces and of conceivable U.S. counteraction as a reason for suggesting caution and thus implying imprudence on the part of those advocating a forward thrust or offensive action. It is not so much that, in such cases, the "moderate" groups in the Kremlin would warn of the likelihood of U.S. counter-intervention, but rather that they would put the onus of proof on the advocates of a Soviet politico-military offensive move to demonstrate that U.S. reaction was definitely precluded. In other words, they would expect to be shown that the odds were at least 99 to 1 or higher that the United States would remain passive. A 5-10 percent chance of U.S. action would, under these circumstances, be regarded as prohibitively high—certainly high enough to smear the more militant Kremlin group with the tag of "adventurism."

The detailed considerations presented here, suggest that the Soviet leadership does apparently operate within a general and consistent framework, especially where major security questions are concerned, such as the dispatch of Soviet military forces and hardware to foreign countries and their subsequent withdrawal. The salient points of the Soviet modus operandi as outlined so far are as follows: In all cases examined, the Kremlin was "acting," not "reacting"—that is, it always took the initiative and never awaited "invitations" or "demands" from client states. (2) In all instances under review, the Soviet decision-making process appears to have been far from monolithic; on the contrary, it seems to have proceeded invariably against a background of factional infighting within the Kremlin, which influenced, indeed shaped, the conditions that had to be met before a final policy was adopted.

The Conventional Interpretation Assessed

The generally accepted version of the Soviet military entry into and exit from Egypt, 1969-72, is strangely inconsonant with both of these saliences of the general Soviet pattern of decision-making; the Kremlin's role as portrayed in that story would render Soviet relations with Egypt a most special, even unique, case, negating all that we know, or have been able to gather about the Soviet modus operandi on military and security questions. On these grounds alone, the 1969-72 story, as presented earlier, must be regarded as highly suspect. Moreover, this version is also contradicted with regard to its basic chronology by sporadic but significant data that gradually have become available. These include, for instance, an approach by Soviet military personalities to the Kremlin with suggestions that Soviet military elements be dispatched to take over a portion of actual combat duties in Egypt, and a subsequent expression of support for this proposal at a Warsaw Pact gathering—both occurring well before the Israeli deep penetration raids and Nasser's subsequent secret visit to Moscow in which he allegedly originated the proposal and "invited" Soviet units to come and help him combat the Israelis.

Another consideration that renders the current version highly implausible concerns logistics and technology: Nasser supposedly dropped from the skies in January 1970 to surprise his hosts in Moscow with his "invitation," and by February 1970, a mere couple of weeks or so later, Soviet combat units and their sophisticated electronic gear had appeared in Egypt. It seems most unlikely that as complex, hazardous, and difficult an enterprise as the Soviet buildup in Egypt, in the face of a hot war with a skillful and daring adversary like the Israelis, a long distance from the Soviet homeland, connected by a slender lifeline that might be cut any time by a hostile U.S. Sixth Fleet, could have been conceived, planned ab initio in all its mass of details—especially considering Egyptian conditions—and actually implemented in a matter of a few weeks.*

---

*As for Cairo's complaints that the USSR allegedly reneged on commitments to deliver certain "offensive" weapons to Egypt, it is difficult to find an item to which this could apply seriously. Prior to the events of midsummer 1972, the Kremlin had sent to Egypt one of the Warsaw Pact's main battle tanks, the T-62, the latest all-weather versions of the MIG-21, practically the entire SAM series (2, 3, 4, 6), the TU-16B, with a sophisticated air-to-surface missile, and other modern weapons. All of these items were emplaced in Egypt, even if some were still operated by Soviet personnel. Both the SAM's and the interceptors, although tactically "defensive," were "offensive" in the Egyptian strategic context. The war of attrition was intended to enable the Egyptians to launch an offensive across the Suez Canal; the Israeli air force was neutralizing the war of attrition; the Soviet SAM's and interceptors were intended to neutralize the Israeli air force and, by so doing, to enable the war of attrition to succeed in its offensive aims. Under these conditions, none of the Soviet hardware mentioned could be considered "defensive." The

MILITARY INTERVENTION AND WITHDRAWAL, FEBRUARY 1970-JULY 1972

These various considerations and data, as well as other evidence that has been sifted in the West, suggest a rather different story, that, if interpreted correctly, sheds interesting light on policy processes within the Kremlin and suggests suitable counterplays by the West.

The Internal Debate

In the summer of 1969, after Nasser's war of attrition along the Suez Canal had been proceeding for some months—but long before the Israeli deep penetration air raids of December 1969 and Nasser's Moscow visit of January 1970—Soviet advisers to the Egyptian forces apparently communicated to their superiors in Moscow their serious misgivings concerning the chances of the Egyptian venture's success. If they did express their concern in this manner, they may have done so for two reasons: (1) because of their unlimited access to and influence on all aspects of Egyptian military activities, from the planning to the operational stage, they were uniquely situated to observe the fundamental weaknesses of their clients, including a lack of improvisation, adaptability, and initiative, at a much earlier stage of the war of attrition than other experts, even the Israelis on the other side of the canal; (2) expecting a renewed Egyptian defeat, they were probably in no mood to take a share of the blame once again. In 1956 and in 1967, the alleged shortcomings and unsuitability of Soviet weapons and advice had been used repeatedly as an alibi for the failure of Egyptian commanders, and it may well be that, on occasion, even some of their superiors in Moscow were inclined to reprimand the military advisers for not having done a better job with their Egyptian charges. It seems likely that, on this occasion, Soviet officers in Egypt may have been determined to forestall such a development, by going on record in good time with predictions of a new setback, plus an analysis of the real causes for their pessimism.

---

only item of significance coveted by Cairo and not delivered by the USSR, although available there, was a tactical nuclear delivery system, the warheads of which, as is well known, Moscow refused to hand over even to its most reliable Warsaw Pact allies. The Egyptians hardly could have expected that they would be given such a weapon or have voiced genuine grievance at its nondelivery.

Military Proposals

In any case, there are reasons for believing that in the summer of 1969, several top Soviet military leaders managed to obtain an invitation to appear before the Kremlin leadership (in all probability before a subcommittee of the Politburo dealing with military and security affairs) and there conveyed the information that yet another defeat in the Middle East seemed in the offing. It appears that they urged the political leadership to avoid a repetition of the debacles-by-proxy that Soviet prestige had suffered through Nasser's military failures of 1956 and 1967, when, as in 1969, Egyptian troops were known by the whole world to have been armed and trained by the USSR. Consequently, Soviet military representatives are believed to have suggested two options to their civilian colleagues: (1) sever all military links with Egypt, publicly, so that the next military setback would be Nasser's alone; or (2) come in directly, in a combat capacity, at least with Soviet crews for sophisticated surface-to-air missiles and Soviet interceptor pilots, to bolster the Egyptian forces in their weakest sector, thereby averting defeat. In all likelihood, this second, rather hazardous alternative was proposed as a carefully planned and supervised probing operation, with Soviet elements first being installed in the Nile Delta, then gradually and carefully being pushed eastward, toward the Suez Canal; if they were able successfully and without untoward reactions to ensconce themselves on the banks of the canal, thus establishing Soviet-Egyptian air supremacy over the canal, its far side and the adjoining edge of Sinai, they might even assist in an Egyptian amphibious operation across the canal to capture the Sinai Peninsula. Apparently, the Soviet military leaders recommended option (2) as their own preference and as the best way of furthering Soviet prestige and influence.

Differing Military Views

We do not know the precise composition of this Soviet military delegation. However, it seems likely that it included some three or four high-ranking officers in all probability Defense Minister Grechko and/or Chief of Staff Zakharov, a representative of PVO Strany (the air defense forces, from the ranks of which the proposed Soviet elements in Egypt were largely to be drawn and, as it turned out, their commander was also to be chosen), and, almost inevitably, Admiral Gorshkov, who had been intimately involved in Soviet-Egyptian relations and Eastern Mediterranean affairs for at least four to five years prior to this development. There is no way of ascertaining whether the recommendation of this group was unanimous and, if not, just who favored which approach. However, the arm of the services least likely to favor Soviet Mediterranean ventures and most inclined to regard them as a wasteful diversion of resources that would have minimal effect on the really decisive military arenas, namely, the Soviet Strategic Rocket Forces, apparently remained unrepresented on this occasion.

Those present, if the guiding assumptions are correct, probably backed the interventionist recommendation with varying degrees of warmth and for somewhat different reasons: 1. Both Grechko and Gorshkov had become linked rather closely with the fortunes of Soviet policy toward Egypt, including (certainly in Gorshkov's case) the fateful decision after the six-day war not to abandon Nasser but to resupply him massively with Soviet hardware, using the Soviet Navy as a protective screen for the operation. Both men had emerged as vociferous protagonists of a new emphasis upon the political tasks of the armed forces, namely, "extending Soviet prestige and influence" —that is, utilizing these forces, by means well short of nuclear war or the danger thereof, for missions that would create a wide periphery of Soviet protectorates and client states in key areas overseas, especially the Eastern Mediterranean and Middle East. Neither officer, therefore, could afford to watch with equanimity yet another Egyptian defeat (and vicarious blow to Soviet prestige in the area), or the complete abandonment of existing Soviet policies and politico-military investments in Egypt.

2. In Gorshkov's case, the Soviet Navy required air cover and other essential services from Egyptian shore-based facilities, at least until a new Soviet aircraft carrier (or carriers) could be launched from Black Sea shipyards in a building program apparently approved in principle shortly before the war of attrition began. While Egypt increasingly had made such facilities available to the USSR during previous years, Soviet advisers seemingly had become convinced of Egypt's incapacity to provide adequate air cover and defense for these valuable establishments, which would be unacceptably vulnerable to air attack and therefore, useless to the Soviet Navy, unless protected by sophisticated Soviet military equipment operated by Soviet crews. At least as an interim measure, therefore, Gorshkov apparently favored the new project, even if, most probably, he regretted the fact that it would place the Soviet presence in Egypt primarily under the supervision of PVO Strany.

3. With regard to PVO Strany, it may be surmised that its representative regarded the interventionist recommendation with rather mixed feelings, welcoming the proposed ascendancy of this arm of the Soviet services in a new arena and, more importantly, rejoicing that a hot-war, "realistic proving grounds" would be provided in Egypt for testing the efficacy of sophisticated Soviet electronic and other equipment and personnel in battle conditions against high-

quality Western aircraft, at a safe distance from the Soviet homeland and without having to depend on Chinese "good will." On the other hand, PVO Strany must have felt great reluctance to emplace valuable equipment in an area where it would be protected on the ground against eager pro-Western hands only by the Egyptian Army.

The Politburo Debate

In any case, whatever the precise considerations of the various military representatives may have been, it is clear that their recommendations were not accepted by the civilian leadership without further evaluation. Several months elapsed before the appearance of the first sign (from the data now available) that the Kremlin had reached a general decision, at least in principle, favoring the military recommendation to build up a Soviet combat capability in Egypt, including interceptor pilots and crews for sophisticated surface-to-air missiles. At a meeting of members of the Warsaw Pact, sometime in November 1969 (four to five months after the military recommendation was made, one month prior to the Israeli deep penetration raids, two months before Nasser's secret Moscow visit, and three months before the first Soviet combat elements actually appeared in Egypt), Soviet representatives mentioned the plan to their allies, possibly because it was originally intended to stage this operation as a Warsaw Pact rather than purely Soviet action, perhaps by including some symbolic East European contingents. (An abortive move in this direction appears to have been made subsequently.)

The reason for this prolonged delay in reaching a decision was almost certainly factional dissension within the Soviet civilian leadership concerning the wisdom of accepting the military recommendation. Diligent studies have been made elsewhere of perceptible differences of emphasis and approach toward Egypt and the Middle East in general, on the part of certain Soviet organs and some figures in the leadership; they leave little doubt that this issue and related Middle East problems have continued throughout the whole of the period, and up until now, to be a subject of contention within the Kremlin. On the basis of the detailed considerations and analogies presented earlier in this chapter, as well as the more directly related analyses mentioned above, it is possible to reconstruct the general form and much of the content of this factional dispute with a reasonably educated guess as to who some of the protagonists* may have been:

---

*Throughout this reconstruction of developments in Moscow, such terms as "interventionists" and "antiinterventionists,"forward" faction, and "moderates" will be employed merely as convenient

The opponents of a potentially risky further Soviet involvement in the Middle East are unlikely to have expressed their antagonism boldly, in frank confrontation with the more interventionist group, especially if, as seems almost certain, the military recommendation from the very beginning, or from an early stage, enjoyed the support of the primus inter pares, Brezhnev himself. As explained earlier, such behavior would not be considered proper "form" among communist leaders in the USSR or elsewhere; instead, as various precedents indicate, opposition of this kind usually is expressed by way of caveats, warnings, misgivings, or even polite requests for the fulfillment of certain conditions (sometimes specifically written into the record), rather than through a futile attempt to veto a proposal supported by the number-one member of the "collective leadership."

The Case Against Intervention

As indicated by the succession of developments that will be analyzed and that apparently had to take place before the interventionist group in the Kremlin could gather sufficient support—or produce sufficiently convincing replies to the misgivings expressed by its more cautious opponents—for the actual dispatch of Soviet combatants, it would seem not too fanciful to speculate that the anti-interventionists made the following points: They would not, of course, stand in Brezhnev's way, if he was really convinced that the military recommendations were appropriate and wise under current conditions. However, for the record (and they would appreciate it if this appeared in the minutes for future reference), they would like to express these caveats and misgivings:

1. Could one really assume that, because of severe constraints on the U.S. administration, the Vietnam syndrome, neoisolationism, and the like, the White House would simply stand by and watch Soviet combat personnel being introduced into a key East-West contest area like the Middle East? Could one ignore the fact that the Soviet buildup would take place in the midst of a hot war, for the specific purpose of helping Egypt to overcome an adversary like the Israelis, who had

---

shorthand to describe the protagonists in the "debate" rather than to imply that there are permanent "hawks" or "doves" in the Kremlin. As we have noted, the various factions in the leadership are perfectly capable of changing sides, of adopting and dropping various issues, as tactical considerations may require; with some exceptions, these issues are exploited as convenient tactical weapons in a struggle for power, rather than being regarded as platforms to which the respective factions must be wedded eternally.

repeatedly proven the quality of their armed forces as well as their utter determination if threatened with extinction and who undoubtedly would regard Soviet intervention as a threat to their survival and would react with desperate measures? Would this not confront the USSR with another "Finland," with all the ignominy that memory recalled, and would it not require a Soviet buildup of some 150,000 Russians in Egypt rather than 20,000, if a victory over Israel really was to be achieved? Would these Soviet troops then not constitute a veritable hostage to fortune, at the mercy of NATO, linked to the far-off Soviet homeland by a slender lifeline passing through NATO territory (Turkey), a line practically begging to be cut by superior NATO elements in the Mediterranean? Was the onus not on those comrades advocating such a venture to demonstrate convincingly that the eventualities mentioned were not merely unlikely but practically out of the question? Was not even a 5 percent chance of a total confrontation with the United States over the Suez area prohibitively high, nay suicidal? Should there not be some practical proof that the chances were less than one in a hundred?

2. If, to avoid such contingencies, the Soviet buildup were of relatively modest proportions, how would that enable Egypt to win? And if it did not achieve that objective, would it not arouse Arab feelings of frustration, disappointment, and ingratitude toward the USSR rather than scoring a Soviet triumph in the Arab world? What kind of leadership, regime, state, army, and society did Egypt have, anyway, to warrant such intimate linking of her fortunes with those of the USSR? Everyone knew how fickle Nasser was and how he had quarreled with Khrushchev; moreover, had not Soviet doctors reported that he was a very sick man? Who would his successors be and how reliable were they? Was it not a typically Middle East habit to play off the various great powers against one another? Would Egypt not continue playing this game, even with Soviet combat elements emplaced there, and would this not endanger them? Was not the whole policy of buttressing military dictatorships in the Middle East unprincipled, since it led to the deliberate sacrifice of the local communist parties and their future? How many times did the Egyptian Army have to be defeated and lose its Soviet equipment, including sophisticated items abandoned and captured intact by the adversary, thus exposing them to eager NATO eyes, and how many times would the USSR have to replace such equipment? Was this not a case of "throwing good money after bad?" Would the proposed Soviet move really make a decisive difference to this situation? If it did, it would have to be of such proportions as to invite U.S. counterintervention, which was unacceptable, and if it did not, remaining within more modest parameters, would not its vulnerability outweigh its utility? In any case, under these circumstances, was not the onus on those

comrades favoring the venture first to produce the necessary conditions that would ensure that Egypt now could be relied upon, that it would act in accordance with Soviet desires, and that it would respond favorably to various essential Soviet demands?

Content analyses and various data indicate that the above points reflect the general lines of contention within the Kremlin concerning Middle East policy during this period, and it is not of primary importance whether or not these precise arguments were voiced at one particular session of the top Soviet leadership. It does seem reasonably likely, however, that caveats and conditions of this kind were advanced by the supporters of a more cautious and anti-interventionist line in the Middle East, as part of an ongoing debate in the Kremlin after military representatives had submitted their recommendations. It may also be presumed that support for a forward policy was not strong enough initially simply to sweep aside these serious objections and that the protracted delay in reaching a final decision in Moscow was due to attempts by the interventionists to meet some of their opponents' conditions, and, by so doing, take the wind out of their sails.

The Impact of the "Rogers Plan"

The most important of these conditions concerned the United States and its possible reactions, should Soviet combat personnel be sent to Egypt and, in its gradual probe toward the Suez Canal and Sinai, clash with the Israelis while providing an umbrella for an Egyptian operation to breach Israeli canal lines. The interesting fact is that, sometime in October 1969, the U.S. State Department, unknowingly, acted in a way that appeared to provide an answer to this vital question—an answer seeming to sustain the contention of the Soviet interventionists that the proposed venture was quite safe and to undermine the argument of their more cautious opponents that U.S. inaction could not be taken for granted and that Moscow should therefore refrain from the dispatch of combat personnel.

Late in 1969 Secretary of State Rogers delivered a policy address on the Middle East in which he presented a detailed blueprint for an Arab-Israeli settlement, subsequently known as the "Rogers plan." In the section concerning Israel and Egypt, he called for the withdrawal of Israeli troops from the Suez Canal and Sinai Peninsula, all the way back to the pre-1948 Egyptian-Palestinian frontier. It is now known that this section of the Rogers plan was the result of prolonged, secret U.S.-Soviet contact (covering the very months when the Soviet military recommendation to send combat personnel to the area was the subject of prolonged argument in the Kremlin) and that Washington submitted it secretly, as its final official offer to the USSR, in <u>October</u>, some <u>six weeks before</u> the secretary of State delivered his <u>speech</u>. It is highly unlikely that the State Department was aware, at that time, of the "debate" in the Kremlin or of the way in which Moscow was bound to interpret the U.S. offer within the context of that "debate." However, there can be little doubt that the Soviet interventionist group waved the U.S. proposal triumphantly in the air, so to speak, as constituting the very evidence its more cautious opponents had requested —that Washington did not regard the Suez Canal as the demarcation line between the Western and Eastern spheres and would, therefore, not regard clashes there as requiring any U.S. reaction. On the contrary, Washington seemed to be ready to concede both Suez and Sinai to the Eastern sphere, at least vicariously via the USSR's friend and clinet, Egypt. Consequently, the dispatch of Soviet combat personnel to Egypt and its gradual probe toward the canal and, possibly beyond it, was not at all "adventuristic" as the more cautious elements in the Kremlin had intimated. This, no doubt, was the reason why Moscow initially encouraged the United States to go ahead with the Rogers plan, although the USSR subsequently reneged because Arab leaders believed that even better conditions could be extracted from Washington. At any rate, it was probably not coincidental that, in November 1969, a bare week or two after the U.S. plan was submitted to Moscow, Soviet representatives informed their Warsaw Pact allies that the recommendation for combat intervention in Egypt had become Soviet policy—in other words, that the Soviet leadership had finally decided in favor of this move and that the forward faction in the Kremlin had overcome the objections of its more moderate opponents. The chronology strongly suggests a cause-and-effect relationship between the transmission of the Rogers Plan in October and the signs in November that ι decision in principle had now been reached.

IMPLEMENTING MILITARY INTERVENTION

At this point, no doubt, planning and essential preparations for the complex logistic and other aspects of the operation were under way in the Soviet Union, and it remained only to meet some of the caveats and misgivings of the Kremlin "opposition" concerning Egypt's future conduct and reliability. With this aim in mind, the Soviet leadership was surely pleased to take advantage of Nasser's military and political plight at the end of 1969, when his war of attrition had turned against him, by suggesting that, if he came to Moscow, he might find the USSR willing to rescue him. This is a far more likely explanation of his secret Moscow visit in January 1970 than the version that he decided, on his own initiative, to fly suddenly to the Soviet capital and ask for help. First, it would have been quite sufficient to transmit a personal letter via the Soviet ambassador; and, in the second place, no one, including friends and clients, drops in uninvited on the Kremlin. The reason why the Soviet leaders may have preferred to bring the Egyptian president secretly to Moscow was to confront him, almost alone, with the full, united, and awesome pressure of the Kremlin, at a time when he was in dire need of Soviet help.

It seems highly likely that Nasser—almost certainly unaware that the USSR, independently of his requests, had decided in principle to send combat elements to Egypt—was presented with a Soviet "offer" of military forces, contingent on certain undertakings from the Egyptian side. Some such understanding was presumably reached during the Moscow visit, whether formally or informally, dealing with Egypt's future policies vis-a-vis the USSR and the United States, and with special military security provisions concerning not only the Soviet combat elements to be dispatched shortly to Egypt but also the facilities already granted there to Soviet naval and air units patrolling the Mediterranean. In this way, the interventionists in the Kremlin apparently were able to meet at least some of the conditions their opponents had laid down for any Soviet combat role in Egypt. By the time Nasser returned home, Moscow's planning and preparations for the new Soviet military presence in Egypt were well advanced, having proceeded for at least three months; consequently, the first Soviet combat elements arrived within a comparatively short period, by February 1970.

THE DECISION TO WITHDRAW

What happened between February 1970, when the "forward" group in the Kremlin seemed to have won all its internal battles against opponents of Soviet intervention and July 1972, when everything apparently turned sour and Soviet combat elements were suddenly withdrawn, amid public signs of conflict between Cairo and Moscow?

The simple fact is that the anti-interventionists in the Kremlin were proven, during this period, to have been right in almost all of their contentions.*

U.S. and Israeli Reactions

Precisely as the more cautious group of Soviet leaders had feared, the United States, despite appearances during February-September 1970, turned out to be very far from a "paper tiger."

---

*The events described here are not presented in chronological order, but rather are grouped according to topic.

After prolonged indecision following the first establishment of the Soviet combat crews in the Nile valley, which lasted even for a short while after these crews pushed forward toward the banks of the Suez Canal, the Nixon-Kissinger team suddenly implemented its new policy of "planned U.S. unpredictability" by pouring sophisticated planes, electronic equipment, drones, and artillery into Israel to balance out the new factor on the other side of the canal. This left little doubt about U.S. unwillingness to permit a Soviet combat role in assisting Egypt to cross the canal and confounded earlier Soviet expectations.

If Moscow still harbored any uncertainty about U.S. determination in the Eastern Mediterranean, it was rapidly dispelled by the reaction of the White House to Soviet behavior during the Jordanian civil war. While King Hussein and his Arab Legion were taking action against the extremist elements that had almost seized power in Jordan, Syrian armor, organized, trained, and, perhaps, even accompanied by Soviet military experts, poured across the Jordanian frontier to assist the king's adversaries. The U.S. Sixth Fleet rapidly moved in the general direction of the Syrian-Lebanese-Israeli coast, and Washington signaled Moscow that, unless Syrian armor were withdrawn immediately, not only might there be action by the Sixth Fleet aircraft, but the U.S. Government would prove quite understanding if Israel decided to unleash its air force to prevent a hostile Syria from establishing a puppet regime on the Jordan. Moscow apparently understood the message, since the Syrians immediately turned around and retreated across the frontier.

The United States had thus shown that it was a far more formidable adversary in the region than the Kremlin's forward group had believed in 1969, and the Soviet antiinterventionists were proven equally correct with regard to the possible complications and dangers of tangling with the Israelis. When Soviet interceptor and missile elements first appeared in the Nile Delta, Israeli Defense Minister Moshe Dayan warned that Israel was prepared to accept the fait accompli, as long as these elements did not move much closer to the canal bank. This was apparently misconstrued in Moscow as evidence that the Israelis were shrinking away from confrontation. Consequently, Soviet interceptor crews, speaking Russian over open communication channels (as a warning to the Israelis that they were not dealing with Egyptians), were sent to fly missions right over the canal. To the amazement of the Kremlin interventionist faction, the Israelis promptly ambushed these advanced Soviet MIG-21 planes and shot down four or five of them. This was but an earnest of Israeli determination when confronted by a real threat, precisely as the more cautious Kremlin group is believed to have warned in the summer of 1969.

The scenario ascribed to the Soviet antiinterventionist faction thus turned out to be all too real: A Soviet combat buildup of the relatively modest proportions first planned would fail to frighten off the Israelis but would constitute enough of a threat to provoke some fairly desperate measures on their part. Therefore, if the Soviet intervention was to have worthwhile results, it would require a massive Soviet buildup in Egypt. However, the U.S. Government had also demonstrated that it was far less impotent and far more unpredictable than the Kremlin interventionists had originally believed, so that a really large-scale Soviet influx into the area might provoke Washington into cutting the slender lifeline linking the Soviet presence in Egypt to its home base via NATO territory. Under these circumstances, it was evident that the Soviet combat elements in Egypt could not become engaged in an offensive across the Suez Canal—that is, could not really help USSR's client militarily, but, at most, could assist Egypt in exerting politico-psychological pressure upon Israel and the United States.

Events in Egypt

With regard to Egypt, too, the more pessimistic, antiinterventionist Soviet view was gradually borne out by events. President Nasser died, and, with him, whatever mutual understanding may have been reached during his Moscow visit. A struggle for succession broke out in Cairo, during which both the contending parties expected Soviet support and were angry when the Kremlin adopted a cautious, neutralist course—President Sadat, because he regarded himself as Nasser's legitimate successor, and Ali Sabri because he saw himself as politically closer to the USSR. After winning, Sadat showed his displeasure with Moscow for its attitude during the struggle and for the communist role during the upheaval in neighboring Sudan, by resorting to the old game of playing off the great powers against one another. He appeared unwilling to implement the commitments Moscow believed it had extracted from his predecessor in January of 1970.

It was to reinforce these commitments and to place Soviet-Egyptian relations on a more permanent footing, irrespective of internal changes in Cairo, that Soviet President Podgorny suddenly appeared in the Egyptian capital in 1971, pulled out of his pocket a prepared text, and told the Egyptians that they would have to sign on the dotted line if they expected sustained Soviet support. President Sadat did sign what became known as the Soviet-Egyptian Treaty of Friendship, but each side chose to interpret it differently. The Egyptian president claimed subsequently that he understood the USSR to have promised certain "offensive" weapons and Soviet direct combat support during a crossing of the canal. The Soviet leaders apparently retorted that the Soviet Union had delivered all of the offensive weapons it could spare and that Egypt was capable of absorbing and utilizing; as for direct combat support in an offensive across the canal, the Kremlin, following its chastening experiences with unexpected U.S. and Israeli toughness, must have informed Cairo that direct involvement of Soviet military elements in such a crossing was now out of the question.

The Moscow Summit Meeting

As long as this Soviet decision remained unknown in the West, as Sadat understood it would remain, the continued Soviet military presence in Egypt was still useful to him, since it constituted a politico-psychological means of pressure on both the United States and Israel. However, even this aspect soon lost its utility when, in the spring of 1972, the Soviet leadership, which had already decided in principle not to intervene directly in the Eastern Mediterranean, felt that it might as well "sell" this decision to Washington in return for certain U.S. concessions. During the Moscow Summit of 1972, therefore, the Kremlin as much as told the U.S. president that it was prepared to undertake to abstain from such interventionist action—as a special "favor" to its guests, of course—if in return it obtained some quid pro quo. Both sides left the summit with a clear impression that such a deal had been made.

From the Egyptian point of view, the Soviet military presence became useless at this point: If both the U.S. and Israel knew that the Soviet forces would not engage in combat their presence could not be exploited even as a politico-psychological threat; moreover, Sadat was bound to, and did, regard Soviet behavior at the summit as a breach of the Soviet-Egyptian Friendship Treaty, which had been interpreted by both parties as an undertaking not to make deals about each other's vital interests behind each other's backs. The Soviet leadership was aware of this aspect, but was unconcerned inasmuch as it was quite prepared to withdraw its combat elements now that their utility was exhausted and that the original plans for the employment of these elements had been proven infeasible.

The Soviets Withdraw

Doubtless, developments involving the United States, Israel, and Egypt had demonstrated that the misgivings and warnings apparently voiced by the antiinterventionists in the Kremlin during the 1969 debate (and duly recorded, in all likelihood, in various official minutes) had been amply warranted. As a result, sometime in

1971-72, the original supporters of the forward move in the eastern Mediterranean must have lost strength, until the Kremlin's decision was finally reversed. This undoubtedly became clear to the Egyptians sometime after the Moscow summit in the spring of 1972, and Sadat was merely "saving face" when he anticipated the obvious next Soviet move by asking the Soviet combat elements to leave. They left, not because they had been requested to go, but because the Kremlin must have already decided weeks earlier that this should be done in the near future.

Protagonists in the Kremlin Debate

In a very tentative and speculative way, partly by means of content analysis* an attempt can be made to establish some of the protagonists in the Kremlin debate.

There is little doubt that Brezhnev originally favored intervention, and although he reversed himself when it had become tactically necessary, he probably did so reluctantly and apparently under pressure. Pyotr Y. Shelest seems to have supported him, although it is not clear whether he participated in the original "debate." His subsequent purge was due to other causes. Suslov appears to have been very skeptical about the proposed military move and is believed to have had consistent reservations about Soviet support for third-world military dictatorships, since it has usually (and especially in Egypt) entailed Soviet sacrifice of the local communist parties—Suslov's own "constituents." Suslov's aide, Ponomarev, has made no bones about his feelings on this matter. Podgorny seems to have joined the "anti-interventionists" at times when the whole venture appeared to be going sour, although he may not have been consistent throughout the whole period. Aleksandr Shelepin may or may not have participated in the original "debate," but he seems to have been sniping fairly persistently at Brezhnev's Middle East policy, if the organ Trud is an indicator of his views. Shelepin's rivalry with Brezhnev is no secret; he had been prominent in conducting Soviet Middle East policy in 1964-65 and then was demoted gradually on all sectors by Brezhnev. There are no indications concerning the views of other members of the leadership.

Brezhnev's own part in this affair seems to demonstrate that he is far from being a "dove," or consistent supporter of detente, as it has become fashionable in the West to claim. He may have reversed

---

*Unfortunately, documentation would be far too bulky, requiring lengthy and tedious interpretation, to be cited here. Moreover, much of this particular type of content analysis has been done by others.

himself in time not to suffer serious political damage, but the episode certainly did not prove helpful to his career and ambitions. Given his well-known persistence, it should have been clear that he would not accept this reversal as the final word, but would attempt to achieve his goals in the region in less direct and overt, and therefore less risky, ways.

A NEW FORWARD POLICY

In retrospect, it would seem that Brezhnev found an opening for the pursuit of such a new, slightly more circumspect, but nevertheless forward and offensive policy, in the Soviet-U.S. "deal" made at the 1972 Moscow summit (which found expression in the joint statement published by the superpowers at the conclusion of the summit). Brezhnev could, and probably did, tell his colleagues in the months following the summit that all the United States had a right to expect after the 1972 meeting was that, in the Middle East, Soviet forces would not be used directly in a combat capacity. He knew, of course (since he and other Soviet officials had done their best to foster U.S. illusions by discreet nods and smiles on appropriate occasions), that his U.S. guests had left Moscow with the distinct impression that Soviet promises not to take unilateral advantage of crisis situations in regions of conflict implied much more. To be precise, Washington now believed that Moscow would not only refrain from direct Soviet combat intervention in areas like the eastern Mediterranean rim but would (1) ensure that other communist states did likewise, (2) discourage USSR's Arab clients from resorting to force, by curtailing rather than escalating Soviet arms deliveries, as well as by employing the required amount of diplomatic pressure, (3) inform the United States immediately of any evidence that these clients were bent upon war, Soviet efforts notwithstanding, and (4) collaborate promptly with Washington, both directly and in the UN Security Council, to prevent, or immediately terminate, armed incursions upon neighboring states by these Soviet clients.

However, Brezhnev presumably argued in the Kremlin that, given an appropriate occasion—such as a U.S. crisis, domestic or other—it would be entirely possible for Moscow to ignore the outer parameters of Washington's understanding of the 1972 summit deal; as long as direct Soviet combat intervention did not actually take place in the Middle East, the U.S. administration probably would be too eager to salvage something from its cherished new achievement of detente to permit itself or others to admit that the USSR was violating most of the spirit and some of the letter of the 1972 superpower "declaration of intent."

It appears that Brezhnev and his forward faction assuaged the misgivings in the Soviet leadership by pointing out the following: (1) a major concession already had been made to the "moderates" in the Kremlin by assuring the U.S. Government that the USSR would eschew direct intervention and by actually withdrawing sizable Soviet combat elements from Egypt; (2) the proposal now was merely to return to Moscow's pre-1969-72 Middle East policy, but to pursue it with greater efficiency, refinement, and sophistication so as to recoup Soviet prestige losses in that part of the world.

BREZHNEV'S PLAN

Brezhnev's new plan, as submitted to his colleagues, seems to have consisted of the following interconnected elements:

1. A further Soviet buildup of the forces of the "progressive" Arab regimes—Egypt, Syria, Iraq, Algeria—starting in the fall of 1972 and escalating gradually during the subsequent year. They would be supplied with highly sophisticated weapons, which the USSR had been unable to spare previously and which would extend even to items not previously given to noncommunist states, but the weapons would be accompanied this time by intensive training of the most thorough kind.

2. As extra insurance against renewed failure or incompetence on the part of the Middle East recipients, some of the most complex new hardware would be handled in combat not by Arab or Soviet or Warsaw Pact personnel, but by military elements from other communist countries, whose death or capture would not involve a NATO-Warsaw Pact confrontation.

3. Most remaining Soviet military advisors would be withdrawn in a demonstrative fashion once the USSR's Middle East clients were ready for combat and had decided to proceed to war, as an "alibi" to convince Washington that Moscow at least had not defaulted on its basic commitment to eschew direct combat involvement.

4. Moscow's clients then would go into offensive deployment, exploiting an occasion when the United States was distracted by a domestic or other crisis, and the Israelis would be forced to mobilize their citizen army, a costly operation that they could not sustain for long or repeat continuously.

5. Once the Israelis had mobilized, their opponents would back off but would return to offensive dispositions as soon as the Israeli alert was called off and the whole maneuver would be replayed until the Israelis, because of economic drain or because their vigilance gradually had become dulled, were caught responding tardily.

6. Moscow's clients would thus, at the least, achieve a tactical surprise, which if it did not suffice for them to gain a final and

decisive victory, would probably be enough to drag Israel into a relatively long, defensive struggle, the attrition of which might prove too much for a small state.

7. To ensure its clients "longer breath" than Israel during such an attrition period, the USSR would rapidly launch a massive air-and-sea lift to its Middle East friends.

8. As long as fighting favored its clients sufficiently for them to seize and hold ground, and to recoup prestige, the USSR would sabotage U.S. efforts, at the U.N. and elsewhere, to terminate the bloodshed.

9. The moment the war turned against its clients, the Kremlin would insist on an immediate ceasefire in place and would embarrass Washington into agreeing to it by suddenly offering at least partial support for the kind of proposals Washington undoubtedly would have made earlier, at the outbreak of war.

10. If the Israelis demurred at being thus robbed of the fruits of victory and insisted that the invading forces be thrown back at least to the preconflict lines, Soviet leaders could always dramatically go through the motions of preparing the dispatch of an intervention force to the area; at which point Washington no doubt would attempt to deter Moscow, but, at the same time would be only too eager to avoid an actual confrontation and, as a "compromise," would agree jointly with the USSR to impose a ceasefire upon the combatants.

11. Such a naked demonstration of Soviet determination and might would have the additional advantage of making Cairo and Damascus say "Thank you, Moscow" rather than "Thank you, Moscow and Washington" for being rescued in time. Moscow's clients would end up with some symbolically important territorial gains and would have caused Israel painful losses, and Soviet prestige would be triumphantly restored—all this without real danger of nuclear confrontation with the United States since Soviet combat forces would not actually be involved. Above all, Washington, frightened by this dramatic orchestration of the Arab-Soviet claim that the Middle East was a "tinderbox," would pressure Israel into a basically pro-Arab settlement to avoid dangers. This would constitute substantial proof that it paid to be a client of Moscow rather than of Washington.

It is not too difficult, ex post facto, to reconstruct these basic elements of Brezhnev's scenario, and it appears unlikely that his opponents in Soviet leadership could have mustered as much political ammunition against it as they did against the far more risky and adventurous intervention plan that Brezhnev initiated in 1969-70.

Various parts of this scheme were, indeed, implemented during the year or so that preceded the "October 1973 war" and during subsequent weeks:

1. The four "progressive" Soviet Middle Eastern clients, whose forces and weapons were to be wholly or partially committed against Israel, received during this rather brief period of 12 or 13 months the following weapons: some 450-500 additional medium, heavy, and amphibious tanks (including a sizable number of T-62, the main Warsaw Pact tank never previously* sent outside the bloc); more than 100 additional self-propelled guns; some 1,400 additional armored personnel carriers; about 1,200 additional medium and heavy field guns and howitzers; several hundred additional antiaircraft guns; at least 150 additional surface-to-surface missiles (including FROG 3 and FROG 7, and, for the first time outside the Warsaw Pact, the SAMLET and SCUD); some 30-35 additional naval units (including more Komar and Osa missile boats); some 350-400 additional warplanes (including Tu-22 bombers, sent for the first time to countries outside the Warsaw Pact, more Tu-16 bombers, a sophisticated version of the MIG-21,† additional Su-7, some large transport planes and many new helicopters, including Mi-6 and Mi-8); and, finally, enough additional SAMs to equip some 35-40 new SAM sites not only more SAM-2 and SAM-3, but at least 120 of the mobile and highly sophisticated SAM-6 batteries,‡ as well as the SAM-7 Strela, which can be employed by infantry against warplanes and which had also never previously been handed over to noncommunist forces). These deliveries did not taper off but escalated as October 1973 approached; for instance, the Tu-22 bombers and massive supplies of SAM-6, as well as surface-to-surface missiles, arrived just weeks before the outbreak of the war. Moreover, major Soviet resupplies of munitions arrived by sea at the very onset of hostilities; in view of the length of the sea journey, this indicates Soviet collusion at least 2 weeks before the war. Thus, there can be little doubt that the USSR was not only aware of the projected date of military operations, but approved and supported them, rather than attempting to restrain its clients.

---

*Except for a small advance shipment dispatched to Egypt for training purposes earlier in 1972.

†These aircraft had been entrusted only to Soviet combat personnel prior to July 1972 and were then withdrawn; but subsequently they were dispatched back to Egypt, although the lack of properly trained Egyptian pilots necessitated that they be flown in many instances by North Koreans.

‡Of the SAM-6, there had been a much smaller number, entrusted only to Soviet combat personnel, in Egypt prior to July 1972; later large quantities were dispatched to Egypt and Syria, although their operation was usually supervised by North Vietnamese.

2. The most complex and sophisticated of the Soviet systems were handed over some weeks prior to hostilities to non-Warsaw Pact communist personnel—North Korean, in the case of the latest versions of the MiG-21, and North Vietnamese (trained during the air defense of Hanoi and Haiphong), in the case of some SAM-3 and many of the SAM-6 batteries. It is inconceivable that this could have occurred without Soviet approval, indeed without the Kremlin's active help and initiative.

3. It was after this development that some of the Soviet advisers were flown out of the Middle East, rather openly and visibly, for political reasons that have already been explained, but reinforcing the evidence that the USSR knew precisely of the projected time of hostilities.

4. Once its clients went into offensive deployment, Moscow not only did nothing to dissuade them but did its best to help confuse foreign intelligence into believing that mere political and psychological maneuvers, rather than military operations, were contemplated.

5. Almost immediately after the outbreak of fighting, the USSR prepared and then implemented its massive air-and-sea lift of arms and supplies to its clients, days before the United States responded by reinforcing the rapidly diminishing Israeli supplies.

6. As long as Egyptian forces were advancing across the canal into Sinai and the Syrians into the Golan Heights, Moscow insouciantly sabotaged all U.S. attempts to activate the UN Security Council and rebuffed other U.S. approaches to bring hostilities to an end.

7. Once it was apparent that the Israelis had regained the initiative and were about to crush their adversaries, the Kremlin summoned Secretary of State Henry Kissinger to Moscow to initiate joint Soviet-U.S. moves to stop the Israeli Army before it could complete its task and while a remnant still remained of the original Egyptian bridgeheads east of the Suez Canal.*

---

*As part of this agreement, and to reward, as it were, the sudden Soviet willingness to "cooperate," the United States was persuaded to approve the imposition, through the UN Security Council, of an almost immediate ceasefire "in place"—despite the fact that the combatants were almost inextricably intertwined, with several enclaves east and west of the Suez Canal, that there was no immediate way of separating them, that there was a glaring lack of reference to any machinery for establishing just where "in place" was. Indeed, all the elements ensuring further bloodshed and dispute were present. Thus, the invading force, at least on the Suez Canal front, was awarded some of the territory it had seized in the initial days of the war. Moreover, the U.S. Government consented to the omission in the proposed new

8. The resulting UN Security Council ceasefire order was followed, under the circumstances, by entirely foreseeable chaos, since it came at the very climax of a military campaign rather than after its natural culmination. Hurried attempts to achieve last-minute advantages were bound to cause escalation rather than diminution of fighting. The Egyptian Third Army, encircled at the southeastern end of the Suez Canal, and equipped apparently with highly sophisticated and sensitive Soviet weapons, accompanied perhaps by supervisory personnel from communist countries, tried desperately to break out, while the Israelis attempted to strengthen their ring of encirclement. It was this situation that Moscow utilized for one of its most dramatic maneuvers. A mere few days after the Moscow "Little Summit" and the joint U.S.-Soviet resolution at the Security Council, the USSR placed six of its seven airborne divisions on emergency alert and dispatched a crudely threatening note to Washington, which warned the U.S. Government to pressure Israel into permitting the trapped Egyptians on the Israeli side of the Suez Canal to reopen their link with Egyptian forces west of the canal—otherwise Soviet troops would have no choice but to come in and crush Israel. This was an obvious psychological maneuver. Had Moscow seriously intended to move in, there would have been no preliminary note and no other visible signals, until Soviet units actually descended from the skies. The note produced the expected U.S. deterrent counterploy of a stern reply and an American military counteralert. However, it also led to the equally expected U.S. attempt to salvage some remnant of detente by offering Moscow compromises in the form of considerable U.S. pressure on the Israelis to permit vital supplies to pass to the surrounded Egyptians, as well as American acceptance of a renewed UN Emergency Force* that, late in October 1973, could serve only as a screen behind which the retreating Egyptians could find shelter and reconsolidate their forces, instead of confronting their Israeli adversaries in truly direct and untrammeled negotiations, free of preconditions.

9. Brezhnev clearly expected the result to be a major recouping of Soviet prestige, which had been seriously weakened in 1967 and again in 1972, and a strengthening of his own power and prestige within

---

Security Council Resolution of any mention of an exchange of prisoners of war, despite the fact that the Israelis had always insisted that this be an integral part of cessation of hostilities and that Kissinger himself had made a final political settlement concerning Vietnam hinge almost entirely on this question.

*The UNEF had been an anathema to the Israelis ever since its pitiful performance in withdrawing precipitously in May 1967 at Egypt's request, thus rendering hostilities inevitable.

the Kremlin "collective," where the developments of 1969-72 had proved less than helpful to him. The Arab world seemed bound to regard new U.S. concessions (to all appearances at the expense of America's friends and, indirectly, perhaps of her own security interests) as a sign of Western weakness, since they followed in the wake of a dramatic Soviet threat. Moscow had persistently told its Middle East friends ever since 1967 that, if they only proved patient and followed the Soviet lead, they would find that the USSR would not only aid them directly to restore their self-confidence but would successfully pressure the United States and, through it, the Israelis, into far-reaching concessions.

IMPLICATIONS FOR THE FUTURE

At the end of October 1973, it appeared that Brezhnev's calculations might be paying off, although of course time alone would tell. If, by some chance, his forward policy in the Middle East should prove a failure once more, his more cautious and noninterventionist rivals in the Kremlin, disparate though their factions and motivations might be, could be expected to combine to clip his wings and to lead Soviet policy toward the Eastern Mediterranean, at least temporarily, into quieter and less disturbing channels. In that eventuality, the West might well discover that it had "bet on the wrong horse" in the Kremlin. However in November 1973 the forward faction in Moscow must be generally satisfied with the results of its policies. Given the record of that faction, including Brezhnev, as analyzed in this chapter, this development should give rise to serious misgivings in the West, rather than engendering premature feelings of relief.*

---

*The author is well aware of the "speculative" flavor of portions of the reconstruction of developments, but he believes the exercise is useful nonetheless, at least by posing serious questions that previously have been bypassed or overlooked, and, hopefully, by shedding some light on the general nature of the Soviet policy and decision-making process.

CHAPTER

12

THE SOVIET UNION AND SADAT'S EGYPT
Robert O. Freedman

At the time of Nasser's death in September 1970, the Soviet Union had reached the pinnacle of its influence in Egypt and throughout the Middle East.[1] The Russians had acquired air and naval bases in Egypt that greatly enhanced their military position in the eastern Mediterranean vis-a-vis the United States, and had obtained port rights in Syria, Yemen, South Yemen, the Sudan, and Iraq, which gave increased Soviet access to the Red Sea, the Persian Gulf, and the Indian Ocean.

The Soviet position in the Middle East had, however, become very expensive to maintain. The Russians had assumed the role of military supplier and financier of economically weak and politically unstable Arab regimes, while at the same time endeavoring to purchase influence in Turkey and Iran. Nor was this position that the Russians had achieved without its risks, since there were a number of Arabs who clearly hoped to involve the Russians in a war against Israel, irrespective of the international consequences. Indeed, one of the reasons for Soviet acceptance of the U.S. ceasefire initiative in July 1970—two months before Nasser died—may well have been a desire to cool its rapidly escalating conflict with Israel, which might otherwise have soon involved the United States. Cooperation with the United States, however, only inflamed anti-Soviet sentiment in Syria and Iraq, Russian clients openly opposed to the ceasefire,

The reemergence of the United States as an active factor in Middle Eastern politics made such Arab disunity even more serious a problem for the Soviets. The "Rogers plan," first announced on

---

The first part of this chapter is a shortened version of an article that was originally published in the Naval War College Review, November-December 1973 and is reprinted with permission.

9 December 1969, was a significant factor in preventing the Arab summit conference, convened at Rabat, Morocco, a few days later, from issuing an anti-American statement as had been rumored it would in early December.[2] The August 1970 ceasefire between Israel, Egypt, and Jordan was an American initiative, and while violated by Egypt, it nonetheless seemed to create a climate for substantive peace negotiations. The strong U.S. support for King Hussein's regime when Syrian tanks invaded Jordan in September 1970, during Hussein's crackdown on the guerrillas, helped restore a great deal of U.S. influence in Jordan and Lebanon as well, and the Soviet Union's disinclination to support one of its erstwhile clients, Syria, against a client of the United States, Jordan, was not lost on the Arab world.

Thus the specter of rising U.S. influence in the Arab world, together with the increasing disunity among the Soviet Union's Arab "clients," dominated Russian thinking in the Middle East when Gamal Abdul Nasser, the man who had been the linchpin of Soviet strategy in the region, departed from the scene.[3]

INITIAL SOVIET POLICY

From the Soviet point of view, the most serious aspect of Nasser's death was that it removed the one man in Egypt so obsessed by his humiliation at the hands of the Israelis that he was willing to give up considerable Egyptian sovereignty in an effort to get revenge. Russia feared that Nasser's successor, not bridled with his mistakes, might prove to be considerably more independent. The presence at Nasser's funeral of a senior U.S. official, Elliot Richardson, did little to ally the Russians' concern.[4]

In addition to a succession crisis in Egypt, the Russians faced government shakeups in Iraq and Syria, in the latter instance involving the ouster of a pro-Russian group of Ba'athist leaders.[5] These domestic changes in their Arab clients led the Soviets to adopt a "watchful waiting" policy. Thus, while attempting to consolidate relations with the new regimes, Russia supported the 22 November UN resolution on the Arab-Israeli crisis. Still, the Russians were far from inactive in the area. Cairo-Moscow traffic was heavy, highlighted by the January visit of Soviet President N. Podgorny to celebrate the opening of the Aswan High Dam, and Syrian leader Hafez al-Asad visited the Soviet capital in February.

As stability returned to the Arab world, the Soviet Union encouraged efforts at Arab unification. A proposed federation of Egypt, Libya, and the Sudan was launched in October 1970, to which Syria adhered on 27 November 1970. The Russians moved quickly to throw their support behind the federation and Syria's decision to join it, since conflict between Syria and Egypt in the past had been one of the main obstacles to Soviet backed "anti-imperialist" Arab unity. At the same time, the Russians were careful to maintain good relations with Iraq, which was highly critical of the proposed Arab federation. Thus, while the Russians agreed to give Egypt a $415 million loan on 16 March 1971, the Iraqis received a $224 million loan on 8 April 1971 for the construction of an oil refinery and two pipelines. The Iraqi loan was not entirely altruistic, however, since it was to be repaid by oil—a commodity the Russians were beginning to find more and more expensive to produce at home. The Soviet loan also served to strengthen the hand of the Iraqi leaders in their bargaining with the Western oil companies.

THE OUSTER OF ALI SABRY AND THE ABORTIVE COUP IN THE SUDAN

As the Soviet-supported Arab federation reached its final stages with a Cairo summit meeting of the member nations on 12 April 1971, serious difficulties arose. One faction of the Sudanese Communist Party came out strongly against the federation, and Sudanese Premier Ja'afar Nimeri had to leave the unity talks to go to Moscow in an effort to get Soviet support in pressuring the Sudanese communists into giving up their opposition. The Russians, however, were either unwilling or unable to bring effective pressure to bear. The end result was that the Sudan was not a signatory to the preliminary agreement of 17 April 1971.

More serious opposition appeared in Egypt. Seizing on the Arab federation as an issue to challenge Sadat, Ali Sabry moved to oust the Egyptian president. Sadat proved too skillful a politician, however, and succeeded in removing Sabry from his post as vice-president on 2 May 1971, three days before the arrival of William Rogers in the first official visit of a U.S. secretary of State to the Egyptian capital since 1953. Consequently, the removal of Sabry, perhaps the most influential supporter of close Soviet-Egyptian relations, was interpreted as a gesture suggesting U.S.-Egyptian rapprochement, if the United States were to bring the necessary pressure on Israel. The follow-up purge of Sadat's other major opponents on 14 May 1971 (including Shaari Gomaa, the Egyptian secret police chief, also rumored to be close to the USSR), intensified speculation in this direction.[6]

In addition to possibly signaling to the United States for an improvement of relations, the purges strengthened Sadat's position vis-a-vis the Soviet Union by making it far more difficult for the Russians to factionalize against him within the Egyptian leadership.

The Russians, though disturbed,[7] proved powerless to undo these changes. Shortly after the second purge, however, Podgorny returned to Egypt and subsequently signed the now famous Soviet-Egyptian treaty. Several commentators assumed that the Russians had thereby spread the mantle of the "Brezhnev doctrine" over Egypt, but the impact of the treaty was far less significant. Egypt merely committed herself to continue regular consultation with the Russians and to eschew any alliance hostile to Moscow. For their part, the Soviets limited their military involvement to "assistance in the training of U.A.R. military personnel and in mastering the armaments and equipment supplied to the United Arab Republic with a view to strengthening its capacity to eliminate the consequences of aggression"[8]—hardly the delineation of master and satellite.

Perhaps the greatest importance of the treaty to the Russians was as a demonstration that the United States had failed in its attempts to "drive a wedge between Egypt and the USSR."[9] Yet the Russians were clearly not satisfied with the progress of events in Egypt or the reliability of Sadat.[10] The sharp limits to Soviet influence in Egypt were revealed the following month by the tumultuous events in the Sudan. The regime of Ja'afar Nimeri, which came to power in a military coup d'etat in May 1969, had received large amounts of Soviet economic and military assistance and appeared to many Western observers to be "lost" to the USSR. However, the opposition of the Sudanese communists to Nimeri's plan to join the Arab federation had become a serious challenge to his power, and on 25 May 1971 Nimeri cracked down hard on the communists. He arrested 70 communist leaders, including nearly all the central committee, and dissolved the unions, which served as the communists' bases of power.[11] While Nimeri was careful to pledge that such actions would not harm Soviet-Sudanese friendship, it is clear that the Russians were not in the least unhappy when Nimeri was ousted on 19 July by a group of army officers.[12]

But the Russians received a severe shock only three days later when, with the aid of Libya and Egypt, Nimeri was able to return to power. One of the Sudanese leader's first actions was to order the execution of the leading communists in the Sudan. The Russians at first adopted a relatively moderate stance to these events, condemning the crackdown on the communists and the plans to execute two key communist leaders. But when, despite Soviet protest, Party Secretary Abdel Mahgoub was executed, on 28 July, the tone became harsher. Pravda commented that " . . . the Soviet people are not indifferent to the fate of fighters against imperialism and for democracy and social progress" and hinted that Soviet-Sudanese relations could not remain close.[13]

Nonetheless, there was no official termination of Soviet economic or military aid to the Sudan. Nor were diplomatic relations broken, although Nimeri recalled his ambassador from Moscow. Perhaps the most severe action that the Russians took at the time was to arrange a demonstration of about 200 Arab students outside the Sudanese Embassy in Moscow. This demonstration was directed against not only Nimeri, but also Anwar Sadat,[14] who had publicly praised Nimeri and denounced the Sudanese communists.[15] Sadat's defiance of the Russians was a clear indication that, treaty or no treaty, Soviet influence in Egypt was quite limited, and this in turn led to Soviet criticism of the Sadat regime.[16]

Relations between the Soviet Union and Egypt continued cool in September. In that month came the visit of Sir Alec Douglas-Home to Cairo, the first visit of a British Foreign Secretary to Egypt since the Suez war of 1956. Douglas-Home's visit to Cairo, which followed by only four months that of U.S. Secretary of State William Rogers, seemed to indicate another move to the West by Sadat's regime—a development not greeted with favor in Moscow, considering the enormous Soviet investment in Egypt.[17]

SADAT'S FRUITLESS JOURNEYS TO MOSCOW

As the date of Sadat's October 1971 trip to Moscow approached, Soviet-Egyptian relations seemed to have hit a new low.[18] While it was one of the goals of Sadat's trip to improve these relations, the primary issue, at least as seen from the Egyptian side, was more specific. Sadat had already committed himself to the thesis that 1971 was to be the "year of decision" in Egypt's conflict with Israel, and he appeared anxious to obtain Soviet support for military operations against the Israelis.[19]

The Russians, however, who almost had been drawn into a military confrontation with the United States in June 1967, were not willing to let themselves be further exploited. Thus, in the official Soviet description of the Moscow talks with Sadat, there were frequent references to "a spirit of frankness" and "exchanges of opinions"—indications that there were a number of disagreements. In his speech of 12 October, Sadat continued his theme that war was the only way to secure Israeli withdrawal and that he expected the Soviet Union to support Egypt in its time of need.[20] By contrast, Soviet President Podgorny emphasized the need for a peaceful solution to the Arab-Israeli conflict, and the joint communique issued at the end of the talks was a clear reflection of Soviet, not Egyptian, priorities. The most the Egyptians were able to extract from the discussions was a somewhat vague statement that the two sides "agreed on measures aimed on the further strengthening of Egypt's military might."[21] Sadat made a similarly unproductive trip to Moscow in February 1972.

The Soviet goal of effective influence in the Arab world had been complicated by the Indo-Pakistani war of 1971. The USSR's aid to Hindu India against Muslim Pakistan was unpopular in Egypt, although Sadat made no official comment. Mu'amar al-Qaddafi, the Islamic fundamentalist leader of Libya, however, openly denounced the Soviet role as "confirming the Soviet Union's imperialist designs in the area."[22] Russian popularity dipped to new lows within the Arab world, and a number of Arab newspapers, once sympathetic to the USSR, began openly to criticize Soviet policy.[23] In addition, Sadat's "year of decision" had passed without a war, and the Egyptian leader openly blamed the Soviet Union for lack of support in Egypt's confrontation with Israel.[24]

Sadat made still another visit to the Soviet capital in April, just before the Nixon-Brezhnev summit talks, which both Sadat and Israeli Prime Minister Golda Meir feared might lead to an imposed Middle East settlement injurious to their interests. According to Sadat, he told Brezhnev during this Moscow visit that Egypt would never agree to a limitation of arms shipments to the Middle East, to a continuation of the "no war-no peace" situation, or to the surrender of "one inch of Arab lands" in a peace imposed by the superpowers. Perhaps even more importantly, however, Sadat once again expressed his desire for advanced weapons, along with Soviet support for renewed hostilities against Israel.[25] The Russians, however, with more important global issues at stake, proved unwilling to sacrifice their relations with the United States and committed themselves to nothing more than "considering measures aimed at further increasing the military potential of the Egyptian Arab Republic."[26]

Far more to the point was the communiqué released after the Soviet-American summit conference, which reaffirmed the two superpowers' "support for a peaceful settlement in the Middle East in accordance with Security Council Resolution No. 242" and declared their willingness to play a role in bringing about a settlement in the Middle East "that would permit, in particular, consideration of further steps to bring about a military relaxation in the area."[27] This emphasis on a peaceful settlement in the Middle East was also prominent in Soviet press commentary.[28]

THE DECISION TO EXPEL THE RUSSIANS

Egyptian disenchantment with the Soviet Union was by no means confined to Sadat. On 4 April 1972 a number of prominent Egyptians

to the right of the political spectrum voiced such complaints in a
memorandum to the Egyptian president.[29] Sadat made this note public
in an interview with the Beirut daily Al-Hayat on 18 May, probably as
a trial balloon to gauge public opinion toward an anti-Russian shift
in Egyptian foreign policy. The Al-Hayat interview was followed in
June and early July by a series of editorials by the editor of the
Egyptian daily Al-Ahram, Hassanein Heikal, who went one step further
by asserting that the Soviet Union, just like Israel and the United
States, was actually profiting from the continuation of the "no war-no
peace" situation.[30]

The lack of Soviet support was compounded by a number of other
serious irritants in Soviet-Egyptian relations. Friction was increasing
between the Soviet military advisers and Egyptian officers, and Egyptian
Defense Minister Mohammed Sadek frequently complained to Sadat
about alleged slurs made by the Russian advisers as to the capability
of the officers and troops under his command. In addition, the Soviet
bases in Egypt had been declared "off limits" to Egyptians, even, on
occasion, to Sadat himself, and this revived unpleasant memories
of the situation that had occurred when the British controlled Egypt
only 20 years before.[31]

Another factor of considerable concern to Sadat during the
prolonged period of "no war-no peace" was that Egypt's position of
leadership in the Arab world, which had once been paramount under
Nasser, seemed to be slipping away. Thus, despite Sadat's bitter
denunciations of the United States in May and June 1972 because of
its support for Israel, the regime of North Yemen, once closely aligned
with Egypt, announced the restoration of diplomatic relations with
the United States on 2 July 1972. At the same time, Sudanese Premier
Ja'afar Nimeri, whom Sadat had helped to restore to power less than
a year before, spoke very warmly of U.S. aid to the war-ravaged
southern section of his country and reestablished diplomatic relations
three weeks later.[32]

Thus Sadat, beset by internal frustration and rising domestic
discontent and under increasing challenge in the Arab world, decided
on a dramatic action prior to the 20th anniversary celebration of the
Egyptian revolution. Following the failure of a final arms-seeking
trip by Egyptian Premier Aziz Sidky to Moscow on 14 July, Sadat
announced on 18 July 1972 the termination of the mission of the Soviet
military advisers and experts, and the placing of all military bases
in Egypt under Egyptian control. At the same time he called for a
Soviet-Egyptian meeting to work out a new relationship between the
two countries.[33]

There is little doubt that these moves were popular among
both the Egyptian masses and the officer corps. Yet a greater degree
of domestic popularity was clearly not the only motive for Sadat's

action. The Egyptian leader was seeking to regain freedom of action in foreign affairs and break out of the cul-de-sac that the relationship with the USSR had gotten Egypt into. His reasoning seemed to be that since the Soviet Union had been unable to get Israel to withdraw from the occupied territories by diplomatic means and was unwilling to expel her by force, Egypt would turn to the United States and Western Europe for assistance.

Despite the close U.S. tie to Israel, the Egyptians had not forgotten that it was primarily U.S. pressure that had forced the Israelis to withdraw from the Sinai in 1957. Indeed, Heikal had editorialized on 21 July in <u>Al-Ahram</u> that "no one can convince Egypt that the United States is incapable of bringing pressure on Israel."[34] High-ranking U.S. officials such as Henry Kissinger as well as President Nixon had made no secret of their desire to get the Russians out of Egypt and thereby weaken the entire Soviet strategic position in the eastern Mediterranean. The weakening of the Soviet presence in the Mediterranean would also benefit Western Europe, and Sadat may have hoped that the Europeans would reciprocate by bringing pressure on Israel by withholding Common Market tariff concessions then under negotiation, as well as by selling Egypt advanced weaponry.

In addition to courting the West, Sadat's new policy involved a move toward union with oil-rich Libya. On 23 July, only six days after the expulsion of the Russians, Libyan leader Qaddafi publicized his offer of a union of Egypt and Libya—something that had been under consideration since Sadat's unsuccessful trip to Moscow in February 1972. In fact, given the strongly anticommunist and anti-Soviet position of Qaddafi, it is quite conceivable that the expulsion of the Russians might have been the condition demanded by Qaddafi before the Egyptians could gain access to Libya's hard currency reserves, estimated by some Western sources as $3 billion.[35] With this money Sadat could afford advanced weapons on the Western market and need not depend on the Soviets. Also, the fact that the United States had major oil holdings in Libya would give Sadat a means of pressure against the United States to weaken its support of Israel.

The Soviet Union, of course, lost heavily by Sadat's decision to expel the Russian military forces from Egypt. Although they were now far less likely to get dragged into a war with the United States— and this fact must have sweetened the exodus somewhat—their strategic position in the Mediterranean was clearly weakened. Without the airfields in northern Egypt, they were unable to give air cover to the Soviet Mediterranean fleet. While the Russians regained the right to visit Egyptian ports, even this was contingent upon a modicum of Egyptian good-will, which could be used as a bargaining chip to assure the continued flow of Soviet economic aid or, at the minimum, the completion of aid projects already under way. A Soviet presence in the vulnerable Egyptian ports also served to deter an Israeli attack, although the possibility of an Egyptian-Israeli war now appeared so remote that on 11 August 1972 Israeli Defense Minister Moshe Dayan was quoted as saying that as a result of the Soviet exodus Israel could now reduce some of its forces and redeploy along the Suez Canal.[36]

The initial Soviet reaction to Sadat's expulsion decision was relatively mild, although as time went on Soviet-Egyptian relations continued to deteriorate and the Russian commentators became more explicit in their criticism of Egyptian policy.[37] Soviet-Egyptian relations worsened further following the Egyptian rejection of a note from Brezhnev to Sadat requesting a high-level meeting. On 19 August 1972 Sadat told the Egyptian People's Council that he had rejected the "language, contents, and type" of the message he had received from Brezhnev. The Egyptian leader further stated that the Soviet Union's refusal to supply the requested arms "aimed to drive us to desperation and the brink of surrender" but that Egypt would, God willing, obtain the needed arms elsewhere.[38] Two days later, Sadat blamed the Russians for not understanding Egyptian psychology and stated that the Western Europeans now owed Egypt a response to the "initiative" he had taken to help them.[39] This was followed by a war of words that broke out between Soviet and Egyptian newspapers in mid-August.[40]

THE EFFECT OF THE MUNICH MASSACRE

The downward spiral of Soviet-Egyptian relations was abruptly ended when a group of Palestinian terrorists killed 11 Israeli athletes at the Olympic Games in Munich and set off a chain of events that greatly upset the pattern of Egyptian diplomacy.

The immediate effect was to strike a major blow to Sadat's hopes in Western Europe and America. The deterioration of Egypt's relations with West Germany, her second leading trade partner (after the USSR) and a potential source of both economic and technical assistance, reached the point in mid-September that Egypt's new Foreign Minister, Hassan el-Zayyat, canceled a scheduled visit to West Germany, part of a planned tour of West European capitals in search of support against Israel. Zayyat did complete a trip to England, but his arrival was marred by the letter-bomb assassination of an Israeli agricultural attache, which inflamed British public opinion against the Arabs.[41]

Egyptian fears of reprisal probably hastened the pace of the Egyptian-Libyan union. On 18 September, Sadat and Qaddafi reached an agreement that proclaimed Cairo as the capital of the union and provided for a single government, a single political party, and a single president elected by popular vote.[42] Egypt was not in fact hit by any Israeli retaliatory strikes, possibly to avoid an Egyptian recall of the Russians. Nonetheless, Sadat was clearly discomfited by the events in Munich. With his attempts to win over Western Europe and the United States having come to naught and condemned both at home and throughout the Arab world for failing to protect Syria and Lebanon from Israeli attacks, Sadat decided to try to stabilize Egypt's relations with the USSR before they deteriorated any further.

Consequently, on 28 September 1972, the second anniversary of Nasser's death, Sadat delivered a major policy address in which he sought to regain some of the momentum in Middle Eastern events. He issued a call for the establishment of a Palestinian government in exile, officially rejected the U.S. proposal for an interim agreement and proximity talks, and, perhaps most important of all, changed his tone toward the Russians. The Egyptian leader declared that he had sent a letter to Brezhnev that was "friendly and cordial in spirit."[43] It was revealed only two days after Sadat's speech that Egyptian Premier Aziz Sidky would undertake a trip to the Soviet Union on 16 October.[44] Nonetheless, the tone in the government-controlled Egyptian press remained quite cool to the USSR until the very eve of Sidky's departure. Thus Sadat himself stated that a peaceful settlement as desired by the Russians meant "surrender to American and Israeli terms" and complained openly, "the Russians had become a burden to us. They would not fight and would give our enemy an excuse for seeking American support and assistance."[45]

Sadat's negative tone changed considerably on 15 October when he stated that Egypt would never have a "two-faced" foreign policy but would always value fully the friendship of the Soviet Union. The Egyptian leader called the Soviet-Egyptian friendship "strategic" rather than "tactical" and warned the United States that it would have to "pay a price" for its support of Israel.[46]

However enthusiastic Sadat may have been, the real accomplishments of Sidky's trip to Moscow were limited at best. In the first place, unlike his earlier trip in July, the Egyptian premier did not get to see Brezhnev but had to be satisfied with meeting Kosygin and Podgorny. Secondly, there was no mention of continued Soviet aid, either military or economic, in the full communique, which described the talks as having taken place "in an atmosphere of frankness and mutual understanding." About the only thing the Egyptians could point to from the talks (assuming that were no secret protocols) was a rather pro forma Russian pledge, frequently found in joint communiques, that the Russian leaders had accepted an invitation to come to Egypt, although no date was set for their visit.[47]

Upon Sidky's return to Egypt, a general debate developed in the top ranks of the Egyptian leadership about the proper relationship toward the USSR. On 25 October 1972 Sidky reported to a mixed Arab Socialist Union-Government meeting that the Russians had promised to resume aid to Egypt, although he did not mention precise quantities. Sadat followed with a speech telling the assembled delegates that "it was up to them" whether or not Egypt should continue to rely primarily on Soviet support, but cautioning that there was little hope in the foreseeable future of replacing the USSR as Egypt's principal supplier of arms. Sadat went on to say that if Egypt should choose continued cooperation with the Soviet Union, its scope would never return to the pre-18 July situation.[48]

The Egyptian leadership apparently decided on continued cooperation with the Russians, and on the very next day Defense Minister Sadek, one of the most anti-Russian of the Egyptian leaders, either was fired or resigned from his position. His ouster was followed by that of the navy commander, Rear Admiral Fahmy Abdel Rahman, another of the outspoken anti-Soviet Egyptian leaders. Sadek was replaced by Ahmed Ismail, Egypt's military intelligence director who, unlike Sadek, had neither alienated the Russians nor possessed sufficient popular appeal to pose a challenge to Sadat himself.[49]

Sadek's fall from the second most powerful position in Egypt gave rise to a great deal of speculation both in Egypt and abroad. Most commentators saw Sadek's ouster as the price demanded by the Russians for a resumption of military aid; the arrival of SAM-6 antiaircraft missiles together with Russian technicians soon after his "resignation" reinforced this belief.[50] But one Egyptian press report stated that Sadek had been dismissed for insubordination and failure to carry out Sadat's orders.[51] Whatever the true reason for Sadek's resignation, the Russians were clearly happy to witness the departure of the most outspokenly anti-Soviet leader in the Egyptian hierarchy. This was evidenced in the extensive coverage given to a speech by his successor, Ahmed Ismail. The new Egyptian Defense minister spoke warmly of Soviet economic and military aid to Egypt and stated that the USSR had fulfilled all the obligations it had pledged to Egypt. In addition, Ismail strongly attacked the United States for its aid to Israel and asserted that "nothing good" could be expected from the United States.[52]

Nonetheless, despite the warmth of the speech toward the Soviet Union, Egypt's new Defense minister also reportedly told Western diplomats, soon after taking office, that the "Egyptian Army Command will never again allow Russian advisors to get key command and advisory posts in the Egyptian armed forces"—a policy goal that Ismail evidently shared with Sadat.[53]

ARAB UNITY AND THE OIL WEAPON

At this point, Sadat made a major shift in his strategy. Having first sought Soviet and then U.S. support against Israel and having failed in both quests, Sadat, under great domestic and foreign pressure to go to war,* decided that the only solution for Egypt was to mobilize the capabilities of the Arab world—including its oil power—against Israel and its supporters. Thus in a major policy speech on 28 December Sadat stated that Egypt "realized the limits of Soviet aid" and that Egypt would take "new initiatives to make the battle a pan-Arab one."[54] The Soviet press commented favorably on Sadat's pan-Arab battle plan, and the Russians utilized the opportunity to remind the Arabs that the United States, Israel's main supporter, was becoming very vulnerable to the oil pressure Sadat had recommended.[55]

A series of events in February and March of 1973 exacerbated the Middle East situation. These included the Israeli downing of a Libyan airliner that had strayed over the Sinai, the assassination of two U.S. diplomats in the Sudan, and the U.S. announcement of the sale of an additional 48 jets to Israel. It was in this context that Sadat fired Premier Aziz Sidky and assumed the premiership himself, along with the post of military governor. In doing so, he stated that war could not be delayed and that Egypt's relations with the USSR had resumed a "correct, friendly pattern."[56]

While the Soviet leaders welcomed the further deterioration in Egyptian-American relations, there is some question as to how friendly they felt toward the Sadat regime. Indeed, in an effort to stabilize his domestic position prior to the battle with Israel, Sadat had embarked on a wholesale purge of the Arab Socialist Union, expelling a large number of leftist intellectuals including Lufti el-Khouli, the only Marxist on the Arab Socialist Union's central committee.[57] This was clearly a reversal of the "National Front" system the Russians had been urging on Syria and Egypt. In addition, Soviet comments on domestic developments in Egypt began to assume an increasingly negative tone as it appeared, through his encouragement of foreign investment and domestic capitalism, that Sadat was embarking on a program to restore Western-style capitalism to Egypt. Nonetheless, the Russians evidently decided not to let these negative factors stand in the way of improved Soviet-Egyptian relations, and by late March arms were again flowing in large quantities to Egypt.[58]

With Egypt now receiving a steady flow of Soviet arms, Sadat turned his attention to developing his relationship with Saudi Arabian

---

*There had been serious unrest in the Egyptian armed forces, and it had spread to the normally docile Egyptian Parliament as well as to the universities and to the media.

King Faisal, whose oil leverage over the United States was a critical factor in the Egyptian strategy against Israel. Faisal's willingness to use the oil weapon may have been partially due to the pressures on him generated by the escalating Arab-Israeli violence in the Middle East. By mid-April, he was threatening the United States that Saudi Arabia would not increase its oil production to meet U.S. needs, unless the United States modified its stand on Israel.[59] Following this warning, the United States, Britain, and France all scurried to sell Faisal modern weaponry, a development further underlining Saudi Arabia's growing importance in the Middle East and the West's growing vulnerability to the oil weapon.

In his May Day speech, Sadat hailed the Saudi warning to the United States as further proof of the growing Arab unity. Indeed, Sadat claimed in the speech that he now had Syrian, Kuwaiti, Algerian, Saudi, Moroccan, and even Iraqi support for the forthcoming battle with Israel. At the same time he pointedly reminded the Russians:

> Regarding a peaceful solution, our friends in the Soviet Union must know the true feeling of our people. From the first moment we believed that what was taken by force can only be regained by force. Our friends in the USSR must know that the peaceful solution which the U.S. has been talking about is fictious.[60]

At this point the Russians appeared not yet willing to back Egypt in a war against Israel. Instead, with the Brezhnev visit to Washington approaching, the Soviet leadership limited itself to supplying weaponry and trying to further discredit Israel and its U.S. supporters in the United Nations and other public forums.[61] As Brezhnev prepared to go to Washington, however, criticism of the lack of sufficient Soviet support for the Arab cause again began to rise.[62]

Thus by the time of the Nixon-Brezhnev summit, the Soviet position was a mixed one. On the one hand the U.S. position had deteriorated sharply, due primarily to its increasing dependence on Arab oil and the apparent willingness of Saudi Arabia to use the oil weapon. On the other hand Egyptian-Soviet relations, while improved, remained tense, as the Egyptians again began openly to complain about the subordination of their interests for the sake of Soviet-American detente.

FROM THE SUMMIT TO THE WAR

As in the 1972 summit, the leaders of the two superpowers appeared to pay little attention to the Middle East in their June 1973 meeting. Indeed, only 87 words out of a total of 3,200 in the final communique issued on June 24 dealt with the Middle East situation, and it appeared as if Nixon and Brezhnev wanted deliberately to downplay the conflict lest it interfere with their pursuit of detente. Thus the joint communique even failed to mention UN Resolution 242—hitherto the basis of the Soviet policy for a settlement of the Arab-Israeli conflict.

As might be expected, the Egyptian reaction to the summit communique was swift and bitter, and the Russians felt constrained to publish a special statement on Soviet policy toward the Middle East. Issued by TASS on 27 June, it reiterated the main tenets of Soviet policy frequently stated in the past, including the need for total withdrawal of Israeli troops to the 1967 borders, a "peaceful solution" based on UN Resolution 242, recognition of the "legitimate interests and rights of the Palestinians," and Soviet support for the Arab states affected by "Israeli aggression" in 1967.[63]

It was perhaps to test the degree of this support that Sadat's security adviser Hafez Izmail made a journey to Moscow on 11 July. In a speech following Izmail's return, Sadat stated that he was not fully "satisfied" with the results of the visit, which the Russians reported as taking place in an atmosphere of "frankness," and that the Russians had told Izmail that detente could be expected to last from 20 to 30 years.[64] Sadat then warned the Soviet leaders that detente would lead to the isolation of the USSR from the National Liberation Movement.

While Soviet-Egyptian relations remained strained, despite the steady flow of Soviet armaments, the Soviet leadership could gain some satisfaction that the Libyan-Egyptian merger project was foundering badly. While visiting Egypt for two weeks in late June and early July, Qaddafi had made little headway in convincing the Egyptians that the advantages of union with Libya outweighed the disadvantages. The Soviet leaders were clearly pleased, since Qaddafi, with his militant anti-Sovietism, had consistently opposed Soviet efforts to gain influence in the region. Nonetheless, the Russians could not have been too happy about Sadat's subsequent choice of Saudi Arabia as his principal Middle East ally. As Egypt's relations with Libya cooled, they became much closer with Saudi Arabia, culminating in a visit by Sadat to Saudi Arabia on 23 August, in which Sadat probably informed Faisal of the coming war with Israel and urged him to use the oil weapon against the United States.

By early September, with the Libya-Egypt merger project having fallen through, Soviet commentators began to express concern over Egypt's relations with Saudi Arabia.[65] But the Russians had increasing difficulty in trying to persuade the Egyptians not to go to war. They argued that through the judicious use of the oil weapon, the Israelis could be forced to withdraw because of pressure from the United States and that this process could be achieved without war—thanks to the existence of the Soviet-American detente.[66]

The Egyptians, however, were evidently not convinced by these arguments, nor by the Soviet decision to allow North Koreans and North Vietnamese to help train the Egyptians and Syrians.[67] Cairo press attacks on detente as subordinating Arab interests to those of the superpowers drew a harsh rejoinder from Moscow.[68] But the Soviet view that East-West detente provided protection for progressive movements in the third world was undermined by the overthrow of the Allende government in Chile in September.[69]

Immediately after the coup there came a major air battle in the Middle East, which resulted in the Israelis shooting down 13 Syrian planes while losing only one of their own. These events may have at least partially undercut the supporters of detente within the Soviet Politburo. They responded by increasing shipments of Soviet weapons such as tanks and antiaircraft missiles to Syria and Egypt, although the USSR still refrained from supplying the Arabs either with fighter-bombers or with ground-to-ground missiles. These shipments of weapons, however, were to prove sufficient for the Egyptians to make a final decision for war. The Russians must have learned about this decision in late September, because they began to withdraw the nonessential technicians and other civilians from both Syria and Egypt well before the outbreak of the fighting on 6 October.

In thus giving their tacit support for the Egyptian decision to go to war—in the viewpoint of this writer it was clearly an Egyptian and not a Soviet decision—the Soviet leaders may have been motivated by a number of considerations. In the first place, it was conceivable that Sadat was again bluffing, as he had appeared to be many times in the past, and that he needed the weapons primarily for domestic considerations. Secondly, should Sadat go to war and be defeated and this was the virtually unanimous feeling of the Western intelligence community and probably of a number of Russians as well, the Sadat regime would very likely fall, perhaps to be replaced by a more pro-Soviet Egyptian regime led by Ali Sabry. At the very minimum an outbreak of war would further inflame Arab feelings against the United States, much as the 1967 war had done, thus weakening the U.S. position still further in the Arab world. This, the Russian leaders may have reasoned, would be fit repayment for the weakening of the Soviet position in Latin America and indeed throughout the third world because of the events of Chile. In addition, since the United States did not inform the Russians about the pending coup in Chile—despite advanced knowledge—the Russians may have decided to equal the score by not telling the United States about the forthcoming Egyptian-Syrian attack, irrespective of the agreement reached at the 1972 summit.[70]

THE OCTOBER 1973 WAR AND ITS AFTERMATH

As the war began, the Soviet leaders had one overriding problem—how to provide aid to the Arabs while at the same time not destroying their detente with the United States. The initial Soviet reaction was a very hesitant one, perhaps because the Russians had some doubts as to the capabilities of the Syrian and Egyptian armies to carry out their offensive successfully—even against an overconfident and poorly prepared Israel.[71] Indeed, if we are to believe Sadat's account of the first day of the war, the Russians tried to get him to accept a ceasefire after only six hours of fighting, by claiming that Syria had requested a ceasefire.[72] Even when Sadat rejected the Soviet ploy, the Soviet media downplayed the war, with Pravda giving far more space to the events in Chile than to the Middle East conflict.[73] When after three days of fighting it appeared that the Arab side was in fact winning, the Russians, perhaps sensing the possibility of finally being able to rally the Arabs into the long-advocated "antiimperialist" alignment, moved to increase their involvement in the war, yet at the same time keeping it within limited bounds. Thus on 9 October Brezhnev sent a note to Algerian President Haouri Boumedienne and other Arab leaders urging them to ". . . use all means at their disposal and take all required steps with a view toward supporting Syria and Egypt in the difficult struggle . . . Syria and Egypt must not be alone in the struggle."[74]

At the same time the Russians began a massive airlift of weapons to Syria and Egypt, thereby demonstrating that while it was to be the Arabs (and not the Russians) who did the fighting, the USSR would provide the necessary supplies.

Several days later, just as he had in 1967, Boumedienne flew to Moscow to ask for more support for the Arab cause; once again he did not receive all the aid he requested. Although TASS published a Soviet pledge to "help in every way" the Arabs to recapture territory seized by Israel in the 1967 war, the communiqué reporting Boumedienne's talks with the Soviet leaders stated that they had taken place in a "frank atmosphere"—the usual Soviet code word for serious disagreement.[75] Nonetheless, the aid provided by the Soviet

Union was sufficient to provoke a massive U.S. military airlift to Israel, as well as attacks on the value of Soviet-American detente.

While the Russians were not ready for a ceasefire on 16 October, the Israeli crossing of the Suez Canal that day and the subsequent enlargement of their salient on the west bank of the canal quickly changed the Soviet leaders' minds. On the same day, Kosygin flew to Cairo to meet with Sadat,[76] and 11 Arab countries, meeting in Kuwait, announced that Arab oil exports to countries "unfriendly to the Arab cause" would be reduced each month by 5 percent until the Israelis withdrew to the 1967 prewar boundaries.[77] Thus the oil weapon, however tentatively, had now been employed. If this action was meant to deter the United States from granting further assistance to Israel, the attempt was a failure, for on 19 October President Nixon asked Congress for $2.2 billion in aid for Israel. Urged on by the Soviet Union, key oil-producing states such as Libya, Saudi Arabia, Kuwait, and the Persian Gulf sheikdoms immediately responded by cutting off all exports to the United States.

One reason for the Soviet emphasis on the oil weapon may have been the rapidly deteriorating position of the Egyptian Army—a development that led to Kissinger's flying to the Soviet Union on October 20 at the Soviet leaders' urgent request.[78] The end result of Kissinger's visit was a "ceasefire in place" agreement—a major retreat for the Russians from their previous position calling for a return to the 1967 boundaries as a price for the ceasefire. The Soviet-American ceasefire agreement, which was approved by the Security Council in the early hours of the morning on 23 October, did not terminate the fighting since both sides, despite agreeing to the ceasefire, continued fighting to improve their positions. The Israelis got much the better of the battle, however, and by 24 October Sadat was forced to appeal to both the United States and the USSR to send troops to police the ceasefire.[79] At this point, with its Arab client about to suffer a major defeat, the Soviet leaders decided to pressure Israel and the United States by alerting Soviet airborn divisions and dispatching transport planes to their bases, while at the same time sending a stiff note to Nixon. The United States quickly reacted to the Soviet threat with a worldwide alert, and it appeared not only that detente had died but that the two superpowers were on the verge of a nuclear confrontation. Cooler heads prevailed, however, and the crisis was defused by the decision to establish a UN emergency force to police the ceasefire, although the two superpowers were later to wrangle about the composition of the UN force.[80]

As the war came to a close, Soviet policy-makers who had been hesitant about the war at the start, were able to total up a number of significant gains for the Soviet Union's position in the Middle East, although many would turn out to be transient. Perhaps the main Soviet gain was the creation of the "antiimperialist" Arab unity they had advocated for so long, and the apparent concomitant isolation of the United States from its erstwhile allies in the region. Not only had Syria, Iraq, Egypt, Jordan, Algeria, Kuwait, and Morocco actually employed their forces against Israel, but even such formerly stanch allies of the United States as the conservative regimes of Kuwait and Saudi Arabia had declared an oil embargo against the United States, while the tiny Gulf sheikdom of Bahrein had ordered the United States to get out of the naval base it maintained there.[81] Soviet press commentary reflected a feeling of exultation with these developments.[82]

Despite certain indirect gains resulting from the war,* the Arab solidarity on an "antiimperialist basis" that the Russians had worked so hard to promote soon began to crumble. Iraq and Libya rejected the Soviet-supported ceasefire agreement (accepted by Egypt, Israel, Jordan, and Syria), with the Iraqis claiming that it was "against the will of the Arab masses." Even more serious for the Russians was the erosion of their influence in Egypt, which had been partly restored by massive shipments of Soviet equipment. By the end of the war, the primary alignment in the Arab world was the Egyptian-Saudi Arabian alliance, with the Egyptians supplying the military power and the Saudis the oil leverage.

U.S. Secretary of State Henry Kissinger was clearly aware of this in negotiating the ceasefire, and he moved quickly to improve relations with Egypt, and many of the oil-producing Arab states as well. His prime weapon in this effort was pressure on Israel to successively agree to the ceasefire, an exchange of prisoners, and a disengagement of forces, including withdrawal from the east bank of the Suez Canal. This provided the basis for a reestablishment of U.S.-Egyptian diplomatic relations, and the entry of U.S. businessmen into Egypt. Not only had Kissinger managed to do this all by himself—thus leaving out the Russians, who had provided the weapons to enable the Arabs to go to war in the first place—he also managed to secure Sadat's help in lifting the Arab oil embargo, thus splitting the "antiimperialist Arab unity," which the Russians had worked so diligently to maintain.

As might be expected, the Soviet leadership was far from happy with these developments. The Russians tried to counter them by emphasizing the direct U.S. military threat to the Arab states and

---

*These included the disarray in the NATO alliance (caused by U.S. action during the war as well as conflicting strategies in dealing with the oil embargo, which also weakened the Common Market) and the discrediting of the Chinese, who had provided no more than verbal support for the Arabs.

the continued U.S. support for Israel as well as by urging the Arabs to keep their unity.[83] Soviet propaganda also played up Israeli Defense Minister Moshe Dayan's visit to the United States in early January as a quest for more U.S. arms, along with U.S. Defense Secretary James Schlesinger's warning to the Arabs on 7 January that they risked the use of force against them if they carried their oil embargo too far. Yet the Soviet leaders were apparently caught by surprise by the Kissinger-arranged Israeli-Egyptian disengagement agreement on 18 January, and Egyptian Foreign Minister Ismail Fahmy had to make a hurried visit to Moscow immediately thereafter to explain the Egyptian position.

The fact that following the disengagement agreement Sadat began to urge the lifting of the embargo was a further blow to the Russians, and they warned that the disengagement agreement might lead to only a partial settlement of the Near East conflict and a weakening of Arab unity.[84] Yet this appeared to be the direction in which Middle Eastern events were moving. Even the decision by the United States to host a conference of energy-consuming nations did not serve to arrest the slow splintering of the facade of Arab unity. While the Russians hailed the decision of the mid-February Arab mini-summit meeting not to lift the embargo, they recognized that differences among the Arabs were increasing.[85]

It appeared to be only a matter of time until the embargo was lifted, since now Sheikh Yamani of Saudi Arabia, as well as Sadat, spoke openly about lifting it. In this atmosphere, Kissinger made yet another journey to the Middle East, procuring from the Syrian leaders the list of Israeli POW's the Israelis had demanded as a precondition for talks with Syria. It appeared that once again Kissinger would be able to pull off a diplomatic coup, and this was too much for the Russians. Having seen the United States replace the USSR as the leading foreign influence in Egypt—however temporarily—the Russian leaders had no desire to see the process repeat itself in Syria. Consequently, Gromyko followed Kissinger to Damascus and worked out a highly bellicose communique with the Syrian leaders that threatened renewed war if Syrian demands were not met.[86] Strengthened by new shipments of Soviet arms and encouraged by Soviet support, the Syrian regime of Hafez al-Assad, less willing (or able) to make peace with Israel than Egypt, adopted a very hard bargaining position and, upon Gromyko's departure, began a war of attrition against Israel that involved daily artillery and tank battles.

If the Russians and the Syrians hoped that by heating up the conflict in the Golan Heights they would be able to prevent the oil-rich states from lifting the embargo, they miscalculated. Despite pleas from Syria and the USSR for "Arab Unity," the oil-producing states, under Egyptian and Saudi pressure, agreed to end the oil embargo against the United States on 19 March, although as a sop to the Syrians, Algeria stated that it would reexamine its embargo policy on 1 June. In any case, Arab unity on the oil embargo issue was clearly broken, as Libya and Syria refused to abide by the majority decision to lift the embargo.

At this point, Soviet-Egyptian relations began to deteriorate rapidly. Just as in 1971, when Sadat refused to support Soviet policy in the Sudan, here again he was strongly opposing a major Soviet policy—despite all the economic and military aid the Soviet Union had given Egypt. The Soviet leaders retaliated by branding Sadat a traitor to Nasser's "legacy," while Sadat attacked the USSR for lying to him on the first day of the war.[87]

CONCLUSIONS

All in all, the Soviet relationship with Sadat's Egypt has not been a pleasant or profitable one for the Soviet leadership. In the period since Nasser's death, the Soviet position in Egypt deteriorated so sharply that not even massive military assistance delivered during the October 1973 war with Israel could restore it. The Soviet military position in the Eastern Mediterranean has been greatly weakened by the loss of air and naval bases in Egypt, and Soviet diplomatic efforts have led to bitter disappointment at high cost, as time and time again Sadat has opposed policies of great importance to the Soviet leadership. The USSR has invested billions of dollars in Egypt, but so far, it has been able to exert little if any real influence over Sadat's foreign or domestic policies.

Ironically, even when the Soviet Union supported Egypt in its war effort against Israel (albeit hesitantly at first), it was not long before Sadat turned to the United States for support, and the end result of the war, in the short run at least, was that the United States emerged in a much better economic and political position in Egypt than it had before the war. Meanwhile Soviet-Egyptian relations have deteriorated sharply despite the massive Soviet aid during the war.

Thus, in the great power game currently being played in the Middle East, there has developed a real question as to who is exploiting whom in the Soviet-Egyptian relationship. At the time of this writing (early April 1974), it would appear that Sadat's Egypt has emerged as the clear victor in the contest.

NOTES

1. For general surveys of the Soviet involvement in the Middle East prior to Nasser's death, see Walter Laqueur, The Struggle for the Middle East (New York: Macmillan, 1969); Aaron Klieman, Soviet Russia and the Middle East (Baltimore: Johns Hopkins Press, 1970); M.S. Agwani, Communism in the Arab East (Bombay: Asia Publishing House, 1969); and George Lenczowski, Soviet Advances in the Middle East (Washington, D.C.: American Enterprise Institute, 1972).

2. An official description of the Rogers plan, calling for the withdrawal of Israeli forces from all but "insubstantial" portions of the territory captured in 1967 in return for a binding peace settlement, is found in U.S. Department of State, United States Foreign Policy 1960-1970: a Report of the Secretary of State (Washington, D.C.: Government Printing Office, 1971).

3. For an analysis of Nasser's role as a "broker" of Soviet interests in the Arab world, see Malcolm Kerr, "Regional Arab Politics and the Conflict with Israel," RM-S966-FF (Santa Monica, Calif.: Rand Corporation, 1969).

4. See the commentary by Uri Glukhov in Pravda, 17 October 1970, translated in Current Digest of the Soviet Press (hereafter, CDSP), 17 November 1970, p. 15.

5. See "Iraq Ousts Vice President Considered Leading Hawk," New York Times, 16 October 1970, p. 4:3; Michael Field, "Iraq— Growing Realism Among the Revolutionaries," New Middle East (London), February 1971, p. 27; and J. Gaspard, "Damascus after the Coup," New Middle East, January 1971, pp. 9-11.

6. For an analysis of the effects of the power struggle in Egypt, see Peter Mansfield, "After the Purge," New Middle East, June 1971, pp. 12-15.

7. See "The Course of the U.A.R.," New Times (Moscow), May 1971, p. 16.

8. The text of the treaty is found in the "Treaty of Friendship and Co-operation Between the Union of Soviet Socialist Republics and the United Arab Republic," New Times, 23 June 1971, pp. 8-9; for a different analysis of the treaty see Nadav Safran, "The Soviet-Egyptian Treaty," New Middle East, July 1971, pp. 10-13.

9. See Podgorny's remarks at a Cairo dinner in Pravda, 29 May 1971, translated in CDSP, vol. 23, no. 22, p. 5.

10. See Yevgeny Primakov in Pravda, 5 June 1971, translated in CDSP, vol. 23, no. 23, p. 7.

11. For a discussion on these events, see Anthony Sylvester, "Muhammed versus Lenin in Revolutionary Sudan," New Middle East, July 1971, pp. 26-28.

12. See Dmitry Volsky, "Changes in the Sudan," New Times July 1971, p. 11. This issue appeared in the brief interval between the time Nimeri was overthrown and the time he returned to power.

13. See commentary by "Observer" in Pravda, 30 July 1971, translated in CDSP, vol. 23, no. 29, p. 5.

14. Cited by Bernard Gwertzman, "Soviet Warns Sudanese as Relations Deteriorate," New York Times, 30 July 1971, p. 2:1.

15. OFNS report from Cairo by Colin Legum, cited in Jerusalem Post, 9 August 1971.

16. See V. Lykov, "New Stage in the Life of the U.A.R.," New Times, August 1971, p. 8.

17. For a suspicious Soviet account of the British official's visit, see "Douglas-Home in Cairo," New Times September 1971, p. 17.

18. See, for example, commentary by Pyotr Demochenko in Pravda, 28 September 1971, translated as "In Moscow There Is a Feeling of Nostalgia for Nasser," New Middle East, November 1971, p. 12.

19. See "War or Peace?—Hussanein Heikal's Guidelines for the Federation," New Middle East, October 1971, p. 39; and Joh Kimche, "Where Do We Go Now?—Agenda for 1972," New Middle East, November 1971, p. 4.

20. Pravda, 13 October 1971, Sadat's speech is translated in CDSP, vol. 23, no. 41, pp. 6-7.

21. Ibid., pp. 7-8.

22. Cited in Jerusalem Post, 17 December 1971.

23. See the examples cited in New Middle East, March-April 1972, p. 63.

24. Sadat's speech explaining his decision not to go to war is printed in "The 'Fog' That Stopped Sadat from Going to War," New Middle East, February 1972, p. 42.

25. For the full text of Sadat's speech to Egypt's Arab Socialist Union, in which he described his visit to Moscow and gave his view of the development of Soviet-Egyptian relations, see the Radio Cairo domestic service broadcast, 24 July 1972.

26. Text of the communique in Pravda, 30 April 1972.

27. The full text, which gave very little space to the Middle East situation, can be found in "Joint Soviet-U.S. Communique," New Times, June 1972, pp. 36-38.

28. See V. Potomov, "A Just Peace for the Middle East," New Times, June 1972, p. 16.

29. A translation of this memorandum can be found in "Egypt and the USSR: Implication of Economic Integration," Radio Liberty Dispatch, 13 July 1972, p. 6.

30. For a survey of the Egyptian press at this time, see Ihsan A. Hijazi, "Domestic Pressure on Sadat Reported," New York Times, 19 July 1972, p. 1:6.

31. For a perceptive view of the domestic situation in Egypt at the time of the Soviet exodus, see P. J. Vatikiotis, "Two Years after Nasser: The Chance of a New Beginning," New Middle East, September 1972, pp. 7-9.

32. For a discussion of the domestic and foreign background to the Yemeni and Saudanese decisions, see John K. Cooley, "Renewed U.S.-Yemen Tie—Diplomatic Toe in Arab Door," Christian Science Monitor, 3 July 1972, p. 1:4; and Henry Tanner, "Sudan Likely to Resume U.S. Diplomatic Relations," New York Times, 18 July 1972, p. 2:3.

33. Text of Sadat's statement in New York Times, 19 July 1972, p. 15:1.

34. Cited in John K. Cooley, "Cairo Looks to U.S. Prod," Christian Science Monitor, 22 July 1971, p. 5:1.

35. For a discussion of the potential benefits to Egypt of the union, see Malcolm Keer, "The Convenient Marriage of Egypt and Libya," New Middle East, September 1972, pp. 4-7.

36. Henry Kamm, "Dayan Says Move by Cairo Reduces His Army's Needs," New York Times, 12 August 1972, p. 1:8.

37. See the Soviet communiqué announcing the withdrawal of military personnel from Egypt in Pravda, 20 July 1972, translated in CDSP, vol. 24, no. 24, p. 18; and the commentary in Pravda, 23 July 1972. See also "Russians Leaving Egypt after Careful Send-Off," New York Times, 20 July 1972, p. 8:1.

38. Cited in John K. Cooley, "Cairo's Anti-Soviet Ire Heats up," Christian Science Monitor, 19 August 1972, p. 1:2.

39. Cited by Flora Lewis, "Ouster Move Long in Making, Sadat Says," New York Times, 22 August 1972, p. 3:1.

40. See John K. Cooley, "Soviets-Arabs Play Out Shadowy Drama in Secret," Christian Science Monitor, 15 August 1972, p. 3:1; Izvestia, 29 August 1972, translation in CDSP, vol. 24, no. 25, p. 5; and "Syrian Seen Mediator in Moscow-Cairo Rift," New York Times, 4 September 1972, p. 3:7.

41. See John A. May, "Fatal Bomb Package Shocks Britons," Christian Science Monitor, 20 September 1972, p. 3:3.

42. "Egypt and Libya Agree on Cairo as Joint Capital," New York Times, 19 September 1972, p. 10:3.

43. Cited in Henry Tanner, "Sadat Rejects Rogers Suggestion of Interim Suez Canal Accord," New York Times, 29 September 1972, p. 18:3.

44. See Henry Tanner, "Egypt and Soviet See Better Tie," New York Times, 2 October 1972, p. 9:1; and the Jerusalem Post, 6 October 1972.

45. See Juan de Onis, "Sadat says Soviet Thwarted Peace," New York Times, 6 October 1972, p. 1:9.
46. Cited in Henry Tanner, "Sadat Stresses Link with Soviet," New York Times, 16 October 1972, p. 15.
47. The text of the communique is found in Pravda, 19 October 1972.
48. Cited in Henry Tanner in the 27 October issue of the Jerusalem Post. See also John K. Cooley, "Cairo Looks Again to Russia as Possible Arms Broker," Christian Science Monitor (Midwest ed.), 24 October 1972.
49. For a description of the possible effects of these leadership changes, see John K. Cooley, "Egypt Still Debates Expulsion of Soviets," Christian Science Monitor (Midwest ed.), 10 November 1972.
50. See William Beecher, "Egypt Is Reported to Get Advanced Soviet Missile," New York Times, 12 November 1972, p. 2:3.
51. Cited by William Dullforce in the Washington Post, 30 October 1972. Interestingly enough, Pravda, on 1 November 1972, gave the same version.
52. Pravda, 1 November 1972.
53. See John K. Cooley, "West's Refusal Leads Egypt to Continue Relying on Soviet Aid," Christian Science Monitor, 11 November 1972, p. 3:1.
54. Cited in Reuters report in the New York Times, 29 December 1972.
55. See Victor Kudryavtsev, "On the Arab Diplomatic Front" New Times, no. 4, 1974, p. 12. The Russians had long been urging the Arabs to use their oil weapon. For an early theoretical treatment of the possible effects of oil pressure, see I. Bronin, "Arabskaia Neft—SShA-Zapadnaia Europa" (Arab oil-USA-West Europe), Mirovaia Ekonomika I Mezhdunarodnaia Otnosheniia, no. 2 (February) 1972, pp. 31-42.
56. Cited in the Washington Post, 27 March 1973.
57. Cited by William Dullforce in the Washington Post, 5 February 1973.
58. See the Egyptian press coverage in Brief: Middle East Highlights (Tel Aviv) (hereafter, BRIEF) no. 54, p. 3.
59. See reports in the Washington Post, 18 April 1973, and the Jerusalem Post, 20 April 1973.
60. Cairo Radio, May 1, 1973, reprinted in Middle East Monitor (MEM), vol. III, no. 10, pp. 3-4.
61. Compare "The International Conference for Peace and Justice in the Middle East," Bologna, Italy, 13 May 1973. For a description of the results of the conference, see New Times, no. 21, 1973, pp. 16-17.

62. See reports by John Cooley in Christian Science Monitor, 20 and 25 June 1973.
63. Reprinted in MEM, vol. 3, no. 14, p. 1.
64. Cited in BRIEF, no, 62, p. 1.
65. See Y. Potomov, "The Egypt-Libya Merger Project," New Times, no. 36, 1973, p. 11.
66. See Dmitry Volsky, "New Opportunities and Old Obstacles," New Times, no. 32, 1973, p. 15.
67. Cited in New York Times, 16 August 1973.
68. See Pravda, 28 August 1973, translated in CDSP, vol. 25, no. 35, p. 18.
69. Compare the confidence expressed by Dmitry Volsky in "Soviet-American Relations and the Third World," New Times, no. 36, 1973, pp. 4-6, with the defensive tone of the lead editorial in New Times, no. 39, 1973, p. 1.
70. For a different view of the Soviet position on the eve of the war, see Uri Ra'anan, "Soviet Policy in the Middle East 1969-73" in Midstream 19, 10 (December 1973): 23-45 and the article by Joh Kimhe "The Soviet-Arab Scenario" in the same journal (pp. 9-22).
71. For an example of the Israeli leaders' lack of psychological preparedness for war, see Defense Minister Moshe Dayan's lecture to the Israeli Command and Staff College on 9 August (reprinted in BRIEF, no. 63, pp. 3-4).
72. See New York Times, 30 March 1974.
73. See Pravda, 7-9 October and the coverage on Moscow Radio for the same dates.
74. Text of the message is found in Radio Paris domestic service 9 October 1973.
75. Pravda, 16 October 1973.
76. Pravda, 20 October 1973.
77. Cited in MEM, vol. 3, no. 20, p. 3.
78. See reports by Murray Marder in the Washington Post, 20 October 1973, and Bernard Gwertzman in New York Times, 21 October 1973.
79. Cited in MEM, vol, 3, no. 20, p. 5.
80. The American alert, and the nature of both the Soviet message to Nixon and the exact nature of Soviet moves are not yet clear. For Kissinger's statement about the alert at a press conference and a description of the alert, see New York Times, 26 October 1973.
81. For a detailed description of the actions of the Arab states during the war, see MEM, vol, 3, nos. 19 and 20.
82. See Dmitry Volsky and A. Usvatov, "Israeli Expansionists Miscalculate," New Times, no. 42, 1973, p. 10; and Georgi Mirsky, "The Middle East: New Factors," New Times, no. 48, 1973, pp. 18-19.

83. See *Pravda*, 27 November 1973, translation in *CDSP*, vol. 25, no. 48, pp. 19-20.
84. See "Commentator" in *Pravda*, 30 January 1974, translated *CDSP*, vol. 36, no. 5, pp. 11-12.
85. See *New Times*, no. 8, 1974, p. 16.
86. Text of communiqué in *Pravda*, 8 March 1974.
87. See reports in New York *Times*, 26 March and 30 March 1974.

CHAPTER

13

THE SOVIET UNION'S QUEST FOR ACCESS TO NAVAL FACILITIES IN EGYPT PRIOR TO THE JUNE WAR OF 1967
George S. Dragnich

This chapter seeks to identify and elucidate Soviet attempts to secure access to naval facilities in Egypt before the June War of 1967. As far as possible, it describes the methods the Soviets employed toward that end. The chapter also seeks to explain why the USSR perceived a need for those facilities and attempts to identify landmarks in the Soviet effort to obtain access to them. Although Admiral Gorshkov's trip to Egypt in 1961 marks the first clear expression of that endeavor, Soviet behavior during three Middle Eastern crises had already shaped Egyptian perceptions of the Soviet Navy. Thus, the chapter also examines the Suez crisis of 1956, the Syrian-Turkish crisis of 1957, and the Lebanon crisis of 1958 for the light they shed on Egypt's subsequent response to Soviet persuasion and pressure for access to its naval facilities. Where relevant to this question, Egyptian views of the Sixth Fleet and other Western naval forces are examined as well.

Soviet-Egyptian dealings on the issue of naval facilities represent a classic example of superpower-third world relations. In this instance, the superpower sought to satisfy its own strategic interests; and the third-world state tried to get as much as it could from the superpower, without having to compromise its sovereignty. Too often, analyses of Soviet attempts to secure military privileges in third-world states dwell too heavily on Soviet interests and intentions, with scant attention being paid to the interests and aims of those third-world states directly concerned. This case study demonstrates that a superpower's persistence and generous military and economic aid are not always enough to overcome a third-world state's reluctance if the latter feels its interests threatened by the superpower's intent.

The June war of 1967 was chosen as the end date for this study because it brought about a whole new set of circumstances that radically affected the relative bargaining positions of the two parties. As this study attempts to show, the Soviet naval presence established in Egypt after that event was not a result of earlier Soviet efforts to attain that objective; rather it reflected Egypt's desperate need to obtain Soviet reequipment of its armed forces in the wake of its disastrous defeat.

1925-55: MINIMAL INVOLVEMENT IN THE MEDITERRANEAN

Western attempts to prove the continuity of Soviet naval policy with the Tsarist period invariably cite the Russian Navy's historical experience in the Mediterranean. Interestingly, the Soviets themselves claim a heritage from much of the Russian Navy's history, especially as it pertains to that sea. For example, the head of the Soviet Navy, Admiral of the Fleet S. G. Gorshkov, has glowingly written of one such period:[1]

> The stay [1769-74] of the Russian Fleet in the Mediterranean Sea is an outstanding example of autonomous operations by a large naval formation completely cut off from its home ports, which increased the international prestige of Russia and evoked warm sympathy toward her by all the peoples of the Mediterranean Sea basin.

Understandably, then, Soviet naval historians reject Western evaluations that downgrade the military effectiveness of Tsarist naval ventures. As they concern the Mediterranean, these negative evaluations[2] are viewed as a deliberate effort by "bourgeois historians . . . to convince the present-day reader of the 'traditional' weakness of the Russian fleet, that it is somehow incapable of justifying the hopes of the Arab nations."[3] In fact, the Soviet Union's first attempt at naval diplomacy in the Middle East ended almost as quickly as it began. In September 1928, the training ship <u>Vega</u> called at Algiers, where its cadets distributed communist literature to Europeans and native Algerians alike and tried to assemble crowds for pro-Soviet rallies. In response, the French colonial authorities there withdrew the Vega's landing permit, and the ship left Algiers two days ahead of its scheduled departure.[4]

Soviet warships first appeared in the Mediterranean during the 1920s and 1930s, in order to visit European ports.* But the Soviet Navy's inability to intervene during the Spanish Civil War—even after a Soviet merchant ship was sunk—demonstrated just how illusory a credible Soviet naval presence in the Mediterranean was at that time. There was no more Soviet naval activity in the region until after Stalin's death in 1953. The first real opportunity to use naval diplomacy in the Middle East came while Stalin was still alive, and it was not taken up.

In January 1952, during a period of acute Anglo-Egyptian tension over the Suez Canal, 33 Soviet fishing trawlers and their depot ship arrived at Port Said for a southbound transit of the canal on their way to Vladivostok. Egyptian demonstrators waiting on the quays welcomed them and then marched through the city carrying portraits of Stalin and shouting pro-Soviet and anti-British slogans. Egyptian authorities, departing from established practice, indicated that they had no objection to the Soviet sailors going ashore while their ships waited to transit the canal.[5]

Less than five months earlier, the Egyptian Government had rejected a British proposal for a naval good-will visit to Alexandria, and, less than three months earlier, the United Kingdom had exacerbated Egypt's Anglophobia by anchoring a cruiser at the northern end of the canal " . . . contrary to all customs and without advising port authorities."[6] That warship was still there and had taken over some of the canal's operations from striking Egyptian workers. Under these circumstances, Egypt would probably have even welcomed a Soviet naval visit. But, although Pravda noted the demonstrations and espoused the Egyptian cause,[7] the Soviet maritime contingent studiously avoided any overt sign of support. The Soviets only allowed the ships' captains and first mates ashore, and these avoided all political activity.

---

*Two Soviet destroyers called at Istanbul and Naples in September 1925. A cruiser and gunboat from the Black Sea Fleet called at Istanbul in October 1928 but did not enter the Mediterranean. In January 1930, the USSR transferred a battleship and cruiser-minelayer from the Baltic to the Black Sea Fleet in defiance of the London Naval Conference (then under way), to which the USSR had not been invited. In September 1930, a cruiser and two destroyers from the Black Sea Fleet called at Istanbul, Piraeus (Athens), and Messina, Italy. A Soviet cruiser and two destroyers called at Naples in October 1933.

Typically, this reflected Stalin's caution about involving the USSR directly in the region. Political considerations may have included the fact that Egypt still had a monarchy and that the anti-British movement was a national rather than class struggle. Moreover, Egypt's outlawed Communist Party was too weak to exert significant influence over that movement. Military considerations may have included the fact that Egypt's borders were not contiguous with those of the USSR and that most of the Royal Navy's Mediterranean Fleet was operating near Egypt.

The first Soviet naval visit in the Mediterranean in the post-Stalin period was to a Soviet bloc state and appears to have served a specific politico-military objective. The cruiser <u>Admiral Nahkimov</u> and two destroyers—the first Black Sea Fleet warships to enter the Mediterranean since the 1930s—visited Albania, May 31-June 4, 1954. Their visit, led by the commander of the Black Sea Fleet (then Vice Admiral S. G. Gorshkov), reassured that isolated Soviet bloc state of the USSR's continued protection. And Albania's leader, Enver Hoxha, thanked the visiting warships for bolstering his country's strength and spirits.[8]

1955-61: THE SHAPING OF EGYPT'S PERCEPTION OF SOVIET NAVAL POWER

The Czech Arms Deal of 1955

In September 1955, President Nasser delivered a sharp blow to Western influence in the region when he announced that he was buying arms from communist Czechoslovakia. According to Muhammad Hasanein Heikal (Nasser's confidant and editor of Al-Ahram), part of the U.S. response was to send Kermit Roosevelt to Cairo, where he told Heikal that the United States might impose a naval blockade of Egypt in order to prevent ships from arriving with arms. At this very time, Heikal has written,[9] Nasser was conferring with the Egyptian ambassador to Washington (then on leave in Cairo). The Egyptian ambassador is reported to have brought up the Sixth Fleet's presence in the Mediterranean, to which Nasser replied, "What can the U.S. Sixth Fleet do to us?" The ambassador answered, "It may prevent the ships from coming to us." Nasser's response was, "there is nothing on our ships. If it stops the ships of others, the Sixth Fleet will have the others to deal with and not us." While this story may be apocryphal, it does illustrate Egypt's apparent perception that the Sixth Fleet's flexibility and capabilities were restricted, that the Soviet Union could curtail the U.S. application of naval power, and that Nasser felt that he could use the USSR as a buffer against the West.

The Czech arms deal, itself a major departure from traditional Soviet foreign policy, opened up an avenue for increased Soviet influence in the area. In the naval sector, the 1955 agreement provided for the delivery of two destroyers, smaller surface vessels, and six submarines. Four T-43-class fleet minesweepers from Poland, and at least 12 P-6-class motor torpedo boats from Czechoslovakia, were delivered to Alexandria in April 1956; two Skory-class destroyers were delivered to the same port in June 1956.

The Suez Crisis of 1956

The USSR was cautiously using middlemen to implement its military assistance policy in the Middle East. In spite of this, the first mention of possible direct Soviet naval influence in the area came that summer, during the Suez crisis of 1956. It is impossible to determine the exact connection, but, as suggested above, the Czech arms deal probably created an expectation in the minds of certain Arab leaders that they could use Soviet military power as a lever against the West.

On August 3, 1956, the Soviet Communist Party newspaper Pravda criticized the British Government's military buildup associated with the Suez Canal dispute.[10] On August 4, the London Times reported the imminent departure from the United Kingdom of three British aircraft carriers for an "unstated destination."[11] That evening, in a broadcast entitled "Gunboat Diplomacy Must Not Be Repeated," Radio Moscow's Arabic-language program warned that the Western military activity was designed to ". . . intimidate Egypt."[12] On August 6, Cairo newspapers banner-lined reports from Damascus that the USSR had asked permission of Arab governments for Soviet warships to visit their ports beginning August 15, the day before the Suez Conference was scheduled to open in London.[13] Although the Athens "Our Cyprus" radio program stated that the Egyptian ambassador to Greece had "confirmed" that Soviet warships might visit Egypt,[14] the Soviet Embassy in Damascus denied all knowledge of such a proposal, and the Syrian Foreign Ministry denied having received any communication from the USSR to that effect.[15]

Very likely, the reports were the result of serious concern brought on by threatening Western naval movements. The Soviet Northern and Baltic Fleets were making a series of naval visits to Northern Europe (the Netherlands, Denmark, Sweden, and Norway) at the time, and perhaps this, combined with the USSR's vocal support of the Egyptian position, sparked speculation that the Black Sea

Fleet might visit Arab ports. But it is unlikely that the USSR would have willingly assumed the considerable military and political risks associated with such a venture. What is important for this study, however, is Egypt's apparent perception that the Soviet Navy might be able to serve Egyptian foreign policy objectives. This is indicated by both the prominence that these reports received in the Egyptian press and by the fact that the Egyptian state radio's international program, "Voice of the Arabs," also carried the Syrian reports.[16]

If, in 1956, that perception remained an unfulfilled expectation, the abortive Anglo-French invasion of Egypt, and the resulting Soviet threats against the United Kingdom and France, did leave many Egyptians with the feeling that the West was no longer free to exercise traditional gunboat diplomacy as it had in the past. The Egyptian Economic and Political Review noted of the U.S. Sixth Fleet in 1957:[17]

> Its super carriers and guided missile cruisers lose their capacity to impress by their very invitation to the Russians, spectacularly accepted by these, to go one better; what with their intercontinental missiles and Sputniks the Soviets seem to have successfully stolen the 6th Fleet's thunder. In the process even the 6th Fleet's role as a somewhat blundering political and diplomatic weapon has been neutralized without the firing of a single shot.

Soviet Bloc Naval Aid to Egypt in 1957, and the Soviet Navy's Transit of the Suez Canal

Beginning very early in 1957, future Egyptian submarine crews began training at the Polish naval base of Oxywie (near Gdynia). About six months later, in June 1957, Egypt's first three Soviet-built submarines—two modern ocean-going "W"s and one coastal "MV"— arrived in Alexandria.

About one week later (June 23, 1957), in a move that may have been timed to occur soon after the above event, two Soviet destroyers and a tanker from the Black Sea Fleet began a southbound transit through the Suez Canal. These ships, which were taken through the canal by Soviet pilots working for the Suez Canal Authority,[18] were the first Soviet naval vessels to transit that waterway since 1924.*

---

*The only Soviet naval transit of the Suez Canal prior to 1957 had been on August 21, 1924, when the combination training and fisheries protection vessel Vorovsky passed through on its way to Vladivostok. (London Times, August 22, 1924, p. 10.)

One month later, after they had reached the Pacific Ocean, the USSR formally closed most of Peter the Great Bay at Vladivostok to foreign shipping. The Soviet Union surely appreciated Egypt's strategic position astride the Suez Canal long before then, but its apparent decision in 1957 to strengthen its naval presence in the Pacific made Egypt's friendship more important.

The Syrian-Turkish Border Crisis of 1957, and the First Soviet Warship Visit to an Arab State

The Soviet warships that transited the Suez Canal in 1957 did not make any port calls in the region; the first Soviet warship visit to an Arab state took place that September, to Syria. On September 21, 1957, the cruiser Zhdanov and destroyer Svobodni from the Baltic Fleet arrived in Latakia. The detachment was commanded by Vice Admiral V. F. Kotov, first deputy c-in-c of the Baltic Fleet, and had just visited the Yugoslav naval base at Split.

Before its arrival in Syria, the Egyptian press exaggerated the size of the Soviet force to include a cruiser, three destroyers, and several submarines equipped with guided missiles.[19] As with the previous year's unfounded reports about the Soviet Navy, this was reported in the context of Western naval movements in the area. According to Al-Akhbar, Soviet Defense Minister Zhukov had written to the pro-Soviet Syrian Defense Minister, Khalid al-Azm, suggesting that Soviet warships pay a friendly visit to Syria. Syria's reported response was that it would welcome such a visit" . . . at any time and in any circumstance."[20] The actual timing and circumstance of the visit are especially relevant because it overlapped a period of Syrian-Turkish border tension.

The Soviet decision to seek Syrian permission for a naval visit appears to have been made only after the detachment had already reached the Mediterranean,[21] by which time the Syrian-Turkish crisis had surfaced. But even if the visit had been planned well in advance, the fact that the USSR maintained a vituperative propaganda campaign against Turkey during the port call indicates that it was willing to have the visit perceived as a gesture of Soviet support for Syria. The Damascus newspaper, Al-Nour, had, early in September, optimistically declared, "Furthermore, there is a big Power which will support Syria and any self-liberated state against foreign aggression. This Power has fleets in the Mediterranean and sufficient intercontinental missiles to wipe out the 85 ships [the Sixth Fleet] which America boast about."[22] Radio Cairo Domestic Service had carried the Al-Nour article, and now some of the Egyptian press saw the Soviet Navy's visit as a clear sign of Soviet support.[23] The

USSR, itself, fostered this impression. Radio Moscow's Arabic language program followed its announcement of the detachment's arrival in Latakia with broadcasts assailing the Sixth Fleet's "demonstrations" in the Mediterranean. And specific broadcasts, such as "The Zhdanov and the Sixth Fleet,"[24] during the ships' visit left no doubt as to the political character of the port call.

The official nature of the visit can be seen in the fact that Admiral Kotov reviewed Syrian cadets at the Home military academy and met with President Kuwatli during his visit. At a dinner in Damascus, Defense Minister al-Azm bestowed decorations from the Syrian Government upon the Soviet naval officers. The Soviet charge d'affaires also gave a reception for Admiral Kotov and his staff.

Most official Soviet naval visits last no more than five days. In contrast, the detachment stayed in Syria for 10 days. Although it was announced before their arrival that they would stay that long,[25] the Soviet units left Latakia just after the crisis appeared to have abated.[26] This may not have been coincidence, since Moscow had been largely responsible for maintaining the crisis at an artificially high level.

In any case, the crisis had only subsided temporarily. The USSR unexpectedly reopened the issue on October 7 and sought to portray it as a major international crisis. During this more internationally tense phase of the Syrian-Turkish crisis, the USSR did not send its naval forces back into the area—even though it used Black Sea Fleet exercises* and a bellicose statement from that fleet's commander, Admiral V. Kasatonov, to demonstrate Soviet concern.[27] Such a move would have been feasible since the cruiser Kuibyshev and two escorting destroyers were in the eastern Mediterranean at the time. (The detachment, which was commanded by a rear admiral, had taken Marshal Zhukov to Yugoslavia. After the warships had put him ashore in Zadar, they had visited the Yugoslav ports of Split and Dubrovnik. Zhukov flew back to Moscow, and the detachment was returning to Sevastopol during this phase of the Syrian-Turkish crisis. Because it was set up to make port calls and because it was unencumbered by Marshal Zhukov, the detachment could have easily visited Syria at this time.)

Thus, while the USSR appeared willing to use naval visits as an instrument of foreign policy in 1957, there appears to have been definite limits governing their utilization in this role. In this specific case, the Soviets may have felt that such visits pass the point of diminishing returns once a regional crisis begins to be played out in the

---

*Radio Cairo Domestic Service (October 24, 1957) did report Red Star's announcement of these exercises.

international arena, and they may have been concerned that the risks attending such a venture might quickly outweigh the limited value that a token force would represent if the crisis escalated. NATO naval forces (including the Sixth Fleet) were operating near Turkey on maneuvers regularly scheduled for that time of year. There was nothing inherently hostile in their deployment, which might explain the USSR's willingness to have made a naval visit to Syria during the first phase of the crisis; but their presence there during the more dangerous second phase might have made another such visit too risky.

The Lebanese Crisis of 1958

The Suez crisis of 1956 had left Egypt with the feeling that the Western powers' ability and resolve to use their naval forces as instruments of crisis management had been irreparably weakened. The Soviet Navy's port call to Syria the following year probably left some Arab leaders with the impression that the USSR would use its navy as a counterforce should the Western powers nonetheless try again. Egyptian perceptions of U.S., British, and Soviet naval deployments during the Lebanese crisis of 1958—prior to the actual U.S. intervention—can be explained in this context.

When the Lebanese internal crisis broke out in May 1958, the United States and United Kingdom sent their naval forces into the eastern Mediterranean. Cairo's Al-Akhbar confidently noted,[28]

> The show of force, on which imperialism relies, has become out of date. It was tried during the aggression on Port Said and ended in disappointment. It was tried in the attempt to destroy Syria and threaten it, and its result was also a complete failure. . . . The appearance of fleets and destroyers no longer frightens the people. The Sixth Fleet appeared before, and a short time ago, but it retreated followed by waves of defeat.

But, as U.S. and British naval forces intensified their operations in the eastern Mediterranean, Egyptians news media evidenced concern that the West might intervene after all. Al-Ahram said on June 19, "The British fleet is preparing and concentrating its forces to proceed to Lebanon. . . . It did not learn a lesson at Port Said. . . . It has forgotten everything."[29] On June 21, Egypt's official newspaper, Al-Gomhouriya, inveighed against the " . . . repeated provocative movements undertaken by America's Sixth Fleet near the eastern coast of the Mediterranean."[30]

The expectation engendered by the USSR's behavior during the Syrian-Turkish crisis that the Soviet Navy would block any future Western attempt at naval intervention surfaced three days later. On June 24, the Egyptian press gave unusual prominence to a UPI dispatch from Copenhagen reporting the westward movement of Soviet naval units through the Baltic, "probably" to counter the Sixth Fleet in the Mediterranean.[31] Included in that day's headlines were, "Russian Fleet Moves as a Countermeasure to Sixth Fleet Maneuvers" and "Guided Missile Carrier Moves with Fleet Units."[32] Radio Cairo Domestic Service commented, "When the Western fleets steam in the Mediterranean, the Eastern fleets move, too, to maintain the balance of power."[33] In fact, the Soviet naval movements in the Baltic appear to have been connected with a Northern Fleet exercise.

When the United States intervened in Lebanon the following month, the Egyptian response could best be described as stunned. President Nasser was visiting President Tito in Yugoslavia when the Marines landed, and he immediately flew to Moscow. Heikal, who accompanied him on that journey, later wrote (1965) that Premier Khrushchev told Nasser: " . . . to be frank, the Soviet Union is not ready for a clash with the West, the result of which would be uncertain." According to Heikal, the Soviet leader offered to announce that the USSR was holding maneuvers (normally scheduled for that time of year) along its southern borders but cautioned President Nasser, "I do not care whether the West imagines that we are preparing for more than maneuvers. But I do care that you yourself should not be led to believe anything. . . . Maybe you expected more from us?"[34]

The Soviet Union's naval response to the U.S. intervention was minimal. The Black Sea Fleet carried on its regularly scheduled maneuvers, albeit with greater fanfare than usual. In late August, the Soviet Union moved four submarines and a submarine tender from the north to Albania; but the transfer does not appear to have been directly related to the U.S. operation in Lebanon. The deployment was unheralded by Soviet media, the vessels' pace was leisurely,[35] and they went directly and unobtrusively to the Soviet base at Vlone, Albania. Had they come one or two months earlier, their politico-military impact would have been substantial; but, coming a week after the U.S. had already begun its troop withdrawals from Lebanon, that impact was virtually nil. In late June, an Egyptian press headline had declared, "Russian Submarines in Albania Supplied with Atomic Missiles."[36] Now, the Arab press completely ignored their presence.

On November 29, over a month after the last U.S. troops had been withdrawn from Lebanon, a Soviet naval detachment consisting of four destroyers, three tugs, and a cargo ship began a southbound transit through the Suez Canal.[37] As with the submarines, their appearance in the Mediterranean a few months earlier would have created quite a stir; instead, their significance lay in the importance that the Suez Canal would eventually assume for Soviet naval deployments east of Suez.

The U.S. resolve in 1958, and the Soviet Union's failure to act effectively in the face of it, left a lasting impression on President Nasser. From that time on, he had to consider the Sixth Fleet a very real factor in foreign policy decisions. Had the USSR been able to create the impression—as it had during the 1957 Syrian-Turkish crisis—that its navy (or its military power in general) had deterred the Sixth Fleet, Egypt would probably have been a good deal more receptive to Soviet efforts in later years to obtain access to Egyptian naval facilities. Instead, the USSR's failure to satisfy the expectations that it had fostered in 1956 and 1957 seriously undermined its credibility in the Middle East.

1959-61: Strained Soviet-Egyptian Relations

The USSR had delivered three more W-class submarines to Egypt in January 1958, and on January 1, 1959 it delivered an additional three Ws—bringing the total Egyptian submarine inventory to nine, all from the USSR. On January 18, 1959, the USSR and Egypt signed an agreement for the construction of a shipyard at Alexandria. That same month, however, the USSR's relations with Egypt deteriorated sharply in response to the UAR's open campaign (initiated in December 1958) against Egyptian and Syrian communists. As a result of this development, Soviet military assistance to Egypt was apparently suspended for a time. In any case, it is doubtful that the USSR had any tangible desire at that time to use Egyptian naval facilities, since its naval base in Albania was certainly adequate.

In September 1960, the Soviet Navy conducted its first major exercise in the Mediterranean. Some 20 vessels, including at least 10 submarines (eight from Albania and two from the Black Sea) took part in maneuvers in the Aegean Sea.[38] Their contingency scenario appears to have been an attack on seaborne NATO reinforcements to Turkey in the event of war.[39] This exercise indicated a probable intention by the USSR to step up its naval activity in the Mediterranean; but, with their base in Albania, the Soviets demonstrated no visible concern about access to Egyptian facilities. Had they foreseen an eventual need for the latter, they would surely have been more generous three months later, when they negotiated a new arms agreement with Egypt.

A Chinese military mission visited the UAR (both Egypt and Syria) in October 1960. While there, it met with the commander of the UAR Navy, Vice Admiral Sulayman Izzat, and inspected Egyptian warships and naval facilities (including a naval training center).[40] This may have encouraged Admiral Izzat to increase his "shopping list" when he went to Moscow as a member of an Egyptian military delegation in December 1960. He is reported to have asked the head of the Soviet Navy, Admiral S. G. Gorshkov, to arrange Soviet financing for the construction of a naval base at Abu Qir Bay (east of Alexandria) and to sell shipborne missiles and other sophisticated weapons to Egypt.[41]

Gorshkov apparently gave him friendly words and some promises; but, in terms of materiel, Admiral Izzat went back to Cairo empty-handed. The USSR sought, instead, to improve Egypt's handling of what it already had. Some 200 East Germans (mostly submariners) were sent to Egypt in 1961 to replace Poles who had been instructing Egypt's Navy.[42] But, if the USSR had not counted on needing access to Egyptian naval facilities, it had not counted on being thrown out of its Albanian naval base in May 1961 either. Even so, the USSR showed no immediate interest in Egyptian naval facilities, probably, because its break with Albania was not yet complete.

The Soviet-Egyptian estrangement remained in effect as Soviet criticism of the UAR's treatment of Arab communists continued unabated. Finally, in June 1961, Egypt opened up a propaganda campaign of its own, aimed directly at the Soviet media assailing Egypt. This, and "progressive" Egyptian economic reforms initiated in July 1961, tempered Soviet criticism somewhat; but the USSR again antagonized Egypt by being the first major state to recognize Syria when the latter seceded from the UAR on September 28, 1961. East Germany and Bulgaria recognized Syria the same day (October 7) as the USSR. In mid-October, a Bulgarian delegation visiting the Middle East dropped Egypt from its itinerary,[43] which included Iraq, Syria, and Tunisia (three of the Arab states most hostile to President Nasser). On December 15, Heikal wrote in Al-Ahram that Soviet aid to Egypt had been given out of self-interest.[44] In this context, Admiral Gorshkov's visit to Egypt in December 1961 was poorly timed.

1961-65: THE GROWING NEED FOR NAVAL FACILITIES IN THE MEDITERRANEAN

Admiral Gorshkov's First Visit to Egypt

That Admiral Gorshkov did visit Egypt at this time is one indication of the loss that the Soviet Navy must have been feeling from its expulsion from Albania. By then, the USSR had obviously given up hope of regaining its base there. On December 11, the day before Gorshkov arrived in Egypt, it announced that it was withdrawing all its diplomatic and trade missions from Albania.

The participation of Albanian-based submarines in the Soviet Navy's September 1960 exercise in the Aegean Sea had demonstrated the value of a Mediterranean base in the event of a NATO-Warsaw Pact war. Similarly, the USSR's desire for basing or access rights to naval facilities in the Mediterranean becomes more understandable when one appreciates the Soviet Navy's concern that its Black Sea Fleet not be bottled up behind the Turkish Straits during such a contingency.[45] The loss of their naval base at Vlone was made even more acute by the fact that the Soviets had apparently decided by this time in favor of forward naval deployments for strategic defense.[46] Warships capable of meeting enemy attack aircraft carriers (CVAs) on the high seas had recently begun to enter the Soviet Navy's inventory.[47] And some Soviet defense planners were also probably concerned about the eventual deployment of Polaris submarines to the Mediterranean, since that sea was a logical area for Western nuclear ballistic missile submarines (SSBNs) to target the USSR. These considerations probably mitigated reservations that some Kremlin political leaders must have had about sending Gorshkov to Egypt at this juncture.*

Once having approved Gorshkov's trip, however, there were certain factors that made that December the best month for the visit. The Egyptian Navy was due that month to receive additional Soviet-built naval units (probably a minesweeper and motor torpedo boats) that would take part in the "Victory Day" celebrations at Port Said later in December. The latter event, which honored Egypt's stand against "Western imperialism" in 1956, held great potential for emphasizing Soviet-Egyptian solidarity. Both factors could be exploited to ease the way for Gorshkov's visit. The actual step of getting him invited may not have been too difficult. Formally, Gorshkov went to Egypt at Admiral Izzat's invitation;[48] but this may have been a standing invitation that Izzat, as a matter of protocol, had probably issued to Gorshkov when he had visited the latter in Moscow.

---

*It should be remembered that, except for Egypt, there was really nowhere else in the Mediterranean for the Soviet Navy to go, Although Syria and Algeria would be wooed in later years for access to their naval facilities, neither offered a realistic alternative to Egypt in late 1961. Syria was too unstable and, having just left the UAR, too much of an unknown. Algeria had not yet become independent, and it was obvious that France would retain a naval base there once it did.

Admiral Gorshkov took along a mission of eight senior officers and flew to Cairo on December 12. The party appears to have been well received. A naval parade was organized in Gorshkov's honor, and Gorshkov hosted a banquet in Alexandria honoring Admiral Izzat, during which the latter thanked the USSR for its "noble help" to the UAR. Gorshkov visited naval training centers and the UAR Naval College in Alexandria, attended naval exercises there, and then went to Port Said, where he laid wreaths on the grave of the Egyptian Unknown Soldier and on the communal grave of Russian seamen who fell during World War I. Following that, he and his mission went to Cairo and then visited a nearby military factory.

This easily met the requirements of a traditional courtesy visit and was accomplished by December 18. But Gorshkov also met separately with Marshal Abdul Hakim Amer (the head of Egypt's armed forces) and President Nasser on December 20. He attended the Victory Day ceremonies at Port Said on December 23, and visited the Aswan High Dam sometime during the latter part of his stay. It is significant that when Gorshkov arrived in Egypt on December 12, it had been announced that he would only be there for 10 days;[49] but he did not leave until December 25.

No Egyptian or Soviet statement was ever made about the purpose of this long visit, but Gorshkov appears to have been laying the groundwork for closer Soviet-Egyptian naval cooperation, which, at a later date, might have facilitated Soviet access to Egyptian naval facilities. To have openly sought the latter then would have been inopportune; Soviet-Egyptian relations were still strained, and President Nasser was firmly wedded to the principles on nonalignment in 1961. When Israeli newspapers charged that Gorshkov was negotiating for a naval base, the UAR Embassy in Washington called the Israeli reports "completely erroneous" and said that Egypt's policy of "positive neutrality" ruled out any kind of alliance.[50] But if Admiral Gorshkov did not openly seek access to Egyptian naval facilities, to replace those lost in Albania, his ultimate goal must have been transparent to the Egyptians. The latter appear to have fully realized the USSR's predicament; on December 13, the day after Gorshkov arrived in Egypt, the UAR ambassador to Albania returned to Cairo for five days of consultations with UAR Foreign Ministry officials about the current situation in Albania.[51]

1962: Increased Soviet Naval Aid

The Gorshkov group apparently agreed to approve credit sales of additional naval vessels. In January 1962, two W-class submarines and two Skory-class destroyers arrived in Alexandria. The latter were the first Soviet destroyers to be transferred to the Egyptian Navy since two others of this class had been turned over in June 1956. Also in 1962, one or two T-301-class inshore minesweepers, one S.O.1-class subchaser, and three Komar-class guided missile patrol boats were delivered to Egypt.[52] The delivery of the Komar boats was especially significant, since these had only entered the Soviet Navy's own inventory around 1959 and marked a quantum jump in the weapons system potential of the Egyptian Navy. As such, they are an important indicator of the degree to which the USSR sought to ingratiate itself with Egypt.

1963: Soviet Forward Deployments, the U.S. Deployment of Polaris Missile Submarines to the Mediterranean, and the USSR's Request For Access to Egyptian Naval Facilities

Admiral Gorshkov wrote of forward deployments in Red Star, February 5, 1963,[53]

> In the last war, naval operations took place mainly near the shore and were confined, for the most part, to operative and tactical cooperation with the army. Today, taking into account the intentions of the aggressors and the role given to their navies in the plan for a nuclear attack against the socialist countries, we must be prepared to reply to them with crushing blows on naval and land objectives over the entire area of the world's seas.

The first concrete steps in this direction were almost unidentifiable as such. Significantly, they appeared at around the same time as Gorshkov's article. In the northern Pacific, Soviet medium-range TU-16 "Badger" aircraft overflew the CVA Kitty Hawk between January 27 and February 3 and the CVA Princeton between February 13 and February 16, 1963. In the eastern Atlantic, TU-95 "Bears" overflew the CVA Enterprise on February 12 and February 13—the first time that these Soviet long-range aircraft had conducted such an overflight. On February 22, four TU-95s overflew the CVA Forrestal southeast of the Azores Islands. On March 16, four TU-95s overflew the CVA Constellation some 600 miles southwest of Midway Island—the first such TU-95 overflight in the Pacific. By late March 1963, some U.S. defense analysts had recognized these overflights to be part of a calculated and deliberate global program and not isolated incidents as had been first thought.[54] However, the fledgling Soviet effort to establish the capability to counter U.S. naval strategic forces far from the USSR suffered a serious setback that same month.

On March 30, the United States announced that a Polaris SSBN was on patrol in the Mediterranean. On April 12, the Pentagon announced that a second Polaris SSBN had taken station there, and that a third would arrive later in the month.[55] This development seriously undermined the Soviet Navy's attempt to establish meaningful forward deployments, since its existing resources were still insufficient to check even the lesser strategic threat posed by the Sixth Fleet's attack aircraft carriers. In order to maintain a permanent naval presence in the Mediterranean to meet this challenge (both SSBNs and CVAs) the USSR needed access to naval facilities in the region itself—at least until its hard-pressed naval expansion program could render such facilities redundant.

Radio Moscow, in several broadcasts that April, warned the Arab World of the dangers inherent in the deployment of SSBNs to the Mediterranean; the "danger of thermonuclear tragedy" remained a frequent subject of Radio Moscow's Arabic-language program for the next few months. The Arabs were told, "Two of these Polaris submarines have arrived in the Mediterranean and are now maneuvering near the Arab coasts."[56] On May 20, the USSR formally proposed that the Mediterranean be declared a nuclear-free zone. When President Nasser told Le Monde that summer that he supported the Soviet proposal, Radio Moscow promptly emphasized the fact to the rest of the Arab World.[57]

The USSR's pressing need to secure access to Egyptian naval facilities may have also been a factor behind its expanded military assistance program in Egypt. In June 1963, the two states concluded their largest arms agreement reached before the June war of 1967. In the naval sector, it provided for the delivery of two submarines, two destroyers, over 30 Komar and Osa missile boats, and various lesser vessels.[58]

The closer state of Soviet-Egyptian cooperation may have led the USSR to believe that Egypt might now permit the Soviet Navy to use Egyptian naval facilities; but it was to be disappointed on this matter. President Nasser told U.S. Ambassador John S. Badeau late in 1963 that the Soviets had approached him, presumably that year, about access to naval facilities in Egypt but that he had rejected their overture.[59]

1964: Premier Khrushchev's Visit to Egypt, and the Soviet Navy's First Sustained Deployment to the Mediterranean

Premier Khrushchev's visit to Egypt in May 1964 reflected a major Soviet attempt to increase the USSR's influence there and in the region as a whole. But, while this appears to have been the visit's main purpose, Khrushchev showed concern for Soviet strategic interests as well. It is very important to note that, while in Egypt, he portrayed the U.S. naval presence in the Mediterranean as a threat to the region itself. In a speech in Port Said on May 19, he declared:[60] "The realization of the plan for stationing submarines equipped with Polaris missiles in the Mediterranean can become a great threat to the security of this area." He went on to say in the same passage:

> The colonialists now want to use aircraft carriers and other warships against the national liberation movement of the peoples, to bring the policies of neutrality and non-alignment into range of their ships' guns and missiles.... The imperialists want, with the aid of aircraft carrier diplomacy, to restore reactionary regimes in the countries of Asia and Africa.

This same theme—indicating Soviet concern about the U.S. strategic threat in the Mediterranean—was reiterated two months later, when the USSR encouraged, and sent observers to, the July 1964 Algiers Conference on the Denuclearization of the Mediterranean.[61] Radio Moscow, in an Arabic-language program lauding the Algiers conference, warned,[62]

> The atomic weapons which the USA and its allies are bringing to the Mediterranean must not only be regarded as fraught with the threat of tragedy, but also as a challenge to the countries of the Mediterranean—including the Arabs.

By characterizing Western naval forces as a common adversary, the Soviets may have been trying to gain Arab acceptance of a permanent Soviet naval presence in the Mediterranean. Such an acceptance would have facilitated Soviet attempts to obtain the use of Egyptian and other Arab naval facilities.

In June 1964, one month after Khrushchev had warned the Arabs of the threat that Western navies posed to their independence, the Soviet surface fleet began its first sustained deployment to the Mediterranean. At the conclusion of its cruise, <u>Izvestia</u> intimated that the Soviet Navy would return to the Mediterranean, "... not for the purpose of saber-rattling and intimidating the peoples, as the United States Sixth Fleet is doing, but to improve its naval and combat skills."[63] In fact, the 1964 cruise marked the beginning of a definite trend, culminating in the permanent Soviet naval presence in the Mediterranean today. That the Soviet Navy's Mediterranean deployment was seasonal then was largely due to its inexperience with replenishment at sea, long a standard feature of the U.S. and British navies. And, unlike the Sixth Fleet, which could count on NATO facilities to make small but necessary repairs, the Soviet Navy had no such facilities available to it in the Mediterranean.

Admiral Gorshkov's Second Visit to Egypt,
and the Soviet Navy's First Port Call
at a Major Egyptian Port

On March 10, 1965, Admiral Gorshkov arrived in Egypt at the head of another Soviet naval delegation. During his 10-day stay, he visited naval installations in Alexandria, the Egyptian air force academy, the Aswan Dam, and the Nasser Higher Military Academy. He also paid a second visit to the latter institution in order to deliver a lecture on modern science and naval operations. He did not meet with President Nasser, as in 1961, but did pay another call on Marshal Amer.

While his visit appears to have been motivated by strategic interests, it is significant that the USSR chose to use his trip for political "state interests" as well. East Germany's Walter Ulbricht had just visited Egypt, and Admiral Gorshkov declared in a speech at Aswan that the Soviet Government and press had clearly defined their attitude toward the ensuing crisis between Cairo and Bonn.[64] Gorshkov's speech, which also praised President Nasser and Soviet-Egyptian cooperation, has been one of his very few ventures into purely foreign policy matters.

Whether or not Gorshkov raised the issue of Soviet access to Egyptian naval facilities during his visit can only be a matter for conjecture. Having been rebuffed before, the USSR probably assumed a more gradualist approach. For example, Gorshkov might have sought Egyptian permission for the Soviet Navy to use the anchorage at the Gulf of Sallum. Located on Egypt's Mediterranean coast near the Libyan border, with a natural but underdeveloped harbor, Sallum had the low visibility necessary to alleviate at least some of Egypt's sensitivity about a foreign naval presence. In April 1966, the New York Times reported that Soviet naval vessels had, sometime in the past, made covert calls at Sallum, where they had allowed their crews a brief chance to stretch their legs ashore.[65]

There is also a good chance that Gorshkov discussed regular port calls, since the Soviet Navy made its first visit to a major Egyptian port that fall. Considering the increased level of Soviet military assistance (and the Egyptian desire to obtain still more), Egypt must have agreed—albeit with some reluctance—to this naval visit in order to help ensure continued Soviet aid. An Egyptian military delegation went to the USSR in August 1965* and returned 25 days later with the last major Soviet-Egyptian arms agreement reached before the June war of 1967. Soviet naval vessels called at Port Said, after the delegation's return, in September.

The visit of two destroyers, two submarines, and a submarine tender to Port Said was low keyed, reflecting Egyptian sensitivity in this matter. No publicity (Egyptian or Soviet) surrounded the visit, and the choice of Port Said was itself probably a conscious Egyptian decision. Port Said was not Egypt's primary port, and the city's population was accustomed to seeing foreign warships there because of the Suez Canal.

In December 1965, then First Deputy Minister of Defense Grechko went to Egypt at the head of a military delegation that included Admiral Sergeyev, then chief of staff of Soviet Naval Forces. During its visit, the delegation saw Alexandria and then went aboard Nasser's presidential yacht, which took them to Port Said and various Red Sea ports. One of the areas on the Red Sea coast that the delegation visited was Ra's Banas, where the USSR had helped develop a fishing port under a March 1964 agreement. Khrushchev had visited the area in May 1964, and it had been subsequently closed to Western diplomats. Grechko's visit there increased speculation that the Soviets wanted to use it to support intelligence trawlers (AGIs) in order to monitor Western fleet movements in the Indian Ocean.[66]

1966-67: INCREASING PRESSURE ON EGYPT

The Second Visit to a Major Egyptian Port

Two months later, Admiral Izzat paid another visit (February 11-23, 1966) to the Soviet Union, at Admiral Gorshkov's invitation. This time he came alone, rather than as a subordinate member of a larger delegation. He met with Gorshkov and Minister of Defense Malinovsky in Moscow and then went to Sevastopol, where he visited warships of the Black Sea Fleet.[67]

The following month, the Soviet Navy made its second overt port call to Egypt. Five vessels from the Black Sea Fleet—the guided missile cruiser <u>Dzerzhinsky</u>, a destroyer, two submarines, and an oiler—visited Port Said, March 20-25. The cruiser docked in front

---

*President Nasser was in Moscow for five days during the delegation's visit. It was his first trip to the USSR since 1958.

of the Suez Canal Authority building, and visits were exchanged between the governor of Port Said and the Soviet commanding officer. Soviet crewmen were allowed to visit the city and engaged in various sports activities with Egyptian naval personnel. This port call received discreet local publicity, and contemporary Western analysis suggested that this reflected Egypt's policy to gradually accustom its people to regard such visits as routine.68

Admiral Gorshkov's Third Visit

That Soviet access to Egyptian ports or naval facilities was still a prominent and undecided issue in Soviet-Egyptian relations at this time, however, can be seen in the inclusion of Admiral Gorshkov in Premier Kosygin's delegation to Egypt in May 1966. Gorshkov's presence becomes all the more conspicuous when one notes that the other members of Kosygin's delegation were the Soviet Foreign minister, the minister of Power and Electrification, and the chairman of the State Committee for External Economic Relations of the Council of Ministers of the USSR.

Gorshkov's business in Egypt appears to have been of the highest order. The Soviet-Egyptian joint communique issued at the conclusion of Kosygin's visit lists Admiral Gorshkov as one of the participants in the Nasser-Kosygin talks; the only Egyptian military officer so listed was Marshal Amer, Egypt's First Vice President and commander of its armed forces.69 Gorshkov's presence would seem to indicate that the Soviet Navy's access to Egyptian ports was still in contention and that Admiral Izzat's visit to the USSR in February had left the issue unresolved.

By sending Gorshkov to Egypt under Kosygin's aegis, the Soviets could be sure that their case would be heard at the highest level. Like Khrushchev before him, Kosygin portrayed the U.S. naval presence in the Mediterranean as a threat to Arab states. In a speech to Egypt's National Assembly, he said, "Warships of states situated far from the shores of the Mediterranean are constantly cruising on the waters of this sea. At present, imperialism is hoping to gain a new foothold in the Near and Middle East."70

Moreover, the delegation's economic weight strengthened Gorshkov's hand.71 Egypt was suffering from a more serious hard currency shortage than usual and needed relief from its debts, as well as additional economic assistance. Contemporary press accounts reported Nasser and Kosygin locked in hard bargaining, with Soviet access to Egyptian naval facilities as one of the possible issues at stake.72

Gorshkov's immediate goal was probably to secure a firm agreement from Cairo that would have allowed Soviet warships to make regular calls at Egyptian ports without having to obtain permission for each visit. The fact that there was only a single publicized Soviet naval visit to Egypt between the March 1966 port call and the June war of 1967 would seem to indicate that the Egyptians were, for the most part, still holding firm. That publicized port call, in August 1966, was, however, the first "official" Soviet naval visit to Egypt. As a necessary step toward winning Egypt's acceptance of a permanent Soviet naval presence in the Mediterranean, it must have offered Gorshkov some compensation.

A simple, but clever, face-saving gesture for Egypt cleared the way for the visit. On June 27, 1966, two Egyptian Soviet-built destroyers arrived at Sevastopol for a three-day good-will visit—the first, and only, such visit in Egyptian history. Thus, the USSR's first official naval visit to Egypt was made less than two months later" . . . to return a friendly visit."[73]

The United States, for its part, had already reacted to the Soviet Navy's March port call. In an apparent attempt to test Egypt's impartiality, the U.S. had requested a port call by two ships of the Sixth Fleet at Alexandria for early May. Egypt rejected the request and cited the previously scheduled visits of President Tito and Premier Kosygin in May as the reason for its refusal. However, Egypt was reported to have suggested October 1966 as an acceptable time for the visit.[74] The fact that the requested visit had only been "deferred," rather than rejected outright, indicated some intention to demonstrate impartiality; but the Egyptian position hardened again in June 1966.

On June 1, 1966, Radio Cairo's "Voice of the Arabs" alleged that the United States and Tunisia had concluded a secret agreement giving the United States base rights at the former French naval base at Bizerte.[75] When an earlier Cairo press report, which had appeared before Gorshkov's visit in May, charged that the United States was negotiating with Tunisia for that base, some Western observers had interpreted this as a possible advance justification for a similar Egyptian concession to the USSR.[76] But such an interpretation does not take into account President Nasser's proclivity to believe the worst of Tunisia's President Habib Bourguiba and his suspicion that the United States was backing pro-Western regimes in order to establish rivals to him in the Middle East.

More serious (and probably more genuine) Egyptian concern surfaced later that same June. Al-Ahram observed (June 21) that at the same time that the United States was extending support to Saudi Arabia's King Faisal and Jordan's King Hussein," . . . we see the U.S. Sixth Fleet sailing about in the eastern Mediterranean because

Syria's attempt at rectifying its stand toward Arab causes has begun to arouse the imperialist forces."[77] In context, a left-wing faction of the Ba'th party had taken power in Syria in February 1966 and was actively supporting Palestinian guerrilla raids into Israel. Apparently, the Sixth Fleet's deployment to the eastern Mediterranean raised the specter of another U.S. intervention (as in 1958) in the region.

On June 24, Heikal wrote his weekly article in Al-Ahram as an open letter to King Faisal, then in Washington. Heikal denied that Egypt had given the USSR naval bases and asserted that " . . . the UAR cannot possibly permit anyone to establish any military or naval bases on its soil or its shores." Moreover, he accused the United States of having focused attention on this issue in order that " . . . the outcry about the myth of the Soviet naval bases will frighten us so that we will hasten to permit the U.S. Sixth Fleet to call at some of our ports, so that we can prove our neutrality." But, according to Heikal:[78]

> We will not allow the U.S. Sixth Fleet to visit our ports. . . . We will not permit it to visit our shores, not because of alignment against you [presumably the U.S. or Saudi Arabia] or in favor of the Soviet Union, but from an attitude springing from the position of patriotism and nationalism and from the concept of freedom and independence.
>
> In our opinion, there is a difference between the Soviet Fleet and the U.S. Sixth Fleet. The Soviet Union has no fleet in the Mediterranean, only occasional units passing through which do not threaten us, but which come flying flags of friendship. But the U.S. Sixth Fleet is stationed in the Mediterranean. From extensive past experience, we have found that the Sixth Fleet moves to threaten Arab nationalism. Furthermore, the main officially declared aim of its presence in the Mediterranean is to protect Israel. Thus, it is an unfriendly fleet. This is how we feel, and so long as this is our feeling, not a single one of its units will be permitted to enter any Egyptian port.

Because of his close association with President Nasser, Heikal's weekly articles in Al-Ahram during the Nasser era were always closely scrutinized. This particular article is especially important for what it says, and does not say, about the U.S. and Soviet naval presences in the Mediterranean. Most obviously, it indicated that Egypt had decided not to permit the Sixth Fleet to visit Egypt after all. But it would be superficial to attribute this decision to Soviet influence. More likely, it reflected President Nasser's chagrin with the U.S. backing of King Faisal's "Islamic Alliance" scheme, which Nasser perceived as a deliberate threat to his position in the Arab world. Added to this, the volatile Arab-Israeli dispute had brought U.S. support of Israel back into the limelight.

Heikal's discussion of the Soviet Navy in the same paragraph, however, was probably related to the upcoming Soviet naval visit to Egypt. As close as Heikal was to Nasser, he must have known of the planned port call. This segment of his article was probably designed to offset unfavorable publicity that would likely arise when Egypt refused the U.S. permission for a similar visit.

But there is no indication in Heikal's article that Egypt had granted the Soviet Navy unhindered access to its ports. On the contrary, the Soviet Navy was acceptable for the very reason that it had " . . . no fleet in the Mediterranean, only occasional units passing through. . . ." Heikal is an astute writer, one who chooses his words very carefully. It is highly unlikely that he would have used the Arabic word abirah (literally, "transitory") to describe Soviet naval activity in the Mediterranean if Egypt had already granted the USSR the wherewithal to maintain a permanent naval presence there.

The First "Official" Soviet Naval Visit

Soviet intentions, of course, were another matter. Strong evidence exists that the USSR attached considerable political importance to its first official naval visit to Egypt. It should be pointed out that the Soviets carefully and clearly delineate between "official" and "business" naval visits. By doing so, they are able to use their naval forces to demonstrate political intent, when they so desire. The Egyptian press, apparently unschooled in this subtlety, had referred to the March 1966 visit to Port Said as an official visit, but the commander of the Soviet naval detachment visiting Egypt in August 1966 said that his was the first official Soviet naval visit to that country.[79] Significantly, the August port call was at Alexandria, rather than at Port Said. The political impact of such a visit is greater at a city like Alexandria, and this is probably one reason why neither of the Soviet Navy's first two regular port calls—at Port Said—were "official."

The importance to the Soviets of the August port call can be seen in the fact that the detachment was commanded by Vice Admiral G. Chernobay, chief of staff of the Black Sea Fleet. The Soviets appear carefully to select politically astute officers to head their official naval visits. Admiral Chernobay had already proven adept at handling important Soviet affairs in the third world, having been decorated by the Indonesian Government in 1964" . . . in recognition of his service in strengthening and developing the Indonesian Navy."[80] He was seen off at Sevastopol by the first deputy commander in chief of the Black Sea Fleet and by the head of the Black Sea Political Board.[81]

Admiral Chernobay left the USSR on August 1 aboard the guided missile destroyer (DDGS) Boikii and was joined by the other units of his detachment in the Mediterranean in accordance with a pre-arranged plan.[82] These other units were the destroyer escort Pantera, the depot ship Magomet Gadzhiev, and two submarines. It may be significant that two submarines were present during all three Soviet naval visits to Egypt. The Soviet Navy's most pressing need for Egyptian naval facilities was probably—as it had been in Albania before—for its submarines' use. Thus, the inclusion of the latter in each visit may have been designed to acclimate Egyptians to their presence. The choice of the Magomet Gadzhiev—from all the available depot ships in the Black Sea Fleet—is interesting because it was named for a Soviet officer who had come from one of the USSR's Muslim republics (Magomet is the transliterated Russian spelling of Muhammad). The Soviets have assiduously exploited their Muslim population in their dealings with the Middle East, and the presence of this ship during the first official Soviet naval visit to Egypt may not have been coincidental.

Admiral Chernobay's detachment arrived in Alexandria on August 6, and left on the morning of August 11. While in port, the ships' crews toured Alexandria and met with Egyptian sailors; a Soviet naval brass band also gave performances in the town. But, in addition to these traditional activities, Admiral Chernobay was an honorary guest at the official opening ceremony of the Alexandria branch of the UAR-USSR Friendship Society.[83] Since the Soviets tend to use their official naval visits for political ends, the two events probably reflect deliberate timing.

The question of timing and circumstance is also very important in that Admiral Chernobay announced at a press conference aboard the Boikii that the USSR was going to maintain a permanent naval presence in the Mediterranean—in order to counter U.S. harassment of Soviet ships there and in order to satisfy a security requirement.[84] This was the first official Soviet statement that the USSR's naval presence in the Mediterranean was to be permanent, and the fact that it was made in Alexandria is highly relevant. Whether or not it reflected Soviet confidence that, at long last, the Soviet Navy would be able to make regular use of Egyptian ports is difficult to say. Having finally passed the milestone of their first official naval visit there, this is certainly possible, but, if so, it reflected misplaced confidence.

As the man most responsible for ridding Egypt of the last vestiges of British colonialism, President Nasser was in no mood to grant the Soviet Navy exceptional privileges in Alexandria or any other Egyptian port. Nasser could not forget that Egypt's diplomatic independence from the United Kingdom in 1936 had been marred for years by treaty stipulations that had allowed the Royal Navy to transfer its Mediterranean headquarters from Malta to Alexandria. While the Soviet naval detachment was still in port, he told students at Alexandria University: " . . . all that has been said about Alexandria becoming a base or supply base for the Russian fleet is nonsense."[85] As their expulsion from Egypt in July 1972 bore out, the Soviets never fully appreciated Egypt's sensitivity on this issue.

Attempts to Balance Soviet Pressure

In what was widely thought to be an attempt to discount Western speculation that it had granted bases to the Soviet Navy, Egypt instituted a remarkable "open port" policy in the period immediately following the Soviet Navy's August 1966 visit. On August 16, three Turkish destroyers arrived in Alexandria for a four-day visit. By itself, this visit was not unexpected; the invitation had been issued soon after the Sevastopol-bound Egyptian destroyers had called at Istanbul in June. Moreover, it could be viewed as a logical outgrowth of Egyptian-Turkish relations, which had been steadily improving since their nadir during the Cyprus crisis of 1964. But, as a port call from a NATO country, the visit took on added significance when, on the day that the Turkish detachment left, the Egyptian press disclosed that a French warship would also visit Alexandria.

That visit (August 25-28), by a single destroyer escort, was low keyed, but the fact that it occurred at all is surprising.* Although Franco-Egyptian relations had been showing definite signs of improvement since the beginning of the year, France's Middle Eastern policy still tended to be pro-Israeli (three French destroyers paid a one-week port call at Haifa in November 1966). It was the first French naval visit to Egypt since the Anglo-French invasion of 1956.†

---

*The Alsacien was returning to Toulon after a tour of duty in Djibouti. (La Revue Maritime, no. 236 [October 1966], p. 1230.) Although it had to pass through the Suez Canal to get to Alexandria, the memories of the Anglo-French invasion of 1956 obviously made Port Said unacceptable as a port of call.

†There have been no French or Turkish warship visits to Egypt since August 1966.

The most startling reversal of Egypt's port call policy, however, came when the U.S. Embassy in Cairo announced on August 22 that two Sixth Fleet destroyers would visit Port Said early in September.[86]

On September 2, slightly over two months after the editor of Al-Ahram had written that Egypt would not permit Sixth Fleet vessels to visit Egyptian ports, the Sixth Fleet destroyers Jonas Ingram and Stribling steamed into Port Said for a three-day visit. It was the first time that a U.S. naval vessel had been in an Egyptian port since the destroyer Soley had picked up an emergency food shipment for Kenya from Alexandria in 1962; it was the first regular U.S. naval visit to Egypt since 1954.[87]

To keep the port call in perspective, it should be noted that it was made at Port Said, not at Alexandria, as had been requested the previous spring. Furthermore, the BBC Monitoring Service failed to pick up any mention of the visit by Radio Cairo.[88] But even if the Egyptian Government was trying to downplay this obvious policy reversal to its own people—as it appears to have been doing—it was still an extraordinary move. As recently as late July 1966, Yemen (which followed Cairo's foreign policy direction at the time) had rejected a U.S. request for a naval visit to Hodeida.[89]

The explanation that Egypt's sudden "open port" policy represented a deliberate effort to discourage Western speculation about Soviet naval access to Egyptian ports falls short by itself. Egypt's stated policy, after all, was to ignore such speculation. As noted earlier, Heikal's June 24, 1966 article in Al-Ahram had specifically accused the United States of trying to force Egypt into just such a policy by spreading stories of a Soviet naval presence in Egypt. Heikal had declared, "If we issued a denial about all the stories fabricated about us, we will spend a lifetime doing nothing but issuing denials."[90] After President Nasser had told students at Alexandria University (August 1966) that talk of Soviet naval bases in Egypt was "nonsense," he went on to say, "It is our policy not to reply to such talk. . . . We shall not reply to such talk. We shall not answer those who spread such talk."[91] Furthermore, the current state of U.S.-Egyptian relations did not seem to warrant the visit.* Thus, there appears to have been no positive incentive to have allowed Sixth Fleet units to visit Egypt at that time.

---

*Diplomatic relations between the two states had been especially cool since the U.S. PL-480 wheat aid agreement with Egypt had expired on June 30, 1966. On August 23, State Department spokesman Robert McCloskey had specifically declined to characterize the upcoming naval visit as a reversal or turning point in the strained U.S.-Egyptian relations. (McCloskey, "Transcript of Press and Radio News Briefing, August 23, 1966.")

It is this writer's conclusion, however, that there was a negative factor—from the USSR—to warrant this dramatic step. John S. Badeau (U.S. Ambassador to Egypt, 1961-64) has written,[92]

> The assumption on which some Arab states have felt free to develop close ties with the Soviets in such vital matters as the supply of arms and large economic assistance has been that if these connections should lead to the brink of Soviet control, the United States would act to prevent the tumble over the edge. Hence the surprisingly muted Arab criticism of the "imperialist" Sixth Fleet in the Mediterranean which is recognized (although seldom publicly admitted) as a final barrier against overt Soviet action.

Although Arab criticism of the Sixth Fleet became a good deal more frequent after the June war of 1967 (Dr. Badeau's book was begun before that event), Badeau's analysis can be applied to this period to obtain a credible explanation for Egypt's behavior in August and September 1966.

It is quite possible that Egypt, itself, may have become alarmed at Soviet expectations, which the USSR may have felt were reasonable now that it had finally been permitted to make its first official naval visit there. The Soviets may have felt that it was time for Egypt to show some tangible gratitude for all of the naval and other military aid it had received from the USSR, especially that year. The Soviet Navy had transferred a R-class submarine to Egypt in February 1966, and had replaced two of Egypt's W-class boats with two Rs in May 1966;* a total of five of these more modern submarines were delivered by the end of the year. In addition, four to five Komar guided missile patrol boats were added to Egypt's existing inventory of three by the end of 1966. More importantly, 10 to 12 of the more advanced Osa guided missile patrol boats were delivered in 1966. Other deliveries that year included rocket assault ships of the Polnocny class, several utility landing craft of the MP-SMB 1 type, and Okhtensky class fleet tugs.[93]

The events of July 1972 showed just how capable the Soviets were of taking too much for granted in Egypt, and this may very well have been the case at this juncture. When Admiral Chernobay announced in Alexandria that the USSR was going to establish a permanent naval presence in the Mediterranean—in direct contradiction

---

*It may be significant that the February 1966 transfer occurred the same month that Admiral Izzat visited the USSR and that the May 1966 transfer occurred the same month that Admiral Gorshkov visited Egypt.

to Heikal's earlier description of the Soviet Navy's Mediterranean presence—the Egyptians must have asked themselves just how the Soviets expected to accomplish that, without using Egyptian ports.

After the first official Soviet naval visit to Egypt, one might normally have expected to have seen more frequent Soviet port calls there. Instead, there were no publicized Soviet naval visits to Egypt until after the June war of 1967, even though there were enough Soviet warships in the Mediterranean during that period to have, technically, justified numerous visits. Turkish Straits data indicate that a cruiser and destroyers from the Black Sea Fleet were in the Mediterranean as late as the middle of December 1966, and back again as early as the end of February 1967.[94]

Two Bulgarian Riga-class destroyer escorts, however, did visit Egypt in late October 1966. In context, the head of the Bulgarian Navy had already flown to Egypt as a member of a Bulgarian military delegation. The warships flew his flag, and their port call appears to have been directly related to the signing of an Egyptian-Bulgarian military-cultural agreement. As such, it is a singularly unusual visit. It has been the only Warsaw Pact warship visit (except for the USSR) to any Arab state—even though there have been a number of military agreements signed between members of this group and various Arab states. Moreover, Bulgarian warships were a rarity in the Mediterranean at the time.*

One possible explanation for the visit is that it might have been made at Soviet urging and represented a surrogate Soviet naval visit. If the USSR and Egypt were stalemated over the future of Soviet naval visits, the USSR might have felt that the Bulgarian port call could serve as a wedge to keep the option of more frequent Soviet visits alive. Because of the impending agreement with Bulgaria, it would have been difficult for Egypt to have refused the visit. But, although the port call was covered by Bulgarian news media,[95] this writer's research has failed to uncover any mention of it by Egyptian news media—even though the latter did give normal coverage to the concurrent visit of the Bulgarian military delegation.

In late November 1966, Admiral Izzat accompanied Marshal Amer to the USSR. Ostensibly, the visit was in response to an

---

*Bulgarian warships have periodically exercised with the Soviet Navy in the Mediterranean since the spring of 1969. But Turkish Straits data indicate that there were no Bulgarian warships in the Mediterranean between this port call and May 1969; and the last time that anything larger than a minesweeper had been in the Mediterranean was in 1957, when a Bulgarian destroyer carried a governmental delegation to Albania.

invitation issued during Kosygin's visit to Egypt the previous May. But the Arab-Israeli dispute had heated up considerably early that November, and a Soviet propaganda program broadcast to the Middle East just before Amer's visit stressed this connection.[96] Thus, one might have expected some new arms agreement to have been reached, but this does not appear to have been the case. It is impossible to tell to what degree Soviet-Egyptian differences might have been a factor, but Egypt's apparent reluctance to allow the Soviet Navy regular use of its ports may have played a part.

Admiral Gorshkov's Fourth Visit to Egypt

In late January 1967, Admiral Gorshkov stopped over in Egypt on his return from Ethiopia. He had gone to the latter country to attend its annual Navy Day's celebration, and it has been suggested that his attendance at that event was a cover for his trip to Egypt.[97] Little is known about his visit beyond the fact that he was met by Admiral Izzat at Cairo airport on the evening of January 28, and was seen off by Admiral Izzat and Air Chief Marshal Mahmud* on January 31.[98] The visit was much shorter than his previous trips to Egypt and is peculiar in other respects as well. It was, ostensibly, a sidelight to another trip; Egypt had been the sole objective of his previous visits to the region. And the USSR made no mention of his presence in Egypt, even though TASS had reported his departure for, and arrival in, Ethiopia;[99] his three previous visits to Egypt had been reported by Soviet news media.

Using the overall pattern of Soviet military personnel visits to third-world states as a guide, Gorshkov's January 1967 trip to Egypt appears to have been a sensitive business visit—probably related to a Soviet-Egyptian stalemate over the future of Soviet naval visits. There is some evidence that the Soviet Navy was planning a major winter deployment to the Mediterranean that February,[100] and Gorshkov may have been trying to arrange for some of the ships to call at Egyptian ports or to use anchorages off the Egyptian coast. The absence of any publicized Soviet port calls to Egypt until after

---

*There is no obvious reason why the Air chief marshal should have helped Admiral Izzat see Admiral Gorshkov off. Perhaps, during his visit, Gorshkov had also sought permission for Soviet naval aviation to operate out of Egypt. It should be remembered that after the June war, Egyptian-based TU-16 Badger bombers—with Egyptian markings but flown by Soviet crews—conducted regular surveillance of Sixth Fleet vessels.

the June 1967 war, however, would seem to indicate that his mission was unsuccessful. Soon after Gorshkov's departure, in an apparent attempt to lessen its naval dependence upon the USSR, Egypt agreed to extend its military cooperation with India to naval matters.[101] And, on February 22, President Nasser again rejected Western news reports that he had given the USSR naval bases in Egypt.[102]

1967: THE JUNE WAR AND ITS AFTERMATH

The Soviet Union Stands Back

Before and after Egypt's "maximum alert" on May 14, the USSR made a number of statements linking Sixth Fleet moves with anti-Arab intentions.[103] But in all of these statements, this writer has not found a single reference to the Soviet Navy—in the Mediterranean or anywhere else. The Soviet Navy did, in fact, react to Anglo-American naval movements in the Mediterranean; but its cautiousness in doing so was in keeping with the Soviet media's omission of any reference to it. The Soviet Navy began shadowing British and U.S. aircraft carriers with small vessels fairly early in the crisis but did not upgrade these "tattletails" with more powerful warships until June 2.[104] On May 22, the USSR notified Turkey that it intended to transit 10 warships through the Turkish Straits.[105] In accordance with Article 13 of the Montreux Convention, the transits could have begun on May 30; instead, only one auxiliary went through, on May 31. Destroyers—the first Soviet warships to go through—did not transit the straits until June 3.[106]

Although the Western press was filled with speculation about their import, the Arab news media appeared considerably less impressed with Soviet naval movements in the Mediterranean. Egyptian rhetoric against Anglo-American naval movements was profuse, but Egyptian reporting of the Soviet naval presence appears to have been limited to a few factual statements that Soviet warships were shadowing the Sixth Fleet.[107] No statements appeared, as they had during the Lebanon crisis of 1958, to declare that the Soviet Navy would deter the Western fleets. It has been observed that the Arab view of the USSR at this time paralleled Arab perceptions of the United States just prior to the latter's intervention in Lebanon in 1958—that the one superpower was capable only of words because it feared the reaction of the other if it attempted military action.[108]

When foreign journalists asked President Nasser, at his news conference of May 28, about reports of U.S. plans to send Sixth Fleet Marines to Israel—and whether Egypt would ask the USSR to intervene—

Nasser replied, "Naturally the dispatch of U.S. Marine units to Israel to protect it when it attacks us will be considered an act of aggression against us and the Arab nation. . . . We will not request any of the friendly states to intervene, but we will leave them to decide for themselves."[109] When war actually broke out, the Soviet Navy's behavior appeared to reflect indifference toward the Arabs' fate. Although the Soviet Navy later (much later) claimed to have had a hand in limiting the conflict,[110] it would have been difficult for the Arabs to have imagined a more awesome defeat.

The Establishment of a Soviet Naval Presence in Egypt

Paradoxically, the June war of 1967 brought the USSR what it had been unable to obtain through years of its own effort. President Nasser, in order to reequip Egypt's decimated armed forces, had to forgo any semblance of a meaningful bargaining posture with the Soviets—even though the latters' behavior during the war did not warrant special privileges. The Soviets were able to demand, and get, concessions that would have been unheard of before the war. They also capitalized on Egypt's fear of continued Israeli raids by making naval visits to Alexandria and Port Said in order to "deter" Israel from attacking those ports.[111] By late December 1967, Red Star was able to write: "Visits of Soviet warships to the UAR have become traditional."[112]

CONCLUSION

Had it not been for the cataclysmic effect of the June war on Soviet-Egyptian relations, it is doubtful that the USSR would have ever obtained the regular use of Egyptian naval facilities that virtually fell into its lap after that event. From Egypt's perspective, the Soviet naval presence that was established after the war shared no meaningful continuity with the prewar period. Admiral Gorshkov's four visits and the Soviet Navy's port calls before the June war had failed to originate any trend that could have logically culminated in a permanent Soviet naval presence there.

While it is not within the purview of this study to examine the origins of the June war of 1967, it is relevant to note that findings of such an examination could be used to support the highly controversial theory that the USSR wanted Egypt to lose that conflict.[113] Specifically, the USSR's abortive attempt before the June war to secure regular access to Egyptian naval facilities provides at least one plausible motive for such an attitude on the part of the Soviets. Certainly, the high degree of Soviet naval oriented activity vis-a-vis Egypt indicates that access to Egyptian naval facilities was a prime consideration of Soviet policy there well before 1967.

In retrospect, Admiral Gorshkov's series of four visits to Egypt (1961, 1965, 1966, and 1967) were unparalleled. He had made only one previous visit outside of the Soviet bloc, to Indonesia, in October 1961.* And, except for his visit to Ethiopia in January 1967, Egypt is the only foreign country that he is known to have visited from October 1961 until his April 1967 visit to Yugoslavia.[114] Moreover, Egypt and Indonesia were by far the largest noncommunist recipients of Soviet naval aid during the period under study; and Egypt especially stood out after Soviet military assistance to Indonesia was cut off in the wake of that country's anticommunist coup in 1965. Soviet naval assistance to Egypt is all the more conspicuous when one remembers that the Arab-Israeli dispute did not acquire a significant naval dimension until after the June war.

In fact, one is struck by the amount of military assistance (naval and otherwise) the USSR gave Egypt, for such a long time, for so little in return. The USSR's military assistance program in Egypt before the June war has been commonly viewed in terms of a superpower struggle for influence in the Middle East; and the 1955 arms agreement and subsequent military aid, up to Admiral Gorshkov's visit in 1961, can legitimately be viewed in this context. But the extent and characteristics of Soviet military aid—especially in the naval sector—to Egypt after that period make sense only if it served goals of much higher priority than those assumed by the simple influence theory.[115] By 1967, Soviet military assistance to Egypt had long since passed the point of diminishing returns from the standpoint of furthering Soviet influence in the Middle East; Egypt itself was taking much more than it was giving.

There were, however, sufficient Soviet strategic interests involved to have justified the military assistance program that had evolved by that time. Foremost among these was the direct threat to the USSR, posed first by the Sixth Fleet's attack aircraft carriers in the late 1950s[116] and then by Polaris SSBNs, as well, in the 1960s. Because existing Soviet naval forces were, both qualitatively and quantitatively, insufficient to meet this challenge, the USSR needed access to naval facilities in the Mediterranean region itself. It is doubtful that the Soviets wanted more (that is, sovereign bases) than this, since their intensive shipbuilding program during this period

---

*Like Egypt, Indonesia occupied an important place in the USSR's "out-of-area" naval strategy.

indicates that they viewed such an arrangement primarily as a makeshift alternative to a larger and more capable navy. But the latter could only be created over time; some indication of the difference that access to naval facilities in Egypt would have made to Soviet naval operations then can be seen in the intensity and duration of the naval activity that the USSR was able to sustain in the Mediterranean after it actually obtained access to those facilities following Egypt's defeat in the June war of 1967.

Because the stakes it was playing for were so high, the USSR proved willing to continue its considerable military assistance program there, even though the odds became increasingly poor that Egypt would satisfy the Soviet Navy's need for access to naval facilities in the region. Whether or not the USSR would have maintained that program at the same level once its own naval expansion program rendered those facilities redundant—or if it had been able to secure such facilities elsewhere in the Mediterranean—is questionable. The June 1967 war so changed the parameters of the Soviet Union's involvement in the Middle East that it is virtually impossible to tell. It is likely, though, that, by the time of the June war, the USSR had come to the conclusion that it was never going to get the full cooperation it sought from Nasser's Egypt in this matter. Egypt no longer represented the only realistic possibility, and the Soviet Navy's attention to Algeria, Syria, and Yugoslavia—especially from 1966 through the first half of 1967*—indicates that the USSR had already begun to seek alternative ports of call.

POSTSCRIPT

On April 3, 1974, in a speech in Alexandria, President Sadat said that Egypt and the USSR had concluded a five-year formal agreement in March 1968 on the Soviet Navy's access to "facilities on the Mediterranean."[117] Ironically, the editor of Look magazine had asked President Nasser in March 1968 if the latter would offer the Soviets naval "bases," and Nasser replied: "That question has never been brought up by their side or ours."[118]

---

*The USSR delivered Komar guided-missile patrol boats to Algeria and Syria during this period. Soviet naval units paid their first visit to Algeria in April 1966, and Admiral Chernobay led a second group of naval units there in November 1966. Admiral Gorshkov led a naval delegation to Yugoslavia in April 1967, less than a month after five Soviet naval units had paid a four-day "informal" visit to Split.

President Sadat had already revealed, in a speech in January 1971,[119] that Nasser had granted the Soviet Navy access to Egyptian naval facilities in 1968, but Sadat left the impression then and at other times that this arrangement was an informal and unwritten mutual understanding. In addition to being drawn up for five years, Sadat disclosed in his April 1974 speech that the 1968 agreement called for the two parties to decide three months before its expiration whether or not to renew it. Accordingly, in December 1972, Sadat had Field Marshal Isma'il call "the Russian general" at the Soviet Embassy in Cairo and "tell him that we had decided, on our part, to extend the facilities for another period." It is clear from the context of Sadat's speech, however, that this demarche carried with it a veiled threat not to renew the agreement if the USSR did not increase the quality and quantity of its arms deliveries to Egypt.

This appears to have had its desired effect, since the first Soviet military delegation to visit Egypt since the expulsion of Soviet advisors in July 1972 arrived in Cairo on February 1, 1973. A new arms deal was concluded that same month and Soviet arms deliveries began to reflect a willingness to meet Egyptian demands for more and better materiel. Thus the Soviet naval presence in Egypt appears to have been a major factor in the USSR's decision to revive its extensive military assistance to Egypt in 1973. This conclusion is reinforced by the fact that the Soviet Navy's use of Egyptian naval facilities was the USSR's most valuable, if not its only, strategic asset in that country after July 1972.

Although President Sadat indicated in his April 1974 speech that the 1968 facilities agreement had been renewed in 1973, he cast doubt on the future of that arrangement three weeks later when he told C. L. Sulzberger of the New York Times that the whole question of foreign naval access to Egyptian ports was under review and that Egypt might grant similar privileges to other foreign fleets (including the Sixth Fleet) as well.[120]

NOTES

1. From the second installment of Admiral Gorshkov's 1972 series, "Navies in War and Peace," in Morskoi sbornik, translated and reprinted in the U.S. Naval Institute Proceedings (February 1974), p. 33.

2. See, for example, Rene Mertens, "The Soviet Fleet in Arab Politics," New Middle East, no. 14 (November 1969), pp. 21-25. A more favorable interpretation of the Russian Navy's Mediterranean experience can be found in Boris Guriel, "Two Hundred Years of Russian Interest in the Mediterranean," New Middle East, no. 2 (November 1969), pp. 35-41.

3. Captain 2d Rank I. Bobkov, Candidate of Historical Sciences, "Presence of Soviet Fleet in Mediterranean Examined," Voyenno-Istoricheckiy Zhurnal, no. 9 (September 1970), pp. 34-37. Translated and reprinted in Translations of USSR Military Affairs, Joint Publications Research Service, no. 51712, November 4, 1970, p. 11.

4. Le Figaro, September 27, 1928, p. 5; London Times, September 27, 1928, p. 4.

5. New York Times, January 9, 1952, p. 6.

6. New York Times (quoting an Egyptian Government statement), October 19, 1951, p. 5.

7. Pravda, January 25, 1952.

8. New York Times, June 4, 1954, p. 2. The detachment returned to the Black Sea on June 6, 1954.

9. Al-Ahram, broadcast by Radio Cairo Domestic Service, April 14, 1967. This article was one in a series that Heikal wrote on U.S.-Egyptian relations.

10. London Times, August 4, 1956, p. 5.

11. London Times, August 4, 1956, p. 6.

12. Radio Moscow, in Arabic, to the Arab World, August 4, 1956.

13. New York Times, August 6, 1956, p. 3.

14. Radio Athens "Our Cyprus," August 6, 1956.

15. New York Times, August 8, 1956, p. 2.

16. Radio Cairo "Voice of the Arabs," August 6, 1956.

17. "Gun Boat Diplomacy or the Impotence of Sea [Power] in the Eastern Mediterranean," Egyptian Economic and Political Review, IV (2d series), no. 1 (December 1957), p. 20.

18. London Times, June 24, 1957, p. 9.

19. Middle East News Agency (quoting Al-Ahram), September 19, 1957; London Times (quoting Al-Akhbar), September 20, 1957, p. 8.

20. London Times (quoting Al-Akhbar), September 20, 1957, p. 8.

21. When the detachment had left the Baltic Fleet on August 31, Pravda had only announced that it would be visiting Yugoslavia. The naval visit to Syria was not announced by Radio Moscow until the night of September 18, when the detachment was already en route there from Yugoslavia. Furthermore, Syrian Defense Minister al-Azm stated that the Soviet naval units had been invited to visit Syria "... on the occasion of their presence in the Mediterranean." New York Times, September 8, 1957, p. 23; Radio Moscow, September 18, 1957; Radio Damascus, September 19, 1957.

22. Radio Cairo Domestic Service (quoting Al-Nour), September 5, 1957.

23. Some of the headlines in the Egyptian press on September 20, 1957 were "Soviet Navy Arrives in Syria"; "Soviet Ship Challenges NATO Navies and Enters Maneuver Area"; "Official American Spokesman Announces That Had the Soviet Ship Been Hit World War Would Have Broken Out." Radio Cairo Domestic Service "Press Review," September 20, 1957.

24. Radio Moscow, in Arabic, to the Arab world, September 22, 1957.

25. Middle East News Agency (quoting Al-Ahram) September 19, 1957.

26. J. M. Mackintosh, Strategy and Tactics of Soviet Foreign Policy (New York: Oxford University Press, 1963), p. 227.

27. Ibid. p. 229

28. Al-Akhbar, May 18, 1958.

29. Al-Ahram, June 19, 1958.

30. Al-Gomhouriya, June 21, 1958.

31. The judgment that the Egyptian press gave "unusual prominence" to the Soviet naval moves is a BBC Monitoring Service evaluation. BBC, Summary of World Broadcasts, Part IV, Daily Series no. 585, June 26, 1958, p. i.

32. Radio Cairo Domestic Service "Press Review," June 24, 1958.

33. Radio Cairo Domestic Service, June 24, 1958.

34. Al-Ahram, January 22, 1965.

35. London Times, August 22, 1958, p. 7.

36. Radio Cairo Domestic Service "Press Review," June 25, 1958.

37. New York Times, November 29, 1958, p. 12.

38. London Times, September 19, 1960, p. 11.

39. Michael MccGwire, "The Mediterranean and Soviet Naval Interests," International Journal, Vol. 27 (Autumn 1972), p. 522.

40. Radio Damascus, October 10, 1960.

41. Izzat may have also asked Gorshkov for three submarines and two large destroyers or light cruisers. Middle East Mirror, December 3, 1960, p. 3; Leo Heiman, "The Soviet Navy and Egyptian Bases," Jewish Frontier, vol. 29 (June 1962), p. 19. Heiman, an Israeli journalist, was an Israeli naval reservist at the time, and his article appears to reflect the Israeli Navy's views.

42. George Lenczowski, Soviet Advances in the Middle East (Washington, D.C.: American Enterprise Institute for Public Policy Research, 1972), p. 148.

43. Yitzhak Oron, (The Reuven Shiloah Research Center), ed., Middle East Record, Vol. II, 1961 (Tel Aviv: Tel Aviv University and Israel Program for Scientific Translations), p. 661.

44. Al-Ahram, December 15, 1961.

45. A statement by the Royal Navy's commander in chief Mediterranean that NATO had plans for preventing the exit of Soviet naval forces through the Turkish Straits in time of war was the subject of a Pravda interview with Admiral Gorshkov published on February 2, 1962. Current Digest of the Soviet Press, Vol. 14 (February 28, 1962), p. 28.

46. Michael McGwire, "The Background to Soviet Naval Developments," World Today, Vol. 27 (March 1971), p. 100.

47. In describing that year's annual Fleet Review at Leningrad, TASS (July 30, 1961) noted, "Also shown, for the first time, were rocket-carrying cruisers [DDGSs], destined for the destruction of large surface ships, such as aircraft carriers and cruisers. These ships are capable of making long voyages."

48. TASS, December 12, 1961.

49. Middle East News Agency, December 12, 1961.

50. New York Times, December 30, 1961, p. 5.

51. Middle East News Agency, December 14, 1961.

52. The figures for Soviet naval deliveries to Egypt in 1962 were drawn from various issues of Jane's Fighting Ships.

53. Admiral S. G. Gorshkov, "The Great Tasks of the Soviet Navy," Krasnaya Zvezda (Red Star), February 5, 1963.

54. New York Times (Hanson Baldwin), March 23, 1963, p. 6.

55. New York Times, April 13, 1963, p. 2.

56. Radio Moscow, in Arabic, to the Arab world, May 15, 1963.

57. Radio Moscow, in Arabic, to the Arab world, July 7, 1963.

58. Lenczowski, op. cit., p. 148.

59. Oral information obtained from Dr. John S. Badeau, April 1974. The author is, of course, very grateful for this information. He should also like to acknowledge his considerable academic debt to Dr. Badeau, his professor for two courses on Egyptian history at Georgetown University.

60. Premier Khrushchev's speech in Port Said, broadcast live by Radio Cairo Domestic Service and Radio Cairo "Voice of the Arabs," May 19, 1964.

61. The Algiers Appeal, issued at the conclusion of the conference, called for the evacuation " . . . of all nuclear weapons and all means of their delivery . . . " from the Mediterranean area. Radio Algiers Domestic Service, July 9, 1964.

62. Radio Moscow, in Arabic, to the Arab World, July 9, 1964.

63. TASS (quoting Izvestia), July 18, 1964.

64. Egyptian Gazette, March 20, 1965, p. 3.

65. New York Times (Hedrick Smith), April 5, 1966, p. 10.

66. New York Times, April 5, 1966, p. 10; London Times, May 11, 1966, p. 8.

67. TASS, February 12, 1966; Radio Moscow Domestic Service, February 12, 1966; Radio Moscow Domestic Service, February 16, 1966.

68. New York Times, April 5, 1966, p. 10.

69. Radio Cairo Domestic Service, May 18, 1966; Radio Moscow, in Arabic, to the Arab World, May 18, 1966.

70. Premier Kosygin's speech to Egypt's National Assembly broadcast live by Radio Cairo Domestic Service and Radio Cairo "Voice of the Arabs," May 17, 1966.

71. It is significant that the only economic aid Kosygin appears to have freely offered was maritime oriented. He reportedly told the director of the Suez Canal Authority, "If you want help in building ships, we are ready to cooperate with you." New York Times, May 16, 1966, p. 8.

72. London Times, May 18, 1966, p. 8.

73. TASS, August 1, 1966.

74. New York Times, May 24, 1966, p. 9.

75. Radio Cairo "Voice of the Arabs," June 1, 1966.

76. New York Times, May 11, 1966, p. 8.

77. Al-Ahram, June 21, 1966.

78. Al-Ahram, June 24, 1966.

79. Le Monde, August 11, 1966, p. 4.

80. TASS, May 16, 1964.

81. Radio Kiev Domestic Service, August 2, 1966.

82. Ibid.

83. A photograph of the event shows Admiral Chernobay in the front row, seated between the governor of Alexandria and the minister of the High Dam. Egyptian Gazette, August 11, 1966, p. 3.

84. Le Monde, August 11, 1966, p. 4.

85. President Nasser's answer to a question from an Egyptian student Alexandria University, August 7, 1966. Radio Cairo Domestic Service, August 8, 1966.

86. Although the United States had made the request the previous spring, State Department spokesman Robert McCloskey stated that Egypt had only formally replied to it " . . . a few days ago." "Transcript of Press and Radio News Briefing, August 23, 1966, 12:34 P.M.," Daily News Conferences, XL no. 165 (Washington, D.C.: Office of Press Relations, Department of State).

87. New York Times, August 8, 1966, p. 2; Le Monde, September 4-5, 1966.

88. BBC, Summary of World Broadcasts, Part IV, no. 2258 September 7, 1966, p. i.

89. New York Times, August 8, 1966, p. 2; London Times, August 18, 1966, p. 8.

90. Al-Ahram, June 24, 1966.

91. Radio Cairo Domestic Service, August 8, 1966.

92. John S. Badeau, The American Approach to the Arab World (New York: Harper and Row, 1968), p. 13.

93. The figures and dates for Soviet naval deliveries to Egypt in 1966 were drawn from various issues of Jane's Fighting Ships.

94. Republique Turque, Rapport Annuel sur le Movement des Navires a travers les Detroits Turcs, 1966 (Ankara: Ministere des Affaires Etrangeres, January 1967); Rapport Annuel . . . 1967. The author gratefully acknowledges his debt to Dr. Harry N. Howard of the Middle East Institute, who generously allowed the former to use his private collection of these reports.

95. Sofia BTA, October 27, 1966.

96. According to Radio Moscow (in Arabic, to the Arab World, November 21, 1966), "The present visit to the Soviet Union by UAR First Vice President Abdul Hakim Amer is especially important if we take into consideration that it is taking place under tense international conditions resulting from the expansion of imperialist and reactionary aggression."

97. MccGwire, "Mediterranean and Soviet Naval Interests," op. cit., p. 525. Although Soviet (and U.S.) naval units traditionally attend Ethiopia's Navy Days, it has been Admiral Gorshkov's only visit to that state.

98. Egyptian Gazette, January 29, 1967, p. 3; Egyptian Gazette, February 1, 1967, p. 3.

99. TASS, January 20, 1967.

100. According to the highly regarded Istanbul newspaper Cumhuriyet (February 20, 1967), which closely covers Soviet naval transits through the Turkish Straits, the USSR notified Turkey that it intended to transit 20 warships (two cruisers, eight destroyers, eight escorts, and two minesweepers) during the last week of February. Turkish Straits data (Rapport Annuel . . . 1967), however, indicates that only one cruiser, three destroyers, and one minesweeper entered the Mediterranean from the Black Sea that week. The USSR's practice of issuing false declarations ("contingency scheduling") in order to circumvent the Montreux Convention is a post-June war development. Thus, the Soviet Navy appears to have genuinely planned, and then rejected, a major winter deployment that year. It is interesting, and perhaps significant, that President Nasser's February 22 statement that Egypt had not granted the USSR naval bases falls between the date that the USSR would have had to have notified Turkey of its intended transits and the date of the actual transits.

101. Indian Express, February 10, 1967, p. 7. Ironically, Egypt had approached India once before in order to lessen its dependence upon another world power (that is, the United Kingdom) for naval assistance. New York Times, February 27, 1954, p. 3.

102. President Nasser's speech at Cairo University on the anniversary of the 1958 Egyptian union with Syria, broadcast live by Radio Cairo Domestic Service, February 22, 1967.

103. For example, for the period from May 14 until the outbreak of hostilities on June 5, this writer has found seven such broadcasts from TASS, five from Radio Moscow Domestic Service, and three from Radio Moscow's Arabic-language program beamed to the Middle East. Similar statements linking Anglo-American naval moves to the Arab-Israeli crisis were broadcast on Radio Moscow's U.K. and North America programs.

104. New York Times, June 3, 1967, p. 31; New York Times, June 4, 1967, p. 4.

105. New York Times, May 31, 1967, p. 16. According to the Egyptian Gazette (May 31, 1967, p. 1) five of the warships were to be of large tonnage, and five were to be destroyers and patrol boats.

106. Rapport Annuel . . . 1967, op. cit.

107. For example: "Meanwhile, more units of the Soviet Fleet pass through the Dardanelles to the Mediterranean to shadow the U.S. Sixth Fleet." Radio Cairo Domestic Service, June 1, 1967.

108. James Cable, "Political Applications of Limited Naval Force," in The Soviet Union in Europe and the Near East: Her Capabilities and Intentions (London: Royal United Service Institution, 1970), p. 57.

109. President Nasser's news conference, broadcast live by Radio Cairo Domestic Service, May 28, 1967.

110. "And if that aggression was soon curbed, if the war in the Middle East did not grow from a local into a big one, a definite part of the credit goes to the Soviet Navy." Admiral of the Fleet V. Kasatonov, first deputy c in c of the Soviet Navy, "The Mediterranean Is Not an American Lake," Soviet Military Review, no. 1 (January 1969), p. 54.

111. In an interview with Radio Jerusalem, then Israeli Chief of Staff Major General Itzak Rabin expressed the opinion that the presence of Soviet warships in Alexandria and Port Said was only a deterrent measure. (Radio Jerusalem Domestic Service, July 21, 1967.) As long as the Soviets took this limited, and purely defensive, posture—leaving the rest of Egypt open to attack—the Israelis appeared to have had little trouble adapting to the Soviet presence.

112. Egypt was the only state mentioned by name and in the context of Soviet port calls. Translation and condensation of three articles written on December 27, 28, and 30, 1967 by a special correspondent to Red Star, "Future Role of Soviet Navy," Survival, 10 (March 1968), p. 80.

113. The leading proponent of this view is Sir John Glubb. His arguments are set forth in his monograph The Middle East Crisis: A Personal Interpretation (London: Hodder and Stoughton, 1967).

114. MccGwire, "The Mediterranean and Soviet Naval Interests," op. cit., p. 525.

115. For a detailed analysis of motivational forces behind the USSR's military assistance program in Egypt, see Gur Ofer's "The Economic Burden of Soviet Involvement in the Middle East," in The U.S.S.R. and the Middle East, ed. by Michael Confino and Shimon Shamir (New York: John Wiley and Sons, 1973), pp. 215-46.

116. Beginning with the CVA Forrestal's deployment to the Mediterranean in February 1957, the Sixth Fleet had the A3D "Skywarrior," a twin-jet bomber capable of carrying nuclear bombs 1,400 miles without refueling. "If Little Wars Come: U.S. Sixth Fleet—The Punch behind the Doctrine," Newsweek, February 11, 1957, p. 49.

117. President Sadat's speech at the conference of the Egyptian Students Federation at Alexandria University, carried live by Radio Cairo Domestic Service, April 3, 1974.

118. Look, March 19, 1968, p. 64.

119. Reportage of President Sadat's speech to a political conference at Cairo University on January 8, 1971. Radio Cairo Domestic Service, January 8, 1971.

120. New York Times, April 22, 1974, p. 7.

CHAPTER

14

SOVIET POLICY IN
THE PERSIAN GULF
Oles M. Smolansky

The reasons for Moscow's current interest in the Persian Gulf are many and varied and are conditioned by the general international situation—that is, by the Kremlin's relations with Washington and Peking, by the requirements of the Russian position and policy in all of the Middle East, and, possibly, by long-range interest in the area's oil and natural gas resources. More precisely, the Indian Ocean, and specifically the countries along its northwestern perimeter, from Somalia and South Yemen People's Republic in the west to Pakistan, India, and Sri Lanka (Ceylon) in the east, has emerged in the late 1960s-early 1970s as one of the main geographic areas of Soviet concern. This trend has manifested itself, among other things, in attempts to establish close working relations (political and, wherever possible, military) with a number of riparian states and in the establishment of a permanent Soviet naval presence in the Arabian Sea.

As I have noted in an earlier work,* this Soviet activity cannot be ascribed to any single consideration but is due to a number of factors, ranging from Moscow's concern about the possible U.S. deployment of strategic nuclear submarines in the Arabian Sea and elsewhere in the Indian Ocean to political competition in Asia with both the United States and the Chinese People's Republic and, finally, to purely domestic political considerations. Since the Persian Gulf is a geographical extension of the Arabian Sea and since, the strategic

---

*"Soviet Entry into the Indian Ocean: An Analysis," in A. J. Cottrell and R. M. Burrell, eds., <u>The Indian Ocean: Its Political, Economic, and Military Importance</u> (New York: Praeger Publishers, 1972), pp. 337-55; reprinted in M. MccGwire (ed.), <u>Soviet Naval Developments: Capability and Context</u> (New York: Praeger Publishers, 1973), pp. 407-24.

aspect notwithstanding, the competition between Moscow and Washington and Moscow and Peking has been conducted mainly along political lines, it could have been expected that great-power rivalry would be extended to the gulf as well. As Chinese influence there has so far been negligible, the main thrust of the Soviet political effort has been directed at undermining the positions of the West generally and of those local regimes that have shown no desire to normalize relations with the USSR. The Russians have approached this latter task, especially, in a very gingerly fashion, seeking accommodation with existing regimes first and resorting to "subversion" only as a last resort.

THE PERSIAN GULF IN THE MIDDLE EAST CONTEXT

Moscow's stand in the Persian Gulf has been influenced by the general requirements of its overall Middle Eastern policy. To appreciate the delicate nature of the Soviet position, it is imperative to distinguish between, on the one hand, the non-Arab, "northern-tier" states—Turkey, Iran, Pakistan—which, in the 1960s, opted for a degree of normalization of relations with the USSR while retaining a basically pro-Western orientation and, on the other hand, the Arab world, which, in the 1950s and 1960s, split into what is typically described as "progressive" or nonaligned and conservative or pro-Western camps. In the former category, the Kremlin has established close working relations with Egypt, Syria, and Iraq, to mention only the most prominent members of the first group, but has failed to make significant inroads into the latter camp. After the death of Gamal Abdel Nasser the Soviet position in Egypt has been substantially weakened, culminating in President Anwar al-Sadat's order of July 1972 for the withdrawal of some 20,000 Russian military advisers. This was followed, in the wake of the October 1973 Arab-Israeli war, by a marked improvement in relations between Egypt and the United States, a process that has obviously disturbed the Kremlin. The Syrians, engaged in perennial delicate internal political maneuvers and thus unpredictable in their behavior, have turned out to be difficult partners to deal with as well. As a result of this deterioration of its position in the Eastern Mediterranean, which, since the 1950s, has been the main focus of Moscow's attention in the Middle East, the USSR has grown progressively more interested in Iraq, whose importance, in the eyes of the Kremlin leaders, has been enhanced by its Persian Gulf location, its oil reserves, and its determination to play an active part in regional politics. Moreover, the Soviet task in Baghdad has been made easier by the relative isolation of Iraq from Arab politics (as manifested by the refusal to cut down its oil production in the wake of the October 1973 war) and by its traditional rivalry with Egypt for leadership of the Arab world. Thus, as Baghdad's influence in the Arab East has declined, accompanied by a marked deterioration of its relations with Iran, Iraq has displayed a growing interest in a closer association with the USSR. Their 1972 Treaty of Friendship was the most significant outward manifestation of this Moscow-Baghdad rapprochement. Unfortunately for the Soviets, like so many prizes easily won, their improved position in Iraq may well prove an empty "victory," for the very decline in Baghdad's own standing in the gulf and elsewhere, which facilitated the growth of the Soviet-Iraqi friendship, has made it a much less useful "tool" in pursuing Russian interests in the region.

Another reason for increasing Soviet interest in the gulf is the fact that the region harbors approximately two-thirds of the world's proven petroleum reserves and the general realization that the noncommunist industrial world will not be able to survive without it until new energy sources are developed and used on a large scale. Since this is not likely to happen soon, enhanced Russian ability to influence the affairs of the Persian Gulf would provide Moscow with enormous bargaining power vis-a-vis its Western competitors, a possibility that is viewed with considerable apprehension by most Western governments. Moreover, it is often argued in the West that the USSR's own petroleum reserves, while significant, may not suffice to guarantee sustained long-range economic development of the Soviet Union itself and its East European satellites. If correct, this requirement may one day make it necessary for the Kremlin to gain a meaningful foothold in one or several oil-producing states of the Middle East. While it is impossible at this juncture to assess either the quantity of Siberia's admittedly vast petroleum and natural gas reserves or the Soviet's ability to exploit them, one may reasonably assume that the USSR is in fact interested in gaining access to some Persian Gulf oil for both its own use and for possible manipulative purposes.

Therefore, it is popular in some Western circles to raise the specter of aggressive Russian designs in the gulf, based on this view of their likely intentions and on their improved military capabilities, including the dramatic buildup of the Soviet Navy. Arguments of this type disregard several crucial considerations. Thus, access to gulf oil will be secured more readily and surely through regular commercial channels than through any effort to take over the region. Similarly, Moscow's ability to interfere with the flow of petroleum to the West would necessitate actions that are bound to raise the risk of nuclear confrontation between the superpowers to an unacceptable level. In addition, in the gulf, the USSR is forced to operate in an environment that is singularly ill-suited for effective, large-scale

Russian penetration. Finally, the Soviet position in the gulf and elsewhere in the Middle East cannot be regarded in isolation from Moscow's total political and economic requirements.

DETENTE AND PERSIAN GULF OIL

Since the early 1970s, the Kremlin has expressed a strong and seemingly genuine interest in a degree of accommodation with the United States. This policy, better known as "detente," lends itself to different interpretations. Whatever else it may mean, however (and it should be noted that detente is an ambiguous and fluid concept whose meaning may change with altering perceptions on the part of both superpowers of both their respective interests and of the best ways of securing them in a rapidly changing world), it would appear that, for one thing, Moscow, no less than Washington, is determined to avoid a general nuclear war. The Kremlin also seems to be genuinely interested in substantially broadening economic cooperation with the industrial giants of the West, above all the United States, in order to secure a massive input of Western capital and technology. This is particularly true of areas in which the Soviets have traditionally lagged behind the West (chemical and computer industries are among the most notable examples) or where large-scale development of natural resources (oil, natural gas) could be undertaken on a solely domestic basis only by diverting capital from other vital sectors, above all defense. This is not to say that the Kremlin seriously believes in the possibility or even desirability of complete normalization of relations with the West, including the elimination of political and economic competition in some parts of the underdeveloped world or the resolution of basic ideological divergencies between "socialism" and "capitalism." However, it does mean that the Soviets must be aware of the fact that they cannot expect Western cooperation in developing their own industrial potential if, at the same time, the USSR is involved in efforts to undermine the industrial capacity of the West.

I am aware of the objections likely to be raised to this line of reasoning. For one thing, the West has until now refused to accede to some of Moscow's basic requests for economic cooperation for reasons that are in no way related to the Persian Gulf. Why, then, should the Kremlin show any restraint in this matter? Might it not make sense to pursue a more active policy in the gulf precisely because it would provide Moscow with the kind of leverage it now lacks to exert additional pressure and make the West more amenable to Soviet wishes? Finally, is it not true that the USSR, in calling for continued Arab embargo against the United States in the wake of the October 1973 war, has done precisely what this writer has argued it would not do?

In tackling the last question first, it must be kept in mind that Soviet pressure (if this, indeed, is the correct term) was limited to verbal calls for the continuation of the embargo, that these urgings were notably ineffective, and that no action whatever was taken to interfere directly with the oil trade. In addition, it was an open secret that some Arab oil was being diverted to U.S. ports in spite of the embargo, and it is very instructive that the Russians have judiciously refrained from making that fact a public issue. Similarly, nothing was ever said about Baghdad's decision to continue full-scale petroleum production. (The irony that this stand was taken only by Moscow's chief client in the Arab East has not escaped many observers.) Moreover, according to some (as yet unverified) reports, the USSR itself was delivering oil to the United States at the very time it was urging the Arabs to uphold the embargo. What all of this amounts to is that encouragement of the petroleum embargo was a gambit that had nothing to do with the Persian Gulf per se but was rather intended to exert political pressure on the United States and Egypt, of whose policies the Kremlin strongly disapproved. Put differently, Moscow's encouragement of a continued embargo testified to Soviet uneasiness over being excluded from active participation in negotiations designed to seek the separation of Egyptian and Israeli forces along the Suez Canal. It was also a relatively "cheap" way for the Russians to lend support to the more radical Arab line and thus, for the moment, counter continuing Chinese jibes about Moscow's ideological heresies.

Of particular importance in the context of this discussion, however, is the fact that the Soviets did nothing to interfere physically with the flow of Persian Gulf oil to the consumer countries. The reason for their inaction is painfully clear to all concerned—any active interference would automatically produce an instantaneous and resolute U.S. reaction resulting in a major international crisis and a possible confrontation between the superpowers. Few would deny that this is the last thing the USSR wants or needs. It is, of course, obvious that the outwardly hardened Soviet attitude is not likely to improve the Kremlin's chances to get Western economic cooperation on the scale the Russians are seeking. But it has also become apparent that the lack of U.S. commitment to accommodate Moscow in this respect (reference is made particularly to the attitude displayed by the Congress) has very little to do with the Soviet stand on the Persian Gulf.

CONSTRAINTS ON SOVIET POLICY

In assessing Russia's policy in the gulf, it should be borne in mind that Moscow's freedom of action in the area is also severely restricted by its general regional requirements in the Middle East as a whole and by local conditions in the gulf itself. In terms of the former, the Kremlin's support of the Arab cause against Israel has traditionally committed the USSR to extensive and close collaboration with Egypt and Syria. Nevertheless, as noted, Moscow's relations with Cairo have deteriorated sharply since 1972, while the partnership with Damascus, built as it is on the shifting sands of Syrian politics, has, from the beginning, rested on a tenuous foundation. This state of affairs has made it necessary to seek closer association with postrevolutionary Iraq. The latter, however, over the years has also presented the Kremlin with a number of serious political problems that do not lend themselves to easy solution. The Iraqis' stanch determination to preserve their country's independence and to pursue a foreign policy of which the Soviets have often disapproved, combined with insistence on conducting their domestic affairs in accordance with their own wishes—including occasionally bloody persecution of local communists and of the Kurds—have in a number of instances severely strained Moscow-Baghdad relations. Nevertheless, the USSR has not allowed these annoyances to lead to an open break with Iraq, for the loss of Soviet influence there would deprive the Kremlin of a needed counterweight to Egypt's periodic efforts to dominate the Arab East while also virtually eliminating any meaningful Russian presence in the Persian Gulf.

All in all, given the basic interests upon which the Kremlin's Middle Eastern policy apparently rests, it is imperative for Moscow to maintain a relatively close association with Iraq. In the process, Moscow may have gained some leverage over the other "progressive" Arab states (Egypt, Syria), but it has also aroused considerable suspicion about Soviet intentions on the part of the "conservative" regimes, above all Saudi Arabia and Iran, which effectively dominate the Gulf scene. The resulting mistrust and occasional overt hostility has not prevented widening economic cooperation between Moscow and Teheran or the establishment of diplomatic relations with Kuwait. The fact remains, however, that, with the exception of Iraq, the USSR has not made any significant inroads into the Persian Gulf: stanchly conservative and openly anticommunist Saudi Arabia, Bahrein, Qatar, the United Arab Amirates (UAA), and 'Uman have refused to have any but the most superficial dealings with the Soviet Union.

AVOIDING FOREIGN ENTANGLEMENTS

Moscow's problems in the gulf have been augmented by the endemically unstable local conditions. The most obvious current example is the war now being fought in the Dhufar province of 'Uman, where local security forces, trained and led by British officers, have with the aid of units of the Iranian Army been battling Marxist-led rebels. The latter, in turn, have been supplied by the Democratic Republic of South Yemen. In line with its ideological condemnation of "medieval" conservative leaders, of whom Sultan Qabus of 'Uman is a classic prototype, Moscow has given political and moral support to the Dhufari rebels, but it has been lukewarm at best. Some Soviet arms and ammunition have reportedly been channeled through Aden and Iraq. The Russians have had little choice but to back the rebels in light of China's initial support of the uprising, combined with Peking's accusations that they were ideologically complacent in the face of a conservative "onslaught" on a Marxist revolutionary movement. It remains to be noted in passing that, in line with its own efforts to improve relations with the Shah, the Chinese People's Republic has discontinued its support of the Dhufari rebels and left the Kremlin "holding the bag."

The rebellion serves as an excellent illustration of the difficulties Moscow has encountered in its efforts to deal with the Persian Gulf. The Soviet Union had nothing to do with the outbreak of the uprising and has never controlled the rebels' actions or policies. Nevertheless, due to circumstances beyond its control, it could not but get involved, no matter how marginally, in a situation that has negatively affected its desire to project a low and moderate profile. As a result, apprehensions about Soviet motives on the part of the Shah, and other conservative rulers with whom the USSR has been trying to establish and maintain friendly working relationships, have been increased.

Soviet difficulties have been exacerbated by the myriad of other disputes and rivalries that abound in the Persian Gulf. There are the periodic armed clashes between Iraq and Iran, accompanied by interference in each other's internal quarrels; territorial disputes between Iraq and Kuwait, Iran and the UAA, 'Uman and South Yemen, Saudi Arabia and the UAA; and "progressive" Baghdad's general hostility toward the region's conservative regimes, to mention but a few examples. Involvement in any of these problems, given the kaleidoscopic shifting of gulf politics, cannot benefit the USSR in the long run, and their generally cautious approach suggests that the Kremlin leaders are aware of the pitfalls awaiting anyone who wishes to become entangled in the region's affairs.

Moreover, their prudence may have been reinforced by the realization that changes in the gulf's present-day political system may not automatically redound to Soviet benefit either. Thus, replacement of conservative by "progressive" regimes, while satisfying to those who take their ideology seriously, may place Moscow in the uncomfortable position of having to choose between active support of the "progressives," who may or may not survive and whose actions would be exceedingly difficult to control, and inaction, which would further tarnish Russia's reputation as the leader of "progressive mankind."

CONCLUSIONS AND PROGNOSIS

On the basis of these observations, it can be concluded that Moscow's ability to influence events in the Persian Gulf states is extremely limited. As noted, this is true not only of such conservative regimes as Saudi Arabia, Kuwait, Bahrein, Qatar, the United Arab Amirates, and 'Uman, but also of Iran and "progressive" Iraq. Teheran has been willing to maintain extensive and mutually profitable economic relations with the USSR, but it has also jealously guarded its political independence, has adhered to a basically pro-Western foreign policy, and has openly persecuted local communists. As long as the shah remains in power, there is no likelihood of relations between the Soviet Union and Iran being anything more than correct and businesslike.

On a relative scale, as noted, the Russians have been more successful in Iraq than anywhere else in the region. This has been due to a number of factors that are outside Soviet control but that Moscow has succeeded in manipulating for its own purposes. As stated above, these include the relative isolation of the Iraqis from their conservative neighbors in the Persian Gulf and the rest of the Arab world, their strained relations with Iran, which could one day lead to the outbreak of large-scale hostilities, and their resulting need for extensive military backing, which the USSR has been willing to provide in exchange for Iraqi oil.

However, even though, at the time of this writing, the Soviet position in Baghdad is relatively more extensive and secure than anywhere else in the Middle East, Moscow's ability to translate these advantages into tangible gains has been rather limited. It is true that the Iraqi leaders are in no position to refuse to listen to Soviet advice when it is given and that, on some occasions, as in the case of the 1973 Iraqi-Kuwaiti border dispute, they responded to Russia's pressure and modified their original demands. Domestically, the Ba'th, in compliance with Soviet wishes, agreed to form a "united front" with the Iraqi Communist Party (ICP) and to grant some concessions to the Kurds. At the same time, it ought to be borne in mind that these "concessions" have been mainly illusory: The ICP remains clearly subordinate to the Ba'th, and the Kurds find Baghdad's interpretation of their demands for autonomy and the "reforms" instituted by the central government unacceptable. Kurdish nationalists, led by Mullah Mustafa al-Barazani and enjoying clandestine Iranian military support, are now poised for another bloody showdown with the Iraqi Army. It appears unlikely that Moscow, though strongly opposed to a serious internal weakening of its chief Middle Eastern client, can succeed in preventing hostilities between Iraq's Kurds and Arabs. Moreover, Baghdad's recent decision to sell and not to barter its petroleum—a move directed mainly against the USSR and its East European satellites—provides another illustration of the limits of Soviet influence in Iraq.

Even this brief and, of necessity, superficial account of the immensely complex political picture in the gulf should suffice to convey the impression of a volatile and potentially explosive area that, while most attractive to the Kremlin because of its economic and strategic importance, presents the USSR with virtually insurmountable difficulties. Since Soviet foreign policy under Brezhnev has been marked by relative restraint and realism and since Moscow, by its circumspect behavior, has demonstrated awareness of the pitfalls awaiting an ambitious outsider, it may be reasonably predicted that, in the foreseeable future and barring totally unforeseen developments, the Soviet Union will abstain from an active, large-scale participation in the affairs of the Persian Gulf except in Iraq.

Nevertheless, the USSR can be expected to continue showing a keen interest in the region. Its concern is partly a function of Moscow's general involvement in the Middle East, reinforced by the growing political and naval competition with the United States in the northwestern Indian Ocean, a competition that may be expected to intensify after the reopening of the Suez Canal. Since Soviet naval units first established a modest presence in the Arabian Sea in 1968, the Kremlin has broadened its base of operations, and Washington, judging by the Pentagon's plans to convert Diego Garcia into a major naval and air base, appears determined to reciprocate in kind. This superpower competition could spill over into the Persian Gulf, although it is difficult to conceive of it as an area where the Soviets can expect in the foreseeable future to score impressive military or political gains.

CHAPTER

15

THE SOVIET NAVAL PRESENCE DURING THE IRAQ-KUWAIT BORDER DISPUTE

Anne M. Kelly

DETAILS OF THE DISPUTE

Following 10 years of unsuccessful efforts formally to demarcate the Iraq-Kuwait border (see Map 15.1)—an issue that had been left unsettled when the Iraqis recognized the independence of Kuwait in 1963—Iraqi military forces attacked the Kuwaiti border post of Al-Samitah in the early morning hours of 20 March 1973. The immediate issue was an Iraqi desire to expand the defense perimeter around the port and naval base they are developing at Umm Qasr.

A word about the geography and strategic importance of Umm Qasr is therefore in order. The Al-Samitah post overlooks Umm Qasr and commands the sea approaches to the port from the Persian Gulf. The Iraqis' objective went beyond this border post, however. Subsequent to the attack, the Iraqis demanded that Kuwait cede a strip of coastline surrounding the border post, and the islands of Warba and Bubiyan as well. These areas lie at the entrance of the channel leading to Umm Qasr.

The ideas expressed in this chapter are those of the author. They do not necessarily represent the opinion of the Center for Naval Analysis, of which she is a member, nor of the Department of Defense and the Department of the Navy.

This chapter was largely completed before the Arab-Israeli war of October 1973. Subsequent reflections suggest that there may be linkages between Soviet actions in the Iraq-Kuwait crisis and preparations for the war. However, there is no firm evidence of a connection between the two. Consequently, a rigorous reassessment has not been undertaken here, although an attempt is made to identify some of the possible linkages between the two events.

MAP 15.1

Iraq-Kuwait Border Dispute

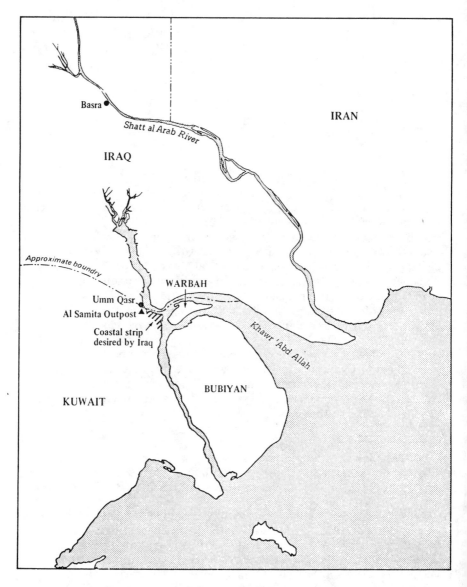

Umm Qasr is being developed as an alternative to Basra, currently Iraq's major port and naval base. For several years, Iraq's movement from Basra to the Gulf—accessible only via the Shatt al Arab River—has been constrained by the Iranian Navy. Teheran disputes Iraqi sovereignty over the eastern shore of the Shatt al Arab and has unilaterally denounced the treaty under which sovereignty was acquired by Iraq. Given the inimical state of relations between Iraq and Iran, Iranian control of the river is a less than satisfactory situation from Baghdad's perspective. Hence the development of Umm Qasr, which will free Iraqi movements from the watchful eyes and potential control of the Iranians.

Tension over the Iraq-Kuwait border had been growing since December 1972, when the Iraqis began building a road through the disputed area to Umm Qasr and augmented their troops along the border.[1] Negotiations in February of 1973 failed to ease the situation. The Iraqis continued to build their road, brought more military forces into the area, and gave every indication of the intention to unilaterally redraw the border. The Kuwaitis protested diplomatically, but to no avail. They then reinforced Al-Samitah and raised the Kuwaiti flag over the post. These moves apparently sparked the 20 March attack. The Iraqis continued to occupy the post until 5 April.

Following the attack, Kuwait asked the other Arab states to mediate. They responded immediately. Arab emissaries visited both capitals during the period 21-29 March. The Arabs generally disapproved of Iraq's methods. The moderate Middle Eastern regimes, particularly Bahrein, Jordan, and the Trucial States, were especially sympathetic to Kuwait. The Saudis, who have their own border problems with Iraq, moved troops to their northern border.[2] Bahrein and Jordan offered troops and military facilities to Kuwait.[3] The Arab mediation effort ended on 29 March and may have prevented an escalation of the conflict; but it failed to get the Iraqis to reduce their demands on Kuwait. In addition to the Arab states, Iran was also sympathetic to Kuwait, and the Iranian press was full of warnings and threats directed at Baghdad.[4]

The two Western powers with long-established interests in the area—the United States and the United Kingdom—adopted a "hands-off" policy. The United States has no leverage in Baghdad. The United Kingdom, according to the London Times, did not feel that the security of Kuwait was seriously threatened.[5]

SOVIET ACTIONS AND POSSIBLE OBJECTIVES

The Soviets immediately gave moral support to Iraq. On 21 March, TASS broadcast the official Iraqi version of the border incident, blaming Kuwait for the attack.[6] Except for this statement, however, Soviet media made no specific mention of the crisis. While the Soviets outwardly assumed a low profile in the crisis, it soon became obvious whose side they were on, and this is important. Despite their patron-client relationship with Iraq, the Soviets have walked a diplomatic tightrope in the gulf. They call it a policy of noninvolvement, but it is more accurately described as a policy of even-handedness toward interstate conflicts in the area. Or so it was until March 1973.

The second public Soviet move came on 30 March when TASS announced that Admiral Gorshkov would pay a "friendship" visit to Iraq "during the first half of April at the invitation of the Iraqi Ministry of Defence."[7] No specific reason for the visit was given; nor did the Soviets announce that the Soviet Navy would also be on hand—but it was. Gorshkov and a contingent of Soviet naval ships visited Iraq 3-11 April. Following is a chronology of the major events.*

20 March	Iraq attacks and seizes Al-Samitah border post
20-29 March	Arab mediation efforts
21 March	TASS blames Kuwait for the attack
30 March	TASS announces Gorshkov to visit Iraq in early April
3-11 April	Gorshkov and Soviet Navy visit Iraq
6-8 April	Iraq and Kuwait hold talks on the border problem
9 April	First anniversary of the Soviet-Iraqi Treaty of Friendship and Cooperation

The visit spanned the first anniversary of the Iraqi-Soviet Treaty of Friendship and Cooperation—so it is tempting to dismiss it as a routine diplomatic visit. Given the conflict setting in which it occurred, on the other hand, it is reasonable to ask if the Soviets were not using the treaty anniversary as a cover for some additional purpose. If so, what was that purpose? Four hypotheses suggest themselves.

The first is that the visit was not intentionally related to the crisis. This requires the assumption that Gorshkov's presence was scheduled prior to the outbreak of the crisis, and it could have been. The Iraqis then presented the Soviets with a fait accompli. The Soviets did not approve of the Iraqi action and wanted to remain neutral—in line with their past performance in the area—but the visit had been

---

*A more detailed chronology may be found in the Annex to this chapter.

scheduled and the treaty anniversary was upon them. They were trapped and could not cancel the visit without alienating their Iraqi client. A variation of this same hypothesis would be that the Soviets were not watching the situation closely and, oblivious to the political implications of their actions, simply went ahead as scheduled.

The other three hypotheses assume that the Soviets intended the visit to be something more than just an expression of good-will.

The second hypothesis is that the Soviet Navy went to Iraq to deter third-party intervention in the dispute. Perhaps they feared actions by the Iranians, or the Saudis, the United States, the United Kingdom, or Australia, all of which have interests in Kuwait and naval forces in the gulf or Indian Ocean.

Third, perhaps the Soviets meant a compelling show of force— to pressure the Kuwaitis to accept the Iraqi demands for the areas surrounding Umm Qasr.

Last, we considered the possibility that the Soviets, because of their treaty commitment, deliberately took steps to express solidarity with Iraq at the time but did not intend this show of force to affect the outcome of the crisis directly.

ANALYSIS OF SOVIET ACTIONS

The first hypothesis asked if the Soviet presence represented a routine good-will visit, which the Soviets either felt they could not get out of, or saw no reason for changing. For a number of reasons, neither potential explanation is convincing.

Could the Soviets have canceled their visit? They probably could have. In the past they have done so when warranted by a change in circumstances. They have canceled or postponed visits to Western European nations when relations suddenly took a downturn. If there were incentives for doing so where the West was concerned, there are even better reasons for doing so when a client is involved. Admittedly, it may be harder to do this, and it may cost something with the client—it may seem to signal a strain in relations. But if the objective is to avoid involvement in a client's squabble and remain above suspicion in the eyes of the third world, then one does not carry out a routine good-will visit in the wake of the client's aggression— when the political and military uncertainties of the situation loom large. There are consequently arguments and precedents for canceling visits when unforeseen events have altered the climate of relations between the USSR and host nations; and there are risks in not canceling, as well as in canceling, a visit under such circumstances.

Secondly, Moscow did not announce Gorshkow's visit until 10 days after the attack. During that time the situation in the gulf was very tense. The Soviets had not only a good reason to cancel a scheduled visit but plenty of time to alter course and devise a proper cover story. They did not do so.

Admittedly, they had a difficult calculation to make. With the treaty anniversary upon them, it could be argued that they could ill afford not to make some gesture of solidarity with the Iraqis. However, Moscow was represented at the treaty celebrations by the Soviet ambassador to Baghdad and a CPSU delegation, but not by Admiral Gorshkov. This is not surprising, given the makeup of Soviet delegations to analogous treaty events in India and Egypt. There is no precedent for a naval representation. Treaty signings and their commemorations are handled primarily by civilian party and diplomatic figures.

Soviet press coverage of Gorshkov's visit and the treaty anniversary contained no linkages between the two events. The closest thing to a connection is Gorshkov's arrival statement, in which he noted that his visit was "coincident" with the treaty anniversary. The Gorshkov and naval visits were, apparently, peripheral at best to the treaty celebrations.

Finally, the argument falls apart if the official Soviet position on the crisis is weighed. If Moscow wanted to remain above the issue, then why did it publicly take the position—which was never corrected— that Kuwait had attacked Iraq? In 1969, when Iran contested Iraqi sovereignty over the Shatt al Arab, the Soviet public position was that of "concerned but objective observer" urging both sides to resolve their differences peacefully. Neither side was accused of wrongdoing or blamed for precipitating the crisis. Moscow's official reaction to the Iraq-Kuwaiti border dispute signaled a different posture.

What about the variation on the first hypothesis—that Moscow slept through the crisis? The evidence suggests otherwise.

First, there is the evidence from the media. Although the press and radio did not refer to the crisis specifically, they were acutely sensitive to Western and Middle Eastern suspicions that the Soviets were masterminding Iraqi moves. They attempted to divorce Iraqi actions from Soviet policy. Particular effort was given to reassuring Iran. The Soviets also wanted to prevent the crisis from creating a bandwagon movement for great cohesion between the West and conservative gulf states. Indications from the media show that the Soviets were clearly aware of and worried about the situation.

The Soviets were in addition working behind the scenes to rein in their client. The day after the crisis erupted, the vice president of Iraq went to Moscow for a two-day visit. According to the Iraqi Foreign minister, the trip was urgent and connected to developments in the gulf.[8] Immediately after the vice president's return, Baghdad moved to cool down the crisis by agreeing to resume talks with Kuwait and withdraw from the border post prior to the talks[9]—meeting a Kuwaiti precondition for the resumption of negotiations. This suggests that Moscow had exerted pressure on Iraq to abandon the hardline policy it had been pursuing. The Soviets clearly had an interest in dampening the crisis. They had been working to pull the Arab world together, telling the Arabs that cohesion was a prerequisite for defeating Israel. To that end, Moscow was attempting to prevent inter-Arab squabbles from tearing apart the emerging superstructure of a united front and diverting Arab attention and resources from the main contest. While the Soviets were apparently sympathetic to Iraqi objectives, they probably did not approve of the methods used to gain the territory in question. So, the evidence from Soviet policy indicators—the press, the Iraqi visit to Moscow, and the effect of the visit on the crisis—is that Moscow was following events closely. The evidence does not support the conjecture that the USSR was duped into a particular course of action.

Finally, the Soviets now have a history of using what appeared to be a routine naval visit to demonstrate and implement commitments to a client in time of crisis. There is the "official visit" to Somalia in 1970, coincident with Somali allegations that they were about to be invaded by the Ethiopians. The Soviets not only stayed beyond the time needed for diplomatic amenities, but dropped all publicity for the visit—curious since such visits usually aim at publicity.[10] There are other relevant examples: a "good-will" visit to Egypt after the sinking of the Israeli destroyer Eilat,[11] and a "business call" to Sierra Leone in 1971.[12] There are precedents for the use of port visits for more than routine good-will and replenishment purposes.

The case at hand appears to be another instance in which a Soviet naval visit went beyond the limits of "good-will." A client was in a conflict situation. The Soviets had arranged for a visit by representatives of the military establishment. The visit came off despite the crisis. The Soviets publicly treated the visit vaguely, permitting the inference that they were quite cognizant that the local situation had added another dimension to their presence.

If the Soviets were not trapped into the visit, or did not simply ignore the implications and possible consequences, what did they hope to accomplish by their naval demonstration? This brings us to the second hypothesis: The possibility of third-party intervention was inherent in this conflict. The Iraqis were faced not only with evident Kuwaiti resolve in the face of their superior military might but also with the strongly negative reaction of the Arab world.* The

---

*The Iraqi Government appeared surprised by the generally negative Arab reaction to its seizure of Kuwait territory. The Iraqi Foreign minister in an effort to downplay the seriousness of the

Saudis sent some troops to their northeastern border. Ath-Thawrah, the organ of Iraq's Arab Socialist Union, singled out Saudi Arabia from among "imperalist and reactionary circles" for exacerbating the situation.[13] Outside the Arab world, Teheran made a gesture of support for Kuwait by offering to come to Kuwait's assistance if requested. The Iranian press was threatening reprisals against Baghdad. The naval forces of the West were also operating in the Indian Ocean and could be of concern. The Iraqis had some cause to be worried.

Whether the Soviets shared this concern is problematical. However, it was probably not the driving force behind their decision to deploy—but the evidence is incomplete and circumstantial, so the possibility that the deterrent motive was an element in the Soviet decision cannot be ruled out. Nevertheless, the following should be considered. We have no record of an instance where the Soviet Navy has deployed against the hypothetical possibility of third-party involvement. Their actions have been in response to manifest threats. There is no evidence that the West or Iran made any military or naval moves that the Soviets might interpret as an intent to become involved.

Verbal threats from the Iranian press must be distinguished from official policy. The Iranian Government took a more cautious approach, making a Kuwaiti request a prerequisite to assistance.[14] Official statements by Iranian spokesmen reflected a reluctance to become involved unless Iraq pushed its attack to the point of threatening Kuwait's existence. However, even if the Iranians were to become involved—or if the Soviets thought they might—it cannot be concluded that the Soviets would have seen this as sufficient grounds to move. There have already been two instances in which the Iranians resorted to gunboat diplomacy—in 1969, when they successfully contested Iraqi sovereignty over the Shatt al Arab River; and in 1971, when they seized the Tunbs Islands in the Straits of Hormuz. Despite the fact that the targets in both instances were Arabs, the Soviets did not react in either, even though the threat to their Arab friends was manifest.[15] Precedent therefore suggests that, where the Iranians are involved, or might become involved, the Soviets are extremely cautious. They have important economic and political interests in Iran, which they have shown they will not lightly risk.

---

dispute and the need for Arab mediation was quoted as saying "the issue does not deserve all this clamor." (An-Nahar in Arabic, 27 March 1973, pp. 1, 12.) Ath-Thawrah accused Kuwait of "magnify[ing] the incident hundreds of times . . ." and Saudi Arabia of "trying by all overt and covert means to exacerbate the situation and magnify the problem as much as possible." ("A Plan Exposed through Its Motives and Aims," 27 March 1973.)

The Saudi diplomatic line was more decidedly one of noninvolvement. A statement by the Saudi Royal Cabinet appealed to both countries to exercise "patience and self-restraint" and to settle the dispute peacefully.[16] Although it was rumored at the time that Saudi troops had entered Kuwait, these reports were denied by the Kuwaiti Foreign minister. Saudi troop movements on their northern borders appeared to be limited to defensive action,[17] motivated by the concern that Iraq might attempt a final solution to its border problems with both southern neighbors. Curiously, however, during the crisis period, Moscow suspended its standing propaganda campaign against the "feudal" and "reactionary" Saudi regime. This would seem to indicate Soviet concern that their own behavior not provoke the Saudis into providing protective support to Kuwait with military force. Whether the Soviets actually perceived Saudi troop movements as a signal of intent to become involved in the crisis is not known. However, the greatest potential for third-party involvement occurred in the week following the crisis. By the time the Soviets arrived on the scene—on the 3d of April—it was clear to everyone that the crisis had abated and that the dispute would be solved through negotiations rather than by further fighting. The Soviet presence coincided less with the conflict itself than with its resolution. If the Soviets were concerned about intervention, one would expect this to show up more quickly in their naval activity. Given the immediate negative reaction of Iraq's neighbors, one would expect Soviet units to have gone to Iraq as fast as possible after the border attack, rather than wait two weeks as they did. Finally, while third-party intervention was a possibility, it was never a strong probability, and the Soviets were in a position to make the same assessment.

Let us go on to the third, or "compellent show of force," explanation. The visits coincided with the resumption of talks between Iraq and Kuwait, 6-8 April. Was it just a coincidence, or did the Soviets want to put pressure on Kuwait? There is no evidence that the Soviets attempted diplomatic pressure on the Kuwaitis. Their presence was probably designed to lend support to Iraq's negotiating position regarding the contested territory, but the Soviets apparently did not cross the line between showing passive solidarity with Iraq and actively pressuring Kuwait. And it is difficult to hypothesize a credible motive for their doing so.

The fourth hypothesis—support of Iraq in this crisis—would be explained by the Peace Program promulgated at the 24th CPSU Congress and by the divergence between Soviet and Iraqi policies that seemed to be building up.

The Peace Program advocates cooperative efforts between East and West designed primarily for averting war—for example, liquidation of war in the Middle East; renunciation of the use of force and threat of its use; final recognition of territorial changes in Europe resulting from World War II; collective security in Europe; dismantling of the NATO and Warsaw military blocs; outlawing of nuclear, chemical, and biological weapons; prohibition of nuclear tests; creating of non-nuclear zones; and reduction of armed force, armaments, and military expenditures by the major states. In contrast to these measures to expand the network of agreements for reducing direct East-West tension, the Peace Program commits the Soviets "to wage a determined struggle against imperialism . . . firmly rebuff intrigues and subversion of aggressors . . . support the struggle of the peoples for democracy, national liberation[18] and socialism." As such, the Peace Program is describing a policy of "active defense of peace and strengthening of international security."[19] If we are interpreting the Peace Program correctly, then there is a cooperative-competitive dyad that governs Soviet foreign policy. At one level, that of direct relations between the blocs, cooperation designed to maintain the strategic balance, avert nuclear war, and promote detente is stressed. Out in the third world, it is competition for political influence with the West that seems to drive Soviet policies. This seems to imply that in the third world, the parameters for superpower behavior are different from those in the area of most direct East-West confrontation: Europe. In the third world, Soviet support for its friends, including that provided by the Soviet military establishment, is not only sanctioned but dictated by the broader Soviet objective of denying to the West any advantage that is detrimental to the interests of the USSR or its third-world clients. During the visit of the Iraqi vice president to Moscow following the border attack, Kosygin seemed to view the propriety of the dispute in terms of the Peace Program formula. In a veiled reference to the conflict, he observed that "inequitable and predatory agreements and treaties" remained in the aftermath of the disintegration of colonialism, which "set third-world nations at loggerheads." Kosygin went on to pledge continuing support for Iraq in "consolidating its national independence."[20] Under this interpretation, the naval presence would be seen as tangible support for border adjustments, which would more reliably insure Iraq's security against her less "progressive" and non-Soviet-oriented neighbors (Iran, Saudi Arabia, Kuwait). By supporting rectification of the territorial status quo between Iraq and Kuwait, the USSR objective was, in part, to lessen Iraq's dependence upon its neighbors' goodwill for the protection of its maritime and petroleum operations. In demonstrating for Iraq, the Soviets were also indirectly demonstrating against imperialism—by challenging a border arrangement left over from the colonial period.

Soviet behavior may also be explained by the divergence between Soviet and Iraqi policies that seemed to be building up in early 1973.

Iraqi domestic and foreign policies were being invigorated by the nationalization of the Iraqi Petroleum Company. The Iraqis had gained a new self-confidence and assertiveness in determining their own future. The immediate result was that Iraqi policies had become more adventuresome—and increasingly independent of Moscow's preferences. As a result, long-standing Iraqi-Soviet differences reemerged more sharply—treatment of the Kurds, a "united-front" government, Iraqi overtures to the West. For the Iraqis, a particular sore point was Moscow's effort to remain in good standing with both themselves and Iran. This led to an erosion of Soviet credibility and prestige in Baghdad. With Iraq showing growing signs of restlessness, the Soviets were no doubt concerned that Iraq might go the way of Egypt. The crisis offered the Soviets a low-cost opportunity to shore up their relations with Iraq—in particular to compensate for their lack of support of Iraq in past disputes in the area. This is not to say that—assuming this assessment of Soviet objectives is correct—the Soviets were successful. There do not seem to have been any post-crisis modifications in Iraqi policy that would indicate that the Iraqis were properly grateful for Soviet support. But demonstrating opposition to a border delineation left over from the "colonial" era was one way of attempting to gain credit with their client, and anything the Soviets could do to refurbish their "progressive" credentials with Iraq would go on the plus side of the ledger.

The generally negative world reaction to Iraq's aggression provided the USSR with an opportunity to demonstrate to Iraq the benefits of strong ties with Moscow; and, given the transitory nature of patron-client relationships, the benefits of such association have to be periodically demonstrated.

The foregoing assessment describes possible motivations for Soviet overt solidarity with Iraq. It is not intended to obscure the equally important inference drawn earlier—that the USSR was apparently discreetly pressuring the Iraqis to cool down the crisis. It may well be that Moscow objected not only (or perhaps, not so much) to Iraqi methods but also to Baghdad's timing. Now that the 1973 Arab-Israeli war is history, too, we know that it was originally planned for the spring of 1973.[21] Under this timetable, the Iraqi offensive could become a critical negative factor in sustaining Arab unity for the war against Israel. The seeming ambiguities in Soviet policy in the Iraq-Kuwaiti dispute may perhaps be due to the pursuit of conflicting objectives—to demonstrate support for the singular cause of a client while, at the same time, preventing that cause from fracturing the momentum that was building toward unity in the Arab world. The following summarizes the evidence for and against each hypothesis examined.

Hypothesis 1: Only a Routine Good-will Visit

Evidence for

- Visit may have been scheduled prior to crisis.
- Visit took place coincident with treaty anniversary.
- For these reasons, Moscow had to carry out the visit, despite intervening events.
- Soviet media carefully divorced Iraqi actions from Soviet policy.
- No evidence of direct Soviet involvement in the crisis.
- The Iraqis may have duped the Soviets into demonstrating more support for them than the USSR would have wished to do.

Evidence against

- There are precedents for Soviets canceling naval visits when a change in circumstances warranted.
- In this case, the Soviets had time and some incentive to change plans.
- If just a good-will visit, it is curious that the ships' visit was not announced.
- Gorshkov did not represent USSR at treaty celebrations.
- No precedent for a naval representation at friendship treaties' celebrations.
- Soviet media did not link Gorshkov's visit to the treaty anniversary.
- Prior to the naval visit, the USSR gave verbal support to Iraq—a departure from their past policy toward crisis in the Gulf.
- Moscow was worriedly watching events closely and not oblivious to the implications of their own behavior or duped into a particular course of action.
  (a) Media sensitivity to political implications for USSR of Iraqi actions.
  (b) Pressure placed on Iraqi Vice President to cool down crisis.
- Precedents exist for the use of "routine" naval visits to demonstrate and implement commitments to clients in time of crisis.

Hypothesis 2: Soviet Naval Presence Intended to Deter Third-Party Intervention

Evidence for

- There is insufficient evidence to conclude that Iran or Saudi Arabia might not have intervened, or that Western naval forces operating in the area might not have come to Kuwait's assistance.

Evidence Against

- No precedence for Soviet Navy's deploying against the hypothetical possibility of third-party involvement.
- Iran and Western powers made no military/naval moves immediately following the border attack that could be interpreted as an intent to become involved.
- Soviets show no verbal concern for Saudi troop movement; Saudi diplomatic line was noninvolvement.
- Soviet visit occurred after the height of the crisis when the possibility for third-party intervention appeared more likely.
- Soviets capable of assessing the credibility of third-party threat to Iraq.

Hypothesis 3: Soviet Naval Presence Intended to Put Pressure on Kuwait to Accept Iraqi Demands

Evidence For

- Visit coincided with the resumption of negotiations between Iraq and Kuwait.

Evidence Against

- No evidence that Soviets attempted to put diplomatic pressure on Kuwait.

Hypothesis 4: Soviet Naval Presence Intended as a Demonstration of Solidarity with Iraq

Evidence For

- There is sufficient evidence that more than a routine good-will naval visit was intended.
- Soviet behavior does not indicate that a compellent or deterrent show of force was intended.
- Soviet presence intended as an additional restraint upon the conflicting parties in order to maintain Arab Unity for the war against Israel.

Evidence Against

- Ambiguities and unknowns in this case preclude judgment.

To summarize, hypotheses 1-3 do not satisfactorily explain Soviet behavior in this case. Something else was intended. Hypothesis 4 is a more reasonable explanation of Soviet behavior. Of course, one could argue that the unknown and ambiguities in this situation preclude judgment or suggest another objective not identified here. But, on balance, there is sufficient evidence that something more than a routine naval visit took place—that is, it <u>was</u> an exercise in crisis diplomacy.

IMPLICATIONS FOR FUTURE SOVIET BEHAVIOR

What are the implications of this case for future Soviet naval diplomacy? In this case, Soviet naval operations had very low military content. Their policy was characterized by discretion and their actions posed little if any immediate threat either to Iraq's adversaries or to the balance of powers in the area. It is consequently tempting to dismiss this operation as a minor occurrence, with no appreciable effects. Indeed, supporting a client on the offensive may not mark a complete departure in Soviet use of their fleet for political purposes. There is the ambiguous case of the West African patrol, which may provide a precedent.* And the case of Iraq does not deny "the rules of the game,"[22] since Kuwait had no Western patron to assist it in this time of adversity. Nonetheless, Soviet behavior was unusual. This is the first time that we have seen the Soviet Navy undertake a demonstration supporting a client who was upsetting—rather than attempting to restore or maintain—the status quo between itself and another sovereign state.† Moreover, if we look at the Iraqi case together with the Soviet transport of Moroccan troops to Syria prior to the 1973 Arab-Israeli war,[23] there appear to be linkages between the two and to the war. Both actions supported the Soviet aim of getting and keeping the Arabs together for the war against Israel. It looks as if, in the third world at least, and outside the context of superpower naval interaction, the Soviets are becoming increasingly activist and

---

*The West African Patrol maintained since 1970 is clearly defensive in the tactical sense; it deters further Portuguese raids on Conakry, Republic of Guinea. Strategically, it is a defensive operation insofar as its objective is to prevent the overthrow of the Toure regime, but an offensive operation insofar as it permits the Guinea (Bissau) insurgency mounted from the Republic of Guinea to maintain and expand its offensive actions against the Portuguese colonial regime.

†Even if one considers the West African patrol to be a strategically offensive operation, it should be remembered that the target is a colonial regime that does not enjoy the respect of the international community.

adventuresome in the exercise of naval diplomacy. In the past, we have suggested that where the danger of the opposing superpower getting involved was nonexistent or minimal, the Soviets would be less inhibited in employing naval force to pursue their interests. These cases from the first half of 1973 tend to confirm this suspicion. The Soviets still seem to be experimenting with the use of naval forces to support their interests in the third world—and apparently there is still room for expansion in that role.

The growing use of naval forces in situations that go beyond the defensive suggests that intensification of Soviet naval diplomacy may be the trend for the foreseeable future. Especially now, in an era of detente, this trend appears anomalous. But, as the Soviet Peace Program implies, detente is a two-edged policy: cooperation and competition. Detente does not seem to encompass a lessening in the spirited competition on the periphery—even though the spotlight is on cooperation in the main arena of great power politics. Soviet behavior in the Iraq-Kuwait crisis provides an example of the Peace Program in operation.

NOTES

1. Selem Lawzi, editorial in Al Hawadith (Beirut), 21 December 1973.

2. Damascus Middle East News Agency (MENA) in Arabic to MENA Cairo 1045 GMT 28 March 1973, Foreign Broadcast Information Service (FBIS), Daily Report V, no. 62 (30 March 1973), p. E1; Paris, Agence France Presse, (AFP) in English 0911 GMT, 5 April 1973, FBIS, V, no. 66 (5 April 1973), p. B1. This is not to say that the Saudi border reinforcement and the Iraq-Kuwait border dispute are necessarily causally related. Incidents on the Saudi-Iraq border happen frequently and the Saudi move might have occurred even without Iraq's attack on Kuwait.

3. Paris AFP in English 1006 GMT 21 March 1973, FBIS, V, no. 55 (21 March 1973), p. B3; Damascus MENA in Arabic, to MENA Cairo 0855 GMT 30 March 1973, FBIS, V, no. 62 (30 March 1973), p. B2.

4. Kayan International in English, 29 March 1973, p. 4; Teheran Domestic Service in Persian 0430 GMT, 4 April 1973, FBIS, V, no. 67 (6 April 1973), p. K1; Teheran Domestic Service in Persian 1630 GMT 12 April 1973, FBIS, V, no. 72 (13 April 1973), p. K1.

5. "The Iraq-Kuwait Border Incident," Times (London), March 21, 1973, p. 15.

6. Moscow TASS in English 1100 GMT, 21 March 1973, FBIS, III, no. 55 (21 March 1973), p. B1.

7. Moscow TASS in English 2309 GMT, 29 March 1973, FBIS, III, no. 62 (30 March 1973), p. B9.

8. Beirut An-Nahar in Arabic, 27 March 1973, pp. 1, 12.

9. Cairo MENA in Arabic 1017 GMT 26 March 1973, FBIS, V, no. 60 (28 March 1973), p. C2; Cairo MENA in Arabic 1055 GMT 4 April 1973, FBIS, V, no. 65 (4 April 1973), p. B1; and Cairo Al Akhbar in Arabic 27 March 1973.

10. James M. McConnell, "The Soviet Navy in the Indian Ocean," CNA Professional Paper no. 77 (August 1971), p. 11, reprinted in Michael MccGwire, ed., Soviet Naval Developments: Capability and Context (New York: Praeger Publishers, 1973).

11. Al-Ahram, 25 October 1967, "Russia Claims It Saved Egypt," Times (London), 27 October 1967, p. 4.

12. Robert G. Weinland, "The Changing Mission Structure of the Soviet Navy," CNA Professional Paper no 80 (November 1971), p. 12. Reprinted in MccGwire, ed., op. cit.

13. Ath-Thawrah, 27 March 1973.

14. Teheran Domestic Service in Persian 1630 GMT 12 April 1973, FBIS, V, no. 72 (13 April 1973), p. K1.

15. For the latter case see James M. McConnell and Anne M. Kelly, "Super Power Naval Diplomacy in the Indo-Pakistani Crisis," CNA Professional Paper, no. 108 (February 1973), p. 2. Reprinted in MccGwire, ed., op. cit.

16. Riyadh Domestic Service in Arabic 0400 GMT 21 March 1973, FBIS, V, no. 55 (21 March 1973), p. B4.

17. Jim Hoagland, "Kuwait Questions Arab Policy,"Washington Post, 31 March 1973, p. A22.

18. Emphasis added. As employed by the USSR, wars of "national liberation" is an amorphous term, difficult for us to define, as it lacks a consistent theory. It seems, however, to cover any movement or conflict that is perceived by Moscow to enhance national independence and security by opposition to "imperialism" and "neo-colonialism"—that is, Western policies or interests in the third world. See Stephen P. Gilbert, "Wars of Liberation and Soviet Military Aid Policy," Orbis 10, 3 (Fall 1966): 839-58.

19. Emphasis added. The Peace Program has been described by Brezhnev in a speech to the 24th CPSU Congress; see "Proceedings of the 24th CPSU Congress," vol. 2, "Text of Brezhnev Report," FBISSOV-71-62-17. See also "Leonid Brezhnev Speech Moscow World Peace Congress" (26 October 1973), FBISSOV-73-208-40, no. 208, supp. 40 (29 October 1973).

20. Moscow Tass in English 1450 GMT, 22 March 1973, FBIS, III, no. 57 (23 March 1973), p. B1.

21. Edward R. F. Sheehan, "Sadat's War," New York Times Magazine, 18 November 1973, pp. 35, 112-20.

22. McConnell and Kelly, op. cit., pp. 7-9. The authors suggest that "rules" are emerging to discipline superpower behavior and expectations in the third world. According to the "rules of the game," it appears permissible for one superpower to support a friend against the client of another superpower as long as the friend is on the defensive strategically; the object must be avert decisive defeat and restore the balance, not to assist the client to victory. It appears that offensive actions (such as the British-French-Israeli invasion of Egypt in 1956) are ruled out, but not defensive, limited intervention (as in Lebanon 1958, Egypt 1970, Operation Linebacker 1972).

23. Washington Post, 22 April 1973, p. A6.

ANNEX: CHRONOLOGY OF EVENTS

Dec. 72: Iraq masses forces on border with Kuwait; Iraqi road-building operations from Umm Qasr penetrate into Kuwaiti territory.

Feb. 26, 73: Kuwait's Foreign minister and seven-man delegation make three-day official visit to Baghdad to discuss border situation. Visit results in no agreement on border, but both sides agree to continue talks. Iraqi Foreign Minister Baqi accepts invitation to visit Kuwait, but no dates are set for visit.

March 1-6: Iraq begins concentration of troops on border with Kuwait.

March 11: Kuwait sends a memorandum to Iraq asking for completion of the demarcation of borders "to prevent the occurrence of problems between the two countries."

March 13: Kuwaiti Foreign minister reveals that the question of extending the Shatt al Arab pipelines to Kuwait is currently under study. Claims Iraq took military measures against Kuwait because latter rejected Iraq's demand for right to build pipeline and extra-territorial port on Kuwaiti territory.

March 17-20: Kuwait reinforces Al-Samitah border post. Moves intended to protest Iraq's building on land Kuwaitis believe to be theirs.

March 20: Iraq attacks and seizes Al-Samitah border post and shells another post overlooking Umm Qasr.

March 20: General mobilization of Kuwaiti armed forces ordered. Kuwaiti National Assembly demands that Iraq withdraw its troops from Kuwaiti territory and stop all military acts against Kuwait immediately. Foreign minister asks friendly governments to use influence to convince Iraq to avoid further "unreasonable" acts.

March 20: Iraqi Interior minister accuses Kuwait of attacking Iraqi forces during maneuvers.

March 20: Leaders of Bahrein, Jordan, and Lebanon call Emir of Kuwait to convey "kind sentiments toward Kuwait."

March 20: Kuwait closes borders with Iraq and the Iraqi News Agency office.

March 20: Qatari Foreign minister denounces Iraqi attack and occupation of Al-Samitah.

March 20: Saudi Arabian Royal Cabinet appeals to Iraq and Kuwait to exercise "patience and self-restraint."

March 20: Kuwait proclaims a state of emergency.

March 20: All Arab ambassadors to Kuwait summoned to foreign ministry.

March 20-27: Iraqi forces remain in place.

March 21: USSR, in only media reference to conflict, broadcasts official Iraqi version of border incident.

March 21: Bahrein reported to have placed its armed forces and airports at the disposal of Kuwait.

March 21: Premier of the Yemeni Arab Republic cables Kuwaiti Crown Prince to express regret and denunciation of Iraq's occupation of Kuwaiti territory.

March 21: Vice president of the Revolutionary Command Council of Iraq, Saddam Husayn Tikriti, begins four-day "friendly" visit to Moscow.

March 21: Arab mediation efforts begin in the persons of Arab League Secretary General Mahmoud Riad, Syrian Foreign Minister Abdal-Halim Khaddam, and Murad Ghalib, special envoy of the president of the Arab Republic of Egypt.

March 24: Tikriti returns from Moscow, reports to President Bakr.

March 24: After three-day hiatus for Ruz holiday, Iranian newspapers feature and condemn Iraqi raid. Editorials pointedly warn Iraq against attacking Kuwait. Kayan International warns Iraq of reprisals.

March 26: Iraq's ambassador to Kuwait states that an Iraqi delegation headed by the Foreign minister will arrive in Kuwait during the second week of April, or earlier.

March 27: Al-Akhbar (Egypt) reports that Arab League Secretary Mahmoud Riad has stated that Iraq's president has ordered the withdrawal of Iraqi forces from Kuwait.

March 27: Ath-Thawrah (Iraq) attacks Kuwait and Saudi Arabia for exacerbating the border dispute.

March 27: Lebanese newspaper An-Nahar reports that Iraq began moving soldiers and equipment. The movements were not the beginning of withdrawal but a repositioning of forces to encircle the road the Iraqis were building inside Kuwait.

March 29: Kuwaiti newspapers for the first time admit the mobilization of Iraqi forces on Kuwait's border last December.

March 29: Saudis move troops towards borders. Lebanese paper An-Nahar says Kuwait has received large quantities of arms. Rumors reported that Saudi troops have entered Kuwait in area adjacent to Iraqi border.

March 29: Kayan International editorial warns Iraq, USSR on the Kuwaiti border issue.

March 30: TASS announces that Admiral Gorshkov, commander in chief of the Soviet Navy, will visit Iraq "in the first half of April at the invitation of the Iraqi Ministry of Defense."

April 1: Kuwaiti Foreign minister denies Saudi troops or other forces have entered Kuwait.

April 3: Iraq announces that an Iraqi delegation headed by Foreign Minister Baqi will arrive in Kuwait on 6 April.

April 3: TASS announces that Admiral Gorshkov departed for Baghdad on a "friendly" visit as the guest of Iraq's Defense Ministry. Soviet naval units arrive in Umm Qasr.

April 4: All Kuwaiti papers affirm that the crisis is about to be solved and that the Iraqi withdrawal will take place tomorrow.

April 4: Gorshkov visits Iraqi President Bakr.

April 4: Iraq announces that it informed Kuwait privately earlier in the week that it will drop its claim to Kuwait border territories if Kuwait cedes Bubiyan and Warba Islands. The Iraqi Foreign minister demands "full" possession of the islands.

April 4: Iraqi Foreign minister, in an interview with the Lebanese magazine As-Sayyad, says the Arab mediation effort is unnecessary; terms the existence of Al-Samitah post a "problem."

April 4: Gorshkov meets with Iraqi Defense Minister Shihab.

April 5: Iraqi troops withdraw from Al-Samitah.

April 5: Gorshkov visits Mosul and Al-Hadr in northern Iraq.

April 5: Kuwaiti paper Ar-Ra'y Al-Amm says Soviet warships as well as other naval vessels believed to be British or U.S. are steaming toward northern part of the Gulf, while Saudi troops are reported to be concentrated close to Iraq-Kuwaiti frontier. Paper also reports intense air activity over the entire northern part of the Gulf West of Kuwait, and intense activity at Dhahran and air bases in eastern part of Saudi Arabia.

April 6: Iraqi Foreign minister and delegation accompanied by Yasir Arafat arrive in Kuwait to begin talks on the border situation.

April 6: The Lebanese newspaper An-Nahar states that "four Soviet naval units have been in Iraqi waters since yesterday."

April 7: Gorshkov arrives in Basra; visits the headquarters of the naval and coastal defense forces.

April 8-9: Gorshkov visits Umm Qasr. Unlike the publicity given to all other portions of Gorshkov's itinerary, no announcement is made at the time of the Umm Qasr visit.

April 8: Celebrations marking the first anniversary of the Soviet-Iraqi Treaty of Friendship and Cooperation begin.

April 8: Iraqi delegation completes two-day visit to Kuwait.

April 9: Gorshkov returns to Baghdad "after a two-day visit to Basra Province."

April 9: Kuwait sends memorandum on border to Iraq.

April 10: Tikriti receives Gorshkov.

April 11: Gorshkov departs Baghdad for Moscow.

April 11: Kuwait Foreign minister states that negotiations with Iraq have failed. Talks described as a "presentation of terms by Iraq." Kuwait refuses to discuss islands of Warba and Bubiyan until borders are demarcated.

April 11 or 12: Soviet naval units depart Iraq.

April 12: Iranian Prime Minister Hoveyda tells London press conference that although Iran has no joint security arrangements with Kuwait, Iran would consider a Kuwaiti request for Iranian aid, should such a request be made.

April 14: Kuwait's Ar-Ra'y Al-Amm says that "Kuwait will soon initiate diplomatic activity . . . in the Gulf area with the aim of establishing diplomatic coordination . . . to prevent any foreign intervention which might affect the area's safety and security."

April 18: Iranian newspaper Ayandegan states that "that the Iraqis did not go too far . . . was due to the fact that they knew they should not force Kuwait to turn to Iran."

April 25: Kuwait says it is prepared to conclude a long-term lease contract with Iraq regarding Warba Island providing Iraq agrees to and recognizes a definite demarcation of the border.

April 26: Kuwaiti forces return to the Al-Samitah post.

CHAPTER

16

SOVIET POLICY IN THE INDIAN OCEAN

Geoffrey Jukes

The initiation of forward deployment of the Soviet Navy away from peripheral waters can in most cases be directly related to a specific combat requirement. In the early postwar years, the projection of submarine forces (not "forward deployment" in the sense of prolonged stationing away from home base) was necessitated by a perceived need to plug gaps such as North Cape-Spitzbergen and later Greenland-Iceland-U.K. against U.S. carrier task forces. Subsequently, the reequipping of the U.S. carriers in the Mediterranean with aircraft of longer range and heavier payload gave the U.S. Sixth Fleet a capability for nuclear strikes against the southwestern part of the European USSR and the southern Warsaw Pact area.* This reequipping began in 1958, and in the same year a Soviet submarine squadron moved into the Mediterranean and operated from a base in Albania[1] until that country espoused the Chinese side in the Sino-Soviet dispute and the Soviet force was withdrawn or, more likely, expelled.

In April 1963, a Pentagon spokesman announced that a Polaris submarine was on patrol in the Eastern Mediterranean,[2] a pilot deployment that became a permanent one after conclusion of the Rota base agreement with Spain in September of that year. By then the Soviets had conducted, with small units, their own feasibility study of the possibility of forward deployment reliant on afloat support[3] and in

---

*The construction between 1952 and 1958 of seven new large carriers for the U.S. Navy, and extensive modernization of three others, probably provided the Soviets with advance indications, before introduction of the Skywarrior and Vigilante aircraft in 1958-61. (Jane's Fighting Ships 1968-69 London: Sampson Low Marston and Co, 1969 , pp. 529-30.)

1964 were able to begin permanent, though not year-round, steady-state deployment on that basis.[4] Year-round basing began in 1967 with acquisition of base rights in Egypt, and the composition and training activities of the forces now maintained in the Mediterranean suggest that their combat missions include both the anticarrier and the antisubmarine role.[5]

The post-1969 deployment in the Caribbean and the abortive attempt to establish a submarine base at Cienfuegos can also be linked closely to the proximity of the U.S. Polaris base at Charleston, South Carolina, though a desire to improve the economics of the Soviet counterpart to the Polaris force probably also played a part.* Deferment of this deployment to such a relatively late date is a factor that need not concern us here; the point is that it has an obvious relevance, when undertaken, to two accepted concepts of Soviet strategy: damage limitation and assured destruction. (Damage limitation is used in the Soviet sense of damage reduction by attacks on enemy offensive systems in order to reduce weight of attack, not in the American sense of limitation by agreed or tacit abstention from attacking certain types of target such as cities. Assured destruction plus damage limitation is combined into a concept of "assured survival.")[6]

The Indian Ocean deployment, which began in 1968, and reached by 1970 a steady-state deployment of four to six surface combatants and two and three submarines on average, augmented to about double that number at times of crisis,[7] does not fit closely with the pattern observed elsewhere, because the U.S. Navy at the time the Soviets began deploying maintained neither carriers nor Polaris submarines in the Indian Ocean. Carriers have paid occasional visits since 1971, but there is still no sign that Polaris/Poseidon boats visit the Indian Ocean; the U.S. Government maintains that they do not, and there is no reason to doubt that this is so—the operating economics of submarines based at Guam would be exceedingly poor (at an assumed 20 knots made good, at least one-third of every 60-day patrol would be spent in transits), and there have been no recorded signs of a locally based presence, such as a submarine tender.

This does not, however, mean that the Polaris/Poseidon factor can be left out in examining why the Soviet Navy deployed into the Indian Ocean in the first place, and why it is still there. Nor can carriers be ignored, because at least in the case of the major strike

---

*The distance from Kola to a Western Atlantic station is of the same order as that from Charleston to Holy Loch, or Guam to Diego Garcia, and even at an assumed 20 knots made good necessitates over two weeks' extra transit time per 60-day patrol compared with a Caribbean base.

carriers, arguments that apply to the Polaris/Poseidon force apply with almost equal validity to them as well.

SOVIET INTERESTS IN THE AREA

Soviet political and strategic interest in the Indian Ocean area was low in the postwar years until the death of Stalin in 1953.[8] The newly independent countries such as India were regarded as bereft of any genuine capacity for politically independent action because of the "bourgeois" nature of their governments and their continued economic links with their former rulers. This political indifference was matched by strategic indifference, except in the case of Iran, the only country that had both a coastline on the Indian Ocean and a common frontier with the Soviet Union. An abortive attempt was made in 1946 to sponsor a breakaway movement in Persian Azerbaijan, and strong forces were maintained along the border, in the Turkestan Military District, as they always had been in the interwar period. Apart from that, the Soviet leadership showed little interest in the area, which, with British withdrawal from the Indian subcontinent and Egypt appeared to be one of declining great-power involvement.

With Stalin's death two factors emerged to increase the Soviet interest. One was that his removal made it possible for those who believed opportunities for influence were being lost to raise their voices, pointing particularly to India's mediatory position over the Korean war as evidence that economic dependence did not preclude politically independent action.[9] The other was that the advent of the Eisenhower Administration, two months before Stalin's death, led to a more forward U.S. policy in the area, through the attempt to involve it in alliances (the Baghdad Pact and the Southeast Asia Treaty Organization) that extended the "ring of containment" eastwards from the NATO area toward the U.S. Far Eastern bases in Korea and Japan. Consequently Soviet involvement in the Indian Ocean area began in the early 1950s with support and encouragement for the nonaligned countries against their aligned neighbors, and has developed through propagation of ideas of Asian collective security,[10] for which treaties signed with three Indian Ocean countries (Egypt, Iraq, and India) have been described as models. These treaties do not amount to alliances but they contain clauses prescribing consultation with a view to concerted action in the event of a threat to the signatories from a third party. It is possible to construe the treaties as at any rate not precluding the offer of base rights, but so far the only country of the three that has provided them is Egypt (in its Mediterranean ports), and it did so four years before the treaty was concluded, so no organic connection between the treaties and the possibilities of basing can or need be assumed. Nevertheless, the fact that the three nations with which such treaties exist are all countries with Indian Ocean shores is sufficient to indicate the strength of Soviet perceived interests in the area, and the fact that Soviet interests there had been actively pursued for 14 years before introduction of naval forces suggests that the navy has hardly been a factor of primary importance in the furtherance of such interests.

Soviet policy-making displayed virtually no interest in the ocean itself, as opposed to the countries along its shores, until 1964. This can be explained on several grounds. Proximity of Soviet ground and air forces in the Transcaucasus and Central Asia made it natural, especially in the army-dominated Soviet defense establishment, to think of those forces first of all as instruments for exertion of pressure or influence, for example by concentration along the Turkish and Iranian borders to deter CENTO intervention in the Iraqi crisis of 1958. The focal point of Soviet policy in the Indian subcontinent was India; India's two main antagonists were Pakistan and China, neither of which threatened India in the naval sense to an extent commensurate with the ground and air threat they could pose. So there was no requirement on India's part for manifestations of Soviet naval support, and the Indian Navy, with mainly British equipment, was the last of the Indian armed services to turn to the Soviet Union for arms, doing so only in 1965 as a consequence of Western embargoes and unfavorable financial terms. Similarly Iraq's fears in respect to Iran and ambitions in respect to Kuwait involved land rather than sea frontiers, and Egypt's problems centered not in the Indian Ocean but in the Mediterranean and Sinai; only after 1964 did the Tiran Straits issue come into prominence as a result of Egyptian action. In any event, it was only with the Mediterranean deployment of 1964 that the Soviet Navy became fully capable of forward deployment and was in a position to put forces into distant areas.

The year 1964 was a watershed in Soviet naval thinking, in that it marked the initiation of a forward deployment based on afloat support. It was also an important year in strategic development of the Polaris system, because of introduction of the Polaris A3 missile. The first version of Polaris had been suitable for deployment only in northern European waters because of its limited range of about 1,200 nautical miles (nm). Introduction of the second version, Polaris A2, with about 50 percent more range, made the Eastern Mediterranean a suitable deployment area, but the range of targets available to an A2-equipped submarine based in the Indian Ocean would have been inferior to either of the previous areas, or to the Pacific, where a number of Soviet cities and bases are close enough to the coast to justify stationing of submarine-launched ballistic missile (SLBM) submarines targeted on them.

THE THREAT FROM SEABORNE STRATEGIC SYSTEMS

Introduction of Polaris A3, and the later Poseidon, with similar range of about 2,800 nm, changed the situation, making the Arabian Sea the second-best deployment area in the world, only slightly inferior in its range of targets to the Eastern Mediterranean. While the U.S. Navy was not actually installed in the Indian Ocean, there were a number of indications that provision was being made for a possible future exercise of an Arabian Sea Polaris deployment option. The extension of the missile range itself pointed to a desire not merely to "stand off" further from the coasts in existing deployment areas, and thus increase the difficulties for Soviet antisubmarine forces (limited by lack of carriers to operations that could be protected by shore-based aircraft), or to attack a wider range of targets from existing offshore distances, but also to complicate further the tasks of antisubmarine warfare (ASW) by utilizing new sea areas for deployment, of which the Arabian Sea was the most obvious "possible." Apart from these general factors, there were other indications more specific to the area. One was that development of the Sino-Soviet rift, which had become public in 1963, was not followed by any reduction in the U.S. Polaris forces stationed in the Pacific, which would suggest to a prudently suspicious defense planner the possibility that they were being retained for use against targets in the Soviet Union (the Soviets by 1964 regarded China as no longer an ally but a potential enemy; America was still reluctant to accept this as fact in 1964, but the Soviets, who knew the true state of affairs, had no incentive to accept the American view as genuine). Development of the multiple reentry vehicle (MRV) and the multiple independently targeted reentry vehicle (MIRV),[11] which had already begun, suggested that even if the Sino-Soviet rift had not already created spare capacity in the Polaris force by removal of Chinese cities from the list of targets, multiple warheads would do so in the next few years. Finally, the North-West Cape agreement with Australia, concluded in 1963, indicated that means for communicating with Indian Ocean naval forces, including submarines, were to be established, while joint Anglo-American surveys of Diego Garcia and Aldabra in July 1964[12] suggested a U.S. search for a base facility close to the Arabian Sea. That the U.S. concern was in fact motivated by a desire to keep the British in the area by shouldering some of the costs of their presence, and as a means of enabling the administration to indicate some support for India against China, without the expense of "taking on another ocean" was not accepted by the Soviets, who probably dismissed it as a cover story similar to ones they have used themselves.

Their immediate riposte was to sponsor a Ceylonese proposal that the Indian Ocean be declared a nuclear-free zone.[13] As the ocean did not offer a range of targets for Soviet missile submarines comparable to those it offered the Polaris force (the main Soviet target systems in Europe and North America being accessible from other areas and remote from all parts of the Indian Ocean), the passing of such a resolution would, as in the case of the Mediterranean, be more gain than loss to the Soviets, while the mere raising of the issue would serve to mobilize opposition among the littoral powers, few of which desired an extension of superpower rivalry to their neighborhood, and serve notice on the United States that any attempt to deploy into the Indian Ocean would be fraught with political difficulties.

At that stage there appears to have been no question of a Soviet deployment into the Indian Ocean. A number of possible reasons can be adduced for this; none would have precluded deployment if political will or operational necessity had required it, but all make sense in a context in which the objective was not to effect a Soviet presence, but to discourage a U.S. one. Firstly, a change of government occurred in Britain at the end of 1964, and while the leaders of the Labour Party were to evince a considerable, though evanescent, enthusiasm for the "East of Suez" policy, within the party as a whole the attachment to a far-flung military role was less than in the party that had preceded Labour in office. Even among those who wished to maintain the East-of-Suez presence, cost considerations were to become more and more important, leading to a conclusion that it could be sustained only at the expense of the contribution to NATO, a conclusion most unwelcome to those in all parties who favored accession to the Common Market. The fifth projected British Polaris submarine, essential if any nuclear presence were to be kept East of Suez, was canceled soon after the new government took office, and despite U.S. pressures to remain in the Indian Ocean, progress of the British debate tended to suggest that the East-of-Suez role would be severely scaled down, and perhaps abandoned altogether, provided no external stimulus were applied to it. Insertion of Soviet naval units would have provided such a stimulus. There is no evidence from Soviet sources as to the possibility that the idea was discussed and rejected, but it is probably not a mere coincidence that the British Government, in January 1968, finally committed itself to withdrawal, and the first Soviet naval visit took place a mere two months later.[14]

The desire not to strengthen the hand of the pro-East-of-Suez faction in British politics was probably the most important single factor in the four-year lag between the Mediterranean and Indian Ocean deployments of the Soviet Navy but will not have been the only one. The urgency that had applied to the Mediterranean was lacking in the Indian Ocean; U.S. submarines and carriers might deploy there, but, unlike the Mediterranean case, had not yet done so, and, insofar as the North-West Cape and Diego Garcia facilities were seen as representing an intention to deploy, that intention would presumably not be realized until the facilities had been built, several years ahead. In any event, forward deployment was still in its infancy in the Soviet Navy. The Black Sea Fleet, the most obvious candidate to supply an Indian Ocean presence until closure of the Suez Canal deprived it of access in mid-1967, was fully stretched in maintaining the Mediterranean squadron. The alternative supplier, the Pacific Fleet, had, in those days, relatively few surface ships and little experience of foreign visiting; the only visits abroad by Pacific Fleet ships had been to China in 1956 and Indonesia in 1969.[15] So, apart from annual visits by one Black Sea Fleet destroyer to Ethiopian Navy Day in 1965-67 inclusive,[16] no voyages east of Suez were undertaken.

DEPLOYMENT INTO THE INDIAN OCEAN

By 1968 the situation had changed. The British had committed themselves to departure, and American involvement in Vietnam had made it unlikely that the U.S. Navy would have resources to spare for assumption of the British role, at least in the near future. Experience in operating the forward-deployed Mediterranean force had made it easier to contemplate dispatch of a small number of ships to the Indian Ocean, and the post-1965 acquisition of Soviet-built warships by the Indian Navy had provided an occasion for a ceremonial visit of goodwill. This took place in March 1968, and an operating pattern had become discernible by 1970. This had two main features, the first being a pattern of six-monthly rotations, with the presence in the winter months, November-April, larger than that of the rest of the year, and a heavy geographical concentration of activity in the western part of the ocean, especially in the Aden-Somalia area. The pattern usually involved departure of a "major" force, of a Sverdlov- or Kynda-class cruiser and one or two destroyers plus afloat support from Vladivostok to arrive in the Indian Ocean in December or January and return from there between late April and July, on replacement by a "minor" force of one or two destroyers plus the same number of small units. Submarines, normally F-class diesel attack boats, but sometimes E-II-class nuclear-powered cruise-missile firers were normally present in numbers varying from two to four. This standard pattern has been varied, as in the Mediterranean, during times of crisis; in 1971 additional ships were sent in following the dispatch of USS Enterprise, and for a brief period there were 20 combatants (13 surface and 7 submarines)[17] present in early 1972, as against the 4 present when the Bangladesh war began in December 1971. When Enterprise withdrew, most of the additional Soviet ships followed suit, and the "presence" reverted to its normal pattern of operations. By mid-1973, for example, it comprised one F-class submarine, one Kashin-class DLG (destroyer leader with SAM), two fleet Minesweepers, and eight auxiliaries (two of them hydrographic research vessels whose functions may not have been solely or primarily naval).[18] In the aftermath of the October war in the Middle East a U.S. carrier and escorts were sent in to the Indian Ocean, and the Soviet presence was stepped up to a maximum of 10 surface combatants and 4 submarines,[19] though it cannot be stated how much of this constituted reinforcement and how much was normal replacement due to be effected in December-January. Following the 1971 pattern, Soviet ship numbers have begun to run down again with departure of the U.S. carrier group; by late March 1974 they stood at four surface combatants and five submarines.

Composition of the Indian Ocean force has been varied, and its numbers normally are too small to enable confident statements to be made about its combat functions. The largest surface unit is normally a visually impressive but elderly Sverdlov-class gun cruiser, or a Kynda. The Kyndas carry the Shaddock ship-to-ship cruise missile, which has a theoretical maximum range of 450 miles,[20] but requires midcourse guidance from an aircraft, helicopter, satellite, or submarine if it is to be used at over-the-horizon range. Abandonment of this missile in favor of the 30-mile SS-N-10 in the latest cruiser (Kresta II), which has also appeared in the Indian Ocean since 1971, may indicate that utilization of its long-range capability required a degree of coordination difficult to achieve in practice, especially in the Indian Ocean theater where the aircraft and satellite options are not available. The E-II class submarine is also equipped with Shaddock, but when the Soviet force level is at its normal low point around midyear, there is quite often no missile-equipped ship in the Indian Ocean. However, since normally there is no target for one either, as major U.S. or British units are only rarely to be found there, operational capability is presumably second to area familiarization for most of the time, and it could be that the cruisers are included in the force not so much for what they can do as for the number of men they can carry and for their visual impact during good-will visits. Only during the 1971 and 1973 crises has there been any apparent correlation between ships and missions, because only during those crises have significant Western forces (British and U.S. in 1971, U.S. in 1973) been present. Given the size of the Indian Ocean, the sporadic nature of U.S. naval visiting, and the small number of ships involved on each side, there has up to now been very little scope for the close mutual shadowing that goes on in the Mediterranean and the

Sea of Japan. Assessment of the combat and other functions of the Soviet Indian Ocean force must therefore be speculative and based more on the areas of activity than on force composition.

UNDERLYING PURPOSES

A total of 162 ship visits was paid to Indian Ocean ports during the first four years of the Soviet presence 1968-71. Of these 96 were to ports in the Horn of Africa/Red Sea/Aden areas, compared with 57 to the Indian subcontinent and Persian Gulf.[21] The pattern has been substantially stable since, with a tapering-off in visits to the Indian subcontinent (partially offset by the presence of a salvage squadron in Chittagong, engaged in a somewhat desultory clearance of the harbor and due to leave in June 1974),[22] and an increase in visits to Iraq, but a continued high level of visiting to Somali ports, especially to Berbera, where refueling and communications facilities have been installed since 1971,[23] and may provide some of the prerequisites of a secure base. This concentration of activity and the laying of buoys in international waters at some points off the East African coast point to two possible objectives, the first being area familiarization with the Arabian Sea against a postulated future Polaris/Poseidon deployment, the second a concern to maintain a position from which continued influence on developments in the Arab world and the Middle East can be exerted.

A further possible rationale for Soviet interest in the Horn of Africa is not connected either with naval participation in matters of strategic balance or with a naval role in the Middle East. At the end of the 1950s, when the possibilities of defense against ballistic missiles were viewed with more optimism than they are now, the Soviets developed a very high-payload missile (SS-9), capable of reaching the United States via the "long way round" route over the Antarctic. This missile was christened the "global rocket" by its owners and the FOBS (Fractional Orbital Bombardment System) by the United States. The mutually agreed limitations on ballistic missile defenses contained in the SALT I agreement may have rendered the "outflanking" characteristic of FOBS redundant, but the missile has been retained in inventory and a successor to it (SS-18)[24] has been developed. It may be that the ability of these two missiles to lift heavy payloads is the reason for their continued use, because the main residual Soviet technological lag in the missile field, the lack of multiple warhead capability on their smaller missiles, is a result of lack of progress in warhead miniaturization. Until this lag is overcome, the mounting of multiple warheads requires missiles capable of carrying large payloads, and it may be this feature, rather than the depressed-trajectory round-the-world capability, that renders SS-9 and SS-18 attractive. But if the depressed-trajectory "back-door approach" feature is retained, and the loss of accuracy inherent in it is to be minimized, there is a requirement for an accurately surveyed point from which midcourse corrections can be applied, and the Horn of Africa is a desirable location for such a point. It could be, therefore, that the Soviet interest in Somalia is not solely or even primarily motivated by naval considerations; settlement of this point can only be attempted by monitoring further development of the Soviet communications facilities there.

In June 1971 Brezhnev referred to "curtailment of cruises by navies in distant waters"[25] as one of the subjects that should be discussed at the renewed SALT talks. The Soviet negotiating position on such an issue would not, of course, be confined to the Indian Ocean, but support for its denuclearization would be popular with the littoral states, whose almost unanimous hostile reaction to U.S. proposals to extend the Diego Garcia facilities has led to Indian and Australian attempts to persuade the superpowers to agree to limit their Indian Ocean deployments. It would also eliminate from the area strike carriers, of which the Soviet Navy as yet has none and ballistic missile submarines. This would be at no cost to the Soviets beyond undertaking to forgo deployments for which they have no present or foreseeable future requirement.

The area in which most Soviet Indian Ocean activity takes place is consistent both with a potential antisubmarine or anticarrier mission and with that described by Admiral Gorshkov as "protection of state interests in time of peace."[26] The two missions may well have coalesced during the 1971 and 1973 crises, when the despatch of cruise-missile equipped units into areas where U.S. carriers were operating expressed not merely a potential to interdict but a readiness to be seen doing so on behalf of countries whose interests and those of the Soviet Union coincided at the time. In that sense the presence was consistent with the prenuclear view, still held by Soviet strategists, that deterrence is an outgrowth of war-fighting capability rather than something distinct from it.[27] But from "protection of state interests in time of peace" as a by-product of a combat role to installation of forces primarily in the interest of such a peacetime function is a long step. The hortatory tone of Gorshkov's 1972-73 articles suggests that this step has not been taken yet in general terms, and the Indian Ocean deployment presents no evidence with which to contradict this interpretation.

NOTES

1. W. E. Griffith, Albania and the Sino-Soviet Rift (Cambridge: MIT Press, 1963), p. 81.
2. M. MccGwire, "The Mediterranean and Soviet Naval Interests" in MccGwire, ed., Soviet Naval Developments: Capability and Context (New York: Praeger Publishers, 1973).
3. Ibid.
4. R. W. Herrick, Soviet Naval Strategy (Annapolis, Md.: U.S. Naval Institute, 1968), p. 154, text and note.
5. MccGwire, Soviet Naval Development, op. cit., p. 314.
6. See J. Erickson, "The Soviet Military, Soviet Policy and Soviet Politics," in "Strategic Review," U.S. Strategic Institute, Washington, D.C., Fall 1973, pp. 24-25.
7. For example, from 4 to 8 and then to 12 in 1971, during the Bangladesh crisis: J. McConnell and Anne M. Kelly, "Superpower Diplomacy in the Indo-Pakistani crisis," MccGwire, ed., Soviet Naval Developments, op. cit., pp. 443-44.
8. I have discussed the reasons for this at length in The Soviet Union in Asia (Berkeley and Los Angeles: University of California Press, 1973), pp. 9-11, 13-16, 99-109, and passim. I do not propose to repeat the arguments here but merely to note that explanations of current Soviet interest that invoke allegedly "traditional" objectives fail to explain the massive indifference that prevailed through most of the Stalin era, or to account for the failure of naval interest to manifest itself until well into the Polaris-Poseidon era.
9. Ibid., pp. 109-10.
10. See V. Mayevsky, Pravda, 22 June 72 and, for a detailed examination of the proposals, I. Clark, unpublished Ph.D. thesis, Australian National University, 1974.
11. The MIRV program was made public in January 1965, but the Soviets probably knew of its existence by mid-1964. The question of surplus capacity is discussed at greater length in G. Jukes, The Indian Ocean in Soviet Naval Policy, Adelphi Paper no. 87, (London: International Institute for Strategic Studies, 1972), pp. 7 and 9-10.
12. P. Darby, British Defence Policy East of Suez: 1947-68 (London: Oxford University Press for RIIA, 1973), p. 265.
13. "Soviet Government Memorandum 'On measures for further easing international tension and restricting the arms race'," Soviet News, London, 15 December 64, pp. 145-47.
14. Australian Federal Parliament, Joint Committee on Foreign Affairs, "The Indian Ocean Region," Commonwealth Government Printing Office, Canberra, December 1971, Appendix L, "Summary of Soviet Naval Deployments in the Indian Ocean."

15. S. Breyer, Guide to the Soviet Navy (Annapolis, Md.: U.S. Naval Institute, 1970), pp. 164-65.
16. Ibid., pp. 165-66.
17. Information supplied by U.S. Information Services, Canberra, March 1974.
18. R. Berman, "Soviet Naval Strength and Deployment," October 1973, tables made up for second meeting of Soviet Naval Studies Group at Halifax, N.S.
19. Information supplied by USIS, Canberra.
20. The Military Balance 1973-4 (London: International Institute for Strategic Studies, 1973), p. 69. Range of the naval version of "Shaddock" may be less than this but is probably at least in the 200-300 nm bracket, so the argument is unaffected.
21. Australian Federal Parliament, op. cit., Appendix L.
22. Financial Review, Sydney, 25 March 74, p. 1.
23. D. O. Verrall, "Soviet Bloc-aided Port Development," compiled for Strategic and Defence Studies Centre, Australian National University, March 1974, p. 1.
24. The Military Balance 1973-4, p. 71, note f.
25. Canberra Times, 14 June 71.
26. Morskoi sbornik no. 2/1967, p. 20
27. See Chapter 26 in this volume, "The Military Approach to Deterrence and Defense."

CHAPTER

17

**THE SOVIET
PORT-CLEARING
OPERATIONS IN
BANGLADESH**

Charles C. Petersen

The surrender of Pakistani Lt. Gen. A. A. K. Niazi in Dacca on 16 December 1971 marked the end of 24 years of semi-colony-hood for East Pakistan. For the 75 million inhabitants of Bangladesh, as the former eastern wing of Pakistan now called itself, there was little else to celebrate, for seldom has a nation begun its independent existence under less auspicious circumstances.

According to a survey completed in October 1972 by the United Nations Relief Operation in Dacca (UNROD), over $1.2 billion in damage was sustained by Bangladesh during the nine-month Pakistani crackdown. Ten million Bengalis, including nearly 4 million able-bodied adults, fled to India; about twice this number drifted about East Pakistan during the "occupation."[1]

Bangladesh's only seaports, Chittagong and Chalna (Map 17.1) had been put out of commission during the fighting. A number of merchant vessels, barges, tugs, and naval craft were sunk in these ports, and their approaches heavily mined as well.[2]

The vital role played in the nation's economy by Chittagong and Chalna made the task of clearing them a first priority for the new government. The job was eventually begun by a fleet of Soviet minesweepers and salvage ships in April 1972 and remains unfinished at this writing. It is the first project of this kind ever undertaken by the Soviet Navy beyond its home waters. This paper includes first, an account of the operation; second, an analysis of the political objectives sought by the Soviets in undertaking the operation; and

---

The ideas expressed in this chapter are those of the author. They do not necessarily represent the opinion of the Center for Naval Analyses, of which he is a member, nor of the Department of Defense and the Department of the Navy.

MAP 17.1

Bangladesh

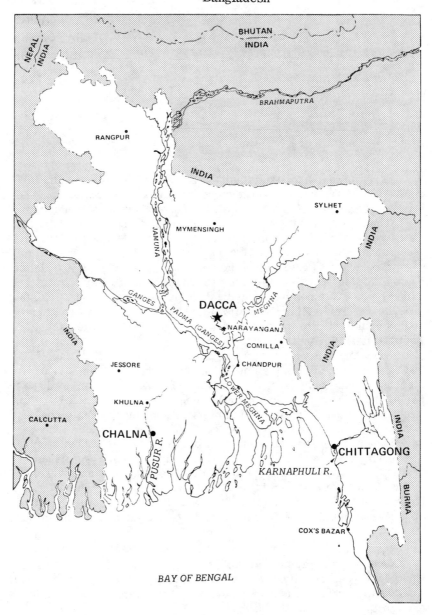

finally, an assessment of the political impact and significance of the operation.

DAMAGE TO CHITTAGONG AND CHALNA PORTS

Sunken or Damaged Vessels

The 1971 war resulted in the sinking of at least six ships within Chalna port limits. (To satisfy the need for a port better suited to the movement of jute [East Pakistan's major export] than the port of Chittagong, an anchorage was established at Chalna in December 1950. Some years later the anchorage was moved about 10 miles seaward to Mangla—Lat. 22°27'N, Long. 89°35'E. Chalna port limits now extend from Lat. 22°34'N to Lat. 22°27'N, covering the area on the Pussur between Chalna and Mangla.)3 Four of these vessels approached or exceeded 5,000 gross tons. Along the Pussur River between Chalna and the Bay of Bengal, the remains of at least six additional vessels have been charted.

The destruction at Chittagong was much greater. To the damage inflicted by Indian carrier-born aircraft and Mukti Bahini guerrillas must be added the toll exacted by previous natural calamities and even World War II: After several months of salvage work, a spokesman for the Soviet port-clearing team stated that 50 vessels of all descriptions had been sunk in or near the port area since 1940!4 Twenty-nine of these have been charted (See Maps 17.2 and 17.3). In the recent war alone 16 vessels were sent down in the Karnaphuli River; three others—the Anisbaksh, Karanphuli, and Al-Abbas (See Map 17.2) were heavily damaged but did not sink. In or near the port's outer anchorage, four more ships went down (Map 17.3). At 11,237 GRT, the tanker Avlos was easily the largest wreck in Chittagong (Map 17.2), and it blocked access to two oil berths.5 But the most critical wreck of all was the 461 gross-ton Sonar Tari, for it sank in the river's main navigational channel less than 700 feet away from the 30-year-old remains of the Golconda (Map 17.2), forcing the Chittagong Port Trust to drastically reduce the maximum permissible length and draft of ships entering the port. (Before the Indo-Pakistani war, the limit was 560 feet overall length and 27 feet maximum draft. Now the limits were, respectively, 505 feet and 21 feet.)

The ceasefire did not bring an end to the shipping casualties. On 4 February 1972, the Indian freighter Vishva Kusum struck a mine and sank north of the outer anchorage; and eight days later, the Esso Ark, a coastal tanker, sank after an onboard explosion, which also destroyed the jetty at which it was berthed.7

MAP 17.2

Wrecks in Chittagong Port

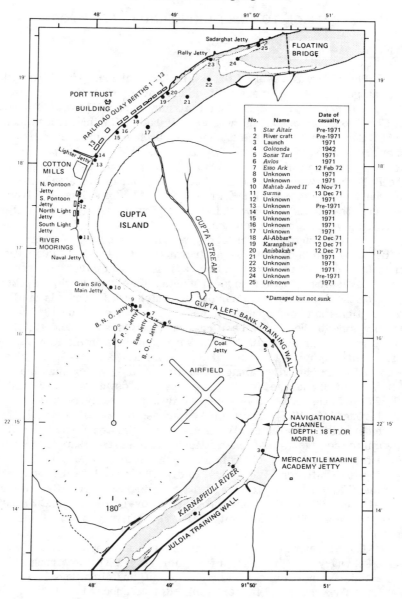

MAP 17.3

Minefields and Wrecks in Approaches to Chittagong

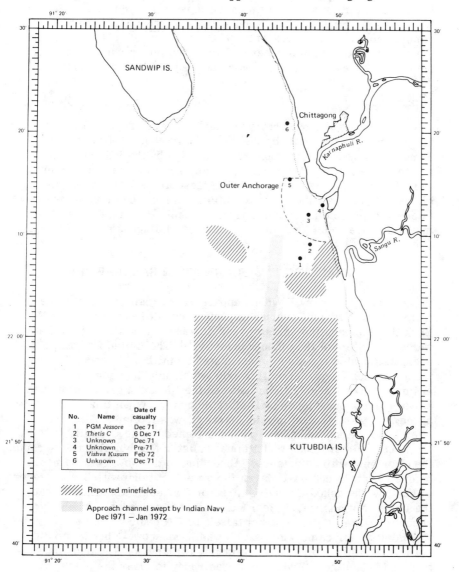

Port facilities also sustained extensive damage during the war. Three berths on the railroad quay had already been destroyed by the November 1970 cyclone. In bombing attacks in December 1971, more berths and storage sheds in this area were destroyed. River Mooring No. 2 was obstructed by the 5,890-gross-ton Surma and went out of commission, as did the BNO and BOC jetties, which did not survive the onslaught of Indian bombers. Moreover, both pontoon jetties and both light jetties were so damaged they required replacement.[8]

Minefields

Large minefields, shown in Map 17.3, were reported in the approaches to Chittagong by the Indian Navy shortly after the war. Somewhat smaller minefields were reported by the Indians at the mouth of the Pussur River. Both Soviet- and Chinese-made mines were later said to have been found in these areas.[9] (The Indians blame only the Pakistanis for the mining. Since the Indians had an interest in preventing supplies from reaching Pakistani troops, they may also have had a hand in the mining of Chittagong's approaches.)

ARRANGEMENTS FOR PORT-CLEARING WORK

In early January UNROD, apparently in response to a request from the Bangladesh Government, informed UN headquarters in New York that clearing the ports of vessels sunk during the "liberation" war was an item of the highest priority.[10] The United Nations, however, was unable to raise the estimated $6 million[11] needed to pay for a clearing operation until nearly two months had gone by.[12] The ensuing contract negotiations between the UN and a group of private salvage firms occasioned still more delays.[13] When finally an agreement was reached in mid-March, it was too late: A Soviet offer to do the job had already been accepted by the Bengalis.

There is evidence that Prime Minister Mujibur Rahman accepted the offer only with reluctance. According to the New York Times, before he left Dacca on 1 March for a week of talks with Soviet leaders in Moscow, he asked UNROD's director for "some assurance" that the United Nations would clear Bangladesh's seaports. Unable to report progress in its contract talks, the UN could promise nothing.[14] The Soviet offer came while Mujib was in Moscow,[15] but Mujib did not immediately reply. Upon returning to Dacca on 7 March, once again he asked the UN whether it was ready to clear the ports. After two days of waiting for an answer, he accepted the Soviet offer.[16] When only 30 hours later the UN announced it was ready, Mujib reportedly tried to cancel the invitation to the Soviets but was unable to do so for unspecified "diplomatic reasons."[17]

The fact that before he left for Moscow Mujib had asked the UN for "some assurance" that it would clear Chittagong and Chalna strongly suggests that he was expecting the Soviet proposal because the Soviets had already shown an interest in the undertaking.[18] Indeed, as if to forestall any further change of mind by Mujib, the Soviets now acted with unusual dispatch. By the time a nine-member Soviet delegation arrived in Dacca on 21 March to conclude an agreement,[19] ships of the port-clearing flotilla were already on passage for Chittagong.[20] The agreement—signed just 34 hours later—committed the Soviet Union to clear Chittagong and Chalna ports free of charge.[21] After a preliminary survey of wrecks in the two ports, the Soviets were assigned to the salvage of 17 wrecks in Chittagong, including two wrecks dating from before the 1971 war.[22] Their assignments at Chalna have not been revealed.

THE INITIAL PORT-CLEARING OPERATION

The lead vessel of the port-clearing flotilla, an Amur class repair ship, arrived in Chittagong on 2 April 1972.[23] Onboard were 100 crewmen[24] commanded by Rear Admiral Sergey Pavlovich Zuyenko, a 54-year-old Ukranian and veteran of 36 years in the Soviet Navy.[25] By 4 May 1972, according to Zuyenko, the flotilla consisted of 22 vessels.[26] Among these were several "small launches,"[27] three salvage tugs (including one Okhtenskiy-class ocean rescue tug),[28] a Sura class buoy tender,[29] a diving tender,[30] floating cranes, and a tanker.[31] The minesweeping component of the expedition was variously reported to contain four and nine "modern minesweepers."[32]

Promise of Early Completion

On 21 April, when the operation had been under way for only four days, Admiral Zuyenko told the Bangladesh minister of Food and Civil Supplies that he hoped the operation would be completed by the end of May 1972.[33] The initial results of the salvage work seemed to bear out his prediction. Three large Pakistani freighters (the Anisbaksh, Karanphuli, and Al-Abbas) were towed out of the port before the end of April and beached about 10 miles north of the Karnaphuli River mouth (Table 17.1). On 6 May, minesweeping was begun.[34] Less than a week later, the river's approaches were pronounced safe for navigation,[35] and during the remainder of May the port cleared nearly six times the cargo tonnage it had averaged during the previous four months.[36]

TABLE 17.1

Vessels Assigned to Soviets for Salvage in Chittagong, March 1972

Name	Type	GRT	Work Begun	Work Completed
Anisbaksh	Cargo	6,273	17 Apr 72	23 Apr 72
Karanphuli	Cargo	6,876	17 Apr 72	25 Apr 72
Al Abbas	Cargo	6,087	17 Apr 72	27 Apr 72
Marine academy launch	Unknown		May 72	20 May 72
Jessore	PGM[a]	143[b]	May 72	23 Jun 72
Harbor patrol boat		52[b]	Jun 72	23 Jun 72
Harbor patrol boat		52[b]	Jun 72	30 Jun 72
Rangamati	Small Barge	Unknown	Jun 72	11 Jul 72
Sonar Tari	Cargo	461	Early May 72	15 Aug 72
Rashid	Harbor Tug	Unknown	Aug 72	22 Oct 72
Esso Ark	Coastal Tanker	899	Aug 72	22 Feb 73
Avlos	Tanker	11,237	May 72	30 Mar 73
Mahtab Javed II	Coastal Tanker	771	Dec 72	7 Apr 73
Dulal	Cargo	360	Apr 72	1 Sep 73
Surma	Cargo	5,890	Mar 73	27 Dec 73
Star Altair	Cargo	6,140	—	—
Golconda	Cargo	5,316	—	—

[a]Motor gunboat.
[b]Full-load displacement.

Sources: Lloyd's Register of Shipping, Register of Ships, 1971-72; Jane's Fighting Ships, 1972-73; People's View, 12 November, 9, 14 April 1973; Bangladesh Observer, 25 May, 11, 27 July 1972, 2 September, 28 December 1973.

These early achievements, however, were deceptive. An approach channel to the Karnaphuli, 26 miles long and one mile wide, had already been cleared by Indian minesweepers in January 1972 (Map 17.3).37 Unfortunately for the Indians, foreign shipping was frightened off when a mine sank the Vishva Kusum on 4 February 1972.38 The accident happened well north of the swept area and therefore could not be blamed upon Indian incompetence. Though it took Russian assurances of navigational safety to restore confidence,39 it was clearly Indian preparatory work that enabled this to be done.

Similarly, the quick removal of the Anisbaksh, Karanphuli, and Al-Abbas was possible only because these ships were still afloat. Fourteen other wrecks, some broken into two or more sections, remained to be cleared. Though the port could now receive traffic, large ships could not enter because of the Sonar Tari, and only 14 of the port's 26 berths were available for docking.40

Unexplained Delays

Indeed, the early successes were followed by delays that, according to one report, were "not fully explained" by local officials.41 Salvage work, said the minister of Commerce on 17 May, would not be finished until "some time more";42 by this, as Port Trust Chairman Gholam Kibria disclosed later that month, the minister had meant the end of the monsoon—October 1972.43 In May and June, salvage work began on several wrecks, but by mid-July the Soviets had raised only the smallest of these. After some two months of effort, they could report no success on either the Avlos or the Sonar Tari. In fact, what appears to be an early attempt to remove the latter failed.*

Moreover, salvage work at Chalna, scheduled for May, had yet to begin. Save for a brief visit by Admiral Zuyenko on 6 May,44 nothing was done there for the next two months. This was hardly because Chalna's salvage could afford to wait. According to a 25 June Bangladesh Observer report, the Pussur River wrecks had become "a serious threat to normal navigation," and Chalna's Port Authority seemed "very much worried" about the failure to remove them. Though the Authority had "brought the problem to the notice of the Government," the latter could not say when salvage would

---

*According to Admiral Zuyenko, in early May the Sonar Tari was "fixed with steel ropes." On 9 July, Gholam Kibria stated that this vessel had been "towed to the other side" of the river. However, no relaxation of length and draft restrictions on vessels entering the port followed this operation. (Bangladesh Observer, 11 July 1972; People's View, 14 April 1973.)

begin. Port officials were complaining that "if no immediate step" was taken to "send the Soviet salvage team to Chalna to clear the wrecks, there will be serious dislocation of navigation which may cause disruption in the normal operation of the port" and "substantial damage to our normal reconstruction programme."[45]

Meanwhile, as progress in Chittagong slowed—with port officials refusing to explain the delays[46]—elements in the Bengali press began to suggest that the Soviets were stalling or engaged in building a naval base.[47] If Admiral Zuyenko was concerned about these rumors, it was hard to tell. His sailors, always in uniform, often toted machine guns and did not mingle with the locals. Their large compound (erected on the railroad quay) was fenced in and well guarded; only Russians could enter.[48] Consequently, many may have suspected that no one really supervised the admiral's activities and that he could get away with what he pleased.

A NEW SALVAGE AGREEMENT

Problems had also arisen between the Soviet and Bangladesh Governments, as a little-publicized visit in mid-July 1972 by a ranking Soviet official, Vasily A. Sergeyev,* indicates. Arriving in Dacca, he met not only with the Communications Minister, but with the Foreign minister and prime minister as well. The acknowledged purpose of these meetings, as well as that of a conversation on 17 July between Mujib, Admiral Zuyenko and the Soviet ambassador was to discuss the clearing operation.[49]

The Protocol of 21 July 1972

Out of these discussions emerged what was essentially a new salvage agreement—although it was called a "protocol"—which was signed on 21 July and revealed to the press in a government communique on the same day. It included the following points.

<u>Extension of Operations to December 1973</u>. It was agreed "that the Soviet [salvage and minesweeping] team would continue to work in [the] Chittagong Port area and its approaches, and would complete the task as expeditiously as possible and not later than December 1973."[50] Additionally, the Soviets had declined to salvage two of the 17 wrecks that had earlier been assigned to them: the <u>Golconda</u>

---

*Deputy Chairman of the State Committee for Foreign Economic Relations of the USSR Council of Ministers.

and the Star Altair, both of which had been sunk before the December 1971 war.51

Establishment of a Joint Committee. It was also agreed that "A joint committee consisting of representatives of the two sides will hold periodic meetings in order to broadly supervise and review the progress of the work."52 It is unlikely that this committee ever met to "supervise and review" the operation. Nothing has ever been reported about its deliberations, if indeed there were any. Moreover, it is hard to believe that the Soviets, who dislike outside interference in the management of their aid projects, would join with representatives of the Bangladesh Government to place Zuyenko's activities under scrutiny. The joint committee's raison d'etre, therefore, was probably mainly cosmetic: a device to quell rumors that the delays in the harbor-clearing work spelled sinister Soviet intentions in Chittagong.

No Salvage at Chalna. The communique also announced that "At Chalna Port, the Soviet [Sergeyev] team examined three sunken vessels and both Governments agreed that these vessels do not at present directly affect the operation of the port [or] its normal handling capacity. Both sides agreed that further work will be the responsibility of the Bangladesh Government to decide."53 The Chalna Port Authority could not have been happy with this assessment. In late July, sources in the Authority complained, according to the Bangladesh Observer, that because no attempt had yet been made to clear the port, the wrecks had "caused the [Pussur] River to silt to the extent that the river has now begun to change its course." Two "diversion channels" were "drawing water from the main stream" and had already "lowered the level of water in the river." Unless the salvage of Chalna was begun soon, there was the "distant [distinct?] possibility" that the country would "lose one of its only two outlets to the sea."54 Indeed, further work would be "the responsibility of the Bangladesh Government to decide" not because it was unnecessary, but because the Soviets were simply unable to do it. As sources in the UN later put it, Admiral Zuyenko was experiencing "considerable difficulties" in his Chittagong operation and consequently could not undertake another one.55 The nature of these "difficulties" will be discussed below.

THE UNITED NATIONS HAS ITS DAY

For the next two months, the Bangladesh Government tried—without avail—to persuade what according to the Communications minister were "a number of foreign countries" to undertake the salvage of Chalna.[56] Finally on 8 September it was forced to turn again to the United Nations, whose help had been refused only seven months before. The UN now seemed determined to avoid delays and decided to finance the project out of available funds rather than go begging for contributions as before. On 24 October, UN headquarters announced that it had awarded an $8 million contract to a consortium of six private salvage firms to begin clearing Chalna in early December, at the end of the monsoon.[57] Less than five months later, the UN consortium had finished removing the six wrecks to which it had been assigned—a week ahead of the target date (16 May 1973) and at "substantially" below the estimated cost.[58]

THE SECOND PHASE OF PORT-CLEARING OPERATIONS

Technical Problems Faced by the Soviets

In this same five-month period, Admiral Zuyenko's team managed to raise only three ships, one of which alone had taken more than 10 months of work. Despite avowals by the Bangladesh Government "never [to] allow any foreign base on our soil," rumors to the contrary persisted.[59] In an attempt to refute charges that his team's work was "deliberately being stalled," Zuyenko complained in early October of the "extremely difficult" conditions under which his men had to work. High temperatures, high humidity, and frequent rainstorms were "hard for us Northerners to get used to." Underwater visibility was so poor that divers had to work "by touch"; strong currents in the fast-flowing Karnaphuli prevented them from going down more than four times a day, and each time for no longer than "25-45 minutes."[60]

Since the salvage firms at Chalna worked under much the same conditions,[61] this explanation fell short of accounting for the remarkable contrast between the two operations. To rebut continuing criticism of the slowness of his work, Admiral Zuyenko claimed during a press conference on 27 March 1973 that the UN consortium's salvage method—underwater blasting of the sunken hulls—could not be used in Chittagong because the Karnaphuli River was too narrow and blasting might endanger the movement of ships within the port. Moreover, "considerable" fish resources would be destroyed and port facilities damaged as well. Though by using his own method, he seemed (as he put it) to be "working very slowly and making little progress," there was far less risk involved.[62]

But underwater blasting, which the UN firms used to break apart sunken hulls into pieces that could be lifted by floating cranes and then beached,[63] is actually a much safer method than Admiral Zuyenko would have one believe. Developed by Western experts in the years after World War II into a highly sophisticated technology, it is far from the messy demolition exercise that Zuyenko implied it to be. With the use of small-shaped charges, underwater explosions can be rigidly controlled and their effects predicted with great accuracy. Sunken hulls can thus be broken apart unobtrusively and with great precision.[64] Admiral Zuyenko's critique of underwater blasting is therefore valid only for much cruder blasting techniques, which may well be the only ones the Soviets at present know how to use. Such methods, which presumably would not involve the use of shaped charges, would indeed have the injurious effects Zuyenko has posited. In short, the admiral's remarks appear to be less a critique of modern underwater blasting than an unwitting admission of the backward state of the Soviet Navy's salvage technology.

Forced for this reason to forgo blasting, Admiral Zuyenko had to rely on an older method—sinking pontoons, attaching them to the wrecks, and floating the pontoons with compressed air, thus bringing the wrecks to the surface.[65] This turned out to be an extremely time-consuming process, since vast amounts of mud and silt had to be removed from a number of wrecks before they could be refloated. For example, it took more than nine months of effort to raise the 5,890 gross-ton Surma, which since its sinking had gathered enough silt to make it weigh some 35,000 tons.[66] The pontoon method, barely practicable for the salvage of recent wrecks, very likely was impossible for the salvage of much older ones like the Golconda, which as we have seen, the Soviets had declined to remove.

Inadequate lifting equipment also contributed to the slow pace of the salvage work, as seems indicated by the arrival of an 800-ton capacity floating crane (made in West Germany) in Chittagong in early October 1973.[67]

Results of Soviet Work by Mid-1973

Though Admiral Zuyenko's operation was something less than a public relations success, and though Dacca could not have been pleased with its early going, by mid-1973 his work had yielded substantial results. Ten vessels of all sizes had been refloated, and three more removed from the port (See Table 17.1). Minesweeping, which compared to the salvage operation was beset by few problems, was completed by 21 October 1972, though not without generous Indian help. (Five Indian minesweepers took part in this operation. By June 1972, the mile-wide approach channel had been widened to three miles. In October 1972 both Soviet and Indian minesweepers began leaving for home. The last Indian minesweeper left in November, but two

Soviet units remained behind throughout 1973 to conduct periodic checks for drifting mines.) Until April 1972, only 15 vessels per month called at the port.[69] During the next eight months, in contrast, 569 ships called, or about 71 per month.[70] Between August and December 1972, the port cleared an average of 406,000 tons per month.[71] Length and draft restrictions on vessels entering Chittagong were relaxed on 15 August 1972 when the Sonar Tari was salvaged,[72] and very large ships could now navigate the river (a 572-foot-long freighter was able to dock on 25 December 1972).[73] In 1973, cargo handling declined some, but at 387,000 tons per month remained well above the average rates that were cleared before the operation began.[74]

Arrangements for More Salvage Work

It is clear that the Bangladesh Government eventually came to appreciate the results of Admiral Zuyenko's efforts. Official praise of his work, noticeably perfunctory throughout 1972, became effusive in the early months of 1973. "The great Soviet Government," said the Communications minister on 14 April, had achieved a "landmark in the economic reconstruction of Bangladesh" with its harborclearing work in Chittagong.[75] At the same time, Gholam Kibria—who as Port Trust Chairman is an official of the Bangladesh Government—publicly expressed the hope that the Soviets would do extra salvage work.[76] At first the Soviet Government, through Admiral Zuyenko, declined.[77] As the operation neared completion in December 1973 however, the Soviets were persuaded to remain; and on 13 December a group of Soviet officials arrived in Dacca to negotiate a new salvage agreement.[78]

The Protocol of 20 December 1973. Following a week of "detailed" talks, a "protocol" was signed on 20 December. According to a government press release, Admiral Zuyenko was to clear six more wrecks by 30 June 1974. "As a friendly grant" of the Soviet Government, "some salvage craft and equipment" would be given to the Port Trust, and "some Bangladesh personnel" would be trained in salvage work as well.[79]

The evidence indicates that the Soviets agreed to do this extra work somewhat unenthusiastically. In March 1972, it had taken less than two days to conclude the original port-clearing agreement; in contrast, this agreement followed a week of negotiations in which the Soviets were careful to avoid overcommitting themselves. The new assignments included only small wrecks: one workshop barge, four lighters, and one coastal vessel.[80] It is likely, moreover, that this was less than what the Port Trust wanted removed; in April 1973 Chairman Kibria had asked Admiral Zuyenko also to salvage the

Star Altair,[81] which at 6,140 gross tons was the largest remaining wreck in the river and, as a glance at Map 17.3 will show, a continuing hazard to navigation. In short, the extra work is hardly formidable; in fact, four of the wrecks had already been refloated by early March 1974.[82] Finally, the fact that the Soviets are to hand over salvage equipment and train Bengali personnel in its use suggests that they will not remain in Chittagong beyond the 30 June deadline to perform any new tasks.

POLITICAL OBJECTIVES AND EFFECTS OF THE OPERATION

Role of the Soviet Navy

"The very first question you should ask" said Admiral Zuyenko in a December 1972 interview with the Los Angeles Times, "is why the Soviet military is doing the work instead of a civilian salvage team." Answering his own question, Zuyenko said,

> Minesweeping was the main work . . . and this was strictly a military job. This had to be done to permit ships to enter the outer anchorage. If the minesweeping were not completed, we could not undertake any salvage work.[83]

Minesweeping, it is true, is strictly a military job. But thanks to the Indian Navy, ships could safely reach Chittagong's outer anchorage well before Admiral Zuyenko arrived there in early April 1972. If he was concerned about mine dangers, then, he clearly did not behave as if there were any: His minesweeping operation began not before but three weeks after his salvage operation, and more than a month after his arrival. No more eloquent proof can be found of his confidence in the Indian Navy's minesweeping performance. When the Soviets began minesweeping, of course, there were still areas that needed clearing. But their help was no longer essential because the "main" task—the removal of mine dangers from the port's approach—had already been finished by the Indians.

However, just as the "need" for minesweeping fails to explain the Soviet Navy's presence in Chittagong, so do other, more sinister, military considerations. There is no reason to suppose, for instance, that Admiral Zuyenko is deliberately "stalling" his work, or playing for time in the hope that the Bengalis will eventually allow the Soviet Navy to use Chittagong for the support of its Indian Ocean operations. The slowness of the work can be blamed, as we have seen, on technical problems alone. If Admiral Zuyenko has been less than candid in accounting for the delays, it is probably because candor would entail admitting the inferiority of Soviet salvage methods.

In any event, hydrographic and meteorological conditions in the eastern Bay of Bengal make Chittagong a poor place for a naval base in the first place. Its waters are much too shallow for submarine operations, and navigational conditions are unpredictable (depths in and around the Karnaphuli River constantly change). The weather is frequently dangerous: In the recent past, a series of violent cyclones caused heavy damage both to the port and shipping in it.[84] There is thus every reason to believe Zuyenko's assertions that the Soviet Navy is not building, and will not build, a naval support facility in Chittagong.[85]

If military considerations do not account for the Soviet Navy's presence in Chittagong, then why is it there at all? The available evidence supports the notion that it is there purely for display. Admiral Zuyenko, for one, has seemed very anxious to advertise its presence by holding frequent press conferences both in Chittagong and in Dacca.[86] In a similar vein, newsmen have been taken on minesweeper cruises in the Bay of Bengal;[87] and the operation's first "anniversary" was marked by three days of ceremonies, which featured the admiral as the guest of honor.[88]

The Soviet Navy's role in Chittagong, therefore, is probably the same role Soviet Navy Chief Sergey Gorshkov has recently outlined for Soviet warships on shorter visits to other ports of the world:

> The friendly visits of Soviet navymen make it possible for the peoples of many countries to become convinced with their own eyes of the creativity of the ideas of Communism. . . . They [the peoples of foreign countries] see warships [as] embodying the achievements of Soviet science, technology and industry. . . . [Soviet navymen] are profitably and convincingly spreading the ideas of the Leninist peaceloving policy of the Communist Party and the Soviet Government.[89]

The specific purpose of this extended "port call" is unclear, but it could well represent a subtle attempt to persuade the Bengalis that the Soviet Navy's presence in the Indian Ocean is a peaceful one and therefore in no way incompatible with the proposal, advanced by the Indians, that the Indian Ocean should be a "zone of peace."

Strengthening of Soviet Influence

Though the Soviet Navy is in Chittagong for display, the Soviets did not undertake this port-clearing operation only to display their navy. In his series of articles on "Navies in War and Peace" Admiral Gorshkov emphasized the "special significance of navies for states as an instrument of policy in peacetime";[90] and it is clear from Soviet statements on the operation that this "instrument of policy" was to be used for the strengthening of the Soviet Union's political influence in Bangladesh. Thus, when the Soviet-Bengali port-clearing agreement was announced to the press on 22 March 1972, the head Soviet negotiator called it "a new step in the development and strengthening of friendly relations between the Soviet Union and Bangladesh."[91] In Soviet parlance, a "friendly" country is one over which the Soviets exert a good deal of influence; and the recklessness with which they entered into commitments they were later forced to abandon shows how important to the strengthening of such "friendly relations" they thought the project would be. It is easy to see why: Bangladesh depended on her seaports for economic survival; were they to remain out of commission much longer, she would starve. By restoring these ports the Soviets could make a crucial contribution to Bangladesh's economic recovery; a contribution for which, as they probably believed, they would be rewarded with political influence.

The Absence of Positive Political Effects

Admiral Zuyenko has claimed that the operation has "set in motion the wheel of economic recovery" in Bangladesh. There is no evidence, however, that the Bangladesh Government has returned the favor with more than minor political concessions. Though it never fails to stress the importance of "friendly relations" with the Soviet Union, on major issues affecting the subcontinent, Bangladesh follows the Indian rather than the Soviet lead. Like India, it has not shown more than polite interest in Brezhnev's "Asian Collective Security" proposal, which envisions the subcontinent's participation in the political encirclement of the People's Republic of China. Nor have the Bengalis taken a hard line against the PRC, despite the latter's opposition to Bangladesh's seating in the UN: Public criticism of China has been confined to regretting this policy.* Finally, Mujib's government continues to endorse the Indian proposal to make the Indian Ocean a "zone of peace" through the withdrawal from it of both superpower navies. (In fact, Mujib has shown little public enthusiasm for the Soviet naval presence in Chittagong. For example, he has conspicuously failed to praise Admiral Zuyenko's work there. So far, only his subordinates have done this.)

---

*The Bangladesh Government's campaign against local pro-PRC elements has been directed at their opposition to its policies and not at their Sinophilia.

Negative Results of the Operation. In many ways the operation's political effect has been negative. The Soviet Union's retreat from its original commitment to clear both seaports could hardly have contributed to the "development and strengthening" of "friendly" Soviet-Bengali relations, and even less to the enhancement of Soviet prestige in Bangladesh. The UN's rapid clearance of Chalna was, as we have seen, a source of embarrassment to Admiral Zuyenko. It is equally clear that the rumors about his activities in Chittagong were intensely embarrassing to the Bangladesh Government, as witness the tone of the following disclaimer by the Foreign minister:

> I would like to declare here categorically that the mischievous propaganda launched by certain interested quarters about the so-called Soviet base, I mean, naval base or any base, on the soil of Bangladesh is malicious and completely unfounded and the Government of Bangladesh will never allow any foreign base on our soil.[92]

Attitude of the Indian Government. By and large, the Indian Government has shown no concern over the Soviet Naval presence in Bangladesh—none that has been reported in the Indian press, in any event. But when Admiral Zuyenko claimed in December 1972 that the Indian minesweeper squadron in Chittagong had been under his operational command,[93] a sharp denial was issued by New Delhi,[94] probably because any implied association between the two navies would tend to make the Indian advocacy of a "zone of peace" for the Indian Ocean seem insincere.

CONCLUSION

The Soviet port-clearing operation in Bangladesh represented a new departure in the Soviets' use of their naval forces for political purposes. Until it was undertaken, the Soviet Navy's "political" operations had been confined largely to one or another variant of coercive diplomacy. The operation in Chittagong, however, is an exercise in influence-building by noncoercive methods. The results, from the Soviet point of view, cannot have been entirely positive. For this reason, the Soviets may well be much more cautious in the future about committing their navy to such projects.

NOTES

1. The People (Dacca), 28 October 1972.
2. It is known that Indian carrier-based aircraft were responsible for most of the shipping casualties at Chittagong and Chalna. Apparently Mukti Bahini frogmen sank a few ships as well. (Lloyd's Weekly Casualty Reports, 19, 16 November 1971; 14, 21 December 1971; 25 January 1972, 8, 15 February 1972; Bangladesh Observer, 3 April 1972; 27 July 1972.) Less certain, however, is the identity of the side responsible for mining those harbors.
3. See M. M. Yusuf, Pakistan Shipping Guide, Karachi, 1971.
4. Bangladesh Observer (Dacca), 24 August 1972.
5. Ibid., 27 July 1972; Eastern Examiner (Chittagong), 9 August 1972.
6. Bangladesh Observer, 16 August 1972.
7. Far Eastern Economic Review (Hong Kong), 26 February 1973.
8. Chittagong Port Trust notices on "Berthing Position and Performance of Vessels in Port" published daily in Eastern Examiner and People's View (Chittagong); Eastern Examiner, 12 June, 9 August 1972; Bangladesh Observer, 27 July 1972, 26 March 1973, 16 December 1973; Ports of the World, 1974, Benn Bros. Ltd. (London, 1974).
9. Bangladesh Observer, 10, 16 March 1972; the Statesman (Calcutta), 3 April 1972.
10. Washington Post, 16 March 1972.
11. Ibid.
12. The estimate was supplied by a Singapore salvage firm (see ibid.). The money was contributed by the Swedish Government at the end of February 1972. Bangladesh Observer, 1 March 1972.
13. New York Times, 31 March 1972.
14. Ibid., 6 April 1972.
15. Washington Post, 16 March 1972; Bangladesh Observer, 10 March 1972.
16. Washington Post, 16 March 1972.
17. Ibid.
18. The Soviets may have shown interest as early as 1 January 1972, when the Soviet envoy in Dacca met with the Bengali Communications (that is, Transportation) minister. According to a Government press release, the two had "discussed the possibility of Russian participation in the rehabilitation and improvement of [the] communication [that is, transportation] system in Bangladesh." (Bangladesh Observer, 2 January 1972.) The "communications" system in Bangladesh, dominated by waterways, has as its vital links the ports of Chittagong and Chalna; therefore it is not unreasonable to suppose that their "rehabilitation" was a subject of discussion.

19. Bangladesh Observer, 22 March 1972.
20. Lloyd's Weekly Casualty Reports, 28 March 1972. The first ships reportedly set sail from Vladivostok on 14 March 1972.
21. Bangladesh Observer, 23 March 1972.
22. Ibid., 27 July 1972; People's View, 22 November 1972, 9 April 1973.
23. A photo of this vessel, the PM-40, appears in the 5 April 1972 Eastern Examiner.
24. Bangladesh Observer, 2 April 1972; Eastern Examiner, 3 April 1972.
25. Los Angeles Times, 26 December 1972.
26. Eastern Examiner, 3 April 1972; People's View, 5 October 1972.
27. Los Angeles Times, 26 December 1972.
28. Eastern Examiner, 5 April 1972; Bangladesh Observer, 6 May 1972; Izvestia, 30 June 1973.
29. A photo appears in the 14 May 1973 issue of Der Spiegel.
30. Bangladesh Observer, 17 April 1972.
31. People's View, 5 October 1972; Izvestia, 30 June 1973.
32. The Statesman, 3 April 1972; Bangladesh Observer, 27 August 1972, 16 December 1973.
33. Eastern Examiner, 16 April 1972.
34. People's View, 14 April 1973.
35. Bangladesh Observer, 24 May 1972.
36. The figures for January-July 1972 are given below (in measurement tons):

Jan.	12,762
Feb.	88,330
March	86,938
April	57,755
May	361,877
June	366,993
July	417,000

The figures for January-May are in the Bangladesh Observer, 12 June 1972. The figures for June are given in ibid., 11-12 July 1972, and for July in ibid., 16 August 1972.

37. Lloyd's Weekly Casualty Reports, 8 February 1972.
38. Far Eastern Economic Review, 26 February 1973.
39. Bangladesh Observer, 8 July 1972.
40. The others were either damaged, blocked by wreckage, or reserved for use by Admiral Zuyenko's ships (Eastern Examiner, 23 February 1972; Bangladesh Observer, 1 July, 16 August 1972; 22 March, 26 March, 16 December 1973).

41. Far Eastern Economic Review, 20 May 1972.
42. Eastern Examiner, 18 May 1972.
43. Bangladesh Observer, 24 May 1972.
44. Ibid., 6 May, 25 June 1972.
45. Ibid., 25 June 1972.
46. Far Eastern Economic Review, 20 May 1972.
47. Bangladesh Observer, 10 August 1972; Los Angeles Times, 26 December 1972; People's View, 30 December 1972; Far Eastern Economic Review, 20 May 1972.
48. Los Angeles Times, 26 December 1972; Far Eastern Economic Review, 20 May 1972.
49. Bangladesh Observer, 12 July, 18 July and 22 July 1972.
50. Ibid., 22 July 1972.
51. Ibid., 27 July 1972; People's View, 22 November 1972, 9 April 1973.
52. Bangladesh Observer, 22 July 1972.
53. Ibid., 29 July 1972.
54. Ibid., 18 August 1972.
55. New York Times, 25 October 1972.
56. Lloyd's Weekly Casualty Reports, 31 October 1972.
57. Ibid.
58. Bangladesh Observer, 17 May 1973.
59. Ibid., 10 August 1972.
60. Interview with Novoye Vremya. English translation appears in 5 October 1972 People's View.
61. Bangladesh Observer, 17 May 1973.
62. People's View, 29 March 1973; Daily Telegraph, 30 March 1973; Bangladesh Observer, 29 March 1973.
63. Bangladesh Observer, 17 May 1973.
64. The U.S. Navy's Salvage Master, Earl Lawrence, has stated that one can stand "right next" to a string of shaped charges as they are set off to break apart a sunken hull, and no harm will result (conversation, 10 September 1973).
65. Los Angeles Times, 26 December 1972.
66. Izvestia, 30 June 1973. Similarly, more than 5,000 tons of silt and 700 tons of crude oil needed removal from the Avlos' stern section before it could be raised. The Esso Ark, which had gathered some 1,000 tons of mud, took seven months to salvage. (Bangladesh Observer, 27 July 1972; People's View, 23 February 1973, 14 April 1973.)
67. Pravda, 12 October 1973; Moscow Radio (domestic service) in Russian, 2200 GMT, 30 August 1973.
68. Bangladesh Observer, 24 August 1972, 25 April 1973; Los Angeles Times, 26 December 1972; Hindustan Times (New Delhi), 5 January 1973; People's View, 30 January, 14 April, 26 April 1973.

69. Bangladesh Observer, 12 June 1972.
70. People's View, 8 January 1973.
71. Ibid.
72. Bangladesh Observer, 16 August 1972.
73. People's View, 27 December, 31 December 1972.
74. Bangladesh Observer, 20 January 1974.
75. People's View, 16 April 1973.
76. Ibid., 14 April 1973.
77. Ibid.
78. Bangladesh Observer, 21 December 1973.
79. Ibid.
80. Bangladesh Observer, 18 February 1974.
81. People's View, 14 April 1973.
82. Bangladesh Observer, 4 March 1974.
83. Los Angeles Times, 26 December 1972.
84. The Statesman, 3 April 1972, Bangladesh Observer, 9 September 1972.
85. People's View, 5 October, 30 December 1972.
86. Eastern Examiner, 3 April, 20 July 1972; People's View, 30 December 1972, 30 January, 29 March, 9 April, 14 April 1973; Bangladesh Observer, 29 March, 16 April, 25 April 1973; Daily Telegraph, 30 March 1973.
87. People's View, 30 January, 14 April 1973.
88. Ibid., 14, 18, 28, 29 April 1973; Bangladesh Observer, 16 April 1973.
89. "Navies In War and Peace," Morskoy sbornik, no. 12, 1972.
90. Ibid.
91. Bangladesh Observer, 22 March 1972.
92. Ibid., 10 August 1972.
93. Los Angeles Times, 26 December 1972.
94. Hindustan Times, 5 January 1973.

CHAPTER

18

THE SOVIET POSITION AT
THE THIRD U.N.
LAW OF THE SEA
CONFERENCE
Robert L. Friedheim
Mary E. Jehn

In the summer of 1974 in Caracas, Venezuela, some 150 states will assemble at the third United Nations sponsored Law of the Sea Conference, where they will attempt to update the regulations which govern men's uses of the sea. The stakes are substantial—how much of the 71 percent of the watery surface of the earth will be reduced to national control; who will control the resources of the relatively shallow continental shelves, slopes, and rises; who will have the right to gather the nodules lying on the deep seabed; and what the rights of fishermen, merchant mariners, scientists, and navies will be. The Soviet Union, its bloc partners, and other friends will be important participants in this multilateral negotiation and presumably will be influential in any outcome achieved by the conference.

The purpose of this chapter is to assess the known probable Soviet positions on a number of major substantive ocean law issues and examine some serious problems the Soviet Union will face in developing its bargaining strategy for the UN conference.

Evaluating the behavior of any state is far from an easy task. It is even more difficult when little is known of the bureaucratic input to decisions guiding such behavior, as it is with the Soviet Union. Some observers use what few clues we are given about such input— a statement in Pravda, an absence from the theater—to make

---

The ideas expressed in this chapter are those of the authors. They do not necessarily reflect the views of the center for Naval Analysis, nor of the Department of Defense and the Department of the Navy.

The authors acknowledge with gratitude the assistance of their colleagues Karen Goudreau, Lorraine French, Karen Young, William Durch, Thomas Anger, and Professor Joseph Kadane.

predictions. We have instead chosen to analyze the present and past decisions themselves—the bureaucratic output. Our raw material is statements by the Soviet delegation to the United Nations Seabed Committee and the First Committee to the General Assembly.

We believe this raw material is appropriate for the following reasons. First, ocean law problems are ongoing problems that have been under continuous negotiation at the UN since 1967, and the Soviet-generated evidence on its own preferences is voluminous and internally consistent. Second, UN parliamentary diplomacy, as public diplomacy, makes it very difficult to alter expressed preferences substantially without deceiving friends as well as enemies. Therefore, we believe that the Soviet Union's "expressed" preferences are reasonably close to their "true" preferences; their statements at the UN are valid indicators of their real positions.

Our analysis begins with some comments on the general pattern of Soviet behavior, followed by the present Soviet positions on the major Law of the Sea (LOS) issues and their evolution. Then the output of one Soviet bureaucracy—the navy—is compared with official Soviet output. Through a quantitative analysis of both, we show the similarities and differences between them. We do not attempt to estimate the extent of the navy's input into Soviet policy. Rather, we simply describe the distance between the two. We also compare these outputs with the positions of both developed and developing caucusing and special-interest groups, in order to put our Soviet analysis into perspective. We conclude with a discussion of the dilemmas facing the USSR in law of the sea bargaining.

BACKGROUND

Before describing the current Soviet bargaining proposals on the issues to be negotiated at Caracas, we must set these Soviet positions in the context of their general views of the laws of the sea and of the bargaining environment—the United Nations system.

Soviet law of the sea has usually been a reflection of Soviet foreign policy. As Soviet interests change, we should expect the international law that Soviet leaders prefer will also change. This is not to say that there are not variations, or that all Soviet international law writing is monolithic—William Butler has demonstrated that this is not so. For our purpose of examining Soviet state behavior, it is appropriate to note the close association between their state behavior, perceptions, interests, and preferred legal doctrine.[1]

Until recently the Soviet Union's maritime law outlook was defensive and protectionist. The Soviet Union viewed the oceans as a place wherein it had few important interests and indeed, a place from which might emanate danger to its land areas. As many have noted, the Soviets and their predecessor regimes were land oriented, a natural outlook considering their position athwart what geopoliticians call the "world-island." Russia historically had no more than a coastal fisheries capability, a minor merchant marine, little interest in oceanography, and a navy geared only to coastal defense.

The law of the sea was viewed by Soviet policy-makers from 1917 to the early 1960s as a set of rules to push others away from Soviet coasts. This defensive stance intensified at the height of the cold war, probably the result of Soviet maritime weakness. The Soviet Union championed the 12-nautical-mile territorial sea to deny Western navies the right to come within the three miles of the Soviet coast. They also espoused "closed seas"; they enclosed waters as internal waters under a "historic" rights doctrine, and they excluded foreign fishermen from access to fishing grounds within 12 miles of their coast.

This emphasis in Soviet maritime law changed in the 1960s when the Soviet Government expended substantial resources to end its maritime weaknesses. As a result, the Soviet Union has "arrived" as a maritime power. The growth of the Soviet Navy to second among world navies, the expansion of the Soviet merchant marine, the creation of a large and sophisticated oceanographic fleet, and the development of world-ranging fleets of fishing vessels with voracious appetites for fish has given the Soviets a concern for maintaining "access" to the world's oceans and resources. The Soviets now want an assured right to get their navy and merchant marine out of their ports on enclosed or semienclosed seas and through straits with as little interference as possible; they want to have access to fishing grounds near the coasts of other states; they want rules that are not discriminatory, administratively burdensome, or prohibitively expensive governing transit of vessels through areas over which other governments claim some rights.

In short, the USSR is now a maritime "have" state. Its interests have forced reevaluation of traditional views. The Soviet Union is now an ardent supporter of "freedom of the seas," a Western doctrine preferred by status quo maritime powers. Support of such a doctrine puts the Soviets in the same camp as most Western developed states, no doubt to their great discomfort. It also increases their concern about "creeping jurisdiction," the supposed tendency of states to lay claims that expand in legal scope and area. Such claims might make some present Soviet naval, maritime, and fishing activities contrary to international law.

To add to the Soviets' woes, these new rules are being negotiated in a United Nations-sponsored conference with 148 to 150 states participating. Since the 1960s, the United Nations has been the preferred forum of the developing state majority, where their numbers can be used to counterbalance (and perhaps tip the balance) against the minority of developed states with real ocean-use capability.

The Soviet Union has a number of reasons to be wary of UN auspices. The history of Soviet involvement with the UN has not been a history of overall diplomatic success. During the height of the Cold War, the United States was able to mobilize majorities to defeat Soviet resolutions and to oppose Soviet actions in the outside world. The United States was even able to push aside the veto, the UN Charter protection counted upon by the USSR to compensate for the minority position of Socialist countries in the General Assembly at the UN founding.[2]

In the late 1950s and early 1960s Khrushchev's ebullient attempt to mobilize the newly admitted developing states into a Soviet-led majority in the United Nations also failed. Analysis of the Soviet voting record indicates that in recent years the Soviets have been voting in the majority more often simply because of Soviet agreement on the developing states favorite issues ("colonial" questions) rather than because the developing states are being led by Moscow.[3] Indeed, what has emerged at the UN has been the growth of solidarity and separate identity of the Afro-Asian group and their successful transformation of the UN into their instrument. The prevailing view of the developing state majority in recent years is that the USSR, despite its claim to a special relationship with the third world because of its revolutionary past or current verbal support for revolutionary regimes, is merely the "other" superpower. This became most evident outside the UN system at the September 1973 meeting of the "nonaligned states" in Algiers, where despite an impassioned defense of the USSR's revolutionary credentials by Fidel Castro of Cuba, many leaders of developing countries such as Colonel Muammar al-Qaddafi of Libya, Prince Norodom Sihanouk of the Cambodian exile regime, and representatives of Mali and Mauritania characterized the Soviet Union as no better than the "other" superpower.[4]

Since the participation of the People's Republic of China in UN activities, the Soviets again must fear that a major rival will use the UN as a collecting point for its enemies. Attacked by a Western-led coalition during the cold war, the USSR remained a UN member out of fear that in its absence the United States might create of the UN a tightly knit formal alliance system directed at the smaller socialist system. Now the Soviet Union must be concerned that China might succeed where the United States did not. The Chinese have, from the first moment of their participation, made it clear that they believe they are the proper leader of the developing group. The representatives of the People's Republic have used virtually every trip to the rostrum, including that of the UN Seabed Committee, as an opportunity to heap vituperation upon the Soviet Union. Peking diplomats have attempted to put the ocean "haves" in general, and the USSR in particular, on the defensive by endorsing many of the most extreme claims made by developing states.

These fears have only further heightened the existing Soviet wariness of the UN. The Soviets have always denied the existence of any kind of "world government," but now they have become increasingly vocal in their denunciation of all attempts to expand UN powers. They have never believed an international organization bureaucrat could be a disinterested party; recently this has taken the form of resisting efforts by the Secretariat to broaden its discretion. They have always preferred negotiating in smaller, more technical, UN agencies. When the seabed item was first being considered by the 23d and 24th General Assemblies, the Soviet Union proposed that it be negotiated in the Intergovernmental Oceanographic Commission of UNESCO, which at that time had a Soviet citizen as its executive secretary. As new issues arise, in the seabed debates, the Soviets continue to seek to move them to smaller agencies. Further, the Soviets have always taken a go-slow posture. Throughout many Soviet speeches, certain themes are evident: "There is ample time," "We ought to gather more facts and data before making a decision," and "There are serious divergences of opinion that preclude quick agreement." Finally, the USSR has always sought an "out"—and needs this now more than ever—voting only by consensus.

PRESENT SOVIET POSITION IN LOS BARGAINING

The USSR as a leading participant in the current law of the sea negotiations has expressed its opinion on virtually all major subjects under negotiation. Soviet positions on five of the major issues that the USSR has entered into the UN record were measured by thematic content analysis of statements by official speakers who expressed for their governments a preferred position on one of the substantive or procedural subjects under negotiation. This provides a systematic record of all major points made by all states in these negotiations since they began in 1967.

Before we begin with what USSR representatives have said, let us mention here the important subjects they have not covered. Soviet negotiators have avoided all discussion of their doctrine of closed seas or historic bays. This absence is no accident; the Soviets are aware than an open espousal of such an unpopular doctrine would jeopardize their other bargaining goals.

The central issue for the USSR in the current law of the sea negotiations appears to be the delimitation of the breadth of the territorial sea. How far from land a coastal state can exercise "sovereign" or general-purpose powers is critical to major Soviet aspirations in transport, fishing, and naval movement. The Soviets do not consider coastal state special-purpose claims over activities beyond the territorial sea as serious a threat as the possible general expansion of the territorial sea. Time has caught up with the Soviet Union. While the claims of many other states have expanded, they have remained essentially in favor of a 12-mile territorial sea since the 1958 Law of the Sea Conference. In 1958 the USSR espoused its "expansionist" 12-mile claim on behalf of all "have-not" states to help thwart the activities of the "imperialists," who preferred to roam freely up to three miles off the coasts of other states.

But as the Soviets too began to roam freely in the open ocean, its stance became more emphatic on the delimitation of the territorial sea. A subtle change in wording in their proposals over time tells us a great deal. In 1958, the USSR proposed that there be a "breadth of . . . territorial waters . . . within the limits, as a rule of three to 12 miles, . . ." which by implication would allow claims to greater distances by those states with different historical, geographic, or security interests.[5] At the second UN Law of the Sea Conference in 1960, the Soviet Union proposed that "every state is entitled to fix the breadth of its territorial sea <u>up to</u> a limit of 12 nautical miles" (our emphasis).[6] In the preparatory phase of a third UN conference in the spring of 1974, the USSR proposed that "each State shall have the right to establish the breadth of its territorial sea <u>at no more than</u> 12 nautical miles . . ." (our emphasis).[7]

In the current negotiations, the Soviets have vehemently opposed the trend toward expansive claims to territorial seas or other zones of jurisdiction. Officially, the Soviet Union's rationale is that these extensions are inconsistent with the customary norms of international law and would increase shipping and air transportation costs, to the detriment especially of developing nations. The Soviet Union and the United States are therefore in the peculiar position of together considering a 12-mile territorial sea the <u>maximum</u> they would like to see enforced by states worldwide.

If coastal states may claim 12-mile territorial seas, it is not difficult to predict ocean-using states would be concerned about their right to freely transit straits of less than 24 miles in width. The Soviet Union is one of those states—again along with the United States—that feel that the right of innocent passage, the transit standard for territorial seas, will not assure their right of "free transit" through international straits. One reason is that under present rules of innocent passage agreed to in 1958, passage of submerged submarines and uncontrolled overflight of aircraft may be prohibited by the coastal state. Moreover, a number of straits states also insist that under the doctrine of innocent passage, they have the right to require that foreign warships provide prior notification of their passage; other straits states insist that foreign warships gain their prior permission before they attempt a transit.

In the recent law of the sea negotiations, the USSR has been a firm supporter of a doctrine of "free transit." They insist that a transiting vessel have the same right for the purpose of transit through a strait as it has on the high seas. But the Soviet Union has indicated a willingness to obey coastal states' designation of corridors for transit. Here, too, the basic position of the USSR parallels that of the United States.[8] However, an important difference must be noted: The Soviets are willing to accommodate their Arab friends by restricting the doctrine of free transit to those straits linking one part of the high seas to another part of the high seas. This would presumably exclude the Strait of Tiran from free transit. The Arab states would like to believe that such an exclusion would legitimatize Arab claims that vessels bound for the port of Elath, Israel, have no right to run the strait without Arab permission.

If the frequency with which a state takes the microphone in UN proceedings to advocate or defend a policy preference is an indicator of the salience of that preference, a victory for free transit is the most important objective the USSR hopes to achieve at the Law of the Sea Conference. They no doubt realize they are especially "geographically disadvantaged"—boxed in by straits; increases in riparian control can only work to their detriment. They were extremely sensitive in March 1972 to a Malaysian and Indonesian statement that the Strait of Malacca was not an international strait, sending L. I. Mendelvitch directly from Moscow to remonstrate with the straits governments. His visits, before tempers cooled, blew up into a mini-crisis, which, at its apogee, had a senior Malaysian official of the ruling party advocating conversion of the strait into a riparian-controlled canal. He estimated that a fee for transit might generate $147 million per year in revenues for the straits' governments.[9]

This is not to say that the Soviet Union is completely intransigent on straits. They are aware of the need to resolve the related special problems posed for archipelago states by their geography. They have gone remarkably far in their apparent willingness to seek a solution for archipelago states. For "ocean archipelagos," they have explored the possibility of granting a special status for each archipelago claim under a criterion of "intrinsic geographic unity," or defining an archipelago under a ratio of land to water scheme. This would not apply to so-called coastal archipelagos, where there is a mainland state with a string of off-shore islands that might be used to extend its jurisdiction. The Soviets are particularly concerned that the Greek Government could do this to control all waters leading out of the Bosphorus into the open Mediterranean.

Soviet fisheries negotiations at the Seabed Committee meetings have not been characterized by flexibility. They have an obvious, clear set of interests to protect, and they have been willing to pay the diplomatic costs necessary in order to protect them. This is not a new pattern of behavior. The Soviet delegation at the 1958 conference voted along with the most conservative Western European states to preserve rights of distant water fishing states in the face of demands of developing states to increase coastal state control of nearby fishing grounds. (The "cluster" these votes represented was the second most important set of conflict issues at the earlier conference as measured by the roll call variance—eigenvalue—of an R-factor analysis of the votes.)[10] Soviet concerns still exist for keeping foreign fishermen away from their coasts, but the Soviets prefer to deal with these problems in bilateral negotiations between themselves and the states concerned, most notably Japan. In Subcommittee II of the Seabed Committee, where fishing allocation proposals have been negotiated, the Soviets have adopted the most traditional possible position—they support the law of capture. They want no exclusive fishing zones beyond 12 miles from any coast. Indeed, they are vehement opponents of a zonal approach—exclusive or not—to the fishing problem. They claim that coastal fishing zones will lead to lower world catches and that enforcement of coastal fishing zones by developing states would harm their own economies.

However, the Soviets do not propose complete laissez-faire in ocean fishing. They have been willing to regulate ocean fishing and, to a much lesser degree, allocate the catch through existing regional fishing organizations. They extoll these organizations highly in the UN debates, and with good reason. The weaknesses of these regional bodies are several: They usually make decisions under a unanimity rule; they usually do not provide for compulsory settlement of disputes; they usually do not have the right to inspect fishing vessels on the high seas to enforce their regulations or allocation scheme. This may seem unfair to the USSR because in the "Moscow Declaration on Rational Exploitation of the Living Resources of the Seas . . ." of July 1972, the USSR and its socialist allies did state that "it is necessary to give international organizations functions of international verification of compliance with fishing regulations. . . ." But the enforcement clause of the fishing proposal introduced by the USSR shows such "international verification" is a way of avoiding national inspection and enforcement even in areas where the USSR might concede some form of coastal state preference.[11]

The USSR will face a difficult dilemma in protecting its fishing rights at Caracas. While the general assault on the freedom to fish is being led primarily by developing states, there is no united opposition among developed states. Indeed, the USSR, most of the Warsaw

Pact states with sea coasts, and Japan stand virtually alone. On the other hand, Canada favors an exclusive coastal fishery jurisdiction zone and beyond that, coastal state competence to "take appropriate measures" to assure the maintenance of the productivity of the sea areas adjacent to the exclusive fishery zone. Even the United States favors a coastal state preferential scheme for coastal and anadromous species in part because of the desire of its coastal fishermen to control and/or limit Soviet fishing off of U.S. coasts.[12]

The Soviet Union's dilemma is in finding a proposal that does not antagonize developed or developing states, does not threaten to break up favorable coalitions on other issues, but still preserves Soviet access to existing fishing grounds or reasonable future alternatives. Its current fishing proposal does not appear capable of doing that. It does, however, indicate Soviet fishing priorities. The greatest short-run fisheries threat they face is the potential expansion of developed states coastal jurisdictions. Thus, the USSR fishing proposal would grant developing coastal states the right to reserve in an area up to 12 miles beyond the territorial sea or fishery zone (or, maximally, 24 miles from the coast) the percentage of the allowable total the coastal state vessels actually can catch. As the developing coastal states' fishing capability expands, therefore, so does their allowable catch. But until it does, distant water fleets can fish as close as 12 miles from the coast. Some observers see this proposal primarily as the USSR's attempt to find some accommodations with the developing states from whom it is growing increasingly estranged on economic and jurisdictional issues. No doubt this partially motivated Soviet decision-makers, but we believe that the most important factor was a hard-headed evaluation of the threat of developed states' coastal expansion. Indeed, since the real "threat" to existing Soviet fisheries interests comes from developed states, it does not seem impossible for the Soviet Union to accept a 200-mile fishing zone as long as the developing majority is willing to create exceptions for existing Soviet fisheries.

A fourth important issue that may be decided at the third UN Law of the Sea Conference is the extent of national jurisdiction on the seabed. The original Soviet stance on seabed delimitation was non-cooperation—they attempted to stall a settlement by a filibuster. The issue, in the beginning of the UN debates, was how to proceed with seabed matters: Whether to decide the principles (and machinery, if any proves necessary) applicable to a regime regulating the deep seabed before deciding its delimitation, or vice versa. The Soviets lobbied heavily for the former. For them, the machinery's configuration determined its territorial scope. Soviet intransigence did help delay the proceedings, but it became evident over time that the negotiations were going to proceed toward a solution to both problems simultaneously. This forced the USSR to state preferred positions on both.

The Soviet seabed delimitation proposal would put under coastal state control all seabed areas within the 500-meter isobath, or 100 nautical miles, whichever is further. A glance at a bathymetric chart of the Arctic Ocean will demonstrate one of the reasons why this is the Soviets' preferred solution. The Arctic, especially on the Soviet side, is quite shallow, with a very substantial portion of it less than 200 meters deep. It also has extensive continental platform areas deeper than 200 meters and insular slopes of greater than 1:40. Because of the extensiveness of the continental shelf and slope in the Soviet Arctic, the Soviets have fewer incentives to claim seabed areas to the end of the continental margin than many other states. Under the 500-meter formula Soviet claims may extend hundreds of miles, while the 100-mile maximum would restrict jurisdiction for states with margins that slope off sharply near shore. By restricting most states' jurisdiction, the Soviets hope to contain "creeping jurisdiction." Under no condition do they want extensive seabed claims to become exclusive coastal state jurisdiction over superjacent waters or airspace. The USSR has so far been very unreceptive to any form of 200-mile economic zone proposals. Nevertheless, it is not impossible for the Soviet Union to accept the 200-mile thesis. While they would worry about encouraging creeping jurisdiction, and other states might benefit proportionately more than the Soviets by adoption of a version of the 200-mile economic zone, the Soviet Union will be able to put very substantial amounts of ocean space under its jurisdiction if it accepts a 200-mile zone.

The Soviet Union has not shown enthusiasm for the creation of a deep seabed regime and machinery (now being referred to as ISRA, or International Seabed Resource Agency). The Soviets' views are consistent with their distaste for world government.

From the earliest seabed debates, they rebelled at the concept of the seabed as the "common heritage of mankind." They felt this implied a common ownership, which in turn would have institutional implications—such as the need for a regulator, or worse still, a universal exploiter. Such joint ownership would lead to a monopoly, the Soviets fear, and one in which they were concerned they would play a small part.

Recently they have been forced to trade off ideology for a degree of pragmatism. The common-heritage concept has become a rallying point for developing nations. Seeing that it was not an easy concept to operationalize, the Soviets also have adopted it as part of their jargon. In addition, they have modified their position on the international machinery. They have apparently come to believe that some international machinery is inevitable—moreover, it appears that they dislike unrestricted competition as much as supranational organization. Here, again, they have chosen not to push for a lassez-faire proposal. In 1971, they proposed a skeleton licensing agency similar to the U.S. Presidential Draft Seabed Treaty.[13]

If this modification in general approach was a gesture of appeasement to the new nations, the attempt probably will be a futile one. The Soviets have retained in this proposal key parts of their traditional outlook—their opposition to "career universal bureaucrats" and their insistence on consensus. Their scheme would create a bureaucracy without individual latitude. Article 23 insists on voting by consensus in the council, and they have further enraged the developing countries by proposing the Eastern European group have an equal number of seats in the council as other regional groups (Article 21). In addition, they have proposed a fixed allocation scheme, resulting in an international agency with only coordinating powers. Their shift in position could hardly be characterized as major.

The issue of property is the key to continued Soviet reluctance to create a powerful new international agency. They refuse to ascertain ownership of the seabed (or high seas): All rights must apply to the entire community of sovereign nations and must be exercised as other freedoms of the seas. The ocean has been a common property resource for them, and only very recently have they acknowledged that there is not in fact an infinity of resources in it. The licensing agency proposal recognizes the need to regulate oceanic resources, but at the same time, it fulfills the Soviet doctrine of interposing sovereign states between international authority and individual exploiters. They do not acknowledge that regulation is in fact a way of assigning property rights—to the regulator. By seeking to create a regime where the component parts are sovereign states, they are merely trying to politicize the regulative act in order to maximize their own access to the ocean while minimizing the property rights of the regulator. The Soviets also propose to limit the latitude of the regulator to exploration and exploitation of the seabed; many LDCs would have its power encompass all oceanic activities. Any proposal suggesting wider powers and scope than theirs is labeled "a tendency to establish a regime for the seabed beyond the continental shelf in the interests of individual countries or groups of countries."[14]

THE SOVIET NAVY'S POSITION

Admiral S. G. Gorshkov, commander in chief of the Soviet Navy, has gone on record[15] as opposing any attempt to abrogate the principle of freedom of navigation by "certain countries." He feels these nations incorrectly view freedom of the seas as favoring imperialist nations to the detriment of developing nations. Rather than discard the principle, Gorshkov proposes stricter observance of it as the only way to curb the imperialists' increasing aggressiveness.

Since Gorshkov speaks for the navy, the numerous installments in Morskoi sbornik provide us with an excellent opportunity for comparing one Soviet bureaucracy's position with the "official" Soviet position on major ocean issues. Where they are very similar or identical, either the navy has had an influential input to Soviet policymaking, or Gorshkov is merely one of the decision-makers' mouthpieces. If they differ, the assumption must be that naval views have been disregarded. Thus, while we cannot characterize the navy's position relative to other ministries, we can show the relationship between it and the official Soviet one.

Our comparison will be facilitated by some of the methods developed by the Law of the Sea project at the Center for Naval Analyses.[16] Positions, or "national scores" of the USSR, and a number of other states on some of the major issues of the Law of the Sea negotiations, will be compared with Gorshkov's "scores" derived by the same method.

The procedure relies on a thematic content analysis of the United Nations records concerning ocean use. Only data from March 1971 to June 1973 were used so that the UN data cover roughly the same period as the germination, writing, and publication of the Gorshkov piece.

For content analysis, a coder extracts a "theme" from the records (UN documents or Admiral Gorshkov's article); a "theme" is a statement indicating an official speaker's preferred position. These themes are then collected into general subject topics. Each set of themes is organized into a model of the policy problem by means of "scaling," or ordering the data according to an underlying conceptual framework. By this device of "artificial measurement" we can compare the preferred outcomes of each nation.

Table 18.1 contains abbreviated versions of the three variables or topics with which we will be concerned, and the "rank" or scale number of the remarks that are used to construct "national" scores" on a topic. National scores are simply weighted averages of the ranks of the remarks made. Since we are interested in comparing Soviet positions, and Gorshkov's expressed preferences with the position of all states eligible to vote at the UN Law of the Sea Conference, it was necessary to estimate positions on the issues for all states. This has been achieved by means of a linear regression model that employs some 31 characteristics (membership in regional and caucusing groups, geographic traits, and economic common-interest groups) as the independent variable and each mention of a theme by

TABLE 18.1

Three Variables
Variable 1: Territorial Sea

Rank	Theme
1	3-mile territorial sea (TS)
3	12-mile TS
4	Might take 12
8	25-30 mile

Variable 2:
Right of Transit Through Straits

Rank	Theme
1	Support free transit through straits
4	Privately sympathize with free transit
9	Free transit with exceptions and limits
12	Moderate innocent passage
18	Conservative innocent passage
24	Riparian control

Variable 3:
Powers of ISRA

Rank	Theme
1	International operating agency
4	Establish joint enterprises
7	Mixed machinery
11	Regulatory agency
13	International licensing system
15	Registry: first-come, first-served

Source: Compiled by the authors.

a state as the depending variable. Coefficients for each characteristic are derived from this equation, and the relevant ones are summed to get a national estimate for each state. For example, the USSR's estimate would be the sum of the positive and negative coefficients of the following groups: the Eastern European caucusing group, major merchant fleet states, Blue Water Navy states, major fishing states, distant water fishing states, straits states, broad shelf states, major mineral-producers, major oil-producers, and offshore oil-producers.

We will compare Admiral Gorshkov's views with the USSR, the U.S., and a number of major groups' views on the three issues upon which Admiral Gorshkov commented: breadth of the territorial sea, the right of transit through straits, and the nature of ISRA. Before we proceed we must enter a caveat and make an important aside.

The caveat first. We believe our comparison is interesting and worth making, but we do not wish to make far-reaching claims for it. The data derived from Admiral Gorshkov's article comprise an extremely small sample size.

The aside covers the absence of data. Our coder could find nothing in the Gorshkov piece that would indicate that the admiral has a preferred position on fishing policy. This is most curious considering the importance of the fishery issue to the USSR and the high probability that within the Soviet bureaucracy, discussions on trade-offs among the major issues have taken place.

The degree to which the navy's (Gorshkov's) "output" agrees with the Soviet external "output" (the USSR's UN statements) can be seen in Table 18.2. In addition, Table 18.2 also reports our "estimates," telling us what the USSR's position would be if it considered only the characteristics we have chosen for our model in making policy. Obviously, we can have no estimates for Admiral Gorshkov. For comparative purposes, we also include the U.S. scores and estimates on the three variables.

On the territorial sea problem, Gorshkov's position is virtually identical with that of the Soviet Union's law of the sea negotiators; the scores are 3.0 and 2.9, respectively. If Gorshkov reflects the Soviet Navy's preferences, it must be satisfied with its country's long-held territorial sea position. Naturally, we cannot conclude that the navy's view was the prime determinant of Soviet policy. But Soviet policy as expressed in the UN negotiations does not show that other bureaucratic forces that might prefer a more expansive territorial sea stance have had any influence on this question.

Gorshkov's position on straits is also to be expected—he favors free transit through international straits with a "score" of 3.3. The difference between Gorshkov's strict view of free transit and the seemingly less committed position of the Soviet delegation can be

TABLE 18.2

Comparison of Gorshkov's Remarks with
USSR and U.S. National Scores and Estimates

Variable		Score	Estimate[a]	Sample
1. Territorial sea	Gorshkov[b]	3.0	—	2
	USSR[c]	2.9	2.9	15
	U.S.[c]	2.7	3.1	10
2. Right of transit through straits	Gorshkov	3.3	—	4
	USSR	4.7	3.9	48
	U.S.	2.1	4.4	41
3. Deep seabed regime	Gorshkov	14.0	—	1
	USSR	11.5	14.7	8
	U.S.	12.7	12.4	16

[a]These numbers are generated from a linear regression of the groups with which states are associated; they are not the substantive positions upon which the states have put themselves on record.

[b]Data coded from Morskoi sbornik no. 2, 1973.

[c]Data from the UN Records contained in Law of the Sea Project files from March 1971 to June 1973.

Source: Compiled by the authors.

accounted for by the need for Soviet representatives to bargain. The weighted average made up of official pronouncements rests at close to rank 5 because it includes not only many endorsements of free transit (rank 3) but also explorations of the acceptability of the Soviet fall-back position of restricting free transit to a limited number of straits (rank 9). Again, it would appear that Gorshkov is not quarreling with his country's position.

One must be cautious in interpreting the comparison of Gorshkov and the Soviet Government on the powers of ISRA because our coder was able to find only one item to extract from the Gorshkov piece on this subject. Nevertheless, that one piece of data allows us, in a very general way, to see that Gorshkov's thoughts and those of his government on the creation of a new international agency are probably very similar. Both clearly are against the establishment of an ISRA that would have broad discretionary powers. It is also interesting to note that Gorshkov's score is very close to what we estimate would be the "proper" position for the USSR, given one view of its interests.

In summary, our data show a very consistent pattern. Gorshkov agrees with his government, his government is taking the positions we "estimate" it should, and given the fact that its interests appear to be similar to the other superpower, it is understandable why its "scores" are similar to the scores and estimates for the United States.

But where do Gorshkov's preferences or the official Soviet positions stand in relation to other conference participants? A "mean bloc preference" sums the national scores, or estimates, for each of the geographical, economic, and political groupings, and the average for each is recorded. Table 18.3 shows a comparison of the Gorshkov and USSR scores with the means of the 9 groups with which they have the most in common.

As can be seen from Table 18.3, both Gorshkov's and the USSR's scores come closest to that of the other major ocean users (the United States, USSR, France, Japan, Netherlands, Poland, and the United Kingdom—states with multiple ocean interests). Eastern European and major merchant fleet countries appear to have the next most in common with the two superpowers. After that, similarities drop off sharply. Thus, one could expect countries with these three characteristics—belonging to the major ocean users, being Eastern European, and having a major merchant fleet—to take positions most similar to those of the Soviet Union and Gorshkov. The dissimilarities with the other six groupings in Table 18.3 is even more striking when one realizes that the extreme scores of the USSR on all three issues are pulling the group scores down (or for regime, up). If the USSR were not part of these six blocs, the mean bloc preferences would be even further from the Soviet score.

Table 18.4 puts the Soviet score in the context of an even broader framework: developed versus developing nations. The final Law of the Sea negotiations will most probably have the most difficulty reconciling these two groups. Representing the "developed" here is an artificial group that had the closest fit with Gorshkov and the USSR in Table 18.3—major ocean users. The "developing" is represented by the Group of 77 (the number of "developing" states in the UN when the caucusing group was formed).

THE SOVIET DILEMMA

With substantive positions that fly in the face of the preferences of a developing majority, interests that would be endangered by major concessions, bureaucrats who have vested interests in resisting change, the Soviets will have difficulty in trying to shape the outcomes

TABLE 18.3

Comparison of Gorshkov and USSR Scores with Mean Bloc Preferences

	Gorshkov	USSR	East. Eur.	Major Ocean Users	Maj. Merchant Fleet	Distant Fishing	Blue Water Navy	Major Fishing	Major Oil Prods.	Off-shore Oil	Broad Shelf
Territorial Sea	3.0	2.9	4.5	3.2	4.8	7.0	5.1	7.0	5.8	6.0	8.2
Straits	3.3	4.7	4.8	3.7	5.6	8.4	8.6	9.7	11.2	11.5	10.2
Regime	14.0	11.5	11.6	12.6	10.8	9.1	7.0	8.0	6.0	6.5	6.5

Source: Compiled by the authors.

TABLE 18.4

Comparison of Developed (Major Ocean Users)
and Developing (Group of 77)
by Mean Bloc Preferences

Issue		Scores	Description
Territorial sea	Developed	3.2	12-mile territorial sea
	(Gorshkov	3.0)	
	(USSR	2.9)	
	Developing	7.1	25-30-mile territorial sea
Right of transit through straits	Developed	3.7	Free Transit
	(Gorshkov	3.3)	
	(USSR	4.7)	
	Developing	10.4	Innocent Passage
Deep seabed regime	Developed	12.6	International Licensing System
	(Gorshkov	14.0)	
	(USSR	11.5)	
	Developing	5.5	Joint Enterprise System

Source: Compiled by the authors.

to be attained at the third UN Law of the Sea Conference. They may have even more difficulty in accepting those outcomes once attained. The Soviets have already served notice that if they do not achieve their major goals, they may walk out of the conference or, even if they stay, they may not sign or ratify the conventions worked out. We do not know if they will or will not walk out or not sign the convention. But their threat of noncooperation may leave them with a major foreign policy dilemma, which we can elucidate.

Four types of outcomes of the Law of the Sea Conference might be envisioned by the Soviet Union: (1) stalemate, (2) complete success, (3) complete failure, (4) partial failure. Only in the case of the fourth alternative would the dilemma have to be addressed by the USSR. But it is our judgment that "partial failure" as seen by the USSR will be the most likely outcome.

If the 10-week bargaining session in Caracas in the summer of 1974 is stalemated, it is obvious that, in the short run, Soviet decisionmakers need not act. But stalemate only defers the problem or moves it to another arena. It is easy to believe that "no decision" will result from the first round of the Third Conference. After all, diplomats have been trying to achieve an overall settlement since 1967, but because the alternative to decision in the UN arena is possible decision by national fiat of 150 states, there are many incentives for most states, the USSR included, to continue the bargaining.

It is possible that agreements that could be characterized as "completely successful" could result from the first session of the conference. Complete success, as seen by the Soviet Union, might result from either of two bargaining strategies: reaching consensus on all main issues or achieving requisite two-thirds majorities in open voting on all main issues. A convention(s) that satisfies, in the Soviet phrase "all main groups" would be, in the language of game theory, a "cooperative solution." If a cooperative solution is achieved, it will be a true diplomatic triumph for the senior negotiators, Soviet negotiators included. However, while possible, we do not believe this outcome probable.

Even less probable is the second strategy that theoretically might lead to a complete Soviet diplomatic success at the Law of the Sea Conference. Forcing issues to open votes is akin to zero-sum games where some players win and some lose. Because of the alignment of forces likely at Caracas, it is obvious why thus far in the negotiations the Soviets have insisted upon "all main groups" being satisfied before a decision is made.

"Complete failure," while not the most likely outcome, is possible. By complete failure we mean the breakdown of a "gentleman's agreement" to bargain toward consensus, the rush to majority voting and a subsequent convention(s) in which the developed states see no language that protects their interests. The developing would use its majority to achieve the maximum "redistribution" of ocean assets possible on paper. The result probably would be that major ocean-using states—East and West—would neither sign nor ratify the conventions, leaving the developing nations with a hollow victory. Confrontational politics could lead to this outcome. There are incentives for some states, especially the USSR, to see complete failure as a more desirable outcome than partial failure. If majoritarianism runs rampant, the Soviet Union is much less likely to be abandoned by the other developed states if it chooses not to sign. More important, the case of the nonsignatories that are not bound by the convention they did not sign is strengthened, perhaps even excusing their use of force to protect their interests.

"Partial failure" is therefore not only the most likely, but the most dangerous from the Soviet point of view. By a partial failure we mean a convention(s) signed that in the main gives the developing majority what it wants but that concedes some of the security, transport, fishing, and resources interests of the developed, but in substantially weakened form. But even "partial failure" could be achieved only if the USSR made a major change in its conference tactics. It seems likely that the conference will not agree to any convention that does not have in it some form of 200-mile economic zone, probably for both fish and mineral resources. For "partial success," it is probably necessary for the USSR to accept the idea of an economic zone, perhaps in return for granting the Soviets some exceptions to complete coastal state control in the zone that would protect existing Soviet distant fishing water interests, especially in relation to fisheries. From the negotiating posture of a number of major developing states it seems likely that Russian concessions on the economic issues is the key to the achievement of a minimally acceptable outcome on the transport and security issues.

This type of mixed-bag comprehensive agreement would put the Soviet Union on the horns of the dilemma. The Soviets might find such an agreement acceptable because it would indicate that this is not likely to be the last assault on the established order of the ocean. Clearly, the magnitude of the Soviet dilemma depends upon how acceptable the developing nations' concessions are to the USSR. If they are even grudgingly acceptable, the outcome would be more like that of complete success, after a certain amount of grumbling. But the Soviets should fear a resolution unacceptable to them. A number of developed states would be prone to accept (sign and ratify) any agreement, even one the Soviets found unacceptable, so that the world could get on with its work. This would set off Soviet fears of abandonment and diplomatic isolation, which might be intense enough to pressure the Soviets to obey the regulations worked out anyway. Moreover, the risks of the use of counterforce against those who claim they have a "legitimate" right under the convention(s) to protect their territory or rights are increased. It is therefore the range of unacceptable "partial failure" outcomes that would force upon the Soviet Union some of its most difficult decisions.

NOTES

1. William E. Butler, The Soviet Union and the Law of the Sea, (Baltimore: Johns Hopkins Press, 1971), p. 3. Much of the following discussion is drawn from the above, and the following other works of Professor Butler: "The Legal Dimension of Soviet Maritime Policy," in Soviet Naval Developments: Capability and Context, ed. by Michael MccGwire (New York: Praeger Publishers, 1973), pp. 109-22; The Law of Soviet Territorial Waters (New York: Praeger Publishers, 1967); and "Some Recent Developments in Soviet Maritime

Law," The United Nations and Ocean Management, ed. by L. M. Alexander (Kingston: University of Rhode Island, 1971), pp. 375-86.

2. George A. Brinkley, "The Soviet Union and the United Nations: The Changing Role of the Developing Countries," Review of Politics 32, 1 (January 1970): 95. Several other works cover the Soviet Union's UN policies during this period fully. They include Alvin Rubinstein, The Soviets in International Organizations (Princeton, N.J.: Princeton University Press, 1964), and Alexander Dallin, The Soviet Union at the United Nations (New York: Praeger Publishers, 1962).

3. Brinkley, op. cit., p. 112.

4. New York Times, September 6, 1973, p. 5; September 8, 1973, p. 13; and September 9, 1973, p. 20 and Section IV, p. 5.

5. UN Doc. A/CONF. 13/C.1/L.80 in United Nations Conference on the Law of the Sea, Official Records, vol. 3 (A/CONF. 13/39), p. 233.

6. UN Doc. A/CONF. 19/C.1/L.1, Second United Nations Conference on the Law of the Sea, Official Records, (A/CONF. 19/8), p. 164.

7. US A/AC.138/SC.II/L.7/Add.1, General Assembly Official Records: 28th Session, Supplement no. 2, (A/9021), vol. 3, Report of the Committee on the Peaceful Uses of the Seabed . . . , p. 1.

8. For a comparison of the two proposals, see SCII/WG Paper no. 4, in GAOR, 28th Session, Report of the Committee on the Peaceful Uses of the SeaBed and the Ocean Floor Beyond the Limits of National Jurisdiction, vol. 5, Supplement no. 21 (A/9021), Section 4.2 p. 14.

9. New York Times, March 13, 1972; Washington Post, March 21, 1972; Economist, March 18, 1972, p. 52; Manchester Guardian Weekly, 106:13 (March 25, 1972), p. 8.

10. See R. Friedheim, "Factor Analysis as a Tool in Studying the Law of the Sea," The Law of the Sea, ed. by L. M. Alexander (Columbus: Ohio State University Press, 1967, pp. 47-70.)

11. For a reprint of the Moscow Declaration and the USSR fishery proposal, see SC.II/WG/Paper No. 4, pp. 6-11 in GAOR, 28th session Report of the Committee on the Peaceful Uses of the Seabed . . ., Volume V, Supplement No. 21 (A9021).

12. For a comparison of the major fishing proposals see Albert W. Koers, "Fishery Proposals in the United Nations Seabed Committee: An Evaluation," Journal of Maritime Land and Commerce 5:2 (January 1974), pp. 183-209.

13. For the Soviet proposal, see GAOR, 26th Session, Report of the Committee on the Peaceful Uses of the Seabed . . .," Supplement no. 21 (A/8421), pp. 67-75.

14. Khlestov, "International-Legal Problems of the World Ocean," International Affairs, March 1973, p. 38.

15. Morskoi sbornik, no. 2, February 1973.

16. For our method see R. L. Friedheim and J. B. Kadane, with the assistance of J. K. Gamble, Jr., "Quantitative Content Analysis of the United Nations Seabed Debate: Methodology and a Continental Shelf Case Study," International Organization 24, 3 (1970): 479-502, and R. L. Friedheim and J. B. Kadane, "Ocean Science in the UN Political Arena," Journal of Maritime Law and Commerce 3, 3 (April 1972): 473-502.

CHAPTER 19

SUMMARY
OF DISCUSSION
IN PART II
Ken Booth

The discussion was opened by Alvin Rubinstein, who suggested that Soviet involvement in the third world had been the most notable new feature of Soviet international behavior since 1953 and more than any other single phenomenon had signified the shift in Soviet strategy from a continental-based one to a global one. He argued that while there is general agreement on what the USSR has expended (in economic, military, and diplomatic terms) in its courtship of key Afro-Asian countries, we do not know nor can we agree on what the USSR has accomplished as a result of its efforts. In particular, we lack any accepted method of establishing the extent to which the USSR has been able to translate its largess into actual political influence. He then made three general observations on the subject.

1. Western analysts have generally exaggerated Soviet influence in the third world. This is especially so in the case of those trained in Soviet studies, who tend to rely on Soviet pronouncements and are not familiar with the domestic politics of the third-world countries that are the targets of Soviet influence.

2. Evidence of Soviet influence is best seen in terms of the actions of third-world countries rather than in Soviet pronouncements. Soviet materials are generally not helpful in studying the question of influence. Furthermore, some of the traditional indicators of "influence" (such as levels of aid, "presence," and trade) are often misleading.

Note: Discussion was based on the presentations of Kelly, Ra'anan, Rubinstein, and Smolansky, and the paper by Kelly and McConnell, "Superpower Naval Diplomacy in the Indo-Pakistani Crisis," in MccGwire, ed., Soviet Naval Developments: Capability and Context (New York: Praeger Publishers, 1973), pp. 442-55.

3. International relations theories are not very useful in operationalizing the idea of influence.

He concluded that the strategic rationale for Soviet policy in the third world had changed since the mid-1950s. It was still important, but today the political rationale was the chief factor. In the mid-1950s, the Soviet Union had entered the third world easily because of the regional problems polarized by U.S. policy. The obvious policy for the USSR had been to support the anti-American groups and/or encourage nonalignment. The denial of forward bases to the United States (for example, for bombers) had been the clear strategic rationale for Soviet policy. Today the Soviet Union needs facilities to support its forward naval deployment in a number of areas. Therefore there is still a strategic rationale in their policy. The question of whether they will then seek base facilities is more difficult. There is an ideological constraint against bases, but ideological gymnastics can be performed if necessary. At present Soviet policy is made up of an untidy package of political reasons. It involves an effort to deny the influence of Western countries in the third world and to increase Soviet influence. It is pushed along by an important but general belief in incrementalism, the hope that something might come off. The interplay of bureaucratic politics also has a role in shaping the policy outcomes.

SOVIET POLICY AND INFLUENCE IN THE THIRD WORLD

There was a measure of consensus on some of the major questions concerning Soviet policy and its influence in the Third World:

Western Assessments

It was agreed that there had been much exaggeration of the likely influence that the Soviet Union had exercised in various Afro-Asian countries. Certainly it was believed that Soviet efforts had not been as successful as the Soviet leaders would have hoped. One view was that U.S. Government reactions to Soviet policy in the third world had been invariably based on what it thought the Soviet Union was trying to do, rather than on what they had the capability of doing. In any case, it was important not to confuse "influence" with "control" and to be specific in analysis—that is, concentrating upon Soviet "influence" in key target countries. It was generally agreed that the Soviet ability to influence important groups in the third world was not impressive.

While it was acknowledged that Soviet success had been modest, two warnings were given. First, a reminder was given that Soviet leaders were believers in "incrementalism"—that is, they believed that small advantages over time might bring great political advantages in the long run, though the outline of possible advantages was not yet clear or even dimly visible. This was thought to be an important rationale for the courting by the Soviet Union of key third-world countries. Furthermore such efforts might deny Western influence in such countries and would certainly complicate Western policy. The "denial" element in Soviet motives had come to be additionally important in recent years because of the Chinese factor. The Soviet Union devotes much attention to the latter, although the Chinese themselves do not yet have much capability. Second, a warning was given against undue complacency. Even if it was valid to argue now that Soviet success had been modest, does this necessarily mean that earlier Western apprehensions were wrong? Furthermore, the trend of the recent past might not be a good basis for thinking about the future. In future, the Soviet Union may become much more successful, as it learns from its experiences: for example, developing better language training. Furthermore, there is evidence of a greater realism in Soviet attitudes; they are more conscious of the limits within which they can operate and in the considerable problems they face. The various institutes that were set up have played an important role in this educative process, especially with regard to the Middle East.

Soviet Assessments

There was wide support for the opinion that Soviet writings on third-world questions had become much more sophisticated in recent years. In the early period of the Soviet penetration of the third world, Soviet journals had talked in universalist Marxist terms about the contribution that aid would make to creating the conditions for the movement of underdeveloped countries toward socialism (in the distant though not unforeseeable future). Since that time Soviet writings had become far more realistic. More recently, writings on third-world subjects had been characterized by more specific articles oriented toward particular countries and often to specific problems.

The question was raised whether the Soviet Union itself thought its efforts had resulted in an important degree of influence. The discussion failed to produce a clear answer, because Soviet writings did not discuss the subject in this way. However, there was thought to be plenty of circumstantial evidence pointing to the conclusion that Soviet participants in this field had a good deal of realism and recognized the narrow limits of their influence and the many problems they faced.

Trade as Influence

The value of trade as an indicator of influence was discussed, and there was agreement that the interpretation of Soviet statistics in this field was unusually difficult. However, it was suggested that the economic factor in the Soviet courtship of the third world was now much more important than previously. In recent years the Soviet Union has been engaged in businesslike deals with third-world countries. Invariably the deals had been on terms favorable to the Soviet Union. In contrast to the second half of the 1950s the Soviet Union is now more cautious in its economic relations with third-world countries; in contrast even to the early 1960s, there now has to be a much better economic rationale for any trading relationship. There was support for the view that the economic instrument of policy was of declining utility for the Soviet Union: Many third-world countries (notably India) did not take up even a half of Soviet credits, and many backed away from Soviet assistance after a time, because it was not found useful and because the Soviet Union did not provide the services required. In fact, one view was that the practical result of some of the deals had been to make certain third-world countries actual creditors of the Soviet Union. It was also suggested that the closer the political link with the Soviet Union, the worse the economic bargain for the third-world country concerned. As a further constraint, it was noted that in a number of important countries, the Soviet Union has to compete with Western suppliers, which placed important limits on Soviet penetration and potential influence.

Treaties as Influence

The question of treaties as indicators of influence was discussed, and once again, the picture was not clear. While it might be agreed that the series of friendship treaties did give a theoretically far-reaching framework for Soviet relations with particular countries, the major consideration is what happens in practice. One view was that the behavior of the government of India was not notably different after its treaty with the USSR, from that which had characterized it before (for instance, India had not taken a strong line over Czechoslovakia in 1968). In this case one could therefore argue that the treaty had not really affected the influence relationship.

Whatever other motives the Soviet Union had for developing its relations with third-world countries, in certain cases there was also a naval rationale. Facilities were needed, especially in the Indian Ocean, for the resting of both men and ships.

NAVIES AND FOREIGN POLICY

Turning to the general question of navies and foreign policy influence, it quickly emerged that there were many difficulties in developing suitable criteria for judgment. Several general considerations were put forward. (1) The need to see the whole range of "maritime" power (including the merchant marine and fishing) as well as just naval activities. (2) The fact that while navies were given much attention these days in Soviet relations with third-world countries, naval units actually seem to have had limited influence in affecting the domestic politics of target states such as Egypt. (3) It was argued that in any particular area "credibility" not "presence" was the key factor. It was agreed that credibility was a subjective and contextual quality, so that it was not possible to generalize. It was suggested, for example, that on matters of little direct concern to the United States the latter would not take any risks against a single Krupny; in some circumstances therefore, a single warship might be judged a "credible force." (4) It was suggested that in any episode of naval diplomacy the important thing was the political situation ashore rather than the naval presence off the shore. In this respect it was stressed that the context of international politics was so important. Neither the United States nor the Soviet Union has control over third parties; much of the thrust of international politics is therefore in the hands of others. Furthermore, influence is a two-way process: The superpowers have to adapt their own policies as well as try to influence others.

The discussion then focused on the question of whether Soviet foreign policy would have looked much different had there been no forward deployment in recent years. There were two conflicting sets of answers to this important but elusive question: One was moderately affirmative; the other moderately negative. The negative case argued that if the relationship between Soviet naval deployments and key events since 1967 was compared, the results did not appear to suggest that various situations had been significantly affected. It was suggested that the kind of military power that had made the difference in critical areas in the six years from 1967 to 1973 was not naval power.

The moderately affirmative case, on the other hand, was argued on the basis of the historical perspective in which the Soviet leadership sees itself and acts. At a time when the traditional powers are thought to be in some decline, the Soviet Union sees itself becoming an increasingly important global power, which needs the appropriate military configuration to play such a role; and naval power is a most important and tangible element of this. It sees Soviet policy in recent years as having been attended with some success. It was suggested that the Soviet Union had used its navy in two stages. First, it had used it to try to deny U.S. influence in local situations in the latter 1960s, by attempting to neutralize the traditional freedom the United States had had in distant conflicts as a result of its preponderant seapower. To the extent that U.S. policy had become somewhat "isolationist" by the end of the 1960s (for example, the "Nixon doctrine") this could be regarded from the Soviet viewpoint as a vindication of its policy of denial. The second stage involved the Soviet effort to use its naval instrument in some local areas to positively improve its own influence. This was the stage they had reached in the Middle East.

While this question could not be easily resolved, there were many interesting things to be said about it. It was surrounded by an inherent obscurity, however, because of the elusiveness of the cause-effect relationship. Until very detailed empirical studies were available, the replies to this question would remain more in the form of hunches than even approximate answers. Running through this discussion was a set of similar questions. What is "influence"? How can it be assessed? What are Soviet interests in the area? These questions were difficult to answer in relation to particular countries. It was therefore even more difficult to arrive at any formula for describing the importance of Soviet interests in the third world in general. Rather than making such an attempt, it was thought much more important to be specific and discriminating in looking at Soviet policy in the so-called third world, an area in which lies a much greater diversity than is conveyed by this blanket term. It was more important to try to understand Soviet interests in, and possible influence over certain key target states, than to look for any general picture.

EXAMPLES OF SOVIET NAVAL INVOLVEMENT

Some discussion concentrated on the specific case studies that were presented.

The Eastern Mediterranean. In this topic particular interest was shown in the reports from the French press that Soviet landing ships had been used to transport Moroccan troops to Syria in April and July 1973, and that these units had sailed as a formed group with a screen of escorts, operating as in a hostile environment. The significance of this transport in force was disputed. One view suggested that its potential significance was considerable and that it was not an insignificant episode even if it does not become a precedent. Attention was drawn to the 24th CPSU Congress, in which it had been stressed that the Arab states should cooperate fully and effectively if they were to deal with the Israeli problem. The transportation of Moroccan troops was tangible evidence of Soviet willingness to support them in their policy. Furthermore, the Soviet Union had been telling the West African countries that they must stand on their own feet but that the USSR would help. In this respect, the precedent of the Guinea Patrol was significant. Against this viewpoint, some argued that the novelty of the episode was being exaggerated. It was pointed out, for example, that the USSR had transported many Arab troops to Yemen in the 1960s and that the only new element in 1973 was the nationality of the passengers. In addition, it was suggested that the episode was probably no more significant than a Soviet response to a Moroccan request for help.

The Indo-Pakistani Crisis, December 1971. An analysis of this episode was presented to the seminar. Among the comments on this Soviet deployment were the following: (1) The configuration of the Soviet detachment sent into the Indian Ocean suggests that Soviet handling of the crisis was restrained and calculating. (2) Even detailed analysis presents many complexities. Too fine a tie-up of dates, for example, might be more misleading than helpful. The possibility of coincidence should be included. The "law of anticipated response" had to be considered. Against this had to be set the fear of "self-fulfilling prophecy"—that is, the level of an anticipatory deployment might be affected by the fear of stimulating a response. (3) There was some disagreement about the importance of the British Task Group in the crisis. One view suggested that it was far fetched to suggest that the Soviet Union was specifically reacting to Britain, because there was little likelihood that Britain would become involved in naval terms even if it did become involved diplomatically. Against this argument it was suggested that the Soviet Union could not be sure about the role of the British forces and so could not just allow its own units to stand passively on the sidelines.

The Persian Gulf. There was some disagreement about the prospect of a more active Soviet intervention in the Persian Gulf. One view was that the situation in and around the Persian Gulf was so complex that it will discourage any large-scale participation in the affairs of the region. It was argued that this had been accepted by the Kremlin, whose policy was active only toward Iraq and Iran (the traditional enemies in the region). Soviet policy would be complicated and constrained by this traditional local confrontation. On the other hand, it was argued that the very complexity of the region, especially in the northern gulf, might result in situations that might well encourage and ease Soviet intervention. The argument was then put forward that even if the Soviet Union tried to control the gulf, the probable outcome would be trouble with the local states, which would attempt to oppose Soviet interference. The parallel was drawn with the history of the British experience in the region, especially toward the end. There was some discussion about base rights, although it did not arrive at a clear conclusion. Is the Soviet Union interested in bases in order to increase its political influence and control, or are bases simply an unfortunate necessity resulting from other compulsions (for example, the strategic need for forward deployment)? Moreover, is the Soviet Union primarily interested in "control" for its own sake or to deny it to others? Answers to these questions were not readily forthcoming. However, taking Soviet policy in the region as a whole, it was clearly felt that there was no justification for any current alarmism. Certainly Soviet policy deserved careful attention, but it should not be forgotten that they acted within many important constraints.

THE WIDER IMPLICATIONS

The group considered some of the wider naval implications of the various case studies.

The postulation of "rules of the game" in naval diplomacy was criticized from a number of standpoints: on grounds of terminology ("rules" have a normative connotation, whereas they are used in the McConnell/Kelly paper to describe empirical behavior); because this was not the way Russians thought about it; and because it is not in Western interests, as the major naval power, to talk of "rules of the game," since this helps the inferior power; and because it was thought that the sample of case studies was much too small to permit such generalizations. In defense of the idea of "rules of the game" it was argued that rules were emerging and that there had been a large enough sample since 1967. In the circumstances of current international politics it was thought to be valuable to feel out the threshold of responses and to have some self-consciousness about "rules" in contested areas (comparable to the "rules" that have developed in areas such as central Europe). We need "rules" in the confrontation below as well as above the nuclear level, in order to clarify situations so that no government is dragged into a conflict against its will.

It was suggested that to the extent "rules of the game" contributed to the stability of the superpower confrontation, regional navies will become increasingly important.

The meeting saw no real incompatibility between a more activist naval policy and detente. In fact, it was suggested that detente encourages the superpowers to meddle in local areas. While the desire for low tension with the United States limits Soviet policy to some extent, it does not stop the USSR from pushing in some local areas.

Finally, the general question of continuity and change in Soviet naval policy was discussed. In particular, there was speculation whether the present moment was a "departure point" in the use of the Soviet Navy in the support of foreign policy. The meeting did not agree about the trend. Against the argument that we are now on the brink of a new era of Soviet naval behavior, it was suggested that the recent past indicated an essentially cautious policy. Others believed that the recent episodes were so different and significant when compared to earlier Soviet naval activity that it would not be safe to extrapolate. In general, it was thought to be important to make a distinction between Soviet naval behavior when the United States was involved and occasions when the USSR was faced by other powers. While it was not possible to arrive at definite conclusions about Soviet motives, it was agreed that there had been a change in recent patterns of behavior: Its magnitude and direction have yet to be determined and assessed.

NAVAL POWER AS AN INSTRUMENT OF PEACETIME POLICY

In concluding the discussion, Ken Booth considered some of the problems involved in assessing the utility of a traditional instrument of coercion (naval forces) in noncoercive roles, within an evolving international system.

He noted the underdevelopment of Western naval analysis in general and the particular shortage of work (in terms of inadequate theories and insufficient case studies) on navies as instruments of policy in peacetime. He argued for (1) greater precision in the use of words such as "power," "influence," and "prestige"; (2) more detailed case studies of episodes where naval forces may have had a significant influence on the course of events; and (3) a keener awareness of the full international context within which naval forces were used as instruments of policy.

Commenting on the state of current analysis, Booth was especially critical of the use of the term "gunboat diplomacy" as a catch-all phrase to describe the political use of naval force short of major war. The term derives from a long-vanished era of international politics, and its historical connotations cannot be escaped. The revival of this term, and its recent subcategorization, reflects the poverty of contemporary analysis; its use should be banned as unavoidably misleading. By comparison with other aspects of strategic studies, the analysis of naval policy remains curiously underintellectualized.

PART III
SOME ANALYTICAL MATERIAL

CHAPTER 20

SOVIET NAVAL OPERATIONS: 10 YEARS OF CHANGE

Robert G. Weinland

The Soviet Navy has now been operating continuously on the high seas for a decade. Although its first postwar cruise took place in 1953, and a contingent of Soviet submarines was based in Albania from 1958 to 1961, it was not until 1964 that the Soviet Navy established a significant permanent presence outside its coastal waters.1 Since then, both the scope and intensity of this presence have increased markedly and the Soviets have begun to make active use of their deployed naval forces—not only in the forward defense of their homeland but in the protection and promotion of their overseas interests as well.

The geographical scope of Soviet naval operations expanded relatively slowly. Their activity was initially concentrated in three areas: the Northeast Atlantic and Northwest Pacific, into both of which they sortied periodically for fleet exercises, and the Mediterranean, where they established a permanent "counterforce" to neutralize the military and political potential of the U.S. Sixth Fleet. Soviet combatants first deployed to the Indian Ocean in 1968, after the British announced their intention to withdraw from "East of Suez." With the exception of the temporary deployment of a contingent of submarines at the time of the Cuban missile crisis in 1962, the initial combatant cruise to the Caribbean came in 1969. A combatant presence was established on the West African littoral in 1970. At present, the Soviets maintain strategic offensive (and also some general purpose) forces in the Northwest Atlantic and Northeast

---

The ideas expressed in this chapter are those of the author. They do not necessarily represent the opinion of the Center for Naval Analysis, of which he is a member; nor of the Department of Defense and the Department of the Navy.

Pacific, and general purpose forces only in the Mediterranean, in the Indian Ocean, and off West Africa. In addition, general purpose forces continue to deploy periodically into the Atlantic, Pacific, and Caribbean for exercises, special operations, and in the conduct of interfleet transfers.

THE EXPANDING SCALE OF OPERATIONS

The scale of these operations has increased significantly over the decade—from under 4,000 ship days in 1964 to over 50,000 ship days in 1973. This represents an increase by a factor of almost 15, although the annual rate of growth has declined markedly in the last few years (from a high of 66 percent between 1965 and 1966 to a low of 3 percent between 1972 and 1973).

Figure 20.1 depicts this growth, showing annual total "out of area"* activity for the Soviet Navy as a whole. Figures 20.2 through 20.6 break down these annual totals by geographic region.

The Mediterranean has clearly been the focus of Soviet naval operations, accounting for over 50 percent of all such activity for 5 of the 10 years—and over 40 percent for four of the remaining five years. The 10-year combined totals for the Mediterranean and Atlantic—the latter being clearly the second focal area of activity—represent almost 80 percent of all these operations. However, this geographical distribution is by no means static. During 1972 and 1973, Pacific and Indian Ocean operations—particularly the latter[†]—have increased significantly, together accounting for 30 percent of current deployments.

The Changing Composition of Forces

The composition of the force deployed forward is also changing. The 10-year ship-day totals show the following breakdown among major ship categories:

Category	Percent
Submarines (strategic and general purpose)	25
Surface combatants and amphibious ships	23
Auxiliaries (and merchant vessels operating under naval subordination)	53

---

*Essentially, outside the Arctic seas, the Baltic and Black Seas, the Sea of Okhotsk, and Sea of Japan.

†Largely reflecting Soviet port-clearing operations in Bangladesh.

FIGURE 20.1

Annual Ship Days by Region
(Cumulative)

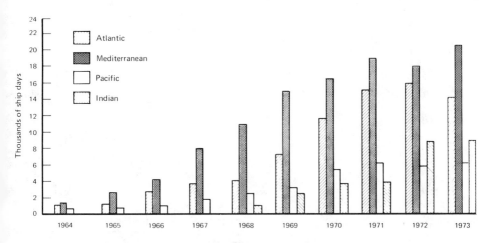

Source: Compiled by the author.

FIGURE 20.2

Regional Distribution of Worldwide Ship Days

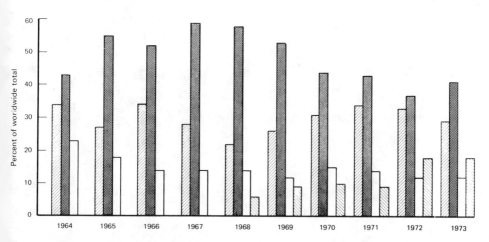

Source: Compiled by the author.

FIGURE 20.3

Ship Days in the Atlantic

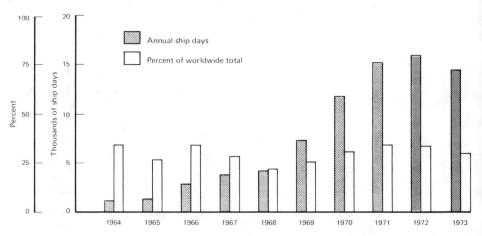

Source: Compiled by the author.

FIGURE 20.4

Ship Days in the Mediterranean

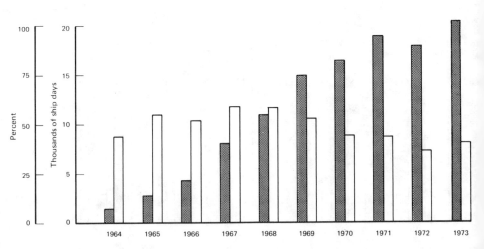

Source: Compiled by the author.

FIGURE 20.5

Ship Days in the Pacific

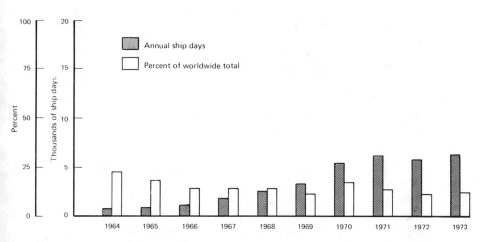

Source: Compiled by the author.

FIGURE 20.6

Ship Days in the Indian Ocean

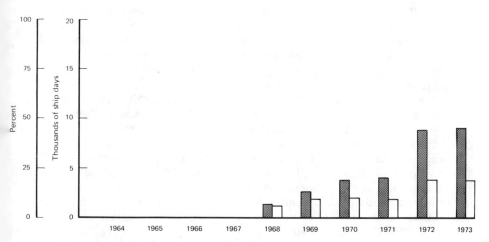

Source: Compiled by the author.

Submarine operations represent a decreasing proportion of the total activity. As recently as 1968, they accounted for over 29 percent of all operations. In 1972, they accounted for less than 22 percent of the total. On the other hand, surface combatant and amphibious ship operations, and auxiliary activity, are on the increase—the former advancing over the last five years from 22 percent to more than 24 percent, the latter from more than 48 percent to more than 53 percent of the total.

As a result of these recent changes in rate of growth, distribution of activity, and composition of the forces deployed, the Soviet naval presence on the high seas now appears to have achieved some degree of maturity. These developments are clearly in line with the changes that have taken place during the same period in the way the Soviets utilize their deployed naval forces.

THE POLITICAL APPLICATIONS OF NAVAL POWER

The Soviets appear to have adopted a "forward deployment" posture for their naval general purpose forces primarily for strategic defensive purposes—to be in position to "counter" U.S. and NATO sea-based strategic offensive capabilities (SSBNs and attack aircraft carriers) whenever those forces operate within strike range of the USSR. This probably remains the principal raison d'etre of these deployments. Since the late 1960s, however, deployed forces have been utilized in an additional capacity: as active instruments of Soviet foreign policy, protecting and promoting Soviet overseas interests.[2]

The scope of their activities in this latter capacity—in which they are carrying out the mission of "protection of the state interests of the USSR on the seas and oceans"—has been consistently expanded. In 1964, when the "forward deployment" posture was adopted, their responsibility probably encompassed only the protection of the Soviet maritime establishment: the merchant marine, the fishing fleet, and scientific research ships. By 1967, this responsibility clearly had been extended to the protection of Soviet interests ashore. Since then, since 1971 at the latest, Soviet naval forces have been utilized not only to protect but actively to promote Soviet interests in the "third world"—and, in the process of protecting these interests, the character of their activities has begun to shift from "passive" toward "active" defense.[3] These more recent changes are in line with the policy of "active counteraction to imperialist aggression," which is an integral component of the "Peace Program" promulgated at the 24th CPSU Congress in 1971.[4]

Table 20.1 lists representative examples of politically oriented operations undertaken between 1967 and 1972. In each of these cases, the Soviets appear to have been seeking to affect the course of events in the international arena—but indirectly, by influencing the actions of another country, rather than by taking direct military action themselves. The United States was the primary, but by no means the only, target of these influence attempts.

The Protection of Established Interests

In the majority of these cases, the Soviet objective was the protection of an established interest that was perceived to be in immediate danger: the territorial integrity and political independence of a client state (such as Egypt), a major economic or political investment in a nonclient state (for example, Guinea), or Soviet men and materiel located overseas (for example, Ghana).[5] Their operations during the major international crises occurring in this period (June 1967) Arab-Israeli war; September to October 1970 Jordanian crisis; December 1971-January 1972 Indo-Pakistani war; May 1972 North Vietnamese crisis) appear to have been intended less to influence the actions of the immediate conflict participants than to constrain U.S. (and in some instances British) involvement.[6] (In the context of each of these crises, significant Western strike forces moved within range of the Soviet Union. Hence, Soviet deployment or concentration of countervailing forces in each case may have had additional, strategic defensive connotations as well.) When conditions warranted, however, the Soviets have not hesitated to move to defend directly a threatened interest—as they did when they provided Egypt with a massive air defense system in the first half of 1970, and when they established their "West Africa Patrol" at the end of that year. Nor have they hesitated to move directly and coercively against a local power—as they did when they established a naval presence in Port Said in 1967 and when they sortied against Ghana in 1969.

The Promotion of Soviet Interests

The operations noted directly above were all essentially "defensive" in character—that is, intended to preserve or restore rather than to alter the status quo—as have been the majority of the politically oriented naval operations undertaken by the Soviets to date. The first major exception to this general pattern may have occurred in April-May 1972, when a submarine tender and a ballistic missile submarine rendezvoused in Cuban territorial waters—an action that contravened a prior U.S.-Soviet agreement on the introduction of strategic offensive forces into Cuba. Soviet objectives in undertaking

TABLE 20.1

Examples of Politically Oriented Soviet Naval
Operations, 1967-72

Directed against U.S.

June 1967	"Shadowing" of U.S. Sixth Fleet carriers during third Arab-Israeli war
Sep. to Oct. 1970	Concentration of countervailing forces in extreme Eastern Mediterranean to deter potential U.S. intervention in Jordanian crisis
Dec. 1971-Jan. 1972	Deployment of countervailing forces to Indian Ocean to deter potential U.S. intervention in Indo-Pakistani war
April-May 1972	Rendezvous of a submarine tender and ballistic missile submarine in Cuban territorial waters
May 1972	Deployment of countervailing forces to South China Sea in response to U.S. interdiction of sea lines of communication to North Vietnam (Operation <u>Linebacker</u>)

Directed against "Third World"

Oct. 1967	Maintenance of continuous combatant presence in Port Said to deter Israeli strikes against Egypt
Feb.-March 1969	Deployment of task group into Gulf of Guinea to effect release of Soviet fishing vessels impounded by Ghana
Dec. 1970 to Present	Maintenance of continuous combatant presence in or near Conakry to deter attacks on Republic of Guinea from Portuguese Guinea
April 1972 to Present	Deployment of minesweeping and salvage forces to Bangladesh for port-clearing operations

Source: Compiled by the author.

this action remain ambiguous. Since the rendezvous took place during the North Vietnamese "Easter Offensive," it may have been only an attempt to assist North Vietnam by applying diversionary pressure to the United States. However, it also took place during the final negotiations leading to the SALT I agreement and may have been an attempt to reinforce the Soviet bargaining position in the talks. Third, it may have been less an attempt to influence U.S. actions than an effort to exploit U.S. preoccupation with the situation in Vietnam, SALT, and the impending summit meeting in order to take an unhindered step toward reestablishing a strategic offensive force presence in Cuba. In neither of the latter two cases could their objective be viewed as "defensive"—they would not have been preserving but altering the status quo; they would not have been protecting, but would instead have been promoting, their interests.

A second instance of the use of naval forces to promote rather than protect Soviet state interests emerged at approximately the same time: the deployment of naval minesweeping and salvage units to Bangladesh to clear the harbor and approaches at Chittagong and Chalna. In contrast to their other major excursions into naval diplomacy, this undertaking had no coercive content. This, however, does not make it any less a political operation.[7]

Further apparent exceptions to the "defensive" or status-quo-oriented pattern of political operations have emerged since then. The following list shows the major politically oriented naval operations undertaken by the Soviets in 1973.

April 1973	Visits of Admiral Gorshkov and naval task group to Iraq during border conflict with Kuwait
April-July 1973	Sealift of Moroccan expeditionary force to Syria
Oct.-Nov. 1973	Concentration of countervailing forces in Eastern Mediterranean to deter potential U.S. intervention (or support potential Soviet intervention) in Fourth Arab-Israeli War

Their actions in the latter stages of the fourth Arab-Israeli war—in which they appear to have been preparing a capability to intervene directly in order to prevent the military and political collapse of Egypt—were clearly "defensive." Their behavior in the six months preceding the outbreak of hostilities, on the other hand, appears to have had the opposite orientation.

The two unequivocally politically oriented naval operations undertaken in the "third world" during this prewar period—the visits to Iraq and the sealift of Moroccan forces to Syria—can both be viewed as adjuncts to the diplomatic campaign then being waged by the Soviets to forge a united Arab front against Israel. In each case, their action served either to suppress discord (the visits to Iraq) or promote unity (the sealift) between radical (Iraq, Syria) and conservative (Kuwait, Morocco) Arab states.8

To say that the Soviets helped to set the stage for the outbreak of hostilities—and employed their naval forces to assist in this undertaking—does not necessarily imply that they wanted (or even anticipated) a war. However, they clearly were bent on enhancing the Arabs' military and hence political power, and in the process enhancing their own position and influence in the region, and they employed their naval forces in direct support of this undertaking.

Whether these apparent exceptions to the predominantly "defensive" or status-quo-oriented character of Soviet naval diplomacy represent aberrations or signify a new trend is not yet clear.* What is clear, however, is that the Soviet Navy must be regarded as far more than one of the several components of the Soviet defense establishment. It is, as noted by its commander in chief, Admiral Gorshkov, "a military factor which can be used in peacetime" and, in that context, "the solitary form of armed forces capable of protecting the interests of a country overseas."9

---

*One additional piece of evidence—which does not, of course, resolve the issue but does lend credence to the hypothesis that a new trend is emerging—surfaced subsequent to this paper's being written. At the end of April 1974, as the United States and USSR were once again engaged in critical SALT negotiations and U.S. attention was focused on both domestic political problems and the attempt to negotiate a resolution of the Arab-Israeli conflict, the Soviets once again sent a ballistic-missile submarine to Cuba. In both cases, the submarines involved were GOLF-class SSBs—older, conventionally powered units, but ballistic-missile submarines nevertheless. In contrast to the previous visit, which was conducted more or less covertly and was publicized by neither the Soviets nor the Cubans, this visit was conducted quite openly—the arrival of the submarine, two surface combatants, and oiler in Havana harbor was covered by the Cuban press (two photographs, one of the entire flotilla and the other showing just the submarine, appear in the May 12, 1974 edition of Granma, an English-language newspaper published in Cuba).

NOTES

1. The establishment of this presence is outlined in Robert G. Weinland, Soviet Transits of the Turkish Straits; 1945-1970—An Historical Note on the Establishment and Dimensions of the Soviet Naval Presence in the Mediterranean (Arlington, Va.: Center for Naval Analyses, Professional Paper no. 94, April 1972); reprinted in Michael MccGwire, ed., Soviet Naval Developments: Capability and Context (New York: Praeger Publishers, 1973), pp. 325-43.

2. The antecedents of this "forward deployment" posture have been vigorously debated in the West. Much of that debate is detailed in MccGwire, op. cit. A recent summary of the debate is presented in Ken Booth, The Military Instrument in Soviet Foreign Policy 1917-1972 (London: Royal United Services Institute for Defense Studies, 1973), pp. 47-50.

3. The evolution of the scope of the "state interests" mission is clearly discernable in the writings of the c-in-c of the Soviet Navy, Admiral of the Fleet of the Soviet Union S. G. Gorshkov. See especially the following articles: "The Great Tasks of the Soviet Navy," Krasnaya zvezda, 5 February 1963; "The Development of Soviet Naval Art," Morskoi sbornik February 1967, pp. 9-21; "Navies in War and Peace," a series of 11 articles appearing in Morskoi sbornik between February 1972 and February 1973 (Translated in U.S. Naval Institute Proceedings, January-November 1974).

4. This aspect of the Peace Program is discussed in depth in: V. V. Zhurkin and Y.M. Primakov, eds., Mezhdunarodnyye Konflikty (International conflicts) (Moscow: "International Relations" Publishing House, signed to press, 15 August 1972), translated in Joint Publications Research Service (JPRS): 58443, 12 March 1973. The military implications of such a foreign policy are outlined with unusual clarity in V. M. Kulish, ed., Voyennaya Sila i Mezhdunarodnyye Otnosheniya (Military force and international relations) (Moscow: "International Relations" Publishing House, signed to press 14 September 1972), translated in JPRS: 58947, 8 May 1973. The latter contains the most straightforward description of this counteractive mission that has yet appeared:

> In connection with the task of preventing local wars and also in those cases wherein military support must be furnished to those nations fighting for their freedom and independence against the forces of internal reaction and imperialist intervention, the Soviet Union may require mobile and well trained and well equipped armed forces. In some situations the very knowledge of a Soviet military presence in an area in which a conflict situation is

developing may serve to restrain the imperialists and local reaction, prevent them from dealing out violence to the local populace and eliminate a threat to overall peace and international security. It is precisely this type of role that ships of the Soviet Navy are playing in the Mediterranean Sea. . . .

The actual situation may require the Soviet Union to carry out measures aimed at restraining the aggressive acts of imperialism. Practical steps towards resolving the problem of regional military opposition to imperialist expansion by expanding the scale of Soviet military presence and military assistance furnished by other socialist states, are being viewed today as a very important factor in international relations. (p. 103.)

5. The interests to be protected and the preferred Soviet means of protecting them are outlined in James M. McConnell, The Soviet Navy in the Indian Ocean (Arlington, Va.: Center for Naval Analyses, Professional Paper no. 77, August 1971), reprinted in MccGwire, op. cit., pp. 389-406. See also Robert G. Weinland, The Changing Mission Structure of the Soviet Navy (Arlington, Va.: Center for Naval Analyses, Professional Paper no. 80, November 1971), reprinted in MccGwire, op. cit., pp. 292-305.

6. The best description to date of Soviet naval operations in an international crisis is to be found in James M. McConnell and Anne M. Kelly, Superpower Naval Diplomacy in the Indo-Pakistani Crisis (Arlington, Va.: Center for Naval Analyses, Professional Paper no. 108, February 1973), reprinted in MccGwire, op. cit. pp. 442-55.

7. This operation is described in detail in Charles C. Petersen, "The Soviet Port Clearing Operation in Bangladesh, March 1972-November 1973" (Arlington, Va.: Center for Naval Analyses, Memorandum (CNA) 504-74, 5 April 1974), reprinted as Chapter 19 of this volume.

8. The visits to Iraq are described and analyzed in Anne M. Kelly, "The Soviet Naval Presence during the Iraq-Kuwaiti Border Dispute: March-April 1973" (Arlington, Va.: Center for Naval Analyses, Memorandum (CNA) 233-74,20, 26 March 1974); reprinted as Chapter 15 of this volume.

9. Admiral of the Fleet of the Soviet Union S. G. Gorshkov, "Navies in War and Peace," Morskoi sbornik, February 1972, p. 23.

CHAPTER
21

**FOREIGN-PORT
VISITS BY
SOVIET NAVAL UNITS**
Michael MccGwire

This chapter tabulates visits by Soviet naval units to ports outside the Warsaw Pact countries. Although such information is inherently unclassified, the wide range of open sources that have to be monitored impede the compilation of a fully comprehensive and up-to-date list. It takes time for all the information to filter through the different channels, which means that the data are increasingly spotty as we come closer to the present. It is hoped that by publishing this list, others will be encouraged to contribute amendments and additions, which will allow us to publish a more complete tabulation next year. Meanwhile, this listing will provide some impression of the pattern of visits, although it is not really adequate to support fine-grained analyses of Soviet naval interests and influence in specific countries.

This tabulation is a blend of many other people's work. Siegfried Breyer in Marine Rundschau, October 1969, pp 343-45, and April 1971, pp. 228-30. Andre Dessens in Problemes et Sociaux, nos. 81-82, 1971, pp. 20-21. James Theberge, Russia in the Caribbean—Part Two (CSIC Georgetown University, 1973), pp. 103-5. Harlan Ullman, "From Despair to Euphoria" (unpublished Ph.D., Tufts University, 1973), pp. 256-66. Australian Federal Parliament Joint Committee on Foreign Affairs, Report on the Indian Ocean Region (CGPO, Canberra December 1970), Appendix L. Foreign Broadcast Information Service, Washington, D.C. Monthly notes on contemporary naval events in Revue Maritime, United States Naval Institute Proceedings; and (most particularly) Marine Rundschau. Current press reports derived largely from the clipping files at the International Institute for Strategic Studies, London.

Visits are listed chronologically for each country. The latter are arranged in geographical order within the following regions:

North and Western Europe
The Mediterranean
West African Seaboard
Central and Southern Africa
East Africa and the Red Sea
Persian Gulf and Arabian Sea
The Bay of Bengal
The China Seas and
 Indonesian Archipelago
Oceania
South America
Caribbean
Central and North America

The information is tabulated in the form: date, port, ship type,* fleet of origin,† class, name and/or side number, senior officer's name. The latter will not necessarily be the senior officer in charge of the detachment. When interpreting the data, remember that this tabulation covers visits by all kinds of naval units, including small survey vessels, fleet tankers, naval tugs, water tenders, and other auxiliaries. The entry "visit" indicates a port call by one or more naval units but gives no indication of the number of units involved or whether it/they were combatants or auxiliaries. Where the date is shown by two months separated by /, this means that the unit(s) visited the port at some time during that period; in certain cases (such as Berbera) it indicates that the <u>port</u> was visited by the unit shown but the <u>number</u> of separate visits is not specified and may have been more than one.

---

*See Chapter 22, list of U.S. naval designators.
† The following abbreviations are used for fleet of origin:

    BF   Baltic Fleet
    BSF   Black Sea Fleet
    NF   Northern Fleet
    PF   Pacific Fleet

NORTH AND WESTERN EUROPE

FINLAND

Date		Port	Type	Fleet	Class	Ship (pennant)	Commander
1954	July 10-14	Helsinki	CL	BF	Sverdlov	Ordzhonikidze	R.Adm. Kutaj
			DD	BF	Skory	Svobodnyj(14)	
			DD	BF	Skory	Smelyj(15)	
1958	Aug 7-11	Helsinki	CL	BF	Sverdlov	Ordzhonikidze	Adm. Ya.Kharlamov
			DD	BF	Skory	Stepennyj(40)	CinC BF
			DD	BF	Skory	Svobodnyj(14)	
1960	Aug 5-9	Helsinki	DD	BF	Skory	Svobodnyj(787)	
			DD	BF	Skory	Statnyj (789)	
1961	Aug 4-8	Turku	PCE	BF	Riga	-	Kuznetsov
			PCE	BF	Riga	-	
1964	Aug 7-11	Kotka	SS	BF	-	-	
			SS	BF	-	-	
1967	Aug 16-21	Helsinki	PCE	BF	Riga	Kommunist Litvy	V.Adm. T.Savelev
1968	Oct 7	Helsinki	DLG	BF	Kashin	Slavnyj	(Conveying Kosygin on official visit)
1969	July 29-Aug 2	Helsinki	CL	BF	Chapaev	Komsomolets(882)	R.Adm. B.Drugov
			PCE	BF	Mirka	-	
1972	Aug 10-15	Helsinki	DD	BF	Skory	Ognennyj	R.Adm. W.A.Lapenkov
			DD	BF	Kotlin	Nastojchivyj	
1972	-	Helsinki	One more visit				

SWEDEN

Date		Port	Type	Fleet	Class	Ship (pennant)	
1954	July 15-22	Stockholm	CL	BF	Sverdlov	Adm. Ushakov(6)	
			DD	BF	Skory	Ser'eznyj(31)	
			DD	BF	Skory	Surovyj(26)	
			DD	BF	Skory	Statnyj(27)	
			DD	BF	Skory	Stepennyj(40)	

389

SWEDEN (cont'd)

Year	Date	Port	Type	Fleet	Ship	Commander
1956	Aug 3/12	Gothenburg	CL	NF	Molotovsk	Adm. A.T.Chabanenko CinC NF
			DD	NF	Otvetsvennyj	
			DD	NF	Ozhestochennyj	
1958	Sept 4-6	Gothenburg	CL	NF	Okt. Revolutsiya	S.M. Lobov
			DD	NF	Otchajannyj	
	Sept 8-13	Stockholm	CL	NF	Okt. Revolutsiya	S.M. Lobov
			DD	NF	Otchajannyj	
1960	Aug	Stockholm	CL	BF	Kirov(961)	
			DD	BF	Svobodnyj (787)	
			DD	BF	Statnyj (789)	
1962	June 12-15	Stockholm	AS	BF	V. Kotelnikov	Kulik
			SS	BF	– (134)	
			SS	BF	– (138)	
			SS	BF	– (100)	
1965	June 28-July 1	Stockholm	CL	BF	Komsomolets	V.Adm. V.Mikhailin
			DD	BF	Surovyj	
			DD	BF	Serdityj	
1967	July 3-7	Stockholm	CL	BF	Kirov(322)	R.Adm. V.P.Belyakov
			DD	BF	Svetlyj(168)	
1972	June 26-July 1	Stockholm	CL	BF	Zheleznyakov	V.Adm. V.M. Leonenkov
			PCE	BF	Barsuk	
			PCE	BF	Kobchik	

DENMARK

Year	Date	Port	Type	Fleet	Ship	Commander
1956	Aug 2-8	Copenhagen	CL	BF	Ordzhonikidze	Adm. A.G.Golovko
			DD	BF	Stremitelnyj(35)	
			DD	BF	Sokrushitelnyj	

390

DENMARK (cont'd)

Year	Date	Port	Type	Ship	Fleet	Commander
1964	May 8-12	Copenhagen	CL	Komsomolets(105)	BF	Adm. A.E.Orel CinC BF
			DD	Spravedlivyj	BF	
			DD	Svetlyj	BF	
	Aug 16-21	Copenhagen	CL	Chapaev	BF	
			DD	Kotlin	BF	
			DD	Kotlin	BF	
			CL	Sverdlov	BF	
			DLG	Kashin	BF	
				Okt. Revolutsiya		V.Adm. Mizin
				Slavnyj		
1973	Aug 24-28	Copenhagen	–	Ugra	BF	V.Adm. I.M.Kuznetsov
			PCE	Riga	BF	
				Gangut		
				Kobchik		
			(Gangut is a naval training ship)			

NORWAY

Year	Date	Port	Type	Ship	Fleet	Commander
1956	Aug 3/12	Oslo	CL	Sverdlov	NF	Adm. A.T.Chabanenko CinC NF
			DD	Skory	NF	
			DD	Skory	NF	
				Molotovsk		
				Otvetsvennyj		
				Ozhestochennyj		
1958	Aug 30-Sept 2	Bergen	CL	Sverdlov	NF	S.M. Lobov
			DD	Skory	NF	
				Okt. Revolutsiya		
				Otchajannyj		
1963	November	Bergen	AGS	–	BF	Capt.1 B.I.Bukovsky
			AGS	–	BF	
				Zenit		
				Azimut		
1964	Oct 17-24	Trondheim	CL	Sverdlov	NF	Adm. S.M.Lobov CinC NF
			DD	Kotlin	NF	
				Murmansk		
				Nastoichivyj		
1966	September	Bergen	One visit			
1970	August	Bergen	Two visits			
1971	Sept 10-15	Oslo	DDG	Kanin	NF	R.Adm. E.Volobuev
			DDG	Kanin	NF	
				Gremyashchij		
				Zhguchyj		

ICELAND

Year	Date	Port	Notes
1966	September	–	One visit
1968	August	–	One visit

ICELAND (cont'd)

Year	Date	Place	Type	Class	Ship	Notes
1969	August	Reykjavik	One visit			
	Oct 24-28	Reykjavik	DLG	NF	Strojnyj	R.Adm. N.V. Solov'ev
			DD	NF	Mosk. Komsomolets	
1970	Feb 3	Reykjavik	DLG	NF	Strojnyj	R.Adm. N.V. Solov'ev
			DD	NF	Mosk. Komsomolets	
	August	Reykjavik	One visit			
	Oct 28-30	Reykjavik	DLG	–	Strojnyj	

UNITED KINGDOM

Year	Date	Place	Type	Class	Ship	Notes
1953	June 7-19	Spithead Review	CL	BF	Sverdlov	Capt.1 O.I. Rudakov
1955	Oct 12-17	Portsmouth	CL	BF	Sverdlov	Adm. A.G. Golovko
			CL	BF	Alex. Suvorov	
			DD	BF	Smotryashchij (52)	
			DD	BF	Sovershennyj (65)	
			DD	BF	Smetlivyj (92)	
			DD	BF	Sposornyj (63)	
1956	April 18-27	Portsmouth	CL	BF	Ordzhonikidze	R.Adm. Kotov (Conveying Bulganin and Khrushchev on state visit)
			DD	BF	Smotryashchij	
			DD	BF	Sovershennyj	
1965	Sept 20-24	London	AGS	–	Nikolaj Zubov	A.I. Rassokho
1969	September	–	One visit			
1970	January	–	One visit			

NETHERLANDS

Year	Date	Port	Type	Fleet	Ship	Commander
1956	July 20-24	Rotterdam	CL	BF	Sverdlov	R.Adm. Kotov
			DD	BF	Sovershennyj	
			DD	BF	Serdityj	
			DD	BF	Surevyj	
			DD	BF/NF	Odarennyj	
1970	Nov 15-21	Amsterdam	AGS	BF	–	Vilkitsky	V.Adm. A.I. Rassokho
1971	Sept 21-25	Rotterdam	DDG	NF	Kanin	Gremyashchij	R.Adm. E.Volobuev
			DDG	NF	Kanin	Zhguchyj	

FRANCE (Atlantic)

Year	Date	Port	Type	Fleet	Ship	Commander
1967	June	–	One visit			
1968	July	–	One visit			
1969	June	–	One visit			
1970	March	–	One visit			
	May 8-13	Cherbourg	CL	BF	Sverdlov	Okt. Revolutsiya	Adm. V.V.Milhajlin
			DLG	BF	Kashin		
	May	–	One more visit			
1971	March	–	One visit			
	Apr 26-30	Le Havre	CLG	BSF	Sverdlov(SAM)	Dzerzhinsky	Capt.1 Dvindenko

MEDITERRANEAN

FRANCE

Year	Date	Place			Ships		Commander
1966	Aug 15-20	Toulon	DD	BSF	Kotlin	Naporistyj	Capt.1 S.S.Sokolan
	October	Toulon	DD	BSF	Kotlin	Naporistyj	Capt.1 S.S.Sokolan
1973	July	Marseilles	CLGM	–	Kynda	Groznyj	R.Adm. E.Volubev
			DLG	–	Kashin	Provornyj	
			DLG	–	Kashin	Krasnyj Kavkaz	

ITALY

Year	Date	Place	Visits				
1966	September	–	One visit				
1968	May	–	Two visits				
1969	May	–	Three visits				
	June	Palermo	One visit				
	September	–	One visit				
	October	–	One visit				
	November	–	One visit				
	December	–	One visit				
1972			Two visits				
1973	Oct 15	Taranto	Two warships				
	End of Oct	Messina	CL	–	Sverdlov	Adm. Ushakov	
			DLG	–	Kashin	Otvazhnyj	

YUGOSLAVIA

Year	Date	Place			Ships		Commander
1954	May	–	CL	BSF	Sverdlov	Adm. Nakhimov(93)	V.Adm. S.Gorshkov
			DD	BSF	Skory	– (10)	CinC BSF
			DD	BSF	Skory	– (12)	

YUGOSLAVIA (cont'd)

Year	Date	Port	Type	Fleet	Class	Ship	Notes
1956	May-June	Split	CL	BSF	Sverdlov	Mikhail Kutuzov	V.Adm. V.A.Kasatonov CinC BSF
			DD	BSF	Skory	Bessmennyj	
			DD	BSF	Skory	Bezukoriznennyj	
1957	Sept 11-18	Split	CL	BF	Sverdlov	Zhdanov(62)	R.Adm. Kotov D/CinC BF
			DD	BF	Skory	Svobodnyj(14)	
	Oct 8-12	Kotor, Split, Dubrovnik	CL	BSF	Chapaev	Kuibyshev(508)	(Conveying Marshal Zhukov on an official visit)
			DD	BSF	Kotlin	Blestyashchij(79)	
			DD	BSF	Kotlin	Byvalyj(75)	
1964	June 18-25	Split, Dubrovnik	CL	BSF	Sverdlov	M. Kutuzov(856)	Adm. S.E.Chursin CinC BSF
			DLG	BSF	Kashin	-	
			DLG	BSF	Kashin	-	
1966	June 5-10	Split	DLG	-	Kashin	-	
			DDGS	-	Krupny	-	
			SS	-	W	-	
			SS	-	W	-	
			AO	-	-	-	
1967	Mar 27-31	Split	CLG	BSF	Sverdlov(SAM)	Dzerdzhinskyj	R.Adm. V.M.Leonenkov
			DLG	BSF	Kashin	-	
			SS	-	-	-	
	June 9-26	Hercegnovi	PCE	BSF	Petya	-	
			SS	-	-	-	
			SS	-	-	-	
	Oct 18-26	Split	DLG	-	Kashin	-	
			DD	NF	Kotlin	-	
			SS	NF	F	-	
			SS	NF	F	-	
			SS	NF	F	-	
			AO	-	-	-	
1968	Jan 29-Feb 3	Kotor	CLGM	BSF	Kynda	Groznyj	Adm. S.E.Chursin CinC BSF
			DLG	BSF	Kashin	Soobrozitelynj	
			SS	-	R	-	
			SS	-	R	-	
			SS	-	R	-	
			AS	BSF	-	-	

YUGOSLAVIA (cont'd)

Year	Date	Port	Type	Fleet	Class	Ship	Commander
1968	Apr 26-May 23	Tivat	AS ASR SS SS	– – – –	Don T-58 R R	F. Vidyaev – – –	
1969	July 23-30	Kotor	DD PCE SS SS	BSF BSF – –	Kotlin Riga – –	– (533) – (687) – –	
	July		One visit				
	Oct 5-10	Split	SS SS	BSF BSF	– –	– –	R.Adm. M.G.Ploskunov
	Dec 23-29	Split	CLGM DLG SS	– – –	Kynda Kashin F	– (854) – (522) –	
1970	May 21-24	Split	DDG PCE PCE	BSF BSF BSF	SAM Kotlin Mirka Mirka	– – –	R.Adm. V.V.Platonov
	Oct 10-14	Split	CLGM DLG PCE	BSF BSF BSF	Kynda Kashin Petya	Groznyj (841) – (368) – (681)	Capt.1 Vasyukov
1971	Mar 20-25	Dubrovnik	–		–	–	R.Adm. Proskonov
	March	Split	One visit				
	Sept 20-25	Rijeka	CLGM DDG SS	BSF BSF BSF	Kynda SAM Kotlin F	Groznyj – –	Capt.1 L.Vasyukov
1972	Mar 5	Split	DD DD SS SS	– – – –	– – – –	– – – –	R.Adm. E.Volobuev

YUGOSLAVIA (cont'd)

Year	Date	Place	Type	Fleet	Class	Ships	Commander	
1972		–	Two more visits					
1973	Mar 23	Rijeka	–	BSF	Group of warships		R.Adm. E.Volobuev	

ALBANIA

Year	Date	Place	Type	Fleet	Class	Ships	Commander	
1954	June	Durres	CL	BSF	Sverdlov	Adm. Nakhimov(93)	V.Adm. S.Gorshkov	CinC BSF
			DD	BSF	Skory	– (10)		
			DD	BSF	Skory	– (12)		
1956	May-June	Durres	CL	BSF	Sverdlov	Mikhail Kutuzov	V.Adm. V.A.Kasatonov	CinC BSF
			DD	BSF	Skory	Bessmennyj		
			DD	BSF	Skory	Bezukoriznennyj		
1957	Aug 30-Sept 3	Durres	CL	BSF	Sverdlov	Mikhail Kutuzov		
			DD	BSF	Skory	Bezukoriznennyj		
1958	August	Valona	Submarine tender and 4 W-Class permanently based					
1959	–	Valona	Second submarine tender and 4 more W-Class based					
1961	May	Valona	Soviet squadron ejected from Valona					

SYRIA

Year	Date	Place	Type	Fleet	Class	Ships	Commander	
1957	Sept 21-Oct 2	Latakia	CL	BF	Sverdlov	Zhdanov(62)	R.Adm. Kotov	D/CinC BF
			DD	BF	Skory	Svobodnyj(14)		
			(Coincided with first phase of Syrian crisis)					
1967	July 17-23	Latakia	DDGS	BSF	Krupny	Gnevynj	Saakyan	
			DD	–	Kotlin	–		
			AO	–	–	–		
	Nov-Dec	Latakia	Five visits					
1968	January	Latakia	One visit					

397

SYRIA (cont'd)

Year	Date	Port	Type	Ships	Name	Commander
1968	Apr 18-23	Latakia	CLGM BSF Kynda		Groznyj	Capt.2 Ushakov
	–	Latakia	Four more visits			
1969	July 15-18	Tartus	CL – Sverdlov DLG – Kashin		– –	
	July	Latakia	One more visit			
1970	Mar 25-28	Latakia	CL BSF Sverdlov DDG BSF		Mikhail Kutuzov –	V.Adm. Sisoev
	August	Latakia	PCE – PCE – AS –		– – –	
	Oct-Nov	Latakia	DDG – SS – SS – MSO –		– – – –	
	–	Latakia	Five more visits			
1971	Dec 14-18	Latakia	CLG BSF Sverdlov(SAM) DD BSF – DD BSF –		Dzerzhinskyj – –	R.Adm. Proskunov
	–	–	Seven more visits			
1972	–	–	Fifteen visits		(Increasing use of Latakia as a base after the rift with Egypt in July)	
1973	Mar 6-12	Latakia	CL BSF Sverdlov DLG – – SS – –		Adm. Ushakov Smetlivyj –	Adm. Sisoev
	April	–	(Military lift of Morrocan troops by Landing Ships)			
	July	–	(Military lift of Morrocan troops by Landing Ships)			

EGYPT

Year	Date	Port	Type	Fleet	Class	Name/Notes
1956	June	Alexandria	DD	BSF	Skory	–
			DD	BSF	Skory	–
1965	September	Port Said	SS		F	–
			SS		F	–
			DD		–	–
			DD		–	–
			AS		–	–
1966	Mar 20–25	Port Said	CLG	BSF	Sverdlov(SAM)	Dzerzhinskyj
			DD	BSF	Kotlin	Plamennyj
			PCE	BSF	Riga	–
			SS	BSF	F	–
			SS	BSF	F	–
	Aug 6–11	Alexandria	DDGS	BSF	Krupny	Bojkij
			PCE	BSF	Riga	Pantera
			SS	NF	F	–
			SS	NF	F	–
			AS	BSF	Don	M. Gadzhev
						V.Adm. Chernobaj COS BSF (Chernobaj announced intended permanent naval presence in Mediterranean)
1967	July 10	Alexandria, Port Said				Soviet warships berthed in Alexandria and Port Said to deter Israeli attacks in the aftermath of the Six-Day War
	Oct 7	Port Said				Permanent presence to stymie Israel attacks
	Oct 7	Alexandria				Submarine Tender and Repair Ship moored alongside to provide base facilities for Soviet Mediterranean Squadron
1968	–	Mersah Matruh				Soviet operational base

LIBYA

Year	Date	Port	Type	Fleet	Class	Name/Notes
1969	Mar 23–29	Tripoli	PCE	BSF	Mirka	– (894)

TUNISIA

1968	February	Tunis	One visit
1969	May	Tunis	One visit
	October	Tunis	One visit
1971	–	Tunis	Three visits

ALGERIA

1966	Apr 24-28	Oran	Group of combatants			
	Nov 10-14	Algiers	DDGS	BSF	Bojkyj	V.Adm. Chernobaj
			AS	BSF	M. Gadzhiev	COS BSF
			SS	NF	–	
			SS	NF	–	
1967	Aug 14-19	Algiers	DLG	–	Kashin	
			AS	–	Don	
			SS	NF	F	
			SS	NF	F	
			AO	–	–	
	November	Algiers	One visit			
1968	January	Algiers	One visit			
	May	Algiers	One visit			
	August	Algiers	AR	–	Oskol	PM-24
			SS	NF	F	(941)
			SS	NF	F	(976)
			SS	NF	F	(973)
	September	Algiers	One visit			
1969	Apr 15-18	Annaba (Bône)	CL	NF	Sverdlov	R.Adm. Golota

ALGERIA (cont'd)

Year	Date	Port	Type	Class	Flag	Commander
1969	July 23-30	Algiers	DD	–	Kotlin	– (383)
	August	Algiers	One visit			
1970	Jan-Feb	Annaba	DDG	–	–	
	May 8-13	Algiers	CLGM	–	Kynda	Adm. Golovko
			DLG	–	Kashin	–
			PCE	–	Mirka	–
			AO	–	–	R.Adm. Leonenkov
	Aug 21	Annaba	DDG	–	–	
			SS	–	–	
			SS	–	–	
			AO	–	–	
	October	Annaba	CLGM	–	–	
			DDG	–	–	
			AO	–	–	
	November	Oran	DDG	–	–	
	December	Annaba	PCE	BSF	–	
			SS	NF	F	
1970	–	–	Seven other visits of which two to Algiers			
1971	–	–	Three visits			
1972	–	–	Eight visits			
1974	Apr 22	–	CL	–	Sverdlov	Sverdlov
			(Other warships)			R.Adm. I.Akimov

GIBRALTAR

1969	June	One visit				
	November	One visit				

MOROCCO

1969	Sept 22	Tangier	DD	-	Kotlin	-	(304)
			DD	-	Kotlin	-	(476)
1970	May	Tangier	Three visits				
1971	-	Tangier	Three visits				
1972	-	Tangier	Four visits				

WEST AFRICAN SEABOARD

MOROCCO

1964	September	Casablanca	AGS	BF	-	Zenit	Capt.1 B.I. Bukovsky
			AGS	BF	-	Azimut	
1968	Oct 15-20	Casablanca	CLGM	BSF	Kynda	Groznyj	V.Adm. B.F.Petrov
			DDGS	BSF	Krupny	Gnevnyj	(Krupny and submarines
			SS	NF	F	-	en-route to Pacific)
			SS	NF	F	-	
1970	May 8-11	Casablanca	DD	NF	Kotlin	-	
			DD	BF	Kotlin	-	
			AO	-	-	-	
1972	Apr 20-24	Casablanca	DLG	-	Kashin	- (484)	Capt.1 I.M.Kapitanets
			PCE	-	Petya	N. Mikhajlovich(688)	
			SS	NF	F	Perepel	
			SS	NF	F	Chirok	
			AO	-	Uda	Kojda	
1972	-	Safi	Two visits				
1973	April	-	(Military lift of Moroccan troops to Syria)				
	July		(Military lift of Moroccan troops to Syria)				

SENEGAL

Year	Date	Port	Type	Ship	Name	Commander
1964	September	Dakar	AGS AGS	BF BF	Zenit Azimut	Capt.1 B.I.Bukovsky
1969	Dec 17-20	Dakar	AGS SS(FS) AO	Fishery Fishery -	- ex-ZV -	Estonia Lira Nevez	Vladimir Zhadanov
1970	November	Dakar	One visit			
1970	December	Dakar	DDG AO	- -	- -	
1971	Apr 16-19	Dakar	DLG SS	BF -	Kashin -	Obratsovyj	Capt. Kasperovich
1972	May 23-27	Dakar	DLG LSV	- -	Kashin Alligator	- (539) - (441)	Capt.2 V.Mabklavsky
1972	-	Dakar	Five other visits			

GUINEA

Year	Date	Port	Type	Ship	Name	Commander
1969	Feb 15-20	Conakry	DDGS DDGS SS AO	BSF BF N BSF	Krupny Kildin F -	Bojkij(976) Neulovimyj(952) - -	Capt.1 V.Platonov
1969	March	Conakry	One visit			
1970	December	Conakry	Visit by combatants			(This was in the aftermath of the Portuguese-supported attack on Conakry and foreshadowed the "Guinea Patrol")
1971	Aug/Oct	Conakry	DLG (2 other naval units)	-	Kashin	-
1971	-	Conakry	Eight more visits			
1972	-	Conakry	Thirty visits			

SIERRA LEONE

Year	Date	Port	Ships		Names	Commander
1971	May 18-23	Freetown	DLG SS	BF -	Kashin - (coincided with threatened coup against President Stevens)	Capt. Kasperovich
	Dec 15-20	Freetown	DDG LSV AO	BSF BF BF	SAM Kotlin Alligator -	Capt.2 Pankov
1972	Mar 1-5	Freetown	DLG AO AO	- - -	Kashin - -	(Transferring to Pacific. Visited Colombo April 28)
1972		Freetown	Two more visits			
1973	Feb 15	Freetown	AGS	BSF	-	

NIGERIA

Year	Date	Port	Ships		Names	Commander
1969	Mar 5-10	Lagos	DDGS DDGS SS AO	BSF BSF NF BSF	Krupny Kildin F -	Bojkij(976) Neulovimyj(952) - -	Capt.1 V.Platonov
	Sept 1	Lagos	DDG DDG LS	- - -	- - -	
1970	May 13-17	Lagos	CLG DDGS AO	BSF - -	Sverdlov(SAM) Kildin -	Dzerzhinskyj Bedovyj(971) -	R.Adm. S.S.Sokolan

CENTRAL AND SOUTHERN AFRICA

EQUATORIAL AFRICA

1970 April	Santa Isabelle	SS	NF	–	(Transferring to Pacific)
		–	NF	–	
1972	Santa Isabelle	One visit		F	

CONGO

1972	Pointe Noire	Two visits			
1973 May 23–29	Pointe Noire	DDG	–	Skrytnyj	(Visited Massawa March 1973)

NAMIBIA

1972	Walvis Bay	One visit			

EAST AFRICA AND THE RED SEA

MAURITIUS

1969 Apr 3–7	Port Louis	CLGM	PF	Kynda	Adm. Fokin	Capt.1 S.E.Korostelev
		DDGS	BSF	Krupny	Gnevnyj	
		AO	PF	Kazbek	Alatyr	
1970 February	Port Louis	AGS	Fishery	–	Estonia	
		SS(FS)	Fishery	ex-ZV	Lira	
		AO	–	–	Nevez	
Apr 19–23	Port Louis	CLGM	PF	Kynda	Adm. Fokin	R.Adm. N.I.Khovrin
		DDG	PF	SAM Kotlin	Blestyashchij (421)	
		AO	PF	Uda	Vishera	

TANZANIA

1969	Dec 16-23	Dar es Salaam	AS	PF	Ugra	I. Kuchnarenko	Capt.1 V.A.Merzlyakov
			SS	NF	F	-	
			SS	NF	F	-	
			AO	PF	Kazbek	Alatyr	
	July 26-31	Zanzibar	DDGS	PF	Krupny	Upornyj	Capt.1 T.A.Lyashko
			AO	BF	Altaj	Egorlik	
1970	February	Dar es Salaam	AGS	Fishery	Fishery	Estonia	
			SS(FS)	Fishery	ex-ZV		

KENYA

1968	Nov 25-Dec 2	Mombasa	CLGM	PF	Kynda	Adm. Fokin	Capt.1 S.E.Korostelev
			DD	PF	Kotlin	Vdokhnovenny(429)	
			AO	PF	Uda	Dunaj	
1969	Feb 17	Mombasa	DDGS	BSF	Krupny	Gnevnyj	
	Summer	Mombasa	AK	-	-	Helicopter-carrying cargo ships supporting Soviet	
			AK	-	-	space programme. Declared as "Survey Vessels"	
1970	Feb 24-28	Mombasa	AS	PF	Don	Kotel'nikov	
			SS	NF	F	-	
			SS	NF	F		
1973	Apr 2	Mombasa	DLG	-	SAM Kotlin	Skrytnyj (447)	

SOMALIA

1968	Apr 17-24	Mogadishu	CL	PF	Sverdlov	D. Pozharskyj(824)	R.Adm. N.I.Khovrin
			DLG	PF	Kashin	Steregushchij(580)	
			AO	PF	-	Polyarnik	
1969	Feb 12-16	Berbera	DDGS	BSF	Krupny	Gnevnyj	
			AO	-	-	-	

SOMALIA (cont'd)

Year	Date	Port	Ships		Notes
1969	August	Berbera	One visit		
	Nov 26–	Berbera	CLGM NF	Kresta I	Vladivostok(532) R.Adm. N.I.Khovrin
			DLG PF	Kashin	Strogij(527)
			AO BF	Altaj	Egorlik
			LSV BF	Alligator	- (424)
	Dec 7-12	Mogadishu	CLGM -	Kresta I	Vladivostok(532) R.Adm. N.I.Khovrin
			DLG PF	Kashin	Strogij(527)
			AO BF	Altaj	Egorlik
			LSV BF	Alligator	- (424)
	Dec 10-11	Kismayu	CLGM -	Kresta I	Vladivostok(532) R.Adm. N.I.Khovrin
			DLG PF	Kashin	Strogij(527)
			AO BF	Altaj	Egorlik
			LSV BF	Alligator	- (424)
1970	January	Berbera	SS NF	F	-
			SS NF	F	-
			AS PF	Don	Kotel'nikov
	-	Berbera	Port facilities being developed to support Arabian Sea operations. Frequent unreported visits by units of the Pacific Fleet detachment from now on, as well as by ships transferring from the West.		
	April	Berbera	DDG PF	SAM Kotlin	-
			SS -	F	-
			AO -	-	-
			LSV PF	Alligator	-
	Apr 17-2nd week in May	Mogadishu	DLG PF	Kashin	- (Coincided with border
			DDGS PF	Krupny	- threat and attempted coup)
	December	Berbera	DDG PF	Kotlin	-
			LSV BF	Alligator	-
			ARS PF	T-58	-
			SS -	F	-
			AO -	-	-

SOMALIA (cont'd)

Year	Date	Port	Type	Type	Class	Name	Admiral/Notes
1971	Feb/May	Kismayu	DD	PF	Kotlin	–	R.Adm. W.S.Kruglyatov
	Feb/May	Mogadishu	CL	PF	Sverdlov	Alex. Suvorov	
	Feb/June	Berbera	DD	PF	Kotlin	–	R.Adm. W.S.Kruglyatov
	Feb/June	Mogadishu	ASR	PF	T-58	–	
	Feb/July	Berbera	LSV	–	Alligator	–	
	July/Dec	Berbera	DD	–	Kotlin	–	
			ASR	–	T-58	–	
			AR	–	Oskol	–	
1972	Feb 16	Mogadishu	CLGM	PF	Kynda	Varyag	(Coincides with visit of United Nations' Secretary General)
			DLG	PF	Kashin	–	
			DLG	PF	Kashin	–	
1973	Mar 15-19	Mogadishu	CLC	PF	Sverdlov(Mod)	Adm. Senyavin	
			MSO	–	–	–	

SOUTH YEMEN (Peoples Democratic Republic of Yemen)

Year	Date	Port	Type	Type	Class	Name	Admiral/Notes
1968	June 25-28	Aden	CL	PF	Sverdlov	D. Pozharskyj(824)	R.Adm. N.I.Khovrin
			DLG	PF	Kashin	Steregushchij(580)	
			AO	PF	–	Polyarnik	
1969	Jan 2-7	Aden	CLGM	PF	Kynda	Adm. Fokin	Capt.1 S.E.Korostelev
			DDGS	BSF	Krupny	Gnevnyj	
			DD	PF	Kotlin	Vdokhnovennyj(429)	
			AO	PF	Uda	Dunaj	
			AK	PF	–	Ulma	
	August	Aden	SS	PF	F	–	
			AS	PF	Don	Kotel'nikov	
1970		Aden	(Regular operational visits by units of the Pacific Fleet detachment from now on)				

408

SOUTH YEMEN (Peoples Democratic Republic of Yemen) (cont'd)

Year	Date	Port	Type	Fleet	Class	Ship	Commander		
1970	Jan/Feb	Aden	DD	PF	SAM Kotlin	Blestyashchij(421)	Capt.1	S.E.Korostelev	
	Mar/June	Aden	DDG	PF	SAM Kotlin	–			
			LSV	–	Alligator				
	Feb/July	Aden	LSV	BF	Alligator	–			
1971	Feb/June	Aden	DD	PF	Kotlin	–	R.Adm.	W.S.Kruglyakov	
	July/Dec	Aden	DD	–	Kotlin	–			
			ASR	–	T-58	–			
			AR	–	Oskol	–			

ETHIOPIA

Year	Date	Port	Type	Fleet	Class	Ship	Commander		
1965	February	Massawa	DD	BSF	Kotlin	Naporistyj			
1966	January	Massawa	DD	BSF	Kotlin	Plamenny	R.Adm.	V.F.Sysoyev	
							D/CinC	BSF	
1967	Jan 21-26	Massawa	DDGS	BSF	Krupny	Gnevnyj	R.Adm.	G.A.Gromov	
1968	June 16-22	Massawa	AGS	–	–	F. Litke	R.Adm.	V.M.Leonenkov	
			AGS	–	–	A. Chirikov			
1969	Jan 9-13	Massawa	DDGS	BSF	Krupny	Gnevnyj			
1970	Feb 1-6	Massawa	DDG	BF	SAM Kotlin	Blestyashchij(421)	Capt.1	S.E.Korostelev	
1971	Feb 20-25	Massawa	DD	PF	Kotlin	–	R.Adm.	W.S.Kruglyakov	
1972	Feb 3-7	Massawa	DLG	PF	Kashin	Strogij	R.Adm.	W.S.Kruglyakov	
1973	March	Massawa	DDG	–	SAM Kotlin	Skrytnyj(447)			

(Except for the survey units in 1968, all visits are for the annual international review which marks Ethiopia's 'Navy Day')

YEMEN

Year	Date	Port	Ships			Commander	
1969	Jan 9–12	Hodeida	CLGM DD	PF PF	Kynda Kotlin	Adm. Fokin Vdokhnovennyj (429)	Capt.1 S.E.Korostelev
1970	Jan 26–29	Hodeida	DDG	PF	SAM Kotlin	Blestyashchij (421)	Capt.1 S.E.Korostelev

SUDAN

1969	Dec 27–31	Port Sudan	DLG LSV AO	PF BF BF	Kashin Alligator Altaj	Strogij (527) – Egorlik	R.Adm. N.I.Khovrin
1970	January	Port Sudan	CLGM DLG LSV AO	NF PF BF BF	Kresta I Kashin Alligator Altaj	Vladivostok Strogij (527) – Egorlik	R.Adm. N.I.Khovrin
1971	Feb/July	Port Sudan	ASR	–	T-58	–	
	Feb/July	Port Sudan	LSV	–	Alligator	–	

EGYPT

1968	June 19–22	Ras Abas	CL DLG AO	PF PF PF	Sverdlov Kashin –	D. Pozharskyj (824) Steregushchij (580) Polyarnik	R.Adm. N.I.Khovrin
	June	Bernice	CL DLG AO	PF PF PF	Sverdlov Kashin –	D. Pozharskyj (824) Steregushchij (580) Polyarnik	R.Adm. N.I.Khovrin
1969	Jan 18–22	Safaga	DDGS AO	BSF –	Krupny –	Gnevnyj –	

PERSIAN GULF AND ARABIAN SEA

Year	Date	Location	Type			Ships	Notes
BAHRAIN ISLAND							
1971	July/Dec	Bahrain	DD	PF	Kotlin	–	
IRAQ							
1968	May 11-19	Umm Qasr	CL	PF	Sverdlov	D. Pozharskyj (824)	R.Adm. N.I.Khovrin
			DLG	PF	Kashin	Steregushchij (580)	
			AO	PF	–	Polyarnik	
1969	Feb 15-18	Umm Qasr	CLGM	PF	Kynda	Adm. Fokin	Capt.1 S.E.Korostelev
			DD	PF	Kotlin	Vdokhnovennyj (429)	
			AO	–	–	–	
	June 14-18	Umm Qasr	DDGS	PF	Krupny	Upornyj	R.Adm. V.M.Leonenkov
			AO	BF	Kazbek	Feder Litke	
	August	Umm Qasr	SS	PF	F	–	
			AS	PF	Don	Kotel'nikov	
1970	May	Basra	DDG	–	–	–	
			AO	–	–	–	
	Dec/Jan	Umm Qasr	SS	–	F	–	
			ARS	–	T-58	–	
1972	Apr 11-16	Umm Qasr	Group of combatants				
1973	Apr 3-11	Umm Qasr	CLC	PF	Sverdlov(Mod)	Adm. Senyavin	R.Adm. W.S.Kruglyakov (Coincides with Iraq/Kuwait border dispute and visit by Gorshkov to Baghdad)
			DD(?)	PF	–	–	
			DD(?)	PF	–	–	

IRAN

Year	Date	Port	Ships			Commander
1968	June 4-9	Bandar Abbas	CL DLG AO	PF PF PF	D. Pozharskyj(824) Steregushchij(580) Polyarnik	R.Adm. N.I.Khovrin
			(Coincides with visit by Zakharov, Soviet Chief of General Staff)			
1969	Feb 7-12	Bandar Abbas	CLGM DD AK	PF PF - -	Adm. Fokin Vdokhnovennyj(429) - Ulma	Capt.1 S.E.Korostelev
	June 28– July 1	Bandar Abbas	DDGS AO AGS	PF BF -	Upornyj Feder Litke -	R.Adm. V.M.Leonenkov
1972	July 10-16	Pahlevi	DD(G) MSO	- -	Escort Minesweeper	R.Adm. Kudel'kin

PAKISTAN

Year	Date	Port	Ships			Commander
1968	May 25– June 2	Karachi	CL DLG AO	PF PF PF	Sverdlov Kashin -	R.Adm. N.I. Khovrin
1969	August	Karachi	SS AS	PF PF	F Don	
					Kotel'nikov	

INDIA (West Coast)

Year	Date	Port	Ships			Commander
1968	April 2-6	Bombay	CL DDGS DLG	PF PF PF	D. Posharskyj(824) Gordyj(981) Steregushchij(580)	Adm. Amel'ko CinC PF
	September	Bombay	AGS	-	Vasili Golovnin	
			(Landed Zond 5 space capsule for return to USSR)			

INDIA (West Coast) (cont'd)

1970	April/June	Bombay	CLG	PF	Kynda	Adm. Fokin	R.Adm. N.I.Khovrin
1971	Feb-May	Bombay	CL	PF	Sverdlov	Alex. Suvorov	
1973	Jan 14-19	Bombay	CLC	PF	Sverdlov(Mod)	Adm. Senyavin	R.Adm. W.S.Kruglyakov
			DDG	–	SAM Kotlin	Skrytnyj(447)	

BAY OF BENGAL

SRI LANKA

1968	July 7-11	Colombo	CL	PF	Sverdlov	D. Pozharskyj(824)	R.Adm. N.I.Khovrin
			DLG	PF	Kashin	Steregushchij(580)	
			AO	PF	–	–	
1969	Jan 5-11	Colombo	AGS	–	–	Chelyushin	Capt.1 A.A.Trofimov
			AGS	BF(?)	–	Vilkitskyj	

INDIA (East Coast)

1968	Mar 27-31	Madras	CL	PF	Sverdlov	D. Pozharskyj(824)	Adm. N.N.Amel'ko
			DDGS	PF	Krupny	Gordyj(981)	CinC
			DLG	PF	Kashin	Steregushchij(580)	
1970	Jan 15-20	Vishakhnapatnam	DLG	PF	Kashin	Strogij(527)	R.Adm. N.I.Khovrin
			LSV	BF	Alligator	–	
			AO	BF	Altaj	Egorlik	

BANGLADESH

1969	Apr 21-23	Chittagong	CLGM	PF	Kynda	Adm. Fokin	Capt.1 S.E. Korostelev
			DDGS	BSF	Krupny	Gnevnyj	
			AO	–	Kazbek	Alatyr	

BANGLADESH (cont'd)

1972	Apr 2	Chittagong	AR	PF	Amur — R.Adm. Zuyenko

(April 1972-June 1974. Harbour clearing operation by naval salvage group. Nine mine sweepers May-Oct 1972, then two until end 1973. About a dozen other vessels)

CHINA AND THE INDONESIAN ARCHIPELAGO

SINGAPORE

1971	Jan or July	–	DD	PF	Kotlin	–	R.Adm. W.S.Kruglyakov

INDONESIA

1959	Nov 17-21	Djakarta	CL DD DD	PF PF PF	Sverdlov Skory Skory	Adm. Senyavin Vyderzhannyj Vozbuzhdennyj	Adm. V.A.Fokin CinC PF
1962	October	–	CL	BF	Sverdlov	Ordzhonikidze	

(Arms supply – renamed Irian)

CAMBODIA

1969	Nov 26– Dec 1	Sihanoukville	DD	PF	SAM Kotlin	Blestyashchij (421)	Capt.1 S.E.Korostelev

CHINA

1956	June 20-25	Shanghai	CL DD DD	PF PF PF	Sverdlov Skory Skory	D. Pozharskyj Vdumchivyj Vrazumitelnyj	Chekurov

OCEANIA

			SS(FS)	Fishery	ex-ZV	Vega	R.Adm. Khomchik
			AO	PF	Uda	Dunaj	
NEW CALEDONIA							
1970	May 14-22	Noumea					
FIJI ISLANDS							
1970	August	Suva	One visit				
1971	November	Suva	One visit				

SOUTH AMERICA

PERU						
1970	June	Callao	One visit			
URUGUAY						
1968	–	Montevideo	Two visits			
1971	–	Uruguay	Six visits			
BRAZIL						
1969	April	One visit				
	May	One visit				
1971		Two visits				

CARIBBEAN

BARBADOS

1969	Aug 10-12	Bridgetown	DDGS	BSF	Kildin	Bedovyj (365)	R.Adm. S.S.Sokolan
			AO	BF		Lena	

MARTINIQUE

| 1969 | Aug 5-8 | Fort de France | CLGM | BSF | Kynda | Groznyj (859) | R.Adm. S.S.Sokolan |

JAMAICA

1970	Oct/Dec	Kingston	CLGM	NF	Kresta I	Adm. Zozulya(?)	R.Adm. N.V.Solov'ev
			DDG	NF	Kanin	-	
			AS	NF	Ugra	-	
			AO	-	Uda	Lena	

CUBA

1969	July 20-27	Havana	CLGM	BSF	Kynda	Groznyj (859)	R.Adm. S.S.Sokolan
			DLG	BSF	Kashin	Soobrazitel'nyj (524)	
			DDGS	BSF	Kildin	Bedovyj (365)	
			AS	NF	Ugra	Tobol	
			SS	NF	F	-	
			SS	NF	F	-	

(The AO Lena and N-Class SSN which were part of this deployment, did not visit Havana)

1970	May 14–	Cienfuegos	CLGM	NF	Kresta	V.Adm. Drozhd(553)	R.Adm. Ya.M.Kudel'kin
	June 2		DDG	NF	Kanin	Gremyashchij(548)	
			SSGN	NF	E II	-	
			SS	NF	F	-	
			SS	NF	F	-	
			AS	NF	Ugra	-	
			AO	-	Merchant	-	

416

CUBA (cont'd)

Year	Date	Port	Type		Ship	Flag Officer
1970	May 25-28	Havana	CLGM	NF	Kresta I	V.Adm. Drozhd(553)
			DDG	NF	Kanin	Gremyashchij(548)
						R.Adm. Ya.M.Kudel'kin
	Sept 9–Oct 10	Cienfuegos	CLGM	NF	Kresta I	Adm. Zozulya(?)
			DDG	NF	Kanin	–
			LSV	NF	Alligator	(454)
			AS	NF	Ugra	–
			AO	–	Merchant	Liepaja
			AO	–	Uda	Lena
			AGS	–	–	–
			ATF	–	–	–
			Buoy-tender			
		(a. The Alligator brought two nuclear submarine support barges for disposal of radio-active effluents				
		b. Kresta, Kanin and Uda sailed Sept 18th)				
	Dec 9-23	Havana, Cienfuegos	DDG	NF	Kanin	–
			AS	NF	Ugra	–
			AO	–	Uda	Lena
			SS	NF	F	–
1970	–	–	Fourteen more visits			
1971	Feb-Mar	Havana, Cienfuegos	CLGM	–	Kresta	
			AS	–	Ugra	
			SSN	NF	N	
			AO	–	–	
	May-June	Havana, Cienfuegos	SSGN	NF	E II	–
			AS	–	Ugra	–
	Oct 31–Nov 9	Havana, Cienfuegos	CLGM	NF	Kresta I	Sevastopol(543)
			DLG	NF	Kashin	(546)
			SS	NF	F	–
			SS	NF	F	–
			AO	NF	–	–
						R.Adm. N.V.Solov'ev
1971	–	–	Six more visits			

CUBA (cont'd)

Year	Date	Port	Type			Name
1972	Mar-May	Havana, Cienfuegos	SS	NF	F	(Sail 17.v)
			DD	–	Kotlin	(Sail 9.v)
			AO	–	–	(Sail 9.v)
	May 4-15	Nipe	SSG	NF	G II	(Sail 9.v)
			AS	–	Ugra	(Sail 15.v)
1972	–	–	Forty-four more visits			
1973	Aug 4 – 9	Havana	DD	–	–	Derzkij
			DD	–	–	–
			SS	NF	–	Admiral Isakov
			AS	–	–	–
1974	Apr 29 –	Havana	DLGM	–	Krivak	Silnyj
			DLGM	–	Krivak	Bodryj
			SSG	NF	G	(K-96)
			AO	–	–	–	R.Adm. Mozarov

CENTRAL AND NORTH AMERICA

MEXICO (Pacific Coast)

1970	November	Mazatlan	One visit

CANADA (Pacific and Atlantic Coasts)

1969	May	–	One visit
	June	–	One visit
1970	January	Vancouver	One visit
	October	Vancouver	One visit
1972	–	–	Three visits

CHAPTER

22

SOVIET NAVAL STRENGTH AND DEPLOYMENT

Robert Berman

The following tables are intended to give some idea of Soviet naval strength in terms of total numbers and their disposition between fleet areas as of the 1st July 1974.

The data in these Order of Battle tables derive from a wide body of source materials, of which Congressional hearings and official releases by U.S. Department of Defense are the most authoritative. To help in identifying the different classes of ship, U.S. naval-type designators have been shown alongside each class name. It must be stressed that these designators are <u>not</u> the same as the terms used by the Soviet Navy to classify various classes, which may be both different and specifically descriptive. For example, the Moskva is designated by the Russians as an "antisubmarine cruiser", and the Kara, Kresta II, Krivak, Kashin, Kanin, and SAM Kotlin are all designated as "large antisubmarine ships" (<u>Bol'shoj Protivolodochnyj Korabl'-BPK</u>); the Kynda is a "missile cruiser."

U.S. NAVAL-TYPE DESIGNATORS
USED IN THE TABLES

AEM	Ammunition Ship, Missile
AGI	Intelligence Collector
AGS	Hydrographic Research Ship
AK	Helicopter-carrying cargo ship (space program support)
AO	Fleet Oiler
AOR	Replenishment Oiler
AR	Ship Repair Vessel
ARS	Salvage Ship
AS	Submarine Supply Vessel
ASR	Submarine Rescue Ship
ATF	Fleet Tug
AW	Water Carrier
CHG	Helicopter (Anti-submarine) Cruiser
CL	Light Cruiser
CLG	Light Cruiser w/SAM
CLG/CC	Light Cruiser w/SAM, Command Ship
CLGM	Light Cruiser w/SSM, and SAM
CVH	Helicopter/STOL Carrier
DD	Destroyer
DDG	Destroyer w/SAM
DDGS	Destroyer w/SSM
DDGSP	Destroyer w/SSM and Point Defense Missile (SAM)
DLG	Destroyer Leader w/SAM
LCT	Landing Craft, Tank
LSM	Landing Ship, Medium
LSV	Landing Ship, Vehicle
LSV/M	Landing Ship, Vehicle, Medium
MSC	Coastal Minesweeper
MSF	Fleet Minesweeper
MSO	Minesweeper
PCE	Patrol Escort
PCEP	Patrol Escort w/Point Defense Missile (SAM)
PCS	Patrol Craft, Anti-submarine
PGGP	Patrol Gunboat w/SSM and Point Defense Missile (SAM)
PGH	Patrol Gunboat (Hydrofoil)
PTF	Patrol Torpedo Boat
PTFG	Patrol Boat w/AA and SSM
PTG	Patrol Boat w/SSM
SS	Submarine
SSB	Ballistic Missile Submarine (diesel)
SSBN	Fleet Ballistic Missile Submarine (nuclear-powered)
SSG	Cruise Missile Submarine (diesel)
SSGN	Cruise Missile Submarine (nuclear-powered)
SSN	Attack Submarine (nuclear-powered)

TABLE 22.1

Order of Battle: Submarines

Submarines		Fleet Areas				Total*
		North	Baltic	Black	Pacific	1 July 1974
P	SSBN	9	-	-	-	9
Y	SSBN	23	-	-	10	33
H II	SSBN	1 + 7	-	-	2	1 + 9
G II	SSB	7	-	-	4	11
G I	SSB	6	-	-	4	10
Z V	SSB	2	-	-	-	2
P	SSGN	1	-	-	-	1
C	SSGN	13	-	-	-	13
E II	SSGN	14	-	-	14	28
J	SSG	12	-	-	4	16
W	SSG	1	3	5	3	5 + 7
A	SSN(?)	1	-	-	-	1
V	SSN	15	-	-	-	15
N	SSN	9	-	-	5	14
E I	SSN	-	-	-	5	5
T	SS	4	-	-	-	4
F	SS	31	14	-	11	56
Z	SS	15	-	-	5	20
B	SS	1	-	1	2	4
R	SS	-	6	6	-	12
W	SS	5	10	15	10	40
Q	SS	-	10	5	-	15

*All ships that have been "delivered" to the navy and that have not been subsequently laid up, scrapped, or otherwise disposed of. Includes ships refitting or undergoing modification or conversion.

TABLE 22.2

Order of Battle: Surface Units

Surface Units		Fleet Areas				Total
		North	Baltic	Black	Pacific	1 July 1974
Kara	CLGM	-	-	3	-	3
Kresta II	CLGM	4	1	-	-	5
Kresta I	CLGM	2	1	-	1	4
Kynda	CLGM	-	-	2	2	4
Krupny	DDGS	-	-	-	3	3
Kildin	DDGS	-	-	3	1	4
Kuril	CVH	-	-	-	-	(1)[a]
Moskva	CHG	-	-	2	-	2
Krivak	DDGSP	5	2	2	-	9
Kashin	DLG	1	4	10	4	19
Kanin	DDG	4	1	-	-	5
Kotlin SAM	DDG	1	1	3	2	7
Sverdlov	CLG/CC	-	-	1 + 1	1	1 + 2
Sverdlov	CL	1	2	4	2	9
Chapaev	CL	-	2	-	-	2
Kirov	CL	-	-	-	-	-
Kotlin	DD	2	4	4	9	19[b]
Skory	DD	6	2	4	5	17
Grisha	PCEP	3	6	7	4	20
Mirka/Petya	PCE	15	14	24	14	67
Riga	PCE	4	8	10	8	30
Kola	PCE	1	1	1	2	5
Nanuchka	PGGP	-	4	4	-	8
Osa	PTFG	25	35	25	35	120
Komar	PTG	-	5	-	-	5
Turya, Shershen	PTF	10	25	10	15	60
P-4,6,8,10	PTF	10	45	15	30	100
Poti	PCS	15	20	15	15	65
SO.1	PCS	15	40	10	15	80
Kronshtadt	PCS	-	10	5	5	20
Stenka, MO VI	PCS	5	20	5	10	40
Pchela	PGH	-	10	10	-	20
Natya, Yurka T-58, T-43	MSF	45	55	40	45	185
Zhenya, Vanya, Sasha	MSC	15	35	25	20	95
Alligator	LSV	3	4	3	4	14
Polnocny	LSM	10	20	20	15	65
MP-2,4,6,8	LSV/M	10	10	10	10	40
Vydra, MP10	LCT	15	40	30	45	130+
Ugra	AS	4	-	2	3	9
Don	AS	3	-	-	3	6
Amga	AEM	2	-	-	-	2
Lama	AEM	3	-	2	2	7
Amur, Oskol	AR	9	1	4	6	20
Dnepr, Atrek, Tovda	AR	6	1	1	5	13
T-58	ASR	5	3	5	9	22
Boris Chilikin	AOR	1	1	-	-	2
Manych	AOR	2	-	-	-	2
Various (ocean-going)	AO	6	1	8	8	23
	AGI	15	10	10	15	50
	AGS	30	25	15	30	105
	ATF	6	1	2	2	11
	AW	1	1	6	4	12

[a] Figures in brackets denote units still in shipyard hands.
[b] Includes the <u>Tallin</u> (if still operational).

TABLE 22.3

Order of Battle: Naval Aviation

Soviet Naval Aviation	1 Jan. 1974	North	Baltic	Black Sea	Pacific	Total 1 July 1974
Badger G	(TU-16: KELT ASM)	150	75	40	35	300
Blinder	(Tu-22)	30	15	9	6	60
Beagle	(Il-28)	–	15	15	10	40
Badger A	(Tu-16: Tanker)	40	20	10	10	80
Badger A,D,E,F	(Tu-16: Recce)	25	10	15	5	55
Bear	(Tu-95: Recce, ASW, Elint)	30	10	10	5	55
May	(Il-38: ASW)	25	5	15	10	55
Mail	(Be-12: ASW)	40	40	10	10	100
Hound	(Mi-4 : ASW)	40	40	15	5	100
Hormone	(Ka-25: A-ASW, B-Radar)	40	15	60	10	130

CHAPTER

23

CURRENT SOVIET WARSHIP CONSTRUCTION AND NAVAL WEAPONS DEVELOPMENT

Michael MccGwire

Current warship construction reflects the requirements of replacing the ships built during the first postwar decade for the traditional tasks of defending the fleet areas and supporting military operations on land; of providing ships for the new tasks that have emerged since then; and of remedying the distortions introduced into the fleet by technological inadequacies and by the decisions taken in 1954. Besides the cyclical process of procurement for replacement, current programs derive from three decision periods: 1957/58, when it was decided to concentrate on nuclear submarines as the means of by-passing Western surface and air preponderance at sea; 1961, when it was appreciated that a nuclear submarine force could not meet the increasing threat on its own but would require a wholesale shift to forward deployment in support; and 1963/64, by when the full implications of this decision had been worked out and accepted, including the need for additional surface vessels if the Soviet Navy was to discharge its newly defined mission.[1] Further decisions may have been made in the aftermath of the 1967 Arab-Israeli war, and there are indications that naval procurement was an important issue when finalizing the Ninth Five-Year Plan in 1971/72.

Since the purpose of this analysis is to provide insight into Soviet naval policy, it consciously sets out to order the available information to provide a coherent explanation of current warship production, well aware of the dangers of working from such fragmentary evidence. Only by making the attempt is it possible to perceive general trends and, equally important, to pick up the anomalies that are such important indicators of changes in plan.[2]

Note: Source data are limited to information that is new in relation to that contained in Chapter 11 in M. MccGwire, ed., Soviet Naval Developments, (New York: Praeger Publishers, 1973).

CURRENT SUBMARINE CONSTRUCTION

The first generation of nuclear submarines was built between 1958-67 at Severodvinsk in the North and Komsomol'sk in the Pacific and was delivered at the rate of five to six units a year, between 1958-67. The decision to place primary reliance on nuclear submarines required that this production rate be at least doubled. The expansion of nuclear construction capacity, and the Y-, C-, and V-classes of submarines, all stem from the same decision period of 1957/58.

In addition to increasing the capacity of the two existing yards, two (possibly three) more yards were converted to nuclear construction during the middle 1960s:[3] Gor'kij on the Volga, and Admiralty Yard in Leningrad; Sudomekh (Leningrad) may also have a nuclear building capability.[4] These extended facilities began delivery of the new family of submarines in 1967/68. Initial Western assessments claimed a building capacity of 20 units a year,[5] but the delivery rate appears to have leveled off at about 11 units a year.[6]

SSBN Construction

SSBN are built at Severodvinsk and Komsomol'sk. The lead ship of the Y-class, carrying sixteen 1,300-nm SS-N-6 missiles, was delivered toward the end of 1967, and 5 were operational by mid-1969.[7] The annual delivery rate built up progressively from 4 in 1968 to 7 in 1971, at the end of which year 22 units had been delivered. Another 7 were delivered in 1972, 3 in 1973, and 1 unit in early 1974, making 33 Y-class in all.[8]

The lead ship of the D-class, carrying twelve 4,200 nm SS-N-8 missiles was delivered toward the end of 1972 and was still on initial sea trials in March 1973.[9] It appears that four units were delivered in 1973 and that another five to seven are expected in 1974 and in 1975.[10] Three D-class were operational by March 1974.[11] There is tenuous evidence that some D-class submarines now under construction are being lengthened,[12] and this may be to increase the missile load from 12 to 16.

In terms of SALT, the number of D-class now under assembly will take the Soviet Union beyond the "base-line" of 740 SLBM and will require that an equivalent number of the "older" ICBM or SLBM be traded in.

There are no indications of the final missile/hull mix, but there are many arguments against building the full 62 SSBN allowed by the agreement. The SS-N-8 can cover the United States from Soviet "home" waters, and with unfilled requirements for other types of nuclear submarines and a limited building capacity, there are obvious attractions in mounting the full number of SLBM allowed under the agreement (950) in as few hull/propulsion units as possible; 25 launchers per submarine would only require 38 units, and it is relevant that the U.S. Trident system will mount 24 launchers in each submarine. However, with 33 Y- and 17 D-class either delivered or building, the Soviets already have 50 of these very large-hull/propulsion units, and it would not be very practical to reconfigure them for another role. But there would seem to be no technical reason why the Y- and D-classes could not be progressively modified (by lengthening) to carry 18 or 19 missiles each—that is, 52 units with 936 SLBM, or 50 units with 950 SLBM. But it would be more simple (and hence perhaps more likely) for them to settle for 59 16-missile units carrying 944 SLBM.

SSGN Construction

The lead ship of the C-class SSGN, carrying the submerged-launch horizon range SS-N-7 cruise-missile system, was delivered in 1968. This class is building at Gor'kij at the rate of some 2 units a year, about 13 units having been delivered by mid-1974.[13]

One unit of the P-class, a somewhat larger SSGN, was apparently delivered in 1971.[14] Its role is not known, but the existence of a single unit suggests some kind of prototype or trial submarine. It might be the vehicle for the new SS-N-13 system discussed below, in which case a series construction program could be expected.

SSN Construction

The lead unit of the V-class SSN was delivered toward the end of 1967.[15] This class must also be building at Leningrad* and initially the delivery rate was about three submarines a year. However, deliveries appear to have been reduced during the period 1971-73.[16] Assuming that this was the result of favoring SSBN construction during the run-up to SALT, we may expect the delivery rate to revert to three per annum during 1974.

This dip in the V-class deliveries may be relevant to the removal of missile launchers from the E I-class and their conversion

---

*Because of their size, SSBN can only be built at Severodvinsk and Komsomol'sk. Gor'kij is building the C-class. If Leningrad is involved in the nuclear program, it must be building V-class.

to SSN. The first family of nuclear submarines was originally configured as SSN (N-class) and SSBN (H-class). The E I-class represents the reconfiguration of the last five of these hull/propulsion units as SSGN, to meet the greatly increased carrier threat,[17] but they were never very effective since they lacked mid-course guidance for their missiles. The opportunity may therefore have been taken to convert these units to SSN, using the sensor and weapon systems that had originally been procured for the V-class program, and perhaps other items as well. This would follow established Soviet practice.[18]

It appears that during the latter 1960s, the N-class of SSN underwent a major modification to effect an appreciable increase in maximum submerged speed,[19] which was formerly about 26 knots.[20] This was probably carried out in the course of the periodic refit/recoring that takes place every four or five years.

Meanwhile the V-class continues to have a speed advantage over U.S. submarines presently in service[21] although it is not superior in other respects,[22] and will be outpaced by the 688 (Los Angeles) class now building.

The Interrelationship of Programs

Taken together, the foregoing analyses would produce the following delivery pattern:[23]

Type	Class	1967	1968	1969	1970	1971	1972	1973	1974	Total
SSBN	Y	1	4	5	5	7	7	3	1	33
	D	-	-	-	-	-	1	4	5	10
SSGN	E II	5	2	-	-	-	-	-	-	7
	C	-	2	2	2	2	2	2	2	14
SSN	V	1	3	3	3	1	1	2	3	17
Other					"A"	"P"				2
Total		6	11	10	11	11	11	11	11	83

Note: The A-class may not be nuclear.

Certain inferences can be drawn from this table:
1. If the SSBN program was in fact accelerated at the expense of the V-class SSN, the decision must have been taken by mid-1968.[24]
2. There should be five V-class weapon/sensor outfits available for fitting to E-class SSN.
3. By comparison with the first decade of nuclear submarine construction, the production capacity for nuclear reactors has been almost exactly doubled.

Diesel Submarines

A new diesel-powered attack submarine is now under series construction and has begun to enter service.[25] It was identified in Sevastopol' on Navy Day (29 July) 1973,[26] and the lead unit was probably delivered in 1972. As Nikolaev has not been associated with submarine construction since the first half of the 1950s, this suggests that the new class is most probably building at Gor'kij on the Volga. It may be relevant that its external dimensions (80 m x 9.5 m) are almost identical to that of the J-class;[27] series production of the latter was planned at Gor'kij for delivery between 1962-67, but only 16 were actually built during that period owing to the reallocation of nuclear propulsion to the anticarrier role.[28] The broad beam of this new class could indicate a teardrop hull; if however the hull dimensions are similar to the J-class, this would allow the extensive use of that program's assembly jigs at Gor'kij, and perhaps equipment and material procured for the original program.

The five-year gap between the completion of the F-class program (60 of which were delivered between 1958 and 1967) and the appearance of the T-class invites speculation. While it most likely reflects the shift in naval policy between 1957/58 and 1963/64, this is not a sufficient explanation in itself.

Inevitably, attention is drawn to the single A-class, which was delivered two years earlier in 1970. This submarine, of about the same dimensions, was probably built at Sudomekh,[29] and there has been unprecedented ambiguity about its propulsion system.[30] The Q-class was built at Sudomekh in the middle 1950s and was intended to have a closed-cycle Walther-type engine, driving the center shaft.[31] It was at Sudomekh that the German scientists from Walther's group worked in the immediate postwar years.[32] And finally, a propulsion system that had a turbine on the center shaft and diesel-electric on the wings would explain the untypical doubts about whether the A-class has diesel or nuclear propulsion. (Nuclear submarines are powered by steam turbines. When these are connected directly to the propeller shaft, it is very hard to silence turbine whine.)

The single "A" may therefore have a closed-cycle system and could be the propulsion prototype of the new class now under series construction at Gor'kij. Closed-cycle drive on the center shaft could account for the beaminess of the Gor'kij class, and the "dawdle and dash" capability provided by such a propulsion system would be a relatively economical way of meeting the Soviet requirement for antisubmarine defense of their fleet areas, and for certain other tasks. A submarine running on batteries should be inherently quieter than a nuclear unit and hence provide a better ASW platform; when operating in "home" waters or in company with its own forces on forward deployment, the batteries can be recharged without undue danger. Meanwhile, the closed-cycle turbine provides a sustained high-speed capability when required.33 If the diesels can be run closed-cycle for recharging batteries when submerged, this would mean increased versatility of tasking.

If the T-class does have closed-cycle propulsion, the design decision would not have been taken much later than 1963/64. If however it has conventional diesel propulsion and relies heavily on J-class design and components, the decision could have been taken as late as 1967/68.

It is tempting to see this new submarine as the direct successor to the F-class,34 replacing the obsolescent Z-class, 28 of which were delivered between 1952 and 1957; these classes have comprised the majority of submarines on sustained deployment in the Mediterranean. However, it will also serve as a replacement for the W-class and its (canceled) successor, the R-class, whose original role was defense of the fleet areas against surface intrusion. The task of fleet-area defense has become of renewed importance with the entry into service of the 4,000-nm SS-N-8 SLBM system, only this time it is defense against intruding submarines.

The B-class remains static at the four units delivered in 1968-69. There are no suggestions that this class is under series construction, and it seems likely that these submarines have a special role, which may include trials and development.

SUBMARINE YARD CAPACITY

There is a general assumption that Soviet nuclear building yards are not working to full capacity,35 but this may partly reflect initial overassessment by Western analysts. Komsomol'sk and Severodvinsk are between them building about eight units a year, of which perhaps six are from Severodvinsk and two from Komsomol'sk.36 Although this is less than the assessed capacity of 12 SSBN p.a.,37 it still represents a 45 percent increase over the first decade. These yards are also likely to have been involved with the modifications to the N- and E-I-classes. Nevertheless, there remain certain anomalies, and these become more interesting when viewed within the overall pattern of nuclear construction.

Despite the impression of underutilization at Severodvinsk, there are reports that this submarine yard, already the largest in the world, is undergoing further expansion.38 This could of course denote an increase in the overall nuclear building rate, but there are other explanations that are perhaps more likely. (1) The assembly of a larger submarine could require the extension of facilities, even if the production rate remained the same. Such would be the case if the Soviets have decided on a 16-20-missile D-class SSBN. (2) Extra facilities would be required if it is intended to modify all the Y-class and the early D's to carry 16-20 SS-N-8, while at the same time continuing at the same rate with new construction. (3) It may be planned to concentrate nuclear construction at Severodvinsk. One reason could be that the size of the C-class successor will prohibit its construction at an inland yard. It is perhaps relevant that the P-class (reputed to be an SSGN) is alleged to be some 15 percent larger than the C-class.[39]

Sudomekh in Leningrad is another enigma. This specialized yard was fully engaged on submarine construction from 1946 through to the end of the F-class program, which delivered six units a year in 1958-67. The single A-class (delivered in 1970) was probably built there, but no new series-production program has been tied to the yard since 1967 although it could be involved in the T-class program. Sudomekh is almost certainly contributing to current submarine construction and one other possibility is its involvement in the V-class program in Leningrad.

Bearing in mind the apparent building rates of nuclear submarines at other yards, there is a certain anomaly in the report that Admiralty Yard in Leningrad is carrying the load of V-class production, particularly since there is also an active civilian shipbuilding program on the main building ways there. (Admiralty recently completed two nuclear-powered icebreakers and is fitting out a third.) As a minimum one would expect that Baltic Yard, which has a long record of collaboration with Admiralty and faces it across the Neva River, is also involved in the program. Baltic's submarine experience goes back to Tsarist days, and it has been continually involved in building or modifying diesel submarines since the war.* Like Admiralty, this yard has been busy with a substantial civilian shipbuilding program since 1955, but it also has separate submarine construction facilities.[40] Meanwhile, Sudomekh Yard lies directly upstream of Admiralty and has collaborated with it in the past (for example, on the K-class submarine both before and immediately after the war). If it is not already involved in some other program, Sudomekh may also be contributing to the V-class program. There are close links between all three yards, and although the arrangement might seem untidy, in terms of the required facilities it is perhaps more likely than Admiralty carrying the whole load on its own, considering

---

*Shchuka V, W-class, and Longbin. The German advisory team (1926) was attached to Baltic Yard, which also built the first Soviet submarine class (Dekabrist, delivered 1931).

the congested circumstances of the Leningrad yards; that is, assuming that the submarines are being fabricated as well as assembled in Leningrad. Another possibility is that components may be shipped to Leningrad by inland waterway for assembly.

The large yard at Gor'kij presents another enigma. This built 36 diesel submarines a year in the middle 1950s and, assuming the full use of the submarine facilities, three nuclear units a year might have been expected.[41] However, Gor'kij appears to have only produced about two SSGN a year since 1968; it is now also involved in the new T-class nonnuclear program.

Surveying the nuclear submarine construction program as a whole, we are faced with a series of anomalies. If we accept that total production is determined by the availability of nuclear reactors, this allows an annual delivery rate of 10-11 nuclear submarines a year. But the way in which the construction of these 11 units is apportioned between the available yards would seem to flout common sense, in terms of the more likely determinants such as yard capacity and the efficiency of production.

Seeking some explanation for this anomalous situation, one could hypothesize as follows. The original decision to double the nuclear construction capacity (1957/58) would have brought Gor'kij into the program, building three units a year, and raised Severodvinsk's building rate from three or four to six units a year. The configuration-mix was planned to be something like three SSGN, five-six SSN, and only two-three SSBN a year, which reflected contemporary assessments of future operational requirements. Gor'kij would build the C-class SSGN only, but the Y-class SSBN and V-class SSN would run as parallel programs at the other yards, as had been the case with the H- and N-classes between 1958 and 1962.

However, the 1961 decisions reflected (among other things) a concern that the United States was trying to build up a first (disarming) strike capability against Russia. This caused the priorities in the allocation of nuclear propulsion to be reversed, and it was now planned to build six Y-class SSBN and two V-class SSN a year. This in turn raised problems of sheer physical capacity, since the Y-class is over twice the displacement and 50 percent as long again as the V-class. The problem was circumvented by shifting the production of V-class to Leningrad (explaining the untidy arrangement at Leningrad), leaving Severodvinsk and Komsomol'sk to concentrate entirely on SSBN construction.

This was the immediate reaction to the 1961 threat reassessment. However, by 1963/64, it had been decided to build the T-class closed-cycle diesel submarine; but such a program could only be accommodated at the expense of SSGN construction at Gor'kij. (Projected nuclear deliveries may have been cut by one third. This would have retained four "notional" diesel-production lines. Gor'kij was still in the middle of the J-class "trickle-delivery" program at this period.) This was acceptable, since the counter-Polaris role had been given priority over the anticarrier role in terms of nuclear propulsion. The planned production rate of C-class was cut from three to two a year, and Leningrad was required to build/assemble a third V-class each year.

This hypothesis has several implications in terms of future possibilities. It suggests that (1) in due course, Gor'kij's full capacity will be available to the T-class closed-cycle diesel submarine program, and that nuclear construction will be moved away (otherwise it would have been simpler to have placed the program at Sudomekh). (2) The construction of nuclear submarines in Leningrad is a temporary expedient; (3) Nuclear construction will once again be concentrated at Severodvinsk and Komsomol'sk, continuing at the overall rate of 10-11 units a year (this would explain the further expansion at Severodvinsk; however, construction of the large SSBN will be complete by 1977 and the "effective" increase in capacity will be equivalent to three C-class a year); and (4) One or more of the Leningrad yards may be drawn into the T-class program (since they will no longer be involved with the V-class).

There is ample precedent for such a reversal of plans as the result of reappraising the threat. The year 1954 saw the downgrading of the naval threat, the wholesale cancellation of series production programs, and the shift to civilian construction. The years 1957/58 saw the full appreciation of the threat from carrier-borne nuclear strike aircraft, the cancellation of the major new classes projected in 1954, and (in the short term) the shift of nuclear propulsion from strategic delivery to the counter-carrier role. It is only to be expected that an even more radical change in threat assessment in 1961 and in consequential policies would bring about changes in operational requirements and the ships needed to meet them. We already have evidence of such changes in the surface warship programs. And compared to the repercussions of the 1957/58 decisions (which reverberated throughout the shipbuilding and naval equipment industries), these changes at Gor'kij seem like a minor adjustment of plan.

Assuming that the T-class is primarily intended to defend the fleet areas and to operate in close company with Soviet surface forces in the extended maritime defense zones, a substantial building program and high production rate would be expected. It is worth recalling that the Soviet Union planned to build 1,200 diesel submarines after the war,[42] and reached an annual delivery rate of some 80 units before the savage cutbacks were ordered in 1954. This included cancellation of the mass-production medium-type submarine programs (W- to be followed by R-class), then running at 72 units a year, of which 36 were from Gor'kij. These units were primarily intended for fleet-area defense, and although the perceived threat then was of a very different kind to what they perceived in 1961-64 and perceive now, it gives some impression of the scale of Soviet reactions.

SUBMARINE WEAPON SYSTEMS

Strategic

An MRV variant of the SS-N-6, with a slightly longer range, is close to entering service and may be deployed aboard the Y-class.[43] There is as yet no evidence of an MRV or MIRV version of the SS-N-8, nor has either system been test-fired on a depressed trajectory.[44] There has been one reference to the range of the SS-N-8 being increased to 4,600 nm,[45] which, if true, would make sense in terms of providing a more comfortable margin for covering the United States from home waters.

Tactical

The Soviet Navy is reported to have developed a new mach-4 surface-to-surface missile (SS-N-13), with a range of 650 km (400 nm) using midcourse guidance from a satellite or aircraft.[46] It is expected to be operational by the end of 1974.

The weapon-outfits of the classes entering service since 1967 demonstrate the Soviet Navy's hard-learned preference for horizon-range SSM systems, and it is most unlikely that they would revert to long-range <u>cruise</u> missiles at this stage. Bearing in mind the alleged speed, one must therefore assume that this new system is a tactical <u>ballistic</u> missile; this makes good operational sense and resolves the <u>anomalous</u> reference to "mid-course guidance," when "initial target location" was intended.

At sea, long-range weapon systems provide the equivalent of high tactical mobility and/or large numbers, particularly where area defense is concerned. This is why in 1954 the political leadership forced the "cost-effective" SS-N-3 cruise missile on the Soviet Navy. Long range demands a homing capability, and cruise missiles have the advantage of only needing to locate the target in one dimension, but there are heavy costs inherent in their aerodynamic flight profile and relatively moderate speed. These include the comparative ease of detecting the missile and its vulnerability to a wide range of countermeasures, the increased liability to failure in some part of the system, and the target's movement during the extended time of flight. These disadvantages in the SS-N-3 caused the Soviet Navy to develop the SS-N-7 and -10 horizon-range systems, which have a short time-of-flight and rely on the firing unit's own sensors for target location data.

It is only recently that developments in ballistic missile propulsion and guidance systems have made their use in the tactical role at sea a practical option. The height of the trajectory allows the re-entry vehicle to obtain a clear plan view of the general target area, with the sea as a homogeneous background. The ability to adjust the final flight path allows the warhead to home in on a specific target. In theory, the ballistic missile can now have at least as good a terminal homing capability as the cruise missile, with none of its inherent disadvantages. Whereas a cruise missile might take as long as half an hour to reach the target, a 400-nm ballistic missile could be there within a few minutes of launch. Failing a mini-ABM system, there would be no way of countering it in flight, and it could be set to burst at any distance above or below the surface of the water, depending on the target. Of course, whatever the missile, all long-range systems have to depend on exogenous target-location data. But the ballistic missile's short time of flight and the large "area-of-look" for initial target acquisition made possible by the trajectory combine to reduce the tactical significance of this factor. Sufficiently up-to-date target location data would have to be ensured on a continuous basis, and for surface formations this could be provided by satellite surveillance.

The role of such a system would be similar to that of the long-range SSM, and against the carrier we already have the precedent of the J-class submarine (and presumably the E-II as well), "maintaining central station" in the Mediterranean, targeting Western carriers.[47] But the appearance of a tactical ballistic-missile system prompts a reevaluation of the general assumption that all Soviet ballistic-missile submarines continue to have a strategic role, long after their weapons have been rendered obsolete by improvements in Soviet delivery systems and U.S. antisubmarine countermeasures.[48] It is perhaps relevant that, whereas Y-class SSBN are deployed to patrol areas in the western Atlantic, north of Bermuda, the G- and H-classes remain east of the mid-Atlantic ridge about level with Nova Scotia,[49] and that the patrol areas of the latter two classes have usually been reported as at least 1,000 miles off the Atlantic seaboard[50]—that is, well-positioned to intercept carrier strike groups sailing from the east coast, but not for striking at strategic targets in North America with their 350- or 650-nm missiles. Meanwhile, the G-class may now be designated as a "cruiser-type" submarine,[51] a prewar category that was not featured in the postwar rebuilding program.*

Of potentially greater significance is the use of this system against Polaris. Although the target location problem is of quite a different order, the Polaris submarine must betray its position when it fires. The rapid reaction time of a 400 nm SLBM offers the possibility of calling down counter-battery fire before all 16 missiles have been launched,[52] thus "weakening the enemy navy's nuclear strikes," which Gorshov lists as one of the three basic missions.[53]

The intention to retrofit the SS-N-13 system into the G- and H-classes would account for the Soviet Union's strong resistance to having these submarines included in the SALT calculations. Ultimately, the United States agreed that the G-class SSB's need not be counted, but the Soviet acceptance of the H-class SSBN's inclusion was only conceded at the very last moment, under the threat that otherwise the whole edifice of agreements would fail.[54]

The in-service availability of SS-N-13 by the end of 1974 suggests that it was intended to retrofit this system to existing classes. The H-class would now seem to be ruled out as the result of SALT, but (besides the G-class) the E II SSGN is an attractive candidate. Such a reconfiguration would extend rather than change its role,† while making the maximum use of existing nuclear-powered hull/propulsion units. Meanwhile the P-class may have been especially designed to take this system.

REVIEW OF SUBMARINE PROGRAMS

The overall impression is one of expectation. Past experience suggests that we may see a new family of nuclear submarines begin

---

*The K-class, which carried 10 torpedo tubes, two 100-mm and two 45-mm guns, began construction in 1936, some 14 being delivered. The K designator stood for Krejserskaya, and this category of submarine was defined in a naval dictionary as being intended for operations against maritime communications, either independently or in company with surface ships and aircraft (K. I. Samojlov, Morskoj Slovar', [Leningrad, 1939]). After the war, a Soviet naval manual stated that submarines were divided into three categories: large, medium, and small, and there is no mention of a cruiser-type (N. A. Shmakov, Osnovy Voenno-Morskogo Dela [Moscow, 1947]). The Z- and F-classes were "large type," with the designator B-Bolshoj.

†That is, its anticarrier capability would be enhanced, and it would also have a counter-Polaris capability.

delivery in 1978, with a lead unit or two being delivered the year before. It is too early to be certain whether or not the Soviet Union will increase its nuclear production rate, but what information there is suggests not. Reactor production capacity is probably the ultimate limiting factor; this was doubled between 1957-62, and while it could again be raised by the same amount (five or six units a year), the information on yard extension does not suggest an increase of this order.

If SSBN construction continues at five to six units a year, some 60 Y- and D-class will have been delivered by 1978, and we can perhaps assume that new-construction SSBN will not figure in the new programs. There are however likely to be major modifications to the modern ballistic-missile units now in service.

Assuming current building rates, by 1978 the Soviet Navy will have some 45 SSN plus 20 of the less versatile C-class SSGN. Because of SALT, they may also choose to reconfigure the H-class as SSN. Nevertheless, in terms of their requirements there remains a shortage of general purpose (rather than task-specific) high-speed submarines for use in the antisubmarine role, both on distant deployment and in defense of the fleet areas. If the new class building at Gor'kij has closed-cycle propulsion, this will go a long way to meeting the in-area requirement. It seems likely that the distant-water tasks will have priority in the allocation of nuclear propulsion; it is relevant that Gorshkov refers to the need for nuclear propulsion in submarines, within the general context of long-range naval operations and sustained high speed.[55]

In view of the major advances in nuclear submarine performance made by the Soviet Navy between 1958 and 1968,[56] it seems likely that there will be further substantial improvements (although relatively not so great) in the new classes that begin delivery in 1978. One can only speculate about the configuration of the new classes, but at the time the final decisions were being made, the most pressing requirement was the task of counter Polaris/Poseidon, and this remains their first priority.[57] If the A-class has nuclear propulsion (as opposed to closed-cycle), it could be serving as a trials vehicle for some such system.

Major modification programs are likely to be of considerable significance in the years ahead and will have a substantial impact on the overall capabilities of the submarine force.

Despite the extensive submarine construction capacity, present programs give the impression of working within strict constraints, and new requirements have to be met by adapting existing facilities.

CURRENT SURFACE SHIP CONSTRUCTION

Carriers

Kiev, the first of the Kuril Class of aircraft carrier, was probably undergoing shipyard trials in the Black Sea at the end of 1973.[58] It is not clear whether she had been "delivered" to the navy by March 1974, but it is forecast that she will not join the fleet until late 1975.[59]

The keel of the second ship (to be called Minsk),[60] has been laid on the same building ways, but progress appears to be leisurely. There is an impression that this program is not proceeding as fast as had originally been forecast. U.S. sources estimate that these ships could carry either 25 V/STOL aircraft or 36 helicopters, and it is considered that a mix of the new V/STOL fighter and Hormone ASW helicopters is the most likely complement.[61] The helicopter cruiser Moskva is thought to have carried out deck-landing trials with this new fighter in 1972/73.[62]

What appears to be continuing dilatoriness over the carrier program, raises questions as to whether the Soviet leaders are fully convinced of its long-term cost-effectiveness in military and political terms. It is not self-evident that they wish to develop the ship-shore role, which has been so prominent in Western practice. They have however demonstrated their requirement for effective air defense and ASW, and in these cases the decision is likely to depend on the various alternatives they can foresee as being available to meet their requirements. For example, developments in laser technology may reduce the potential importance of ship-borne fighters in defense against air attack at sea. On the other hand, they may find that ship-borne fixed-wing aircraft are essential to support their counter-Polaris operations.

Cruiser-Size Ships

The third Kara-class cruiser was probably delivered by mid-1974 and construction is continuing.[63] The present delivery rate is about one unit a year from Nikolaev North Yard, and there is still no suggestion that construction is taking place elsewhere.

Five Kresta II were delivered from Zhdanov Yard, Leningrad, by mid-1973, but it appears that no more were completed during the next 12 months.[64] The shipbuilding pattern suggested that 12 Kresta hulls were originally programed for delivery (that is, four Kresta I and eight Kresta II), and this was supported by the report that by 1972, six or seven Kresta II were in service or fitting out.[65] It is not clear whether the program has in fact terminated at five units, or whether there are up to three more Kresta II waiting to be completed. The pattern of deliveries continues to support the assessment that the delays in fitting out this class resulted from the modification in characteristics decided on in 1961.* The most likely bottleneck is the availability of SS-N-10 systems, at present being fitted in both Kara and Krivak.

The Kara class, and Krestas I and II, are all designated by the Soviet Navy as large antisubmarine ships (bol'shoj protivolodochnyj korabl'—BPK).[66] This contrasts with Kynda, which is referred to as a missile cruiser.[67]

At least two Sverdlov-class cruisers have been converted to command ships, with extensive new communications arrays, including satellite.[68] Admiral Senyavin (in the Pacific) traded both her after-turrets for a helicopter hangar and an SA-N-4 point-defense SAM system. Zhdanov (operating from the Black Sea) traded X-turret for an SA-N-4 system. Both ships now have four "gatling"-type 25 mm mountings abaft the for'd funnel, Senyavin having another four mounted atop the new hangar; each pair of mountings has its own control system.[69] It is possible that a third Sverdlov is under conversion,[70] and assuming that they prove operationally successful, it seems likely that further units will be completed to fill an obvious need. There should be a lot of life left in these hull/propulsion units, which were delivered between 1952 and 1956 and have done relatively little sea time.

It appears that one gun cruiser has been retired during the last 12 months.[71] The solitary prewar Kirov would seem to be the most likely candidate.

Destroyer-size Ships

The first Krivak, built at Kalingrad, was at sea in the Baltic in December 1970[72] and presumably entered service during 1971. In April 1972, one or two units were reported as being "active",[73] three

---

*The projected Kynda program was canceled during the 1957/58 decision period, only four units (out of 12 planned for delivery between 1962 and 1965) being completed. At the same period, the design of the follow-on Kresta class was modified to incorporate organic target location for the long-range SSM, in the form of helicopter-borne radar. Between 1961-64, the Kresta design was further modified to replace the long-range SS-N-3 system with the horizon-range SS-N-10, which also allowed an ASW helicopter to be carried.

being "active" one year later.[74] In the 12 months from 1 July 1973, at least three more were delivered,[75] implying a sharp rise in the production rate. This uneven pattern can be explained if we assume that after a slow start, Kalingrad achieved a delivery rate of two a year in 1972,* and Kamysh Burun began deliveries in 1973. This would imply an ongoing delivery rate of four units a year.

Five Krupny-class "missile ships" have been converted to Kanin-class "large antisubmarine ships." The remaining three Krupny are in the Pacific, and it is not clear whether they too will be converted.

It is reported that a substantial surface modernization program is under way, involving the addition of new missiles, antisubmarine systems, and communication equipment.[76] Within the destroyer size, units of the Krupny and Kotlin classes have already undergone major conversions, and the two remaining candidates are the Kashin and Kildin. The Kashin is comparatively modern, having completed delivery in 1969/70, but the class began construction in 1960, since when there have been substantial advances in Soviet ASW. The Kashin is already classified as a "large anti-submarine ship," and this class may be fitted with improved sonar systems (including VDS) in the course of periodic refit.

The four Kildin had already been passed over for conversion to antisubmarine ships in favor of the older Kotlin hull/propulsion units. This class is now being modified by removing the single (and obsolete) SS-N-1 system mounted aft and replacing it with two twin 76 mm turrets (as on Kynda), and an Osa II outfit of four SSM horizon-range missiles.[77]

The Soviet Navy still has 36 gun-armed destroyers in service,[78] but the number of Skories (about 17 at present) is expected to drop rapidly in the next few years. One SAM Kotlin was handed over to Poland in 1970 and it is possible that other units of this class have been supplied to Warsaw Pact allies, whose Skories and Rigas are now obsolescent. It may also be intended that these ships should operate in company with Soviet units on forward deployment.

Escorts and Smaller

There are relatively few comments to be made on the smaller classes, construction and replacement proceeding apace. The Turya,

---

*The delivery rates of Kynda, Kashin, and Kresta I, were each two a year, using four "notional" production lines. It is this precedent, rather than the capacity of Kalingrad, that prompts such an assumption.

successor to the Shershen-class torpedo boat, had begun to enter service in 1973. Both classes have about the same-sized hull and carry four tubes, but instead of the Shershen's two twin 30-mm turrets, the Turya mounts a twin 25 mm and a twin 57 mm.[79]

The "gunning-up" of this torpedo boat tends to confirm the impression that the Soviet Navy has had to cut down on the number of different types of surface ship. The Grisha class (850 tons, one twin 57 mm and one SA-N-4), which began delivery in 1970/71 can be seen as the single successor to both the Mirka escort (1,050 tons, two twin 76 mm) and the Poti large subchaser (650 tons, two twin 57 mm). All three classes carry two 12-barreled ASW rocket-launchers, but the Mirka has 10 torpedo tubes, whereas both Grisha and Poti only have 4. The Turya's 57 mm turret will partially compensate for the loss of the escort's gunfire.

The Nanuchka, a 200-foot, 800-ton missile patrol boat carrying six SS-N-9 launchers, provides the all-weather surface attack capability that used to be supplied by the cruiser/destroyer group, operating within range of shore-based air support. The SS-N-9 system, which requires external target location and midcourse guidance, fires to 150 nm, the optimum range being 50 nm.[80] The Nanuchka program appears to have difficulties, the numbers in service remaining virtually constant since the class was first reported in 1971;[81] it is not clear whether these problems have been solved and if the program is now proceeding normally.

Afloat Support

There continues to be a steady, if slow increase in the number of units that are purpose-built for afloat support.

Boris Chilikin, a gun-armed, fast-replenishment ship for solids and liquids (AOR) was delivered in 1971. This was the first of its kind in the Soviet Navy, being a modification of the 16,300-deadweight-ton (dwt) Velikij Oktyabr' class of civilian tanker, which had been building at Baltic Yard since 1967. This single unit has been followed by the Manych, a 7,500-dwt AOR, which entered service in 1972 and is probably in series production.[82]

The Amur class of repair ship/tender (AR), which began delivery in 1970, does not represent a new departure. However, at 6,500 tons, it is twice the size of its predecessors Dnepr' (1958-62) and Oskol (1965-70), both of which displace 3,000 tons, and the Amur is better suited to sustained forward deployment. (All three classes are designated as "repair ship"—plovuchaya masterskaya—and often carry the side designator PM.) Eight Amur were reported in 1973, suggesting a delivery rate of about two a year.

The first of the Amga class of missile-support ship had entered service by early 1973 and is probably in series production.[83] It succeeds the Lama Class, five of which were delivered 1962-67, and is expected to provide support to ballistic missile submarines.[84] The Lama class was originally intended to provide support to detachments, dispersed in the fjords of the Murman coast. With the ballistic-missile force rising toward 60 submarines, it seems likely that this will also be the Amga's primary role.

There is no information on any successor to the Ugra class of nuclear submarine tender, 10 of which have been delivered since 1962. (One of these was supplied to India. A second is being used as a training ship by the Soviet Navy.) This class, which was designed in the middle 1950s, was intended to provide support to submarines in sheltered berths. A new type of ship designed specifically for open-ocean work might be expected.

SURFACE YARD CAPACITY

There continue to be no indications that surface yard capacity has been switched from civilian to warship construction. Rather, the impression persists of the Soviet Navy having to work within the restricted limits of what it already has. The Kuril-class carrier program is confined to a single way at Nosenko Yard, the cruiser-sized Kara has replaced the destroyer-sized Kashin at Nikolaev North, and each category of construction has been shunted down one "type-yard."

Construction is steady rather than forced, and Kara and Krivak appear to have reached their full production rate at the yards presently involved in these programs. One Kara (compared to two Kashin) a year from Nikolaev, and probably four Krivak from the two former escort yards. Meanwhile, Zhdanov, which can build two Kresta a year (but isn't doing so) remains something of an enigma. In the five years 1969-73 it appears to have built 5 Kresta II and converted 5 Krupny to Kanin; it may still be involved in the terminal stages of fitting out up to three more Kresta. On past experience it appears underemployed and bearing in mind the apparent squeeze on yard capacity, one would expect it to be making some more substantial contribution to the surface program. One plausible explanation for the apparent hiatus is that Zhdanov will build Kara, and this may require substantial alterations to her assembly ways.

SURFACE WEAPON SYSTEMS

It is interesting that despite the general shift to horizon-range SSM systems, the Nanuchka's SS-N-9 has a maximum range of 150 nm.

There would seem to be three reasons for this. The Nanuchka's size virtually dictates that it will operate within range of shore-based air cover and other forms of support; it can therefore rely on external target location data and midcourse guidance, whereas the Kara and Krivak must be self-sufficient on forward deployment. The second reason is spatial. Nanuchka is intended for area defense, and there is an inverse relationship between the number of units required to cover a certain area, and the range of their weapons. Kara and Krivak are intended to mark large enemy warships (particularly strike carriers), in which circumstances there is no relationship between weapon range and the number of units required, the latter depending on the number of targets to be marked. The third reason is the different types of engagement. Nanuchka, whose role (along with many other types of naval unit) is to defend the offshore zone against intruding surface units, will inevitably be involved in some kind of closing encounter. In such circumstances, the greater the missiles' range, the better the chance of surviving to launch them. In contrast, the exposed Kara and Krivak must rely on being within missile range of their target before conflict erupts; otherwise they are fairly certain to be sunk by submarine and carrier-air attack before they can launch their weapons. For units on forward deployment, self-contained, rapid-reaction cruise-missile systems are to be preferred. But within the fleet areas, given external midcourse guidance and a coordinated attack by several different types of weapon system, the advantages of longer range outweigh its drawbacks.

REVIEW OF SURFACE PROGRAMS

The range, accuracy, and payload of cruise missiles has meant that the minimum size of hull for a specific task can now be determined by the hostility of the operational environment and by seakeeping requirements. This means that the fleet area tasks that were formerly discharged by cruiser-destroyer groups can now be carried out by relatively small, missile-armed units, supported by shore-based weapon systems and sensors. Meanwhile, the cruiser- and destroyer-size ships become available for forward deployment.

The Soviet Navy has followed this course, and the classes now entering service show a much clearer distinction between the types of ship designed for distant deployment (out-of-area), and those intended for defense of the fleet areas (in-area). Furthermore, they have been able to drop the escort-size (1,000-1,500 tons) from their inventory of ship types, by upgrading the capabilities of three other categories; the large subchaser (Grisha), the missile patrol boat (Nanuchka), and the torpedo boat (Turya). The escort ship had been built for 35 years, but it was not large enough to be really effective on distant deployments; it could also be argued that it was larger than absolutely necessary for the in-area task. Dropping this type served the purpose of releasing the "escort building ways" and allowed destroyer-size ships to be built at these yards.

For surface ships on distant deployment, the need to rely on a ship's own sensors for target location places a limit on the useful range of the SSM, which means that both the destroyer- and cruiser-size ships can mount the same "caliber" of main armament, although the Kara carries twice as many launchers as Krivak, and more reloads. But there are limits to what can be packed into a destroyer-size hull, and the Kara's larger displacement (about 2.5 times greater) allows her to carry the long-range SAM systems and associated radars, whereas Krivak only has the point-defense system. The long-range SAM can be seen as a "force" weapon system (that is, for the protection of the whole force) and for similar reasons the Kara could be used to carry very large hull-mounted sonar arrays.

While this explanation of the rationale underlying the new classes now entering service may be close to the official Soviet line, one suspects that it represents the best deal that the navy could get in the circumstances, rather than their "preferred solution." The growing divergence between "in-area" and "out-of-area" capabilities must reduce operational flexibility, and it is relevant that on average, one in three of the surface combatants in the Mediterranean throughout 1968-69 were escorts,[85] the ship type that has now been dropped from their inventory. It is true that the progressive shift in emphasis to antisubmarine surface forces, heavily armed for self-protection, has increased flexibility in other ways. But while the Kara and Krivak are impressive ships, one can imagine that naval leaders would have preferred to pack rather less into each hull. One suspects that Gorshkov's comment that "attempts . . . to build general-purpose combatants to carry out all (or many) missions have not been successful,"[86] was not idly made.

Meanwhile, they are still faced with the problem of insufficient numbers. If we take the evidence of canceled programs and major changes in design specifications as a measure of the Soviets' own assessments, it would seem that only the 9 Kresta, 19 Kashin, and 5 Kanin are considered fully adequate for distant deployment. They are now adding to these ships at a rate of one Kara and four Krivac a year. By 1980 (when the first Kanin hull will be 20 years old), they will have about 75 of these units, plus two Moskva and perhaps three Kuril helicopter and/or V/STOL carriers. Bearing in mind their basic strategic requirements,[87] this would seem a rather austere building program and does not suggest that the navy has been allocated additional resources to support a purely peacetime role.

OVERVIEW

If present trends persist, by 1985 the Soviet Navy will have some 200 nuclear submarines, of which about 60 will be SSBN, and the non-nuclear submarine force may have been built up to over 200 modern units.* Ocean-going surface forces will comprise more than 100 ships of destroyer size and above, of which at least three-quarters will carry SSM;† long-term intentions on carrier construction remain unclear.

There is a growing distinction in characteristics and capabilities between ship-types intended primarily for distant deployment and those designed to operate within range of shore-based air-cover and other forms of support. Since the interchange between tasks can now be only one way, there is a loss in operational flexibility, which will increase the difficulties of sustaining out-of-area operations. Meanwhile, afloat support is growing slowly, but at a rate that will take several years to relieve the Soviet Navy of its dependence on overseas facilities.

The heavy bias toward submarines has been maintained and perhaps increased. This, coupled with the emphasis on shore-based naval aviation, means that the Soviet Union is not building a balanced fleet in the Western sense.[88] A large proportion of their forces are still hampered by task-specific design; the new tactical ballistic-missile is a step back in this direction, and reflects the continuing importance of the counterdeterrent role. However, they are now building into their major surface units a high self-defense capability as well as a strong offensive punch; this greatly increases the navy's ability to operate in a hostile maritime environment and hence the flexibility of its employment.

Despite the steady improvement in the quality of Soviet ships, the niggardliness of the economic response to the major reappraisal of the maritime threat that took place in 1961 is striking. The navy has been required to shift to forward deployment and to develop a whole range of new operational concepts to meet a serious threat to the homeland, but there has been no reallocation of shipyard facilities to meet these new requirements. It was fortunate that in the early 1960s nuclear building capacity was already in the process of being doubled, but otherwise the navy had to make do with what it already had. By manipulating the available facilities, the navy was

---

*Comparable figures in mid-1974 were 130 nuclear submarines (of which 52 were SSBN) and 210 diesel units (of which 90 were obsolescent).

†Comparable figures in mid-1974 were 110 ships of destroyer size and above, of which about 30 carry SSM.

able to gain one cruiser a year, but only at the expense of eight escorts; and in terms of tonnage, the output has remained constant. The extra oceangoing units needed to sustain forward deployment will only become available through handing over the task of fleet-area defense to smaller surface ships.

The economic response is in contrast to the "foreign policy" one that reversed a long-established policy on overseas bases, so as to enable the shift in naval deployment. In the short term, this was unavoidable, but it is significant that at this late date the Soviet Navy still lacks the kind of afloat support that might indicate a decreasing dependence on facilities in foreign countries. It would seem that the political costs count for less than do the economic ones.

NOTES

1. For a justification of these decision periods, see "The Turning Points in Soviet Naval Policy," Ch. 16 in Soviet Naval Developments: Capability and Context M. MccGwire, ed. (New York: Praeger Publishers, 1973). See also Chapter 28 in this volume.

2. The data being analyzed are Western assessments and not the actual building rates in Soviet yards. Although it is often possible to identify the detailed figures on which generalized official statements are based, new evidence may change these original Western assessments. Official information is therefore not always internally consistent from year to year, nor between different sources. Analysis of these inconsistencies is a further aid to understanding the actual course of events.

3. Admiral Rickover's testimony to U.S., Congress, Joint Committee on Atomic Energy, Naval Nuclear Propulsion Program 1972-73, p. 249.

4. Norman Polmar, Soviet Naval Power (New York: National Strategy Information Center, 1972), p.70.

5. U.S. Congress, House, Committee on the Armed Forces, Seapower Subcommittee, Status of Naval Ships, report dated 19th March 1969 p. 419. This assumed that all seven yards with previous submarine experience (including Nikolaev) would be involved in the program and assessed that Severodvinsk and Komsomol'sk could build 12 SSBN a year between them. In 1972 it was claimed that the Soviet Union had built 15 nuclear submarines of all types during the previous year; see p. 4 of Admiral Zumwalt's statement to the Senate Committee on Armed Services concerning fiscal year 1973. By 1973 all assessments had been revised downward; see MccGwire, "Soviet Naval Programmes" in Survival (London: IISS, Sept./Oct. 1973), p. 221.

6. Rickover, op. cit., p. 249, speaking in March 1973: ". . . in the past twelve months the Soviet Union . . . produced about 10 nuclear powered submarines." Judd, U.K. Minister for the navy, speaking in June 1974: ". . . one nuclear submarine being commissioned every five weeks . . ." (London Times, 6 June 1974). About 54 nuclear submarines were delivered 1968-72 (inclusive), which supports this annual production capacity.

7. Military Review, November 1969.

8. U.S. Secretary of Defense, Annual Defense Department Report FY 1975, p. 47. Also chairman, U.S. Joint Chiefs of Staff, Annual Posture Statement FY 1975 (hereafter, Moorer), p. 22. Admiral Moorer credited Soviet nuclear submarines with 666 SLBM by mid-1974: $(3 \times 10H) + (16 \times 33Y) + (12 \times 9D) = 666$.

9. Rickover, op. cit., p. 250.

10. Schlesinger (U.S. Secretary of Defense, note 8) stated that 18-19 D-class had been launched or were being assembled by the end of 1973. Five units had been delivered by this date, which leaves 13-14 hulls in the yard. Figures given in previous years for "under assembly or fitting out" have turned out to be equivalent to about 2.5 years of delivery. Moorer (op. cit., p. 21) estimates future annual production at five to seven for the next few years.

11. Moorer, op. cit., p. 8.

12. Ibid., p. 21. However, in January, Defense Department spokesmen were more dogmatic and claimed that the Soviets were building a lengthened D-class, which would carry 16 missiles firing to 4600 nm (London Times, 12 January 1974).

13. U.S. Naval Review, May 1972, p. 278.

14. The P-class was referred to during the 1972 hearings, but there was no reference to it in 1973 or 1974.

15. Early information on both the Y- and V-classes indicated their lead units were delivered to the Northern Fleet before the winter ice in 1967—for example, Spectator, 4 October 1968, Jane's Fighting Ships 1968/69, addendum p. 522.

16. Since 1972, there have been anomalies in the figures quoted for SSN and "attack submarines," and they do not conform to the trends established by statements during the preceding four years. In particular, the same number of SSN (25) was given for 1971 and 1972 during the Senate Appropriation Hearings for FY 1974 (pp. 754 and 587, respectively); and Admiral Zumwalt gave the same number of SSN (28) for 1972 and 1973 in letters to Senator William Proxmire (2 June 1972) and Representative Les Aspin (April 1973). The initial rate of three a year tends to be confirmed by the figure of 17 SSN for 1968 (p. 587 of Senate Hearings), which breaks down to 13 N + 4 V, one of which delivered in 1967. The cutback in SSBN construction suggests that the V-class delivery rate will be restored.

17. See MccGwire, Soviet Naval Developments, op. cit., pp. 185-86.

18. See "Soviet Naval Procurement," in The Soviet Union in Europe and the Near East, (London: Royal United Services Institute, 1970), p. 87.

19. Rickover, op. cit., pp. 128-29.

20. Time Magazine, 23 February 1968; New American, 15 November 1968.

21. Senate Armed Forces Committee, Appropriation Hearings FY 1972; March 1971, Admiral Zumwalt (p. 910) and Senator Margaret Chase Smith (p. 946).

22. Zumwalt's letter to Proxmire, 8 June 1972.

23. The apparent production rate of about 11 nuclear reactors has been used to determine annual totals, this number being supported by the overall available figures. The Soviet Union had about 55 nuclear submarines at the beginning of 1968 (Time Magazine, 23 February 1968), 100 in early 1972, and 110 in early 1973 (Rickover, op. cit., pp. 125 and 148). There is fairly detailed information on annual SSBN deliveries (see MccGwire, "Soviet Military Programmes," op. cit.), and what is left must be divided between the C- and V-classes.

24. See note 10 for evidence that the time spent in assembly and fitting out is at least 2.5 years.

25. Moorer, FY 1975, op. cit., p. 74. Moorer says "A new diesel-powered attack submarine is now entering the submarine force. . . . It is believed to be . . . a . . . replacement for the . . . Foxtrot." This implies series construction.

26. Marine Rundschau, February 1974, p. 119.

27. Ibid. Jane's Fighting Ships gives the J-class as 85.5 m by 9.5 m. For comparison, the F-class is 90.5 by 7.3 and the R-class 75.5 by 7.3. The beaminess of the J-class and this new class is noteworthy. In the case of the J-class, it could perhaps be explained by the requirement to mount missiles on the hull. This does not apply to the new class, and other explanations must be sought.

28. See MccGwire, Soviet Naval Developments, op. cit., pp. 185-86.

29. There is no other obvious place for the single A-class to have been built, and its delivery fits in to the end of the F-class program at Sudomekh in 1967. The amount of detail on this enigmatic unit appearing in Jane's Fighting Ships favors its origin from the only yard open to direct public observation.

30. For example, Jane's 1973/74: "Her form of propulsion is by no means certain." All other new classes have been identified as nuclear or diesel at the time their existence was first announced.

31. S. Breyer, Guide to the Soviet Navy (Annapolis, Md.: United States Naval Institute, 1970), pp. 150, 284. The system was fitted in

several of the Q-class but was not successful. The Walther system, which was under development by the Germans at the end of the war, used high-test hydrogen peroxide (HTP) to provide the oxygen necessary for closed-cycle combustion. HTP is unstable, and the closed-cycle Q-class was known in the Soviet Navy as "zazhigalka" (cigarette lighter).

32. Reports from returned German scientists, who worked on developing a closed-cycle diesel. The British secured Walther himself and in due course built two experimental submarines, Explorer (1956) and Excalibur (1958), which were used as high-speed targets. In addition to diesel-electric drive, these units has a turbine driven by a mixture of steam and carbon dioxide, produced by burning diesel oil in an atmosphere of steam and oxygen, the latter formed by the decomposition of HTP (Jane's 1959/60, p. 42).

33. Explorer and Excalibur were credited with a speed in excess of 25 knots, Jane's 1959/60.

34. For example, Moorer, op. cit.

35. Rickover (op. cit., p. 249) is quite explicit on this.

36. Between 1958 and 1967, Komsomol'sk and Severodvinsk built five or six nuclear submarines a year. The three issues of Jane's published 1962-64 indicate that E-I construction only took place at Komsomol'sk, building two units a year; therefore Severodvinsk was building three or four units a year. In 1972, eight SSBN were delivered from Komsomol'sk and Severodvinsk. At least two must have come from Komsomol'sk, at most six from Severodvinsk. In view of the substantial increase to Severodvinsk's capacity (see note 37), the proportion 6:2 seems more likely than 5:3.

37. House, Status of Naval Ships, op. cit., p. 419. The capacity at Severodvinsk was reported to have been doubled; London Times, 12 October 1971. Most of the Y-class were built at Severodvinsk; Rickover, op. cit., p. 165.

38. Rickover, op. cit., pp. 165 and 249.

39. Enclosure to Admiral Zumwalt's letter to Senator William Proxmire, 2 June 1972, gives P-class displacement as 5,000 tons.

40. Breyer, op. cit., p. 232. The design bureau at Baltic Yard was responsible for the majority of prewar classes. For example, D, L, Shch, P, S, and it set up the M-class project at Nikolaev.

41. The 1968 assessment was that the Soviet Union could build 20 nuclear submarines at seven yards, of which 12 could be SSBN (note 5)—that is, 12 at Severodvinsk and Komsomol'sk and 8 at the other five yards. Assuming that this was based on the previously established diesel submarine capacity, Gor'kij would have taken 3 of these 8. Similar results are obtained by applying the crude conversion factor of 0.5 to derive "notional" nuclear-production lines, and then applying the nuclear assembly rate.

42. Mentioned by Admiral Wheatley in testimony to Congress; New York Times, 4 February 1959. This figure is substantiated by the building rates actually achieved in 1952-54 and by the output of the truncated program carried on after the change in plan.
43. Moorer, op. cit., p. 8.
44. Schlesinger, op. cit., p. 47.
45. Times (London), 12 January 1974.
46. Air et Cosmos, 24 November 1973; Janes 1973/74.
47. Rickover, op. cit., p. 128.
48. I am indebted to K. J. Moore for this thought.
49. Washington Star, 20 April 1974.
50. Missiles and Rockets, 4 April 1966; Senator Stuart Symington quoted in Washington Post, 29 May 1966.
51. Granma, (Havana) 12 May 1974, reporting a visit by a Soviet naval group, showed a picture of a G-class SSB and gave its number as K-96.
52. I am indebted to Harlan Ullman for this point. See also his "The Counter-Polaris Task," Chapter 31 of this volume.
53. Morskoi sbornik, February 1973, p. 21. He uses "oslablenie."
54. See J. Newhouse, Cold Dawn: The Story of SALT (New York: Holt, Rinehart and Winston, 1973), pp. 254-55.
55. Morskoi sbornik, February 1973, p. 22.
56. Rickover, op. cit., p. 127.
57. Norman Polmar quotes discussions with the Soviets on how they would try to counter Polaris/Poseidon, and their opinion that these methods would be effective. U.S. Congress, Senate Armed Services Appropriation Hearings, FY 1974, p. 6048.
58. Marine Rundschau, January 1974, p. 56; Navy International, October 1973, p. 37.
59. Moorer, op. cit. p. 70.
60. Marine Rundschau, May 1973, p. 307. It is perhaps noteworthy that the names of both Kiev and Minsk were generally known when the ships were still under construction and in the case of Minsk, while still on the building ways. Could this be a form of insurance by the navy, against a change in policy affecting this program?
61. Moorer, op. cit., p. 70.
62. Jane's Fighting Ships, 1973/74.
63. Moorer, op. cit., gives 17 missile cruisers by mid-1974 (FY 1975, p. 69) and gave 16 missile cruisers in mid-1973 (FY 1974, p. 37).
64. Kresta are referred to as "missile cruisers." Moorer refers to more Kara joining the fleet in the next few years but makes no reference to Kresta (p. 71). Hence the assumption that one Kara was delivered but no Kresta II.
65. Flottes de Combat, 1972.

66. Krasnaya Zvezda, 23 February 1974; Morskoi sbornik, June 1973, frontispiece.

67. Morskoi sbornik, March 1972.

68. Norman Polmar, p. 6048 of Senate Armed Services Appropriations Hearings, FY 1974.

69. Marine Rundschau, January 1974, pp. 39-40.

70. A 1972 release stated that three cruisers were undergoing conversion. Zhdanov was said to be in hand at Sevastopol and Adm. Senyavin at Vladivostok (New York Times, 30 April 1972; Seapower, May 1972).

71. Moorer, op. cit., gives 14 gun cruisers by mid-1973 and 13 mid-1974. See also note 63 above.

72. Rickover, op. cit., p. 134; Revue Maritime, July 1971.

73. Enclosure to Zumwalt's letter to Proxmire dated 2 June 1972. This gives 11 SSM destroyers—that is, 5 Krupny, 4 Kildin, and 2 Krivak.

74. Enclosure to Zumwalt's letter to Representative Aspin, April 1973. This lists three Krivak.

75. Moorer, op. cit., gives 40 missiles destroyers in 1973 and 43 in 1974; see note 63 above. If no SAM Kotlin were given away during that period, this implies three Krivak were delivered.

76. Moorer, op. cit., p. 69.

77. United States Naval Institute Proceedings, June 1974, p. 81.

78. Moorer, op. cit., p. 69.

79. Marine Rundschau, January 1974, p. 56.

80. Jane's Fighting Ships, 1973/74.

81. Jane's Fighting Ships carried six units each in 1971, 1972, and 1973.

82. Moorer, op. cit., reported that two new classes of replenishment ship had become operational (FY 74, p. 41). The wording implies series production.

83. Moorer, op. cit. note 82.

84. Moorer, op. cit., FY 75, p. 55.

85. MccGwire, Soviet Naval Developments, op. cit., p. 383.

86. Morskoi sbornik, February 1973, p. 21. This was the final section of his 50,000-word statement, in which he sums up the Soviet Navy's present-day requirements.

87. That is, to deter Western attack on the Soviet Union by demonstrating the capability to fight and win the subsequent war. Gorshkov lists three main tasks in this context: (1) contribute to strategic nuclear strikes; (2) weaken the nuclear strikes of opposing naval forces; and (3) participate in military operations within the continental theaters. This last task covers the whole range of traditional naval operations. Morskoi sbornik, February 1973, p. 21.

88. Gorshkov (<u>Morskoi sbornik,</u> February 1967, p. 20, footnote) defines a well-balanced fleet as one that by reason of its composition and armament is capable of discharging the tasks assigned to it, whether in nuclear-missile war, in non-nuclear-missile war, or in protection of state interests in time of peace. But this is a tautology. When discussing "balanced forces" in late 1971/early 1972 with Western officers in Moscow, Gorshkov commented that while it was easy to defend the requirement for submarines, it was much harder to justify the need for more surface ships.

CHAPTER 24

COMPARATIVE CAPABILITIES OF SOVIET AND WESTERN WEAPON SYSTEMS
Nigel D. Brodeur

ANALYTICAL PROBLEMS

In the days of wooden ships and hand-operated guns, capabilities comparison was a simple matter. By and large, "like fought like"; moreover, the range and destructiveness advantages of one piece of naval ordnance over another were insufficiently significant to offset completely such factors as the "weather gauge," the experience and skill of the master gunner, and the maneuvering skill of the captain in gaining a tactical position that maximized his firepower concentration and minimized that of the enemy. Consequently, estimates of comparative capability were based largely on numbers of "ships of the line" and the numbers of guns they carried. The concept of "outgunning" an adversary or being "outgunned" by him was born—and persisted as a basis for comparison long after it had been made obsolescent by new forms of maritime warfare.

Steam power and armor plating brought dramatic changes. Heavy calibered guns could now be installed and readily loaded, trained, and elevated. The newer weapons significantly outperformed guns of lesser caliber—both in range and destructiveness. The ship with the heavier gun could not only outrange an adversary but also, with the addition of suitable protective armor in vital areas, remain immune

---

The concepts and conclusions expressed in this paper are those of the author and should not be interpreted as necessarily reflecting those of the Maritime Command or the Canadian Armed Forces. The author is indebted to his staff at the Canadian Forces Maritime Warfare School for their research and assistance in preparing this paper, the work done by LCDR R. C. Withinshaw, RN, LCDR H. R. Waddell, CAF, and Lt. L. L. Fletcher, USN, deserving special mention.

from a "chance vital hit" within certain range brackets of the enemy's gun. Thus was born the concept of the "immunity zone." Heavy guns and superior speed would now enable the ship so endowed to punish an outgunned, slower adversary at will. Now capabilities comparisons became based on caliber and number of guns, strength of protective armor, and speed capability. The problem of making comparisons became more complex, but, even following the introduction of shipborne torpedoes, ships still fought ships and the weapons comparison process—though somewhat more technical—was still straightforward.

It must be remembered, however, that earlier comparisons of weapon systems capabilities were founded principally on an examination of very basic adversary situations. The capabilities of naval ordnance in the context, for example, of its comparative superiority to field ordnance was a secondary consideration unlikely to preoccupy unduly the designers of naval ordnance.

The infinite variety of adversary situations in modern warfare can, however, make straightforward comparisons of comparative weapon systems capabilities either impossible, or alternatively, misleading. To be sure, the anti-air (AA) weapons on one side may be compared to those on the other side—but the fact that such comparisons inherently disregard the adversary situation may lead to erroneous conclusions concerning the supremacy of one side over the other.

In considering then, the example of a comparison of AA systems capabilities, it will be apparent that a meaningful comparison cannot be arrived at without considering the adversaries—in this case, the respective air threats. In such a comparison of apples with oranges and apples with bananas, the only realistic basis for comparison is the degree to which each side's appetite has been satisfied—in other words the degree to which each side's AA systems has met the challenge of the other side's airborne weapons and platforms.

Having arrived at a satisfactory basis for comparison from the conceptual viewpoint—namely the degree to which the types of combat systems on each side meet the threat against which they are intended; the analyst then finds himself in a very sensitive area indeed—an area hardly conducive to complete revelation and hence open an informed debate.

The dilemma resulting from the need to safeguard information, which by its nature is highly sensitive, extends not only to the adversary situation but even to the side-by-side comparison of related weapons systems and can make such comparisons even more misleading. The maximum in-flight range of two missiles, for example, may be postulated in unclassified literature; however their effectiveness, in output terms, is gauged not by their maximum trajectories but by their respective single-shot kill probabilities against specific targets within specific range and altitude envelopes—probabilities that result both from the characteristics of the missiles and those of their associated fire-control systems. Such information is again very sensitive.

Notwithstanding the previously described problems and limitations, an attempt will be made in this paper to conduct both side-by-side and adversary-situation comparisons of Soviet and Western weapon system capabilities. The "output terms" and conditions on which the comparisons are based will either be stated or will be implicit in the "formulae" adopted to reduce situations to fundamental problems. Thus, a reasonably useful portrayal of comparative standings can be inferred—bearing in mind that specific details of most systems cannot be disclosed, that the adversary situations have been highly simplified, and that certain system parameters not incorporated in the analysis method could, in some cases, significantly affect the comparative standings.

SUBMARINE DESIGN AND CONSTRUCTION

The relationship between a submarine's displacement, configuration, propulsion power, and maximum speed is too technical for detailed discussion in this paper. Suffice to say that a well-designed submarine of a particular displacement will encounter less drag moving through the water than one that has a less optimized hull form or too many drag-causing appendages.

Deviations from the ideal hull form are, however, closely related to volumetric problems associated with machinery size and systems design. Thus a designer takes an in-production power plant of X horsepower, projects it into an "ideal" submarine of Y tons displacement intended to attain Z knots maximum speed, then finds that he needs more cubic space—either for the powerplant or for other systems in that submarine. In the quest for space he could increase either the beam, the length, or both. Increasing only the beam or the length causes a departure from the ideal hull form—and the drag increases. Increasing both, so as to retain the "ideal" form, increases a surface area—again drag increases. The designer now finds he needs a more powerful power plant to achieve Z knots, which in turn etc., etc.

Steel technology and design depth also may force compromises in hull form. As design depth (hence pressure) increases, hull strength must be increased. This can be achieved by increasing steel thickness, by changing the hull shape, or some combination of both. Eventually, however, a limiting depth is reached for any given hull shape and material where the overall weight starts to exceed that of the volume of water displaced. The solution then is to use stronger steel or make the hull more spherical thereby improving the displacement to weight ratio. Both alternatives have practical limitations— on the one hand the designer is constrained by the technological state of the art in high-yield steels and the exorbitant cost of alternatives, such as titanium; on the other hand he must depart once again from the ideal hull form.

Submarine design is therefore a complex and iterative process that involves not only the above factors but also other critical considerations such as stability, noise reduction, and life-support capability. As a consequence of all the factors involved, submarine design is considerably more complex and precise than that for surface ships, submarine equipment is highly "systems-engineered," and completely successful major retrofits or conversions are much less likely than may be generally believed.

From the preceding, it can be seen that a graphical presentation of submarine maximum speeds against power/displacement ratios should provide some useful "state-of-the-art" comparison in the aggregate area of hull form design, steel technology and hull construction, power-plant design, and (possibly) overall systems engineering. Such a presentation is given in Figure 24.1. It can be seen that the "envelope" encompassing Western submarines is closer to the theoretical ideal—the Von Kharman-Gabrielli line[1]—than the "envelope" encompassing Soviet submarines, but that the USSR is decidedly closing the gap.

It must be pointed out, however, that the calculation process inherent in Figure 24.1 presupposes that both Western and Soviet submarines have the same depth capability. If the Soviet's submarines are much deeper-diving than their Western counterparts, then the "gap" would be smaller, and indeed, might not exist at all.

ANTI-AIR WARFARE

Assuming that other missile operational characteristics are equal, graphical analysis of missile range versus launch weight can provide an insight into comparative SAM technologies in the fields of aerodynamics, propellants and component design and systems engineering.

The "envelopes" circumscribing the Western and Soviet families of SAM's in Figure 24.2 indicate that the West leads both in the fields described above and in the development of long-range surface-to-air missiles.

The calculation process does not, however, take into account whether or not the Soviet missiles have greater speed, kill probability,

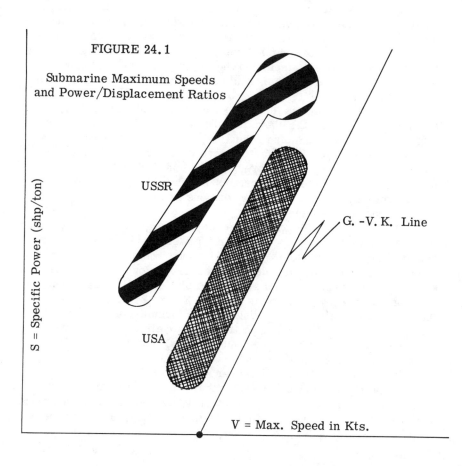

FIGURE 24.1

Submarine Maximum Speeds and Power/Displacement Ratios

S = Specific Power (shp/ton)

USSR

G.-V.K. Line

USA

V = Max. Speed in Kts.

or countermeasures capability than their Western counterparts. Any of these factors could account for increased launch weight.

The form of presentation also does not incorporate the adversary situation. The weapon-release ranges of the respective Soviet and Western aircraft are not shown; consequently comparisons of achievement versus need are not evident. A criticism that the Soviet Navy has inadequate defense against aircraft delivering long-range stand-off missile attacks presupposes that the Soviets would be attacked by aircraft that, at this time, only they possess—a somewhat unlikely situation.

In considering comparative anti-air capability the approach given in Figure 24.3 is more profitable. This figure is based on the overall capabilities of the best anti-air ships of similar vintage against

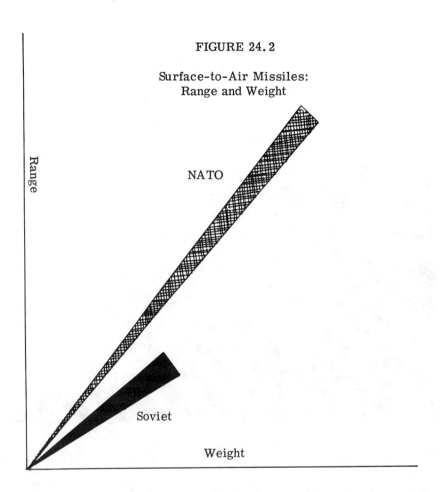

FIGURE 24.2

Surface-to-Air Missiles:
Range and Weight

identical representative closing air targets. The calculations incorporate all the AA systems installed in each ship and their respective capabilities to detect, engage, and reengage. Assuming again equivalent systems reliabilities, the "best in the West" has the capability to commence an AA engagement at much greater range and can engage nearly three times as many targets as the Soviet contemporary in the long and medium ranges combined. The Soviet ship capability improves rapidly however in the shorter-range bracket to the point where the total number of targets ultimately engaged is only slightly less than that engaged by the Western ship. Again, however, it must be remembered that the comparison applies only for the selected representative air target. Comparative capabilities against, for example, a very fast sea-skimming missile could be considerably different.

FIGURE 24.3

Engagement Capability: Surface to Air

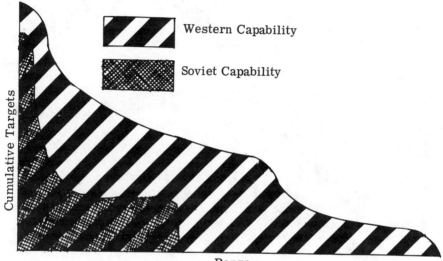

SURFACE-TO-SURFACE WARFARE

The comparative surface-to-surface warfare engagement capabilities are shown in Figure 24.4. The representative ships selected are those assessed as having the best antisurface capabilities in the Western and Soviet navies.

The engagement spectrum portrayed commences with long-range surface-to-surface missiles and incorporates medium range antisurface missiles, surface-to-air missiles fired in an antisurface mode, and guns. The sharp cutoffs signify termination of engagement with a specific system owing to lack of reload capability. In this instance the number of targets engaged is not shown cumulatively because repetitive engagement may be necessary before an opposing warship is ultimately destroyed.

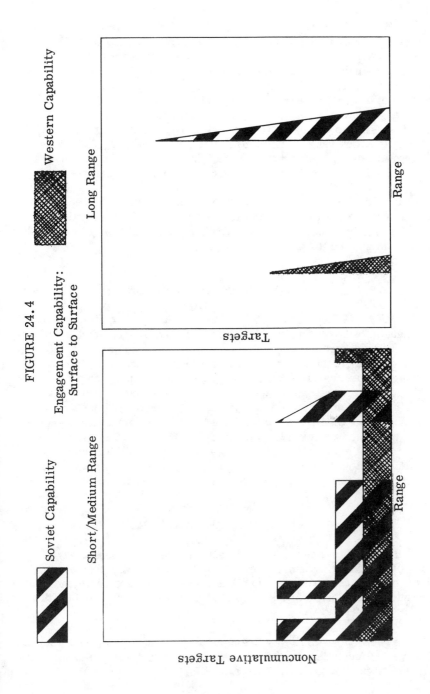

FIGURE 24.4 Engagement Capability: Surface to Surface

This analysis implies third-party assistance for long-range missile firings but does not incorporate carrier strike capability against surface targets. The latter is often advanced in the context of surface-to-surface capability. However, careful consideration leads to the premise that carrier strikes are principally surface projection of delivery systems as opposed to weapons systems; hence their engagement capability may be less a function of the range to which airstrikes can be delivered and more a function of aircraft stand-off attack ranges and numbers and effectiveness of weapons delivered. In this context it is argued that carrier airstrike capability should be considered in the air-to-surface warfare comparison.*

In examining Figure 24.4 it will be noted that the Soviets have the capability of engaging up to twice as many surface targets at up to three times the range of anything in the West. In the medium- and short-range brackets the Soviets are outranged but have the capability of engaging more targets simultaneously.

AIR-TO-SURFACE WARFARE

Figure 24.5 compares the number of surface targets that can be engaged by Western and Soviet air launched weapons and the range brackets from which the attack can take place. The analysis approach assumes that the number of attacking aircraft is identical in each instance; consequently Figure 24.5 is based principally on aircraft payload, weapon release range, and weapon systems characteristics.

It will be noted that the Soviet aircraft considerably outrange the West in weapon range but that the Western aircraft can engage twice as many targets in the closer-range bracket. The overriding factor in this situation will be the degree to which either side can establish and maintain local air superiority using fighter aircraft— a factor too complex to be incorporated in this analysis.

ANTISUBMARINE WARFARE

The combinations of forces, tactics, and environmental factors possible in the ASW confrontation situation makes it extremely difficult to conduct capability comparisons in the context of a generalized and unclassified paper. There is also considerable danger that the

---

*These are admittedly argumentative points—they are intended however to focus thinking on the comparative surface-to-surface capabilities of carriers versus long-range SSM ships.

FIGURE 24.5

Engagement Capability:
Air to Surface

 Soviet Capability

 Western Capability

advocate of a given position will postulate the scenario that most aptly supports that position, disregarding the fact that his scenario will constitute but one of several hundred combinations of opposing forces, tactics, and environmental conditions. Unless detailed parametric analysis is conducted encompassing a wide range of scenarios, there is danger that conclusions may hold true in real life only to the extent that the situations postulated actually occur. It is particularly important to avoid planning forces on the basis of optimizing the outcome of a single representative situation—the fact that the outcome is optimized in one side's favor virtually guarantees against its repetition in real life since a competent adversary generally will not make the same mistakes twice.

Reduced to its most simplistic terms, the ASW problem can be expressed as two objectives in descending order of desirability: (1) to detect, localize (pinpoint), and attack the submarine before it detects the presence of targets; and (2) to attack the submarine before it can attack its targets. Since the ASW forces could involve any combination of air, surface, and subsurface ASW forces, the problem is fundamentally seven-sided. As such, it defies simple analysis— all the more so since it should, for comparative purposes, be portrayed simultaneously in the context of Soviet ASW and Western ASW.

The problem of analysis can, however, be simplified by ascribing the West a baseline capability and seeking where the Soviets stand in detecting, localizing, and attacking submarines using ships, aircraft, and submarines in comparison to the West. Similarly the same baseline can be used to see how the Soviet submarine compares to the Western one in attacking ASW forces before it first is attacked.

Figure 24.6 illustrates such a comparison—it will be seen that detection and localization capabilities of Soviet ships, aircraft, and submarine against a "standard" submarine is considerably less than that of Western ships, aircraft, and submarines but that their attack capability is only slightly less. It will be seen also that the Soviets have a better capability than the West to attack ASW surface forces before being attacked by them (because of their cruise missile submarines) but fare less well against hunter/killer submarines.

It should not be inferred from Figure 24.6 that Western ASW forces have 100 percent overall capability to detect, localize, and attack submarines before the submarine detects, localizes, and attacks them. Barring a major breakthrough in ASW, the submarine will continue to hold the overall advantage in these respects.

ELECTRONIC WARFARE

Just as sonics has become the single dominating factor in undersea warfare, so has electromagnetic radiation in above-water warfare. Deprived of it, communications—hence command and control—rapidly deteriorate, detection ranges are circumscribed by optics and visibility, and guided weapons have no guidance. It is the Achilles heel of modern naval warfare.

Electronic warfare possesses two inherent characteristics not always recognized. First, it ultimately is directed against only a small segment of an enemy's overall technological or command capability—against, for example, 4 subsystems (or people) in an overall system comprising 50 subsystems (or people). Yet in thwarting the 4, it renders the remaining 46 ineffective. It is thus a very cost-effective form of warfare. Second, electronic warfare is the only form of maritime warfare that can be fully and legitimately practiced against one's potential enemies in peacetime. Its more ardent practitioners consequently enjoy an immediate advantage at the outset of hostilities.

Electronic warfare is frequently confused with "eavesdropping," an occupation fundamentally abhorrent to Western concepts of individual privacy. The relationship exists to the point that electronic eavesdropping provides information that enables electronic warfare to be effectively applied. However, the distinction from conventional eavesdropping can be allegorically illustrated. There is a major difference between spying on what someone says—in order to learn

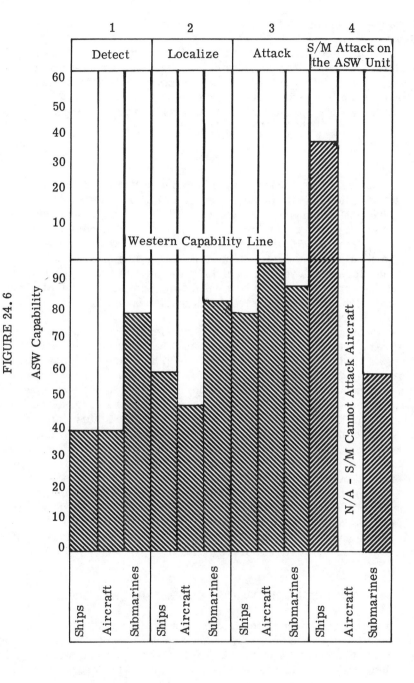

FIGURE 24.6

secrets—and spying on how he speaks—in order to determine how best to paralyze his vocal cords.

The impact of electronic warfare is such, therefore, that it can render all the above-water weapons systems' comparisons previously discussed completely invalid. In this connection, the Soviet Navy's overall capability in this science is assessed as being well ahead of that of the West[2], although the gap has closed in recent years.

MINE WARFARE

The naval commander for the landings at Wonsan during the UN intervention in Korea complained in a historic message that his force had lost command of the sea to a nation without a navy using weapons that were obsolete in World War I laid by vessels of a type that were in service in the time of Jesus Christ. Albeit somewhat exaggerated—these were the latest type of Russian mines and the landings, though delayed six days during mine-clearing operations, nonetheless took place—this chagrined statement remains an eloquent summation of both the offensive and defensive potentials of mine warfare and its cost-effectiveness.

The mine is generally viewed as a defensive weapon when laid in proximity to one's own waters and an offensive weapon when laid in proximity to another's. The mine tends, however, to be notoriously impartial in discriminating between foe or neutral and between combatant or noncombatant. This, together with other aspects of mine warfare such as declaration or nondeclaration of the purpose and existence of the minefield, internationally accepted status and customary use of the waters concerned, and nationalities of ships likely to be affected can have consequences so that a minefield, irrespective of its location, may be either tactically defensive or offensive and at the same time be either strategically offensive or defensive.

Inasmuch as mine-laying equipment continues to be installed in nearly all Soviet warships whereas it appears in few Western warships, it is concluded that the Soviet capability in mine warfare, if only on the basis of general emphasis, awareness, and overall training, is probably superior to that of the West.

TREND SIGNIFICANCE: SOVIET SHIP WEAPONS

It is generally contended that the Soviet Navy exists primarily to defend Russia from attack by submarine-launched ballistic missiles and from nuclear attack by the strike fleet and that these tasks largely deny the USSR the "disposable surplus" of seapower needed for flexibility.

It is important, therefore, to examine the recent trend in Soviet shipborne weapons systems in the context of their relevance to the stated primary roles of the Soviet Navy.

Defense Against Submarine-Launched Ballistic Missiles

Defense in this instance must entail either destruction of the submarine before missile launch or destruction of the missiles in flight or some combination of both.

As missile ranges increase, the problems inherent in the first task become insurmountable. The threat area, consequently the area requiring continuous surveillance and close pursuit, becomes immense and the missile submarine's position is no longer a useful indicator of the imminence of attack. Consequently, the defender is faced with two options—either of which may be politically and militarily unacceptable:

Preemptive attacks on the high seas against those missile-firing submarines that have been detected, an escalatory measure almost certain to invite nuclear retaliatory attack from those missile submarines still undetected; or

Retaliatory attacks following the initial firing(s)—in other words, "damage limiting" in protection of what logically may be expected to be lesser targets than those against which the initial missiles are launched.

In considering the forces that can usefully be applied in exercising either option, it must be remembered that the modern nuclear submarine enjoys both a detection advantage and, except under the most ideal conditions, a speed advantage over most surface ships. It must also be remembered that the strategic missile submarine, unlike the antishipping submarine, is under no compulsion to approach warships in the execution of his mission. If we remember also the size of the threat area, it will be evident that the Soviet naval units capable of providing the most meaningful capability against the Polaris/Poseidon threat would be long-range hunter-killer submarines and long-range ASW aircraft.

Since, however, such units cannot provide complete assurance that no submarine will be able to launch any missiles, it then follows that a navy seriously committed to antimissile defense would have placed great emphasis on airborne and submarine-borne systems designed to detect and destroy submarine-launched ballistic missiles in their comparatively slow, hence more vulnerable, early stages of flight. The fact that such systems are now precluded under the Strategic Arms Limitation Agreement would not be pertinent in the 1960s time frame, and to date, there is no evidence that the Soviet Navy had developed such systems prior to SALT.

Assertions that the Soviet Navy surface ship program (including even ASW carriers, whose area ASW capability is not unlimited) is primarily intended for defense against the Polaris/Poseidon threat, cannot, therefore, be supported from hardware examination. Moreover it is inconceivable that the proponents of ship programs would have been so ignorant of Soviet SLBM potential developments at that time that they would have failed to realize that an expensive surface ship program could only be achieved at some cost to other vehicles likely to be more efficacious in anti-Polaris defense.

There are some grounds, therefore, to postulate that the Soviets may, for some time, have perceived their first line of defense against Polaris to be the deterrence[3] provided by their own strategic missile capability. Surface ship deployments to the Indian Ocean are more logically explained as resulting from the "disposable surplus"[4] arising from such a realization than from any serious pretensions of countering a proclaimed Polaris threat from that quarter—notwithstanding Soviet political, or politico military, propaganda to the contrary.

Defense Against the Strike Fleet

In this case, defense entails either destruction of the carriers before aircraft launch, destruction of the aircraft before weapon release, destruction of the weapon in flight, or some combination of all three.

The dilemma inherent in the Polaris/Poseidon problem exists also to some extent in connection with the strike fleet—namely whether or not to conduct a preemptive attack against the carriers—since flying operations are more or less a continuous occurrence in carrier forces and consequently a strike launch may be well under way before it is recognized as such. The consequences of delay are, however, less serious than in the SLBM case observing the flight times involved before the strike reaches the target zones and the impressive Soviet air defense they must penetrate.

Antisurface missiles are basically a logical extension of the "outgunning" principle—namely the "stand-off" weapon. They are, however, very expensive and, for any given size of warhead, the missiles and their associated systems increase greatly in size and complexity as their ranges increase. It consequently makes little sense to have a "stand-off" capability greatly in excess of an enemy's weapon range. Consequently, it is logical to conclude that Soviet long-range antiship missiles were developed primarily for standoff attacks against carrier forces, the weapon range being predicated in this instance by carrier aircraft rather than by guns.

Missile attack at long ranges, however, poses several problems. The radar horizon and acoustic tracking limitations will force the firing ship or submarine to rely on target data from a third party, for example, a reconnaissance aircraft or a shadowing submarine—a more difficult process than it sounds, and one that can be counterproductive in that it may place a key player in the "standoff" scenario within the retaliatory range of carrier borne aircraft. Target identification and selection will also be a problem requiring a very discriminating missile to ensure that the target hit is the carrier and not a lesser warship, or worse still, a wandering neutral. Finally, the longer the missile flight time, the more it is susceptible to early detection, and consequently the greater the likelihood of countering it.

The Soviet temptation to use nuclear warheads in long-range missiles in order to compensate for target selection problems and close-in countering must be very strong if not irresistible. In the case of a warship, however, this may reduce its status in the utilitarian sense in that the threat it poses may not be entirely credible outside the context of nuclear warfare unless, of course, the credibility is restored by creating doubt as to whether or not the warhead is necessarily nuclear.

The disadvantages of long-range surface-to-surface missile systems could, of course, be overcome if the "standoff" range requirement could be reduced. This could be realized if the carrier air threat could somehow be discounted. If, for example, the Soviet Navy were to view the strike fleet as a strategically offensive but tactically defensive instrument that would not prejudice its primary mission by premature preemptive attacks on Soviet warships, then the likelihood of attack before a Soviet "crushing blow" would be diminished. If Soviet warships possessed an adequate anti-air capability, then the carrier air threat might also be discounted.

If then the Soviet surface fleet's primary role really is to defend the USSR against the strike fleet, then any shift from long-range surface-to-surface missiles to shorter-range systems in combination with improved surface-to-air systems must also be considered in conjunction with a shift from a tactically defensive to a tactically offensive posture, since the fleet's effectiveness against the carriers must rely on the "attack first, defend second" principle. Such a posture would presuppose a degree of Soviet aggressiveness not conceded by all Western observers. It is undeniable, however, that warships with good SAM and shorter-range SSM capabilities would be far more utilitarian in other roles—particularly if their effectiveness did not depend on nuclear warheads.

If then the range trend in Soviet air-to-surface missiles is counter to that in their surface ships; it must be questioned whether or not the combination of the Soviet nuclear deterrent[5], their impressive territorial air defenses, and their long-range maritime aircraft do not constitute a far more effective and realistic defense against strike fleet attacks on the USSR than do the Soviet surface fleets.

Conversely, however, such a weapons trend in Soviet warships would not only make them more utilitarian vessels for peacetime political purposes but would also make them far more capable of surviving in the hostile maritime environment that might be encountered in "wars of fraternal aid",[6] which raises the question of whether or not they might be designed more for countering strike fleet attacks on Soviet maritime interests than for countering Strike Fleet attacks on the USSR.

CONCLUSION

The recent trend in Soviet warship armaments is not deemed to be wholly consistent with primary roles of defending the USSR against nuclear attack by ballistic missile submarines and the strike fleet. The recent trend in air and submarine weapons does, however, seem entirely consistent with that role.

NOTES

1. The V. Kharman-Gabrielli line shown is an approximation of the limiting line for submarines discussed in G. Gabrielli and The Von Karman, "What Price Speed—Specific Power Required for Propulsion of Vehicles," Mechanical Engineering, October 1950.

2. Admiral Zumwalt's testimony during Senate Armed Services Committee hearings, reprinted in Aviation Week and Space Technology, 21 February 1972. See also quote from Sokolovsky's Soviet Military Strategy, reprinted in same edition of Aviation Week and Space Technology.

3. "Deterrence" in the sense used presupposes a close association with "defense." The interrelationship is discussed by Geoffrey Jukes in Chapter 26.

4. The concept of "disposable surplus" originally advanced by Michael MccGwire; for discussion of surplus of capability over minimum requirements, see MccGwire's "The Turning Points in Naval Policy Formation," December 1972, reprinted in MccGwire, ed., Soviet Naval Developments: Capability and Context (New York: Praeger Publishers, 1973).

5. Ibid.

6. The term is explained by Peter H. Vigor, "The Soviet View of War," October 1970 reprinted in MccGwire, ed., op. cit.

PART

IV

THE SOVIET UNDERSTANDING OF DETERRENCE AND DEFENSE

CHAPTER 25

THE SEMANTICS OF DETERRENCE AND DEFENSE

Peter H. Vigor

In the Western world, most of the analysts of Soviet defense problems have English as their mother tongue; the material on which they work is largely written in Russian. Hence there arise considerable, and often extremely important, problems of translation from Russian into English and vice versa, of which the most intractable at present appear to be those related to the concepts of "deterrence" and "defense." This paper is an attempt to make a modest contribution toward solving them.

Problems in semantics are always tricky; and it is usually wise to begin to attempt their solution by equipping oneself with an underpinning whose integrity cannot be questioned. In the case of the English language, such an underpinning would be the 12 volumes of the Oxford English Dictionary; in the case of the French, the six tomes of the Larousse; while in the case of Russian this function, I would suggest, is fulfilled by the 17 volumes of the Slovar' Sovremennogo Russkogo Literaturnogo Yazyka (A dictionary of the modern Russian written language), published by the Soviet Academy of Sciences between the years 1950 and 1965.

In what follows, therefore, I propose to take the definitions of that magnificent dictionary (hereafter referred to as SSRLYa) as being my underpinning. In other words, although I have consulted a total of 15 of the best-known of the dictionaries of the Russian language (from which in many cases I have received valuable additional information), I have taken the definitions of SSRLYa as being unimpeachably authoritative at that moment in time at which the volumes were published that contain them.

But time, of course, rolls on; and words that had a given, universally accepted meaning at one moment can acquire a very different one at a later period. While SSRLYa in the 1960s defined ustrashat as vyzyvat' strakh, pugat', it is still possible that in 1973

471

its meaning has become quite different. But before we can accept that its meaning has changed, we must, I suggest, be given evidence in favor of that supposition by those who wish us to adopt it. This point is likely to be of importance when we consider the word oborona.

It is generally agreed that there are two Russian words that correspond to the English "defense"; and it is often supposed, moreover, that the two Russian words are synonymous. This, however, is wrong.

The word zashchita, one of the two words for defense, is derived from the basic schchit, which means a "shield." The kind of defense that is represented by "zashchita" is therefore that of a shield or some other protection inserted between the threatening object or person and the person or thing that is threatened. Thus, a gas mask is a "zashchita" against gas; civil defense however (the organ supplying the gas mask) is grazhdanskaya oborona.

This is because "oborona," to quote SSRLYa, has as its first meaning "actions directed to defense (zashchita) from enemy attack." Civil defense, like AA defense, is an activity; and therefore "defense" here must be translated into Russian by "oborona." The same thing is true of battles. "The Defense of Stalingrad" is Stalingradskaya oborona, not Stalingradskaya zashchita; while the Ministry of Defense, which is supposed actually to do something to ensure the safety of the country, is Ministerstvo Oborony.

"Oborona" has as its second meaning, according to SSRLYa, a "system of defensive (zashchitnykh) works", or, in other words, "defenses" as distinct from "defense," while its third meaning is given as "the sum total of the means for ensuring the safety of the country," which is illustrated by a quotation from Saltykov to the effect that every state must have its "military defense and its budget." SSRLYa gives no other meanings for "oborona," nor indeed do any of the 15 other dictionaries, not even those that have been published as recently as 1972. If indeed, as some allege, "oborona" has recently extended its meaning to include "deterrence," strong supporting evidence will need to be provided.

The result of this has been that "oborona" has come to be used preeminently for military kinds of defense, while "zashchita" tends to be reserved for more pacific varieties. One's defense in a law court, for instance, is a "zashchita," and so is "the defense" at football. The case is most clearly put, I think, by an old Tsarist dictionary, Malyi Slovar' Russkago Yazyka, which, dealing not with the nouns but with the related verbs, defines zashchishchat' as "to ward off an attack, to protect, preserve"; while oboronyat' is "to defend ("zashchishchat'") someone or something with arms in one's hand" (emphasis mine).

There are, of course, exceptions to this general rule. Soviet writers can be as idiosyncratic as English writers; and one can find instances—in SSRLYa, as well as in less august publications—where "zashchita" and "zashchishchat'" are used in a military context, while "oborona" and "oboronyat'" are employed for nonmilitary defense matters. Nevertheless, these are undoubtedly exceptions; and the basic rule remains as stated above.

The English word "to deter" and its associated "deterrent" and "deterrence" cause many more difficulties for the Russians. There exists in Russian no word that adequately covers the full meaning of "to deter." Thus "to deter" means "to discourage or hinder by such means as fear, dislike of trouble, etc." (Concise Oxford Dictionary, 1964 ed.). Webster's Seventh New Collegiate Dictionary of 1963 gives a similar definition—"to turn aside, discourage, or prevent from acting (as by fear)." But the oldest English-Russian dictionary that I have been able to consult, Alexandrov's English-Russian Dictionary, published in St. Petersburg in 1899, can do no better than otvrashchat', sovrashchat', and finally, as a last resort, uderzhivat', meshat'.

Subsequent English-Russian dictionaries, whether published in London or Moscow, follow the same line. Thus, Segal (London, 1948) has "otvrashchat'" and "uderzhivat'";* the 1943 edition of Muller (published in Moscow) has "uderzhivat'" or otpugivat'; while the authoritative Bol'shoi Anglo-Russkii Slovar' published in Moscow in 1972 merely adds ostanavlivat' to the preceding. None of the other 14 dictionaries that I have consulted have anything further to add to the above suggestions, except for the Anglo-Russkii Voenno-Morskoi Slovar' (Moscow, 1962), which gives sderzhivat' and predotvrashchat' as additional options, and the Anglo-Russkii Voennyi Slovar' (Moscow, 1968) with a further suggestion of ustrashat'.

The trouble starts, however, when one embarks on that tedious process that in one's salad days at university was forced on one by one's supervisor. "Do not," I remember well being told by that wise old Russian who supervised me, "Do not accept as gospel what one half of the dictionary tells you, any more than a man of sense would accept as truth what the wife alone, or the husband alone, may tell you of a domestic incident. Check your story. If Alexandrov's English-Russian tells you that the correct translation of "to deter" is "otvrashchat'," "sovrashchat'," what does his Russian-English tell you is the meaning of "otvrashchat'" (to say nothing of "sovrashchat'") ?"

It is a fair question; but the answer to it is highly inconvenient for that same Alexandrov, who tells you that the Russian for "to deter"

---

*As well as a brilliantly idiosyncratic otsovietovat' strakhom ("to dissuade by fear"); but this had no successors.

is "sovrashchat'," and then proceeds to tell you that the English for "sovrashchat'" is "to corrupt, to debauch" (and as a matter of fact it is). Similarly, Muller's Anglo-Russkii Slovar' (Moscow, 1953) gives "uderzhivat'" for "to deter," while its Russian-English complement (Smirnitsky ed., Moscow, 1953) renders "uderzhivat'" into English as "to retain; to hold back from; to suppress (laughter etc); to deduct from (esp. wages)." The reader will note that "to deter" is not one of the meanings that are attributed to it.

The very latest English-Russian dictionary to come out of Moscow, the Bol'shoi Anglo-Russkii Slovar' of 1972, declares that "to deter" is to be rendered into Russian by "uderzhivat'" (a word we have just been dealing with) or by ostanavlivat', or by otpugivat'. But the very latest Russian-English dictionary, the Oxford Russian-English Dictionary of 1972, declares (quite rightly) that "ostanavlivat'" means "to stop" or "to restrain"; and that "otpugivat'" means "to frighten off," "to scare away." As to "uderzhivat'," which is the first meaning for "to deter" given by the Bolshoi Anglo-Russkii Slovar', the Oxford Russian-English Dictionary agrees with Smirnitsky both in its understanding of the various meanings properly to be ascribed to "uderzhivat'" and also in the fact that "to deter" is not to be considered one of them.

If one turns from the basic "deter" to the associated "deterrent" and "deterrence," the picture gets worse, not better. To begin with the latter, not a single English-Russian dictionary published prior to 1960 that I have been able to consult contains so much as an entry for it; so we may assume that, before 1960, the concept of "the act of deterring" was considered to be of but little importance in Russia.

Even in 1960, what was given was an explanation, rather than a translation. The dictionary published in that year was the Anglo-Russkii Voennyi Slovar', issued by the Soviet Ministry of Defense. It has no entry for "deterrence" as such, but it does have one under the letter "N" for "nuclear deterrence"; and this is rendered into Russian as either ustrashenie protivnika yadernym oruzhiem ("the frightening of an enemy with a nuclear weapon") or uderzhivanie protivnika ot napadeniya ugrozoi yadernogo udara ("the keeping of an enemy from committing aggression by the threat of a nuclear strike"). By 1968, no advance had been made; for that year saw the publication by the Soviet Ministry of Defense of a new edition of the same dictionary, where exactly the same explanation of 'nuclear deterrence" was given (no entry for "deterrence," as before), and where again no translation of the concept was even attempted.

Perhaps this was just as well; for in 1972 the Bol'shoi Anglo-Russkii Slovar' (not a military dictionary, but published by Sovietskaya Entsiklopediya) attempted a translation, and made a hash of it. For it gave either uderzhanie (which conveys the idea of "holding back," but no idea of "fear") or, as alternatives, otpugivanie and ustrashenie (which convey the notion of "inspiring fear," but not of "holding back"). And that, with "deterrence," is as far as we have got to date.

If we now turn to "deterrent," things are different. (Before I go on, I should make it clear that throughout this section I am omitting any consideration of "deterrent" in its adjectival form—as in a "deterrent appearance"—and considering it only as a substantive, in the sense of "that which deters.")

"Deterrent," then, is to be found quite early in Soviet dictionaries, though not in the Tsarist ones. Taube's Voennyi Anglo-Russkii Slovar' (Moscow, 1942) translates it as predokhranitel'noe sredstvo ("preventive means") or sderzhivayushchee vozdeistvie ("a restraining or containing influence"). But the following year Moscow put out another dictionary, edited by Muller, in which "deterrent" was translated as sredstvo ustrasheniya ("a means of inspiring fear"). This, it will be noted, was before the dawn of the atomic, let alone nuclear, age.

By 1960, however, the Soviet Ministry of Defense's dictionary was translating "deterrent" as sredstvo uderzhivayushchee protivnika ot napadeniya ("a means of restraining the enemy from attacking one"), though it retained as a third alternative the sderzhivayushchee vozdeistvie of the Taube dictionary. Despite this, however, the 1963 edition of Muller stuck, like the 1943 and 1958 editions, to sredstvo ustrashenie; though the Muller, of course, is not one of the dictionaries published by the Ministry of Defense.

The latter, however, with that doggedness (some might even say "obstinacy") that characterizes Ministries of Defense the wide world over, was not put out by what Muller had said concerning the correct translation of "deterrent." As a result, when it produced a new dictionary in 1968, the phrases it used to render "deterrent" into Russian were exactly those it had used eight years before.

We may ignore that leg of the dichotomy whereby "deterrence" and "deterrent" can be translated into Russian either by the appropriate compound of the verb for "to frighten" (strashit'), or else by that which is based on the verb "to hold" (derzhat'); neither, however, are good translations. Nonetheless, we are still left with the problem of why it is that the English-Russian dictionaries (including all those published in Moscow) prefer to translate these two important words by the appropriate forms of "uderzhivat'" (uderzhanie, uderzhivayushchii, and so on), when actual Soviet books on military and political matters prefer to use the appropriate form of sderzhivat' (sderzhivanie, sderzhivayushchii, and so on). Thus A. E. Efrimov's Evropa i Yadernoe Orzuhie (Moscow, 1972) is a history of Western policies in connection with nuclear weapons. Naturally, in the course of it he frequently quotes Western speeches and articles on the subject; and whenever he does, he uses "sderzhivanie" to translate the original "deterrence."

Now is this an exceptional, quirkish usage? A number of Western authorities have remarked that "sderzhivanie" is an equivalent for "deterrence" that is widely accepted today among Soviet writers and speakers; and Marshall Shulman reminds us that this is not just modern trendiness, but that as early as the late 1940s "sderzhivanie" was used for "deterrence"; though in those days, it seems, it was followed by putyom ustrasheniya.

In the course of the years, the phrase "putyom ustrasheniya" has dropped out of usage. Nor is this very surprising. In any language neologisms make their appearance in a much more self-explanatory, more long-winded format than later generations will tolerate, or indeed find necessary. The "wireless receiving apparatus" used in the late 1890s to describe Marconi's invention has been shortened to "wireless," which in turn has been shortened to "radio."

By a similar process we have reached the situation where Soviet military and political writers use "sderzhivanie" to express the English "deterrence," while Soviet lexicographers do not. Why this difference?

Any explanation must reckon with the fact that the lexicographers' preference, "uderzhanie," is a noun derived from the perfective aspect of its verb, while sderazhivanie derives from the imperfective. This, together with the slight differences in meaning between sderzhivat'/sderzhat' on the one hand, and uderzhivat'/uderzhat' on the other, might indicate that the military and political writers think the deterrent to be less of an effective deterrence than do the lexicographers, though it must be borne in mind that the deterrent to which this refers is the Western deterrent, the one directed against the USSR.

We can therefore conclude that the basic essence of deterrence, as expressed in the verb "deter," has no real equivalent in the Russian language. As a consequence, every attempt to translate it, whether made by a Russian or a Western dictionary, has so far unquestionably failed.

This failure to be able to express in Russian the essential notion of the concept "to deter" has naturally been reflected in a similar failure to express correctly in Russian either "deterrent" or "deterrence." In other words, the Russian mind is particularly ill-equipped to apprehend the notion of "the act of deterring" and not much better to apprehend that of "the thing that deters."

From here we can go on to say that in any case the Soviet Ministry of Defense, if we are to judge it by its dictionaries, recognizes no deterrence that is not nuclear, though it appears to be able to conceive of a nonnuclear deterrent.

We can go further. As I remarked before, the phrase used for these two words in the Ministry of Defense are not translations; they are explanations. These dictionaries were not compiled to assist Westerners to learn military Russian, but to help Russian officers to learn military English. If there existed in Russian proper equivalents for "deter," "deterrent," and "deterrence," the Russians would use them; as there does not, they concoct phrases to explain their meaning to those who require to know it.

And from this it followed, until very recently, that, by the sheer, clumsy mechanics of the phrases they have been forced to use, real debate on "deterrence" between Russians was bound to be a remarkably cumbersome business. One cannot think clearly with tangled knots of verbiage; and ordinary human usage is that, whenever a concept has to be seriously debated, the language has to be simplified to allow the debate to continue. Without such simplification there will be little meaningful debate.

But the Russians need to have meaningful debate about deterrence; so those concerned (chiefly the military) have had to produce for their own professional use an expression that their lexicographical compatriots have so far refused to recognize. This expression is "sderzhivanie"; and so widely is it accepted now that one can find it coupled with yadernoye ("nuclear") in yadernoye sderzhivanie. In view of the difficulties of translating "sderzhivanie" correctly, to which more than sufficient allusion has been made in the preceding pages, I will refrain from attempting a translation of this neologism; but I am prepared to say that, when read in the context of the only example of it that is known to me, it appears to represent with considerable accuracy our concept of "nuclear deterrence."

I hope that this rambling, and somewhat pedantic, discussion will have the result of gaining general acceptance for at least the following points:

1. "Zashchita" is a "defense" in the sense of some sort of shield or protective device inserted between the enemy and the intended victim.

2. "Oborona" is basically the activity involved in defending oneself, in inserting the shield between oneself and the enemy.

3. There is not, even today, a Russian word that properly expresses the two meanings of "inspiring fear" and "holding back an aggressor" that are contained in the English "deter," and for many years the Russians had to make do with clumsy paraphrases. This must inevitably have hindered the evolution in the Soviet Union of a proper discussion, and hence of a proper understanding, of the concept of "deterrence."

4. The Soviet lexicographers have remained in this unsatisfactory position; but the Soviet military, impelled by the needs of the practicalities of everyday life, appear to have adopted by mutual consent the word "sderzhivanie" as their own expression to translate the English "deterrence." This custom is by now so well established that the phrase yadernoye sderzhivanie is even to be found; and it seems to be the more or less exact equivalent of the English "nuclear deterrence." Further than that one cannot safely go at the present moment.

CHAPTER

26

THE MILITARY APPROACH TO DETERRENCE AND DEFENSE

Geoffrey Jukes

In attempting to outline Soviet use of the terms that correspond most closely to our "deterrence" and "defences," two pitfalls have to be avoided. One is that of laboring the obvious, the other that of not laboring it enough, which usually results from wrong assumptions about what is obvious. I propose to risk the first pitfall rather than the second, and to begin by restating some of the most obvious ways in which the context of Soviet defense decision-making differs from that to which we are accustomed, in the belief that to do so will help dispel any idea that our use of the term is in any sense the norm from which others "deviate." And because the terms are to be considered particularly in the naval context, I shall emphasize the respects in which the status and circumstances of the Soviet Navy differ from those that some of us have been conditioned by history to regard as normal.

THE CONTEXT OF MILITARY DECISION-MAKING

The Soviet Navy is not, and never has been, the "Senior Service." The head of it does not operate within a Chiefs of Staff framework in which all service heads enjoy hierarchical parity under a chairman or a chief of Defense Staff who may be drawn from any service, and a civilian minister or secretary of Defense. On the contrary, its head is a member of a larger body in which the leading positions (minister, c-in-c Warsaw Pact, and chief of General Staff) are now, and always have been, held by soldiers, or politicians with army experience.

It is not even called a "navy." The words normally translated as "Soviet Navy" are actually "Soviet Military-Sea Fleet," or "Soviet War-Sea Fleet," and there is no word to distinguish the navy as a whole from the "Fleets" of which it is composed. In fact, until 1937 it was known by the title of "Military-Sea Forces of the Workers' and Peasants' Red Army." This is not to suggest that the Soviet leaders regard admirals as simply "Generals at Sea," but that the Russian words "military" and "army" primarily mean "not civilian." There are, in fact, three ways to describe the totality of the Soviet armed forces. These are

    1. Vooruzhennyye Sily (Armed Forces), the term most often used by the career officers themselves.

    2. Armiya i Voyenno-Morskoi Flot (The Army and the Military-Sea Fleet), as for example in the title of the body charged with political supervision of the armed forces—The Chief Political Directorate of the Soviet Army and Military-Sea Fleet.

    3. Armiya (The Army). Brezhnev, for example, frequently uses this term in contexts that clearly include the navy, air forces, and so on.

Without laboring the terminological point unduly, it should be recalled at least that the terms imply less exclusivity than the English terms used to translate them.

There is no tradition of public debate on defense issues, in the media, the academic world, or the elected political bodies, nor are there interest groups such as "lobbyists" for competing arms firms— though limited competition exists between some of the aircraft design bureaus, for example, there is no evidence to indicate that members of elected bodies are enlisted in furtherance of it; in any event their ability to confer or receive favors is limited by comparison with their counterparts in some Western countries.

There is, therefore, no alternative forum or "court of appeal" for any service that finds its requests opposed by the army-dominated apparatus, except by recourse to the very summit of the political leadership. This recourse was occasionally open under Stalin and Khrushchev, but the practice of the Brezhnev-Kosygin regime up to now has been for detailed public pronouncements on military matters to be made by the appropriate member of the government, namely the minister of Defense, Marshal Grechko. This implies a stricter division of responsibility than was customary under previous leaders, and a diminution in service lobbying.

Assessment of strategic options, of the weapons systems needed to make them available, and the implications that technological developments may have on them, is essentially conducted by the military themselves (including defense scientists whose leaders, at least, hold high military ranks), subject to overriding control by the political leadership, which is now rarely exercised except on economic grounds.

Because of a tradition of secretiveness on military matters, which dates back well beyond 1917, the lack of a licensed vocal opposition, and the unavailability of "indirect" sources of support for the schemes of individual services, Soviet defense decision-making is more insulated than that of most other nations from extraneous influences and therefore more open to domination by a relatively small number of senior politicians and officers.

These men are overwhelmingly soldiers, or politicians who served in the army during World War II. Because of the very orderly career progression of armed forces officers in peacetime and, except at the very highest levels, in the Communist Party itself, these men tend to be older than their Western counterparts, evolutionary, rather than innovatory, in their thinking, and, because of their insulation from extraneous factors, under no compulsion to emphasize real or imaginary novelties in their decisions.

In this decision-making milieu there is no incentive to conceptualize defense as containing discontinuities between past experience and present reality, present and future, or peace and war. In this regard it is of particular importance to note that there is no motivation to regard "deterrence" of war as anything other than an outgrowth of war-fighting capability. An officially sanctioned definition of military doctrine of the Soviet Union views it as "a system of scientifically substantiated, officially approved views on preparation for and victorious conduct of war . . . compiled by the state political leadership with the assistance of the higher military bodies."[1] Note that this formulation does not mention deterrence at all. In this it expresses continuity with the prenuclear era, in which deterrence of attack was effected by manifest ability to fight and defeat an attacker—that is, was a by-product of perceived war-fighting capability.

In the naval policy context it should also be noted that there can be no independent "naval strategy" as such. However much Soviet naval thinking is influenced by that of potential adversaries, it cannot be translated into ship procurement and deployment unless it is acceptable to the "soldiers," and hence the most that can be expected is a naval contribution to an overall strategy formulated as mentioned in the preceding paragraph.

SOVIET MILITARY CONCEPTS

The effect of the various factors mentioned above is to produce patterns of thought and action that can be seriously misinterpreted unless we constantly bear in mind the underlying "continuity" between prenuclear and postnuclear thinking, between deterrence and defense, between peace and war. Some of the main features of this pattern are as follows.

By (Soviet) definition the Soviet Union is incapable of an act of aggression (some acts that do not conform to this definition have been rationalized away), but this does not mean it must be unprepared for war. Indeed, Stalin is blamed for having enunciated the proposition (itself a rationalization to explain the disasters of 1941) that "unaggressive nations must inevitably be in a state of low preparedness for war, the theory of active defence (i.e., retreat) . . . of the counteroffensive as an indispensable form of strategic operations."[2]

The stress on preparedness applies not merely to the strategic deterrent forces (the strategic rocket forces and the missile-firing submarines), but to all the other armed forces as well. They must be offensive-oriented from the outset.

Soviet military writers[3] recognize a theoretical possibility that Soviet forces could be involved in wars below the level of general nuclear war but do not attempt to elaborate categories of "limited war" and have greeted with derision Western attempts to devise "rules" for them.

This reluctance to contemplate watertight categories for war shows itself also in training programs and exercises, which attempt to integrate nuclear and nonnuclear operations rather than distinguish between them,[4] and which train troops for large-scale war rather than the counterinsurgency actions (East Germany, 1953, Hungary, 1956, and Czechoslovakia, 1968) that have been their only actual operations since 1945.

The desirability of eschewing certain types of procurement or deployment in order to affect adversary behavior is not publicly discussed. For example:

1. Initial deployments of strategic missiles (MRBMs, IRBMs, and ICBMs) were soft-sited, and soft-site deployment continued for longer than in the United States, no attention being paid, at least in public, to Western suggestions that soft-site deployments indicated first-strike intentions (because the soft sites could not survive an attack by an adversary).

2. A weapon claimed to be a mobile intercontinental ballistic missile (ICBM) was displayed in 1964. No response was made to Western suggestions that such weapons, theoretically easy to conceal and difficult to "count" could have a serious effect on the prospects for arms limitation, and though the missile was not deployed it was never suggested that it had been abandoned in deference to these arguments, despite the contribution that such a suggestion could have made to the Soviet image as a "peace-loving" state.

3. Antiballistic missile (ABM) deployment similarly continued for several years despite Western arguments against it as a destabilizing factor, and agreement to limit it came only in SALT I, by which time it had probably become clear in the USSR as in the United States that the technological and economic arguments against ABM were too strong to make its deployment worthwhile, except in limited

numbers for protection of some headquarters or sites against "accidental" launch of one or a few missiles.

4. Surface-to-air (SAM) missile systems and new interceptor aircraft continued to be developed and deployed to protect Soviet cities despite the downgrading of the manned bomber as a strategic nuclear weapon vector in both Western and Soviet air forces.

5. Construction of shelters against blast, fire, and fallout continued as a standard feature in new apartment blocks until the beginning of the 1960s.* When it ceased, the fact was not publicized, and its place was taken by intensified civil defense training programs, which gave prominence to rapid evacuation of cities in crises. This was despite the fact that a Soviet "peace offensive" was in progress and that the giving of publicity to the ending of basement shelter construction could have served to counter Western arguments that the shelter and training programs were evidence that the Soviet Union was preparing to wage aggressive or preemptive war.

6. The military is uniquely privileged within Soviet society in the extent of its freedom to express reservations about the government's conduct of policy. Even after due allowance has been made for the tendency of the military in all countries to be more "hawkish" than the population as a whole, the frequency with which Soviet military media denounce governments with which the Soviet Government is negotiating arms control agreements or general detente, as actively planning aggressive war against the Soviet Union, is remarkable.[5] So is the fact that denunciation continues despite the ammunition it provides for those in the West who believe the Soviet leaders to be using detente as a vehicle with which to "disarm" their adversaries.

7. In discussion of deterrence emphasis is laid on ability to retaliate in overwhelming strength and in a number of ways, rather than on the ability of the Soviet missile force to render attack pointless through manifest ability to survive it. This applies even where such statements contain preemptive overtones and can hence be interpreted as covering a possible intention to strike first, as, for example, a statement by the then Defense minister, Marshal Malinovskiy, in 1962: "The best means of defense is a warning to the enemy of our strength and preparedness to defeat him <u>at the very first attempt to commit an act of aggression</u>."[6]

---

*Withdrawal of the Civil Defense authorities, from the list of those authorities required to approve plans of new buildings actually took place in 1958; "pipeline inertia" was responsible for continued basement shelter construction in some areas until 1961.

DETERRENCE AND DEFENSE

In summary, Soviet usage does not distinguish rigidly between deterrence and defense. Soviet description of future war lays great emphasis on damage limitation—by swift action to occupy peripheral areas from which enemy attacks could be mounted, by nuclear attacks on targets of all types (enemy missile sites, airfields, troop concentrations, ports, transport bottlenecks, industrial, administrative and population centers, and by mass civil defense action in urban and rural areas to counter enemy nuclear, chemical, and biological weapons.[7] As Malinovskiy's statement indicates, manifest preparation for successful conduct of war is considered to make an essential contribution to deterrence. In this, Soviet practice differs considerably from that general in the West, where deterrence is seen to depend on making it clear that sufficient nuclear force would survive a Soviet first strike to eliminate the Soviet Union as an organized (and hence war-capable) society. Convincing the Soviets that NATO conventional forces could sustain a long conventional campaign, advancing to Warsaw or Moscow in the course of it, or that damage would be limited by Air and Civil Defense plays no part in Western deterrent doctrine.

The lack of a rigid distinction between deterrence and defense is apparent also in Soviet terminology, where no word for "deterrence" exists. For their deterrence of the West, the Soviets most commonly use the word sderzhivaniye (restraining); for Western deterrence of them they use the word ustrasheniye, which comes very close to meaning "intimidation." The Russian language, like the English, borrows foreign words freely, and the fact that so far they have not created a neutral word "deterrentsiya" to cover deterrence of either by the other suggests they do not feel a need for it. In normal usage they manage quite happily without a definite or indefinite article, nor do they distinguish between "foot" and "leg." On the other hand, the language is extremely well-equipped with words for expressing gradations—for example, between emotional states, and we should be careful not to strait-jacket their behavior to fit our own love of categorizing. Soviet practice and terminology suggest that deterrence is regarded as part of a spectrum of defense, rather than that defense is something resorted to when deterrence "fails"; that the maintenance of a high and manifest overall war-fighting capability forms an essential part of the deterrence concept; and that deterrence "shades off" into damage limitation in its literal sense, not in its special U.S. interpretation as a reciprocal process by which each undertakes to avoid certain types of target—for example, cities—if the other does the same. Such a concept emphasizes continuities—for example, between past, present, and future war, or between the various stages of the process deterrence-war-victory and the interdependence of the various armed services more than is normal in Western discussions of present and future strategy.

In attempting to evaluate the trends in Soviet naval development it is advisable to bear these points in mind; the alternative, which is to interpret them by transference from Western experience or by analogy from Western naval history, can lead the analyst very rapidly into error.

NOTES

1. Colonel P. Sidorov in Soviet Military Review, no. 9, 1972, pp. 14-15.

2. Marshal V. D. Sokolovskiy, ed., Voyennaya Strategiya (Military strategy), 2d ed. (Moscow, 1965), p. 9.

3. For example, Colonel-General Shtemenko in Red Star, January 3, 1963; Marshal Sokolovskiy and Major-General Cherednichenko, in Red Star, August 25-28, 1964; Marshal Yakubovskiy, in Red Star, July 21, 1967; and Colonel-General Povalii, in Red Star, March 13, 1968.

4. See J. Erickson, Soviet Military Power (London: Royal United Services Institution, 1971), pp. 65-73, for discussion of the concept of "coordination" in the context of the land battle.

5. The denunciations were particularly virulent in the mid-1960s (for example, Marshal Rotmistrov in Red Star, April 25, 1964, and Colonel Rybkin in Armed Forces Communist, no. 17, 1965, pp. 50-6), and still occur (for example, in M. Avakov, and S. Chernichenko, "The Vital Force of the Leninist Principles of Peaceful Coexistence," Armed Forces Communist, no. 23, (1973). In the latter example, the "imperialists" are said not to have given up their "traditional" foreign policy practices: "interference in the internal affairs of socialist states, economic blockade, aggression and all possible means from the Cold War arsenal."

6. Malinovskiy, Vigilantly Stand Guard in Defence of Peace, (Moscow: Voyenizdat, 1962), p. 25.

7. Sokolovskiy, ed., op. cit., pp. 364-401 and 440-43 (2d ed., 1965), pp. 332-68 and 400-403 (3d ed., 1968).

CHAPTER
27
SOVIET STRATEGIC
WEAPONS POLICY,
1955-70
Michael MccGwire

If we are to believe Soviet pronouncements, the Soviet Navy's shift to forward deployment in the early 1960s was in response to the West's increasing investment in seaborne strategic delivery systems. There is a great deal of material in this vein by political as well as military writers, appearing in major works such as <u>Military Strategy</u> as well as in newspaper and journal articles. Admiral Gorshkov stated the requirement very clearly in February 1963 in an article that concluded that the maritime defense of the Soviet Union would henceforth depend on naval engagements fought far from their shores and that the Soviet Navy must develop the capability for sustained all-weather operations on distant deployment.[1] A 1966 editorial in <u>Morskoi sbornik</u> is quite specific as to the reasons for this adjustment of policy:

> The world oceans . . . are now being converted . . . into a vast arena containing the launch points for highly mobile, secretive vehicles carrying strategic missiles and for mobile aircraft carrier forces. Hence, the struggle with the main forces of the enemy fleet is . . . taking on an immeasurably greater significance than it had in any of the past wars. The main content, and the main purposes of this struggle will be to destroy the aggressor's strike forces in the shortest space of time and to eliminate his ability to deliver his nuclear strikes from the sea.[2]

Admiral Kharlamov in an article discussing the problems involved in this kind of warfare, the importance of getting in the first strike, and the requirements for command and control, makes the point that as Minuteman and the Polaris take over, the aircraft carriers will be transferred to the "second wave" of the primary strategic delivery forces[3]—that is, they will be held back from the initial exchange for deferred strike.

Most recently we have it on Gorshkov's authority when talking directly to the officers of his fleet, that one of the navy's three main tasks in nuclear war is "to weaken the nuclear strikes by enemy naval forces." Over the last decade there has been a notable consistency in these statements, and in large measure their claims are borne out by the extending pattern of naval deployments between 1961 and 1970 and by the characteristics of the warships emerging from their new construction and major conversion programs.

Despite this body of evidence, there has been considerable reluctance in the West to take these statements at face value or to accept that the United States' seaborne delivery capability could be the primary determinant of this shift in Soviet naval policy. Much of this stems from difficulty in moving from Western to Soviet viewpoint, but there is one argument that must be taken seriously. The United States strategic ballistic missile inventory comprises 1,054 ICBMs and 656 SLBMs. Why should the Russians devote special efforts to countering the Polaris submarine when they cannot defend themselves against the ICBM? Similarly, why should they take special steps to counter the aircraft carriers in distant sea areas, when those aircraft are only part of the manned bomber threat to the Soviet Union, which is already the responsibility of PVO Strany?

The answer to these objections lies in the meanings the Soviets attach to such concepts as "world war" and "deterrence." Starting from the Soviet understanding of these terms, the purpose of this chapter is to examine the policies that would logically flow from them, and the consequential weapons requirements, and then to compare the postulated requirements with what the Soviets have actually produced, built, and deployed.

TWO BASIC ASSUMPTIONS

World War

Marxist-Leninist theory lays down that the initiation of war can only be justified if (1) the Soviet Union is virtually certain of winning, and (2) the gains outweigh the costs. World war is seen in communist theory as a global fight to the finish between imperialism and socialism and meets neither of these criteria.[4] Excepting the possibility of a preemptive strike, this implies that the Soviet Union will not deliberately initiate a world war and in fact their theoretical definition of such a war assumes that it will be started by the imperialists.

Communist theory and Russia's national interests coincide in this matter, and it is generally acknowledged that the prevention and avoidance of world war is a prime objective of Soviet foreign policy. But the possibility of world war remains and the penalty of losing such a conflict would be the extinction of socialism and, of course, Russia. Therefore, despite the low probability that the imperialists will initiate global conflict, the catastrophic consequences of defeat require the Soviet Union to cover this contingency and to be able to fight and win this type of war should the need arise. In a fight to the death, victory is synonymous with survival.

Defense Through War Fighting

Without going into details, it is fair to assert that in the West there has emerged since the war a strategic concept known as "nuclear deterrence," which has developed its own vocabulary, body of theory, and consequential policies. It started from the relatively simple idea of defending Europe against a conventional attack by Russia by demonstrating the capability and resolve to inflict "unacceptable damage" on the Soviet Union. This "deterrent" concept has become increasingly complex as on the one hand Russia developed her own strategic nuclear capability and, on the other, the theories were spun out to their logical ends. There is a clear conceptual distinction between what is involved in "nuclear deterrence theory" and the old traditional idea that there is a strong element of deterrence inherent in defense. And it is also fair to say that "deterrence theory" has conditioned strategic thought and attitudes in the West, particularly as concerns the possibility of war with Russia.

The Soviets do not have an equivalent concept to Western "nuclear deterrence," and they even lack a proper term for it. On the occasions when they do refer to the concept, it is as something particular to the West.[5] The reasons for this divergence in approach are interesting in themselves and help illuminate the important differences underlying Soviet and Western strategy. But in this chapter, we must concentrate on what I will call the Soviet Union's "strategic weapons policy," which fills the place held by "deterrence policy" in the West.

Soviet military doctrine does not separate out the idea of "deterrence" from the general concept of defense. Defense of the Soviet Union depends on the capability to repel (or at least absorb) any attack and then to go on and win the subsequent war. The Soviets are not looking to be attacked and obviously hope that their defense capability will be sufficient to discourage or hold back (sderzhivat') an aggressor, which is of course deterrence in its traditional sense. But the crucial distinction between this and "nuclear deterrence" theory is implicit in the Western comment that "if the deterrent is used, it will have failed." The Soviets do not entertain such ideas. While they would much prefer to avoid general war, should war come, their defense will only have failed if their armed forces are unable to recover and go on to victory.

The Soviet emphasis on defense through war fighting is central. While the West saw nuclear weapons primarily as a means of threatening "unacceptable damage" to Russia, the Soviet Union saw them as adjuncts to their war-fighting capability. Where the West thought in terms of credibility, argued about the merits of counterforce or countervalue, or worried about stabilizing and destabilizing developments, the Soviet Union thought in terms of achieving victory in war. It was not that the Soviet leadership wanted war any more than the West, and in fact both Malenkov and Khrushchev advocated "minimum deterrence" policies as a means of cutting back on defense expenditures. Neither were successful because the collective leadership was not prepared to rely for their defense on a theoretical construct, particularly when weapons' inventories in the West did not always match Western theory.

In his book Military Strategy, Sokolovskij did not discuss the Soviet strategic delivery capability in terms of "deterrent forces" but in terms of war fighting. He states that "strategic operation of a future nuclear war will consist of coordinated operation of the services of the armed forces and will be conducted according to a common concept or plan. . . . The main forces of such an operation will be the strategic nuclear weapons. . . ."[6] In discussing such nuclear missile attacks, he says that "the basic aim of this type of operation is to undermine the military power of the enemy by eliminating the nuclear means of fighting and formations of armed forces, and eliminating the military-economic potential by destroying the economic foundations for war, and by disrupting governmental and military control."[7] This is war-fighting with nuclear weapons.

IMPLICATIONS: OBJECTIVES IN THE EVENT OF WAR

World war, if it comes, would be a fight to the finish between the two opposing systems with victory synonymous with survival. For victory to have any meaning, this would require the continued existence of some kind of governmental apparatus, and of an economic and social base on which to rebuild a Socialist society. In terms of strategic policy these requirements have the following implications:

1. A high value is placed on reducing the amount of damage inflicted on the Soviet Union, in protecting the center of government and in preserving the population.

2. Western Europe assumes a crucial importance as an alternative economic base on which to rebuild society.

3. A high value is placed on destroying the West's war-fighting potential as well as her existing forces.

These national objectives generate the following military requirements:

1. Destroy Western strategic nuclear delivery systems at the outbreak of war in order to:
   a. Reduce the amount of damage inflicted on Russia.
   b. Deny the West the option of holding back strategic missiles, which could be used to influence the subsequent course of the war and to destroy Europe as an economic base.

2. Provide sufficient active and passive protection to the locus of central government to ensure its ability to function at an acceptable level. To provide more widespread passive protection so as to ensure the survival of an adequate proportion of the population, and a skeletal framework of national government.

3. Develop a concept of operations for the seizure of Europe at the outbreak of war, which will spare Europe's industrial and agricultural capacity, to the greatest practicable extent.

4. Destroy the West's war-making capacity, both the forces in being and its war potential.

OPERATIONAL CONCEPTS AND PRIORITIES

Given the above objectives and military requirements, we can postulate the policies that would flow from them, remembering always that the main aim is the survival of the socialist system and the destruction of the opposing one.

Defense against Air and Ballistic Missile Attack

Moscow is the center of government, and we would therefore expect it to have overriding priority in the provision of defenses against air and missile attack. Western-type concern for the theoretical merits of stability will not be a factor in Soviet decisions.

This is in fact what has happened. Deployment of the ABM-1 complexes around Moscow began in 1962,[8] within a year of the Soviets' deploying their first operational ICBMs. This is their standard reaction to any threat to Moscow and involves massive expenditures on the development of defense systems that are at or beyond the limits of contemporary technology. This same syndrome can be seen in the vast concentric complexes of SA-1 SAM sites that defended Moscow against free-fall bombers in the middle 1950s, and in the Griffin/Tallin system, which was developed against stand-off weapons like the canceled Skybolt.

It may well have been planned to extend some kind of ABM cover to other important areas (as they did with their SAM systems), but SALT has removed that option for the time being.

One would not expect ABM protection to be given to ICBM sites, since the Soviet Union has no theory of "mutual deterrence," which requires her to be able to ride out a U.S. strike. Considerable emphasis is given to the importance of getting in the first blow in nuclear war and, given the war-fighting orientation, a policy of "launch on warning" would be expected and is indeed implicit in some of their doctrinal statements.

Civil Defense

We would expect high priority to be given to civil defense and shelter programs. This is in fact the case.

Civil defense used to come under the Ministry of the Interior, and shelter facilities were included in all new apartment blocks until about 1960. In mid-1961 there was a major reorganization instituting a nationwide system with a central headquarters within the Ministry of Defense, headed by Marshal Chuikov who held that post until he retired in 1972.[9] He was replaced by the young and very promising Colonel-General Altunin, and the post was raised to deputy minister status.[10]

A large number of organizations are involved in the very active civil defense program, with heavy military participation both in terms of staff appointments throughout the country and of specialized military civil defense forces. Problems of civil defense are continuously publicized in DOSAAF's monthly journal and in various pamphlets and books, and at regular intervals in the national press.

The Seizure of Europe

There are only two reasons that could justify the Soviet seizure of Western Europe in existing circumstances. The one postulated in this analysis; or the belief that the benefits of conquering Europe would outweigh the costs of general nuclear war. The latter idea flouts both common sense and Marxist theory and is not really tenable.

We should also note that the seizure of Europe is only a self-evident requirement if it is intended to make subsequent use of the territory and its resources. Otherwise it would be simpler, more certain, and more cost-effective to ravage Europe with nuclear weapons as it is planned to ravage North America.

We would therefore expect the Soviet Army's war plans to center on the seizure of Europe using tactics and weapons that will limit the devastation of war. The concept of operations would avoid "meat-grinder" tactics except for destroying enemy forces on the main battle front. The emphasis would be on high mobility and deep penetrations, the ability to seize and hold areas deep in the enemy's rear, the extensive use of chemical weapons and emphasis on conventional rather than nuclear weapons away from the main battle front.

What we know of the Soviets' operational concepts accords with this general picture.[11] Their exercise scenarios usually start with defense against (Western) assault, moving over to the offensive as soon as possible. Nuclear weapons will be used to destroy enemy ground and air forces and will be accompanied by the simultaneous advance of high-speed strike forces moving round the clock, with armor and airborne forces striking deep in the enemy's rear. Rates of advance of 50-70 miles a day are planned and the ground forces are highly mobile and hard-hitting. They include seven airborne divisions with over 50,000 men, and chemical weapons are organic at all levels of the army.

Destruction of the United States War-Making Capacity

Given the concept of a fight to the finish, we would expect the Soviet Union to plan on destroying the United States' war potential as well as weapons and installations already in existence. Military targets located within the United States fall into two major categories: (1) U.S.-based nuclear delivery systems that can strike directly at Russia; and (2) the more general targets such as political and military command centers, communications systems, rear services, war industry, and the like. And we know from Marshal Sokolovkskij and other Soviet writers that these are indeed the targets they intend to attack.[12]

As long as the two categories of target are located within the same general area and are liable to destruction by the same type of weapon, we could expect the Soviet Union to adopt the single, most cost-effective weapons policy to handle both categories. But if the U.S.-based nuclear delivery systems are resited away from the other military targets, and are so designed that they are not liable to damage by the same weapons, then we would expect to see a shift in weapons' policy reflecting this new development. Exactly such a process can be seen in the first half of the 1960s.

1955-61. Between 1955 and 1961, with few exceptions, all U.S. targets of both categories were unhardened or "soft." Such targets are highly vulnerable to the radiation effects of a nuclear explosion, particularly heat. The very great majority of them either lay within the great conurbations, or were in or near major cities. Given such a situation, "area devastation" is the most attractive strategic weapons' policy, because the requirement for warheads is directly related to a known geographical area; the requirement can therefore be calculated in advance and, most important, it is finite and not dependent on the varying number of enemy targets. A further advantage is that the larger the warhead, the smaller the number of missiles required to cover a given area. And since in these circumstances, the increase in yield is roughly proportional to the increase in the area of devastation, there is no penalty to be paid for having fewer missiles with larger warheads.

(One has to distinguish between blast and other kinds of damage, and between radius and area. The radius of blast damage increases as the cube root of the yield, y; the area of blast damage increases as $\sqrt[3]{y^2}$. Unhardened targets are highly vulnerable to the other "prompt" effects, particularly the incendiary effect; the range of these effects increases approximately as the square root of the yield, or the cube root at large distances because of various attenuating factors. Thus, the incendiary effect increases roughly in proportion to the yield.)[13]

The first family of ICBM comprised the SS-7 and SS-8, of which 209 were deployed between 1961 and 1963 (See Table 27.1). These missiles carry a 5-megaton (MT) warhead, which was probably proved during the series of tests that preceded the moratorium in 1958.[14]

They were followed into service by the SS-9, whose 25-MT warhead would have been tested in the series that took place in September 1961.[15] At that time the Soviet Union claimed that she was developing missiles with warheads in the 20-100-MT range, which was lent substance by the size of the devices that were tested.[16] Both their rocket and their nuclear weapon development reflected a clear trend in policy toward ever larger warheads.

If we think back to the situation in 1953-58, which is probably the period when the relevant decisions were taken, such an area-devastation policy makes good sense. The manned bombers of Strategic Air Command (SAC) were the United States' primary means of nuclear delivery, and there were strong indications that these aircraft would continue to play a major role for many years to come.

TABLE 27.1

Pattern of ICBM Deliveries

	Soviet					Totals		U.S.				Totals		
End of Year	SS 7/8 5 MT	SS 9 25 MT	SS 11 1 MT	SS 13 1 MT		Annual	Cumulative	Atlas	Titan I	Titan II	Silos MM I	MM II	Annual Delivery	Cumulative	End of Year
1960								18					18	18	1960
1961	59					59	59	81	54				135	153	1961
1962	75					75	134	18		54	150		222	375	1962
1963	75					75	209				250		250	625	1963
1964		75				75	284	(-81)			100	200	300	844	1964
1965		75				75	359	(-36)	(-54)			100	100	854	1965
1966		75	5			80	439					100	100	954	1966
1967		75	175			250	689					100	100	1,054	1967
1968		13	175	60		248	937								1968
1969			250			250	1,187								1969
1970			250			250	1,437								1970
1971			75			75	1,512								1971
1972			25			25	1,537								1972
1973			25			25	1,562								1973
1974			25			25	1,587						(-171)		1974
Total	209	313	1,005	60		1,587		-	-	54	500	500	1,054		Total

Sources: Soviet figures in this table represent a realistic "production schedule," which has been derived from the following data: (a) "First Deployment" dates given in The Military Balance 1973-74 (London, IISS), p. 69; (b) the SALT information that SS-7 and -8 were deployed before 1964; (c) type totals given by chairman, Joint Chiefs of Staff in U.S. Military Posture for FY 1975, p. 10; (d) ICBM midyear totals for 1972-74 given in SALT press releases and the U.S. Secretary of Defense's Annual Report FY 1975, p. 50. Actual deliveries may not have run exactly as shown, but the table probably portrays the general pattern, assuming the correctness of the information on when each type of missile was first deployed.

U.S. missile development had concentrated primarily on the medium and intermediate ranges, which exploited the geographical advantages of their alliance structure. Prior to the jolt of Sputnik in 1957, the United States appeared to be giving relatively low priority to the development of intercontinental-range rockets for emplacement in North America, and those that were under development were intended for unhardened sites. And finally, there was the geographical concentration of targets referred to above.

Once having chosen an area-devastation policy, the use of very large warheads offered certain advantages: (1) the turbulence created by large-yield weapons would knock down a substantial proportion of those SAC bombers that were airborne over the United States, as well as destroying aircraft caught on the ground; (2) missile accuracy would be a relatively insignificant factor; and (3) the number of missiles required to devastate a given area was inversely proportional to the size of the warhead; hence fewer missiles would be needed. Given the extensive demands on the Soviet Union's aerospace industry, large warheads provided the maximum "defense" from a given production capacity.[17]

1961-70. The early months of Mr. Kennedy's presidency saw a reorientation of U.S. defense including a major adjustment to its strategic weapons policy. This included the sharp acceleration of the Polaris program,* and a doubling of the planned production rate of the new Minuteman solid-fuel ICBM, which was then completing development. Minutemen and the earlier Titan II were to be deployed in underground silos remote from existing centers of population. At the beginning of 1962, it was announced there would be 800 Minutemen in service by 1967, but in practice 800 missiles were deployed by 1965; a further 200 missiles were added by mid-1967.

This buildup of the U.S. strategic missile forces was remarked on by General Sokolovskij, writing in the fall of 1961 for the 1962 edition of Military Strategy, in which he quoted from President Kennedy's message to Congress on 28 March 1961.[18] In the 1963 edition Sokolovskij sets out a table showing the sharp jump in ICBM from 200 at the beginning of 1963 to 1,190 by the end of 1966, and he notes that Minutemen will be mounted in silos.[19]

Given the Soviet Union's war-fighting objectives, these changes in U.S. policy could not but generate a consequential change in Soviet requirements. The shift in emphasis toward seaborne systems and

---

*Between 1958 and 1960, fourteen Polaris had been authorized. On taking office, President Kennedy accelerated the program, and 15 additional units were authorized in 1961 alone.

the latter's increased capabilities was a separate problem that
generated its own responses. But in terms of the present discussion,
the new emplacement policy for U.S. ICBM had major implications
in terms of the Soviet's existing strategic weapons program.

Targets hardened to withstand pressure of 300 pounds per square
inch (psi) can only be destroyed by the blast of a nuclear explosion,
the effective range of which is some 30 times less than that of the
range of radiation effects of a 1-MT weapon against "soft" targets.
The range of the blast effect only increases as the cube root of the
warhead's yield, which means that when planning to destroy hardened
silos set 5-10 miles apart, each silo must be individually targeted.
In such circumstances the optimum size of warhead (which determines
the range of blast effect) will depend on the missile's accuracy. But
because of the cube root relationship, it is more cost-effective to
invest in improved accuracy than in increased yield.

We would therefore expect to see a clear-cut shift in Soviet
strategic weapons' policy from area devastation using large warheads,
to one of individually targeted small warheads. We would also expect
there to be some correlation between the number of the smaller Soviet
missiles produced, and the U.S. ICBM inventory, and to see attempts
to improve the missile's kill capability by some method or another.

Meanwhile, the more general category-two targets would still
require to be attacked. Since there has been no change in this respect,
the former policy of area devastation could continue to apply and we
would therefore expect to see the larger missiles being retained in
the inventory and in due course being replaced by a second generation.

This is very much what has happened. The trend toward larger
warheads was reversed with the deployment in 1966 of the SS-11
carrying a 1-MT warhead. Deliveries of ICBM rose sharply from
about 75 a year to about 250 a year. Production continued at this rate
for four years (1967-70); by then, over 900 1-MT weapons had been
deployed. Deliveries then dropped abruptly, and between 1973-74
only 25 missiles a year were deployed.

This pattern has two features that are of interest to the postulated
weapons' policy:

1. The annual delivery rate of 250 SS-11 is about the same as
the average delivery rate of Minutemen between 1962 and 1965. This
may only be a coincidence. But it would also be the logical corollary
to a weapons' policy that requires each U.S. ICBM to be targeted.
Unless the Soviet production rate could match the U.S. one, the United
States would have the option of outbuilding her and hence ensuring
that some U.S. missiles would survive the initial exchange.

2. The mid-1974 total of SS-11 and SS-13 (the 1-MT solid-fuel
ICBM, of which only 60 were deployed) was about 1,053[20] and SALT
allows 1,096 in all. This is very close to the U.S. inventory of 1,054
ICBM.

If we allow that the Soviet leadership appreciated the new requirements by mid-1961, this implies five years' lead time before the deployment of the new missile. This would seem reasonable in view of the radical shift in the development priorities from range and payload to terminal accuracy, which was a consequence of being forced to rely on blast effect only. They were not fully successful in their efforts, and at present only the first and second modifications of the 25-MT SS-9 have the combination of yield and accuracy needed to attack hard targets effectively.[21] But this requirement may provide an explanation of what are seen in the West as anomalies in the development of Soviet MRV and MIRV.

Because a missile's payload is finite, the yield of a single reentry vehicle (RV) is always appreciably more than the total yield of an MRV or MIRV. The latter's radiation effects will therefore cover a smaller total area than a single RV, and except for the purpose of foiling ABM, multiple RVs are not appropriate to an area devastation policy. However, if MRVs are closely clustered in the impact area, the blast effect for a given payload is increased. This is partly because of the cube root rule and partly because of the compounding effects arising from the interaction of the separate explosions. In view of the inadequacy of the SS-11 and -13 as counter-silo weapons, this might explain the attempt to MRV the 25-MT SS-9. It might also suggest that MIRV-ing the successors to the 1 MT SS-9 and -11 (SS-X-16, -17, -19) may have as much to do with increasing their lethality against missile silos as any other reason.

Overview. The available evidence would seem to support the hypothesis that between 1955 and 1961, the Soviet Union's targeting policy against the United States was one of area devastation. After 1961, reacting to the buildup, dispersion, and hardening of the U.S. inventory, the Soviet Union set about building up its own counterforce, with each U.S. ICBM individually targeted.

The hidden effects of the 1961 shift in policy are not known, but it is possible that it was originally planned to deploy even larger warheads than the 25 MT SS-9.

(The SS-10 ICBM, which was never deployed, was somewhat larger than SS-9. This missile was displayed in a Moscow parade in 1964. It may have been a "competitive development" with SS-9, or it could possibly have been a canceled program.)

It seems likely that present policy is a combination of area devastation for the soft, category-two targets, and individual targeting for ICBM. It is relevant that the Soviets have developed a successor to the SS-9, one variant of which carries a single large RV.

The Threat of Deferred Nuclear Strikes

Strategic weapons that are held back from (and survive) the initial exchange for purposes of "deferred strike" could well determine the outcome of the struggle. To some extent, this would depend upon the progress of the war and whether any kind of reinforcement or logistic resupply was feasible once battle is joined. But a deferred strike could be the final blow that tipped the balance between the troops locked in combat, between government control and chaos, between survival and obliteration.

These considerations apply to both sides. However, for Russia, the implications of deferred nuclear strikes are immeasurably more serious because of Europe's central role in strategy. In the initial stages of the war, the extent to which Europe was devastated would largely depend upon the policies adopted by the Soviet Union. By choosing a selective weapons policy, limiting military operations to essential areas, and using the diplomatic tools of bribery, blackmail, and coercion to their fullest extent, in some measure she should be able to control the extent of devastation in Western Europe and ensure that certain parts are left largely unscathed apart from fallout.

It is the availability of Europe that allows Russia to plan on the assumption that a general war need not imply mutual suicide. The United States has no comparable option. She cannot count on moving into Canada because it will be Russia's decision whether or not Canada is to be spared. Canada is a member of NATO and NORAD, and a potential base for the revival of Western society. There would seem ample reason for including her within the North American target complex.

As we have seen, the targeting policy for Soviet ICBM is designed to ensure, among other things, that no strategic weapons emplaced in the United States survive the initial exchange. Shorter-range, land-based systems in Europe are likewise targeted. Those air-delivered weapons that manage to get airborne will only be available for as long as the aircraft have fuel to fly. It is in such circumstances that seaborne strategic nuclear weapons assume a unique significance. They are the only systems that the United States can withhold from the initial exchange with some certainty that they will be available to be used for deferred strikes. Or rather, such certainty will exist unless and until the Soviet Union takes positive measures to counter these forces and to deny them the option of withholding.

The immediate implications in terms of Soviet naval requirements are obvious. But one final and important point should be noted. To achieve the objective of "damage reduction," it is necessary to prevent Western maritime strike units from launching at least some of their weapons at Russia. In operational terms, this is very demanding. Paradoxically, the more important objective of denying the West the option of withholding nuclear weapons for use at a later stage of the war is somewhat simpler. It is only necessary to pose a level of threat that is sufficient to persuade the West that it is more prudent to use its weapons while it may.

CONCLUSIONS

This chapter set out to examine the validity of the Soviet Union's assertions about the reasons underlying its navy's shift to forward deployment. Starting with two of the key Soviet assumptions about the defense of Russia and the nature of "world war," we deduced the policies that logically should flow from them, and then looked to see whether the evidence supported the existence of such policies. In general terms the answer must be "Yes," and the range and diversity of the evidence that points to that conclusion are impressive.

It has long been realized that Soviet defense policy was under intensive debate during 1960-61; it has been generally assumed that this stemmed from the army's opposition to the new emphasis on strategic rocket forces given by Khrushchev at the end of 1959. While this was undoubtedly one reason, the present analysis suggests that at the root of the debate lay the new strategic capability being developed in the United States, which was thrown into sharp relief by the abrupt acceleration of programs following Kennedy's inauguration. Now, in addition to the general suggestion of discussion and change within the Soviet armed forces, we have three significant indicators that 1961 was a major turning point in the Soviet Union's strategic weapons' policy.

The most clearly defined is the major reorganization of civil defense in 1961, its subordination to the Ministry of Defense, and the appointment of Chuikov as its head in August that year; but this only becomes really significant when viewed in conjunction with the other two indicators. The second is the navy's shift to forward deployment. And although the argument continued until 1963/64 about what the navy needed to be able to carry out these new tasks, there is clear evidence that the decision in principle was taken before the end of 1961. The third and most significant indicator is the evidence of a major shift in targeting policy, in reaction to the new weapon's policies adopted by the United States after Kennedy's inauguration. Until 1961 there is a clearly established trend toward larger missiles and warheads, suitable for area devastation, and the projection of this trend can be seen in the nuclear tests carried out in 1961. This trend is then abruptly reversed by the introduction of a relatively small missile suitable for individual targeting, with a production rate and final inventory very similar to the ICBM force built up by the United States since 1961. It is not satisfactory to explain this sudden shift of policy by reference to Cuba or to Khrushchev's demise, since the "objective" factors of the strategic situation had not changed. The only thing that had changed, and changed radically, was the number, nature, and location of targets that had to be covered. The decision must therefore be dated back to 1961.

What changed in 1961 were the Soviet policies concerning the targeting of strategic weapons, civil defense, and naval countermeasures; but the principles and objectives of Soviet defense policy remained unchanged. The continuity of that policy since the war has been striking, and the "objectives" outlined in the early part of this chapter could have applied to any period since 1945. And unless Soviet policy has changed drastically as the result of SALT, these objectives are likely to still apply today.

The vital role of Europe in Soviet contingency plans in the event of world war would seem to be central to their strategic weapons' policy. While "damage reduction" is an important task in itself, it is secondary to the requirement that Western Europe be preserved as a safe haven in which to rebuild the socialist system. Strategic resources must therefore be assigned to the function of limiting the U.S. option of withholding nuclear weapons for deferred strike; this can be done by reducing to very low odds the probability of these weapons' survival. This requirement applies to land-based as well as seaborne systems, and Western efforts to develop a "second strike" capability merely ensure further Soviet endeavors to erode the certainty of this capability. In terms of nuclear "deterrence theory," this would seem to imply an attempt to achieve a "first strike" capability, with all the aggressive connotations implicit in the term. In terms of Soviet strategic (war fighting) doctrine it is a logical defensive response.

Inherent in a U.S. policy of dispersed and hardened ICBM silos is a Soviet strategic requirement for sufficient weapons to target each silo independently plus additional weapons to dispose of the remaining (category two) military targets in the United States by area devastation. In such circumstances, sufficiency for the Soviet Union of necessity implies some form of numerical superiority. But this is for objective military reasons and not the result of some deep emotional Russian response.

This chapter has outlined the strategic weapons' policy the Soviet Union had adopted to meet its strategic requirements. A large gap often exists between these doctrinally derived requirements and the actual Soviet capability to fulfill them, and this might seem particularly true of the need to counter Polaris. But we should bear in mind Soviet willingness to expend substantial resources in attempting to meet such requirements, even when the likely returns are low and the chances of success seem small. This glaring discrepancy between Soviet capabilities and their stated objectives frequently leads the West to dismiss the latter as unreal and to describe Soviet objectives and intentions in terms of Western strategic concepts that bear little relation to Soviet requirements. Above all we should remember that Western scepticism has been confounded in the past and that on several occasions the Soviet Union has gone a long way toward achieving what were seen in the West as unrealistic strategic aims.

NOTES

1. Krasnaya zvezda, 5 February 1963.
2. Morskoi sbornik, July 1966, pp. 3-7.
3. Ibid., January 1966, p. 33.
4. Peter Vigor, "The Soviet View of War," in M. MccGwire, ed., Soviet Naval Developments: Capability and Context (New York: Praeger Publishers, 1973), pp. 17, 18, 22. I have drawn on my contribution to that chapter (pp. 26-29) in the following paragraph.
5. See Chapters 25 and 26 in this volume.
6. V. D. Sokolovskij, Voennaya Strategiya (Moscow, 1968), pp. 346-47.
7. Ibid., p. 349.
8. U.S. Congress, Senate Armed Forces Committee, Military Authorization Hearings, FY 1974, p. 240.
9. E. L. Warner in "The Institutional Setting of Soviet Defence Policy," draft Chapter 2 of unpublished Ph.D. dissertation, note 88. Chuikov was appointed 17 August 1961 but retained his existing title of C-in-C, ground forces, until August 1964; he retired at the age of 72. This brief survey of civil defense draws heavily on Warner, who in turn draws on Leon Goure, Soviet Civil Defence Revisited 1966-69, RM 6113 PR (Santa Monica, Cal.: Rand Corporation, 1969).
10. Altunin took command of the North Caucasian Military District in 1968 with the rank of lieutenant general at the age of 47. Two years later he was promoted to colonel general and transferred to the Ministry of Defense as chief of the Central Directorate for Personnel, a thoroughly political position. A. O. Ghebhardt and W. Schneider, "The Soviet High Command," Military Review, May 1973.
11. See J. Erickson, Soviet Military Power (London: Royal United Services Institute, 1971), pp. 65-73, and his "The Soviet

Concept of the Land Battle," in The Soviet Union in Europe and the Near East (London: RUSI, 1970), pp. 26-32. Also G. H. Turbiville, "Soviet Airborne Troops," Military Review, April 1973.

12. V. D. Sokolovskij, Voennaya Strategiya (Moscow, 1968), pp. 244-45, 342, 349-50. These explicit statements about the targets that are to be attacked by strategic nuclear weapons have remained unchanged in the 1962, 1963, and 1968 editions.

13. H. F. York, Arms Control (San Francisco. W. H. Freeman, 1973, pp. 130-31.)

14. There was an accelerated test schedule in the fall of 1958, immediately preceding the test ban talks. This included three explosions in the MT range in six days, followed by six more nuclear tests in nine days. A. Kramish, Atomic Energy in the Soviet Union (Stanford, Cal.: Stanford University Press, 1959), p. 126.

15. Pravda, 31 August 1965.

16. Explosions up to at least 25 MT were detected, and these included the testing of a triggering device for what was reputed to be a 100 MT warhead. See R. M. Slusser, The Berlin Crisis of 1961 (Baltimore: Johns Hopkins Press, 1973), pp. 183-203.

17. It may also be "cheaper" by Soviet economic accounting. In Western terms the cost of a missile increases in proportion to its weight, which is a function of range and payload. See York, op. cit., p. 132.

18. V. D. Sokolovskij, Soviet Military Strategy, Rand Corp., Translation (Englewood Cliffs, N.J.: Prentice-Hall, 1963), pp. 172-77.

19. V. D. Sokolovskij, Voennaya Strategiya (Moscow, 1963).

20. U.S. Secretary of Defense, Annual Report FY 1975, p. 50.

21. Chairman, Joint Chiefs of Staff, U.S. Military Posture for FY 1975, p. 12.

PART
V
ASPECTS OF SOVIET
NAVAL POLICY

CHAPTER

28

THE EVOLUTION OF SOVIET NAVAL POLICY: 1960-74

Michael MccGwire

The analysis of Soviet naval policy is best seen as an ongoing hypothesis. The depth of our understanding depends largely on the extent of our hindsight, and as time goes by, we gain perspective, fragmentary evidence begins to accumulate in meaningful ways and trends become clearer; even more important, we begin to perceive the operational and hardware outcomes of decisions taken several years before. During the years 1970-73, a great deal of new evidence has become available in the shape of routine and crisis operational deployments, new classes of warship entering service, and substantial public pronouncements. Fresh analytical approaches have shed new light on past events. Meanwhile, it has become increasingly clear that 1961-64 was a period of fundamental decisions concerning Soviet defense policy, which had a major impact on other branches of the armed forces besides the navy. It would also seem that 1970-74 has seen a heated discussion on the role of the navy in war and peace as part of a wider debate on the future shape of Soviet foreign policy and the part to be played by the Soviet armed forces. The purpose of this chapter is therefore to offer an updated hypothesis of Soviet naval policy to take some account of the new material that is available.

BACKGROUND

In 1945, the Soviet Union had established her military perimeter across the narrower part of Europe but her maritime flanks were uncomfortably exposed. The Baltic coast bared the lines of communication to her forces in the west, while the Black Sea gave directly on to her industrial hinterlands, outflanking her natural defensive barriers and the advantages of space. In neither area did Russia have a battleworthy fleet, while her likely opponents were now the 'traditional maritime powers," which had recently demonstrated their capacity to project and sustain continental-scale armies over vast distances of sea. The likelihood of maritime invasion was considered substantial; naval requirements were carried over from before the war, as Russia embarked on rebuilding a large, mainly conventional navy with heavy emphasis on "medium-type" submarines.[1]

In 1954, as a consequence of the post-Stalin reevaluation, the Soviet leadership downgraded the threat of seaborne invasion and gave first priority to the dangers of surprise nuclear attack. This engendered a radical reappraisal of naval requirements and the decision to place primary reliance on long-range cruise missiles, to be carried by surface ships, diesel submarines, and aircraft. The operational concept relied on the reach and payload of these weapons (which had still to be developed) to substitute for tactical mobility and mass; this would allow resources to be released from warship construction to the domestic economy.

Khrushchev brought 45-year-old Gorshkov to Moscow to implement these decisions, which had been strongly opposed by the commander in chief of the navy (Kuznetsov), whom he replaced.[2] The building of cruisers was halted in midcourse, the mass-production of medium-type submarines was sharply brought to a halt, and while the destroyer, escort and subchaser programs ran their full course, their successor classes were put back four years. At this same period, the Soviet Naval Air Force was stripped of its fighter elements, which were transferred to the newly formed National Air Defense (PVO Strany). The new concept of operations was predicated on engaging the enemy carrier groups within range of shore-based air cover, and envisaged a coordinated missile attack by strike aircraft, diesel-submarines, and light cruisers. These newly designed units would begin to enter service after 1962.[3]

If the 1954 decision had been fully implemented, the result would have been a task-specific, defensively oriented navy, more firmly tied to home waters than at any time since the 1930s, with strategic delivery submarines as the sole exception. However, by 1958 the basic premise that shore-based air support would be available over the encounter zone, had been falsified by the increase in aircraft range; this allowed U.S. carriers to strike at Russia from the Eastern Mediterranean and South Norwegian Sea. To meet this threat from distant sea areas, it was decided to rely on nuclear submarines and plans to double their production capacity were put in hand for deliveries beginning in 1968. The cruise-missile diesel submarine programs were canceled (that is, the J-class and Longbin SSG), and as an interim expedient, the J-class missile systems were used to reconfigure the second generation of ballistic-missile units to SSGN.[4] The ill-founded decision to gamble on long-range surface-to-surface missiles (SSM) as the primary armament of the fleet (a mistake the Soviet Navy is now finally working out of its system) was reversed, and the development of horizon-range systems with organic target-location was put in hand. At this same period, the requirement for a sea-going helicopter platform to extend the range of airborne ASW coverage in arctic waters generated the decision to build the Moskva class, which the Soviets classify as an antisubmarine cruiser.[5]

1960-1964

1960-61: The New Strategic Threat

The new defense policy announced by Khrushchev in January 1960, placed heavy emphasis on nuclear delivery systems and tended to downgrade the role of conventional ground forces. Repercussions on naval policy were few. It reaffirmed the navy's contribution to the strategic strike forces. It also confirmed that the submarine would provide the main defense against attack from distant sea areas, and (by implication) that the Soviet Navy was not intended to challenge the West's worldwide maritime capability.

Within 18 months, the situation had changed dramatically as the result of a thoroughgoing reevaluation of the strategic threat. At the root of the Soviet defense debate was the new nuclear delivery capability being developed by the United States in the form of Minutemen ICBM and Polaris submarines. Already a cause for concern, it was thrown into sharp relief by the abrupt acceleration of the programs ordered by Kennedy on taking office in January 1961,[6] raising serious questions as to the U.S. intentions.* Soviet military doctrine does not separate out deterrence from the general concept of defense, and defense of the Soviet Union is premised on the capability to throw back any attack, and then go on to win what would inevitably be a world nuclear war between the two antagonistic social systems.[7] Since "world war" is seen as a fight to the finish, victory is synonymous with survival, and it is in such circumstances that Western Europe assumes a very special place in Soviet military strategy. In the

---

*The Soviets were aware of the U-2 flights and would have assumed that U.S. authorities knew that the "missile gap" did not exist. From the Soviet viewpoint, therefore, the United States must have been embarking on this buildup for reasons other than "balancing" or "deterrence."

initial stages of such a war, the extent to which Western Europe is devastated will largely depend upon the policies adopted by the Soviet Union, and it is the potential availability of this partly undamaged area on which to rebuild the socialist system that allows Russia to plan on the assumption that if "world war" is forced on her it need not imply mutual suicide.[8]

With the deployment of the SS-7 ICBM in 1961, the Soviet Union had at last acquired a sure means of striking directly at North America, her weapons' policy giving priority to targeting all nuclear weapons that could be launched against Russia. Now she was faced with the new problems of the dispersal and hardening of U.S. ICBM sites, and a major shift in emphasis toward seaborne nuclear delivery systems. The latter were particularly suitable for withholding from the initial exchange and would then be available to deny Russia the use of Europe.

If the seaborne component of the West's nuclear delivery capability was to be countered, the Soviet Navy would have to move forward in strategic defense. Gorshkov explained the requirement in February 1963;[9] commenting on the West's increasing investment in seaborne systems, he concluded that the maritime defense of the Soviet Union would henceforth depend on naval engagements fought far from her shores. As he said in 1966, one-third of the U.S. nuclear strike weapons were seaborne, adding (nine months later) that by 1970 this proportion would have risen to one half.[10] The task of Soviet naval forces on forward deployment would be to prevent the launching of nuclear strikes against Russia to the extent possible.[11] But even more important, it would deny the United States the option of withholding such systems from the initial exchange, for use at a later stage of the war.

The fleet was ill-equipped and ill-prepared for such a radical shift in operational requirements, and there can be little doubt that the decision to shift to forward deployment reflected strategic imperatives rather than long-stifled "traditional" naval aspirations. There are no indications that the Soviet Navy had been straining at the political leash, while preparing itself for an oceanic role through local operations and training. It is evident from the elementary nature of the sea-keeping capabilities that the navy was now exhorted to acquire[12] that the opposite was the case. In fact Gorshkov later admitted that the new requirements demanded the "organic restructuring of the fleet and the reorientation of traditional naval policy and operational habits."[13]

Short-Term Measures

Whatever might be decided in the longer term, something had to be done meanwhile. By January 1961, the first Polaris submarine had only recently sailed on its first patrol, and the strike carrier was still seen as the primary maritime threat to Russia. Four of the Forrestal class had entered service between 1955-59, and two of the Kittyhawk (improved Forrestal) class were due to enter service in 1961 as well as the nuclear-powered carrier Enterprise, while a third Kittyhawk had already been authorized in 1960. The Soviet Navy was very conscious of the threat inherent in Polaris, but the military was less easily persuaded, perhaps because of the inadequacies of the Soviet SSBN, and continuing Western doubts about Polaris, which were orchestrated by the U.S. Air Force.

The area needing most immediate attention was the South Norwegian Sea and Greenland/U.K. Gap, which harbored both threats, although the Polaris one was still embryonic. In this area it was practicable to plan for the progressive extension of the Northern Fleet's defense zone to beyond the Iceland/Faeroes gap, and for the sailing of missile units to mark the carriers whenever they came within strike range of the Soviet Union. Such an approach would not be possible in the Mediterranean, and if the Sixth Fleet carriers were to be countered by ships (as opposed to land-based missiles or aircraft), some form of continuous presence would be necessary.

In 1961, the Soviet Navy was at a low ebb as the result of the cutbacks in naval construction resulting from the 1954 decisions, and the further disruption arising from the reversal of plans in 1957/58. After a four-year hiatus, surface ship deliveries were due to recommence in 1962, with the Kynda-class missile cruiser and the Kashin-class SAM-armed destroyer each building at the rate of two a year. In the submarine field, technological difficulties had exacted their toll. The first generation of submarine-launched ballistic missile (SLBM) and of nuclear submarine had both fallen short of the escalating operational requirements. Five of the first and all of the second generation of nuclear submarines had been reconfigured to carry long-range SSM, and from 1962 five to six of these E-class SSGN would be entering service each year. Diesel submarine deliveries were running at eight units a year, of which two would be J-class SSG.

About the only thing going the navy's way was the 1957/58 decision to double the production capacity of nuclear submarines, which meant that 11 units a year would begin delivery in 1968. But that was still seven years ahead and in the meantime it was necessary to maximize the utility of what was already available.

The 1961 Decisions

Concern that the United States was trying to build up a first-strike (disarming) capability against Russia, coupled with the apparent success of Polaris, highlighted the potential of an SSBN force as a relatively invulnerable element of the Soviet strategic delivery capability.14 It seems likely that the new generation of nuclear submarines, due to begin delivery in 1968 at the rate of 11 units a year, was originally planned in the proportion of three SSGN, five-six SSN and two-three SSBN; this reflected the 1957/58 assessment of future operational requirements. The new concern over the dangers of surprise attack generated a reversal in the priority given to SSN over SSBN, resulting in the 1961 decision to program the annual production of six Y-class, two V-class, and three C-class.15

Second, in view of the inadequacies of existing Soviet ballistic-missile submarines in their designed role of strategic delivery, and the inadequacies of the long-range cruise-missile system as an anti-surface weapon, it was decided to transfer the H-class SSBN and G-class SSB to the anticarrier role.16 This would be effected as these classes were retrofitted with the submerged-launch 650 nm SS-N-5 SLBM system. (This system was originally intended for the second generation of SSBN, which was reconfigured as SSGN; it was retrofitted in H-class between 1963 and 1967, and thereafter in a dozen or so of the G-class. Patrols were initiated in 1964.)17

Meanwhile there was a pressing requirement to improve the capability of Soviet surface forces to carry out sustained operations in a potentially hostile environment;* on the other hand, with the E-II SSGN program in hand, there would be no shortage of SSM launchers, particularly in relation to the target location capability. It was therefore decided to cancel the Kynda program, completing only 4 of the 12 ships planned. This would allow (1) the major conversion of 8 Kotlin-class destroyers to SAM-armed ASW ships, using the weapons procured for the Kynda program; and (2) the use of Zhdanov Yard's building ways to provide a two-year boost in the delivery of Kashins, the only class that had the required capabilities. It was decided not to cancel the Kynda's successor, Kresta, whose ASW capability was somewhat better than Kashin's, and her fully automatic gun systems were a considerable improvement. The Kresta's characteristics were however redesigned to incorporate an organic target location capability for its long-range SSM, in the shape of helicopter-borne radar. The provision of helicopter facilities must have necessitated some rearrangement of the weapons' layout; this may have involved the sacrifice of a second SSM system but may also have allowed the fitting of a second SAM system and increased magazine capacity for all self-defense systems.

---

*In the early stages, target-location information on the carriers had to be provided by Soviet destroyers maintaining close station on them.

Meanwhile, other plans were held in abeyance until the longterm policy had been formulated. Of the ships under construction, the only one affected was the Moskva, and work on her was stopped until her future usefulness was clear. But in 1961, initial procurement planning would have been well in hand for the successors to these classes then building and due for delivery between 1962 and 1969. These successor classes would begin delivery in 1970 and procurement for their long-lead items would have to come within the current seven year plan (1959-65). It seems likely that at this time it was not intended to provide a successor to the Kynda/Kresta missile cruiser since it had been decided in 1957/58 to place primary reliance on nuclear submarines. It is however fairly certain that the need for a destroyer-sized unit to succeed Kashin (and replace the Skorys and Kotlins) was accepted as a routine requirement, although the details of the armament had probably still to be finally agreed.

The Long-Term Requirements

The operational requirements were daunting. To be able to attack Western strike units at the outbreak of war, Soviet forces had to be in weapon-range contact with them at the vital moment. In dealing with the carrier, there would be little difficulty in maintaining contact; if surface ships were used, the main problem was how to survive in the potentially hostile environment. Remaining in contact with the Polaris submarine was another matter, and in 1961 the difficulties of detection and location might have seemed insuperable.

Whatever operational concepts were developed to meet these new requirements, they would have to be introduced in successive stages over the succeeding 15 years or so, as the necessary weapon systems and forces became available. Initially the Soviet Navy would have to make do with what it already had in service or building, but gradually as major conversions, followed by modified new construction and finally new development and design began to enter service, it would be possible to adopt more ambitious concepts. Counterforces did not have to be limited to the navy, and from the earliest stages it seems likely that shore-based systems, including ballistic missiles and satellites were included in the plans.

Carriers could be marked continuously by surface ship or nuclear submarine, or periodically relocated by reconnaissance aircraft or satellite systems. The kill capability could either be organic to the marking unit or lie with some other vehicle or vehicles, which might be surface ship, aircraft, submarine, or land-based missile.

The fundamental difficulty with Polaris was detection, location, and continuous tracking. If that could be achieved, killing on command was relatively simple. Theoretically, the submarine could be detected by various forms of fixed systems such as bottom mounted or moored arrays; by different forms of airborne surveillance systems carried by satellite or aircraft and including the use of large mid-ocean fields of drifting or floating sensors; by various types of trapping barriers established at natural choke points or outside Polaris bases; and by continuous tracking by submarine, surface ship, or aircraft. All these methods presented massive problems, which in the Soviet case were exacerbated by the relative backwardness of their ASW capability and submarine technology and by their geographical location. This did not deter the Russians, who have been making strenuous efforts in this field and outwardly express confidence that the problems can and will be overcome.[18]

The Debate about Means

While there are clear indications that a shift to forward deployment was decided in principle before the end of 1961, it would seem that this only reflected agreement on the immediate short-term response to the new threat, while the longer-term measures remained in dispute until 1963/64.

There appear to have been several strands to the debate. To some extent it was an argument between those who had a penchant for radical solutions and believed in the likelihood of technological breakthroughs, and those who thought it best to persist with established methods, while trying to develop new ones. It was also an argument between those calling for increased naval expenditure and those who were trying to hold down defense costs.

Much of the naval debate appears to have centered on the value of the surface ship against Polaris submarines. There were some who argued that the most fruitful approach was to use all available means and that the surface ship had an important role in the combined ASW team. Others argued for primary reliance on the submarine backed by long-range aircraft, with shore-based missiles as a supplementary strike weapon.[19] But this still left unanswered the question of support to submarines on distant deployment and the extent to which such ships would need protection. And it seems that there was also contention between those who claimed that Western naval forces would not dare to interfere with Soviet vessels and those who argued that local war at sea was a very real possibility and that the Soviet Navy must be able to protect itself against such attacks. The all-arms solution had the disadvantage of requiring that the maritime defense zones be pushed out beyond their existing limits of 200-300 miles from Russia's shores, in order for the Soviet Navy to apply its existing concepts of antisubmarine defense. It also meant that Soviet forces would have to conduct sustained operations in distant sea areas, which were then dominated by their opponents.

The international circumstances that helped shape the outcome of the debate are suggested by events such as the U-2 incident in May 1960; the announcement of the 2,500-nm A-3 Polaris missile in July that year; the expulsion of the Soviet submarine force from their Albanian base in May 1961; the Berlin crisis of that year, which evoked the Western concept of "Maritime Countermeasures," whereby Soviet ships would be seized as an asymmetrical response to pressure on Berlin; the successful launch of the A-2 Polaris missile in October 1961; the authorizing of a fourth Kittyhawk attack carrier in the spring of 1962; the Cuban missile crisis in the autumn of that year, which involved the detection and close tracking of all six Soviet submarines sent to the area, and the turning back of Soviet merchant ships; Kennedy's proposal for a multilateral nuclear force (MNF) comprising ballistic missiles mounted in merchant ships;[20] and the deployment of Polaris submarine on patrol in the Eastern Mediterranean in March 1963.

If there was any one factor that caused a resolution of the debate, it may have been this Polaris deployment. With the range of modern weapons, the Eastern Mediterranean is now of greater defensive concern to the Soviet Union than her arctic seas, for while Moscow is equidistant between the two, to the south and east of the capital lies a large part of Russia's industrial strength. In any case, naval pronouncements took on a more satisfied tone from about the middle of 1963;[21] there was a pilot deployment of small surface units into the Mediterranean that summer; and at some time during the year the Soviet Union approached Nasser about the use of Egyptian base facilities.[22] The 1963 edition of Military Strategy gave equal priority to the tasks of countering the carriers and countering Polaris, both being described as "most important," and by 1964 the primacy of the Polaris threat had been formally acknowledged in the military press.[23]

The full implication of what was decided on threat priorities and the means of countering them have yet to be fully perceived. As so often, the outcome of the theoretical debate seems to have been a pragmatic compromise, making the best use of what would be available in the near future, while seeking more radical solutions in the longer run. An important element of the basic concept was the extension of the maritime defense zones to take in the South Norwegian Sea up to the Greenland-Iceland-U.K. "Gap," the Eastern Mediterranean and the East China Sea. The Soviet Navy was progressively to establish a presence in these areas and thereby contest their unhindered use by the West for the deployment of strategic delivery units; they were to become areas of "no command."[24] This would represent an extension of the national defense perimeter, but the Soviet Navy would also have more distant tasks. It would need to develop the capability to sustain deployments in the vicinity of U.S. naval bases, to cover the main routes from North America to the Mediterranean and the Norwegian Sea, and to establish the infrastructure in other potential areas of threat, which would allow the rapid establishment of countermeasures, should the need arise.

Surface ships would play an important role in this new concept of operations, but submarines were to remain as the "main striking arm" of the fleet. In order to limit the number of ships (combatant and auxiliary) needed to sustain this forward deployment, reliance would be placed on gaining access to shore facilities in the area of operations and on expedients such as relieving crews on station. Meanwhile research into new methods of submarine detection and new sensors and weapon systems would be pushed ahead.

THE 1963-64 DECISIONS

Several decisions concerning warship building programs appear to have been made at this period.

Surface Programs

It was concluded that despite the changed operational scenario, the helicopter-carrying ASW cruiser was still a useful concept, but that Moskva could not operate as many aircraft as would be required for sustained distant deployments. It was therefore decided to complete the two units already under construction, which could be used to evaluate the concept and develop operational procedures and techniques, but that the remainder of the program should be canceled.[25] At the same time, design and procurement for a ship roughly twice the size was put in hand (Kuril), although the final details of its characteristics were probably left open until some experience had been gained with Moskva, and more was known about the prospects for VTOL aircraft.

Cancellation of the Moskva program made available the improved SA-N-3 SAM systems and associated air surveillance radars that had been procured for these ships. The Kresta class was shortly to begin construction at Zhdanov Yard, and it was decided that the first four units should be built as planned (Kresta I) but that this new capability should be incorporated into the design of the remaining ships in this program. This in turn released the SA-N-1 systems procured for those units, which allowed the conversion of the Krupny missile ship into the Kanin class of SAM-armed ASW ship.

It was accepted (reluctantly it would seem) that a cruiser-sized ship was needed in addition to the destroyer-sized successor to the Kashin. The requirement for the Kara class derived from the decision that Krivak must carry the SS-N-10 SSM and dual-purpose gun systems, which limited its capacity in other directions.* A larger ship was therefore necessary to carry long-range weapons and sensors such as the SA-N-3 SAM system and associated radars for force air defense, and large hull-mounted sonars; and an ASW helicopter to attack submarines at maximum detection range.

The preference shown in Krivak's characteristics for the SSM over an improved self-defense capability suggests that within the extended maritime defense zones, it is intended that the anticarrier role will largely fall upon the surface units, allowing the submarines to concentrate on the task of countering Polaris.

No additional yard capacity was allocated for building the Kara class, but the navy got round this by placing the Krivak program in what were traditionally "escort" yards, and the cruiser program in a "destroyer" yard. This involved some far-reaching decisions concerning the future composition of Soviet surface forces, including the dropping of the escort-size ship from their type-inventory. In the building programs that emerged from this decision period, the Soviet Navy appears to have been forced to accept a clear-cut distinction both in terms of sea-keeping qualities and combat qualities between those forces intended for distant deployment and those that will operate within range of shore support in defense of the home fleet areas.[26]

Submarine Programs

Submarines were involved both with strategic defense and with strategic delivery. During the early 1960s, there was a genuine Soviet concern that the United States might launch a surprise nuclear attack. The very rapid buildup of U.S. strategic systems in 1961-64 could be interpreted as an attempt to achieve the capability of launching a

---

*Krivak only carries the SA-N-4 point-defense SAM system. She also lacks the two after-facing six-barreled ASW rocket launchers that are standard fitting in Kashin, Kresta, and Kara and may provide protection against homing torpedoes. In terms of contemporary technology, Krivak would seem to have a lower air-defense capability than Kashin had on entering service; and apart from a second set of torpedo tubes, her ASW weapons fit seems little improved over the Kashin.

disarming first strike against Soviet ICBM, of which there were some 284 at the end of 1964 compared to a U.S. inventory of 844.[27] In such circumstances, the comparative invulnerability of sea-based systems would seem to offer considerable attractions to the Soviet Union, except that the first generation of Soviet SSBN had established a very poor record of evading U.S. submarine detection systems.

It was therefore decided to develop an SLBM system with sufficient range to be able to strike at North America from the comparative safety of the home fleet areas. This system (the 4,200 nm SS-N-8) would be fitted in the D-class, which would begin delivery in 1973. (The D-class is a "mid-term" modification of the basic Y-class hull/propulsion unit.)

Meanwhile a 1,300-nm SLBM (the SS-N-6) was already due to be fitted in the Y-class SSBN, one of the new generation of greatly improved nuclear submarines that was due to begin delivery in 1968. The much longer missile range made it unnecessary to close the U.S. coast, and the Soviet Navy could further improve the invulnerability of these units by maintaining only a small number of SSBN on patrol.[28] The majority would be held back in home waters; here they could hope to survive the initial exchange, and a subsequent surge deployment against disrupted Western defenses would have a high success rate. Production of this new generation of submarines was programed for 11 hull/propulsion units a year, and it would seem that in 1961 the proportion to be configured as Y-class SSBN was raised from perhaps as low as two-three units a year up to six a year. (This assumes that it was originally intended to build three a year at Gor'kij.)[29]

In the context of strategic defense, it appears to have been concluded that whatever new developments might ultimately become available for use against the Polaris threat, in the foreseeable future the nuclear-powered attack submarine was likely to provide the highest immediate return. Outside the extended maritime defense zones, it was indeed the only practicable antisubmarine vehicle. Any submarine-based solution to the Polaris problem would require a large number of units, and given the limitations on reactor production capacity, nuclear propulsion would have to be reserved for those roles where it was absolutely essential.

It was therefore decided to develop a closed-cycle diesel submarine, which could be used for the defense of the home fleet areas against intruding submarines and would also be able to work with Soviet surface forces in the extended maritime defense zones. This program would be placed at Gor'kij, which was already committed to the C-class SSGN program, with deliveries due to begin in 1968. However, the higher priority that was now accorded to Polaris over the carrier threat, and the reassignment of the counter-carrier task back to SSM-armed surface ships meant that they could afford to drop the production rate of C-class from three to two. This would release production facilities to the new closed-cycle program, the lead ship being scheduled for delivery in 1972. The third nuclear reactor would be reallocated to the V-class SSN program, increasing the annual delivery rate to three a year.

It seems likely that it was at this period of "working out the operational concepts and tactics needed to meet the new requirements"[30] that it was also decided to follow up the reassignment of the G- and H-classes to the counter-carrier role, by initiating the development of a 400-nm submarine-launched tactical ballistic missile system, to be fitted with some form of terminal homing.[31] Besides being extremely effective against an aircraft carrier, the system had a potential role against Polaris. It could be used in the damage-reduction role to bring down counter-battery fire as soon as the Polaris submarine exposed its position by launching its first missile; it could be used to sanitize high-probability areas with barrage fire; and if or when the detection problems were solved, it would serve as an excellent counter-Polaris strike weapon.[32] It has been reported that this system will be operational by the end of 1974.[33] If it is to be carried by the P-class, this implies that the latter is now under series construction. It may also be intended for retrofitting into some existing class in order to upgrade its present capability; if that is the case, the E-II SSGN would seem a most worthy candidate.

And finally, it was most probably at this period that the decision was taken to modify the N-class so as to improve its performance as an SSN.[34]

Longer-Term Plans

During this same period, the longer-range requirements for research and development would have been mapped out and the design objectives for the third generation of nuclear submarines would have been specified,* both in terms of the performance characteristics of the hull/propulsion units and the required configurations. On past precedent, this new family of nuclear submarines is not due to begin delivery until 1978, with a lead unit or two at the end of 1977; hence

---

*Each 10-year period has seen a distinct advance in nuclear submarine technology, a new "generation" of nuclear submarines. The five-year mark has so far seen the introduction of a changed configuration and some "improvement" to the basic hull/propulsion unit, but not sufficient to justify the term "generation."

one can only speculate on the nature of the decision that will have shaped them.

However, the new closed-cycle diesel program, the mounting of SS-N-10 in Krivak, and the fact that SSBN/SLBM will have reached the SALT limits by 1977, combine to create an impression of clearing the decks so that nuclear propulsion can be devoted exclusively (or at least primarily) to the task of countering the threat from Polaris/ Poseidon.

1964-1968

The Shift to Forward Deployment

The first significant Soviet naval exercise in the Norwegian Sea took place in June 1961 and became an annual event. Their major theme was the simulated passage and attack of a Western carrier strike force. They progressively increased in scale, scope, and complexity, involving the Baltic and Northern Fleets in a series of related exercises within the general scenario of defending an extended fleet area. During the NATO Strike Fleet Exercise in 1964, Northern Fleet destroyers mingled with the Western forces, marking the carriers closely, while a missile cruiser stood back within range. It became standard practice for Northern Fleet units to deploy whenever any significant Western naval forces operated in the Norwegian Sea. Meanwhile the Soviet Navy's own operations in the South Norwegian Sea were progressively increased, with rising emphasis on ASW. The perimeter of the extended maritime defence zone was pushed out to the Greenland-Iceland-U.K. Gap, with submarines stationed further forward in the Atlantic approaches.

The first sustained deployment of surface warships to the Mediterranean was not attempted until 1964 and had to rely upon open anchorages for support. Thereafter there was a gradual buildup in numbers and the length of deployment was extended each year; but lacking local base facilities, it was not possible to sustain the deployment without breaks or throughout the winter months. Operational activity was low and was mainly in the eastern basin. It became increasingly clear that lacking effective afloat support and without access to shore facilities, it would not be possible to build up an effective naval presence in the Mediterranean. The pressure on President Nasser to allow the use of Egyptian ports was increased.[35]

A New Assertive Policy

Between the fall of Khrushchev and the middle of 1965, there was a distinct hardening of Soviet attitudes toward the United States,[36] and the latter's overseas' involvements. This would have been the product of many factors, which probably included the following developments, relevant to this analysis: the availability of the 2,500-nm A-3 Polaris systems for deployment in the Eastern Mediterranean;* U.S.-U.K. discussions about establishing a base at Diego Garcias in the Indian Ocean;† failure of the Soviet Union's proposal to the United Nations in December 1964 that the Mediterranean and Indian Oceans should be declared nuclear-free zones; public announcement of the Posiedon program in January 1965;‡ the increasing U.S. involvement in Vietnam; the bombing of Hanoi during Kosygin's visit in February 1965; the Soviet decision to supply Haiphong by sea; the potential situation in Cyprus; overestimation of West European resentment of U.S. domination; the Soviet leader's success in resolving the Indo-Pakistan conflict; the imminent buildup of Soviet ICBM; and the continued rift with China.

These developments had two kinds of implication. On the one hand there was a continued growth in the strategic threat to Russia from distant sea areas, both in terms of the range of the sea-based nuclear weapons and the areas from where they could be launched. And on the other hand the Soviet leadership was moving (willingly or not) into a position of having to react more positively against U.S. overseas involvements, which were "smothering national liberation

---

*On 5 September 1964, President Johnson announced that the A-3 system would be deployed later that month.

†Knowledge of these discussions, which were not formally announced until later, is shown in the Soviet U.N. proposal for a non-nuclear zone. For the Soviet Union, this development would have been seen in the context of the 1963 U.S. agreement to build a VLF (very low frequency) communications station at North West Cape in Australia. The latter could only be interpreted as developing the capability to communicate with submerged submarines in the Indian Ocean. Submarine tenders could be moored inside the lagoon of Diego Garcias. The Arabian Sea provides better coverage of Russia and China than any other sea area. The extra 500 miles of the Poseidon missile would bring the whole of Russia within range of a submarine deployed in this area except for the northeastern part of Siberia and the Kamchatka Peninsula.

‡The Soviet Union would have been aware of this program before this date.

movements."37 From the naval viewpoint, these implications were mutually reinforcing. The shift to forward deployment had as its primary purpose the countering of U.S. strategic delivery units. This was to be achieved in part by extending the maritime defense zones to take in such areas as the Eastern Mediterranean, in order to contest their unhindered use by Western navies. But these same forces would also be available to impede the traditional use of Western naval power and, where possible, to provide support to Soviet proteges. However, while the locational requirements were similar, there was an element of inherent conflict between the two different types of mission and their related tasks. One mission was intended to reduce the likelihood of "world war" and the task was to maintain the operational posture best suited to achieving victory, should war come. The other mission would tend to lead to direct U.S./Soviet confrontation, which would if anything increase the likelihood of war. And it was at such times of high tension that Soviet naval forces would have to concentrate on their primary, war-related tasks. This would not affect those units assigned to mark the carrier groups, since the latter would be directly involved in the confrontation. But it would restrict the employment of counter-Polaris forces in support of this secondary, "peace-time" mission. Doubtless, Soviet leaders recognized this conflict, but they must have concluded that benefits could still accrue from the assertive use of a naval presence in circumstances when the dangers of war were low.

This line of reasoning led to a significant elaboration of the functions of forward deployment. In certain limited areas, besides "marking" nuclear strike units, Soviet naval forces were to contest the West's unhindered use of the seas for the projection of military power. This new departure may have been outlined at the 23d Party Congress in March 1966.38

Once again, the area of most immediate concern was the Mediterranean. Soviet naval forces were to adopt a more assertive (even truculent) posture, but if this was to have any effect there must be more ships on station and continuous deployment. However, with the forces then available, this could only be achieved through access to a base in the area of operations. Pressure on Egypt for the use of such facilities was redoubled, and Gorshkov was a member of the high-powered team that accompanied Kosygin on his visit to the UAR in May 1966.39

Several new classes of warship were to begin delivery in 1967/68, and this may have determined when the new policy could be implemented. But its purpose was made explicit in August 1966 during an official naval visit to Alexandria; the Soviet Union was going to maintain a permanent naval presence in the Mediterranean to counter the harassment of Soviet ships there and to satisfy the requirements of national security.[40] In April 1967, Brezhnev demanded the withdrawal of the Sixth Fleet from the Mediterranean,[41] and this signaled a sharp rise in the navigational intransigence of Soviet warships in the Mediterranean and the Sea of Japan.* But the problem of bases had yet to be solved.

The exact origins of the Arab-Israeli war in June 1967 remain obscure, but certain dates and consequences are clear. At the end of January, Gorshkov once again visited Egypt; in April he headed a naval delegation to Yugoslavia, which had also been pressed for naval facilities. The six-day war took place 5-11 June; on 9 July, Soviet warships berthed in Alexandria and Port Said to provide "protection" against Israeli attack. On October 7 Soviet naval support facilities were established in Alexandria and Port Said, including a submarine tender and other support ships. During the next 12 months, the average length of individual deployment in the Mediterranean almost doubled, a naval presence was maintained throughout the winter for the first time, and the number of naval units on station more than doubled. It was also in October 1967 that Gorshkov was promoted to Admiral of the Fleet of the Soviet Union. (Equivalent to marshal of the Soviet Union, and the first appointment to this rank in the Soviet Navy.)

By luck rather than good management and with considerable help from Western commentators, the Soviet Navy emerged from the June war and its aftermath with its international reputation established and its operational capability in the Mediterranean greatly increased. Others were less fortunate. The Soviet Army's prestige took a heavy blow; and the Merchant Fleet, whose Black Sea ports were the main supply points for Vietnam and Pacific Russia, was forced to use the Cape Route and to increase its foreign charters at a substantial cost in hard currency earnings.[42]

1967-68: Program Modifications

It is probable that the postmortem of the wider implications of the June war generated various decisions that have yet to be perceived. However, certain adjustments to existing plans may date back to the general period, and some may stem from that particular experience.

It is very likely that the requirement to remain in close company with the Sixth Fleet throughout the 1967 crisis highlighted the need for a fast under-way replenishment capability, which they had been

---

*This led to collisions in the Sea of Japan within three weeks and in the Mediterranean within three months.

able to sidestep until then. Both the Boris Chilikin merchant-tanker modification and the Manych class now building in Finland probably stem from this period.43

It appears that there have been delays in the Kresta II program, and it has been suggested that these are related to the provision of SS-N-10 SSM systems for these ships. If this is in fact the case, the decision to replace the original SS-N-3 with the short-range system must have been deferred until about this time, in order to have produced this effect. This design modification may be related to the experience gained in the June war, but that was also the period when the first Kresta I was evaluating the helicopter-borne organic target-location capability for its long-range SSM. Either or both of these reasons may have forced the decision to delay the final completion of the Kresta II's, until short-range systems were available to be fitted.

(It is assumed that production was set up to meet the Kara and Krivak requirements, and that Kresta had to be supplied as fully as possible. There appears to have been a delay in reaching the full Krivak delivery rate, which may have allowed the fitting of the earlier units.)

The decision to provide the Kuril with an angled through-deck that would permit the operation of V/STOL aircraft could have been taken in 1968 or perhaps even later. If the decision was related to the June war (as opposed to V/STOL development prospects), it might reflect a new awareness of the threat posed by the local air forces such as Israel's, which had shown itself to be highly effective and much less inhibited in its responses than the Sixth Fleet. (The Israelis sank the USS Elint ship Liberty. There are suggestions that this was deliberate and was intended to prevent the United States from eavesdropping on the course of the battle.)

The fourth adjustment was directly concerned with the strategic balance. On the information available, it appears that deliveries of Y-class SSBN during the years 1971-73 were increased above the previously established rate, at the expense of the V-class SSN program.44 If this was the case, the decision to readjust the production schedule must have been taken by mid-1968 at the latest.45 It is relevant that it was in May 1968 that the Soviet Union finally indicated its willingness to embark on SALT, having spent 15 months debating the initial U.S. proposal.

1968-1971

After the June war, there was a marked increase in the political exploitation of the presence of Soviet naval units in distant waters.

However, it appears that the actual areas of naval activity continued to be largely determined by long-term war-related strategic requirements rather than by other foreign policy interests. In fact it is increasingly clear that the latter have at times been sacrificed to the strategic requirements. It is therefore necessary to distinguish between the "war-related" aspects of Soviet naval policy and the other ways in which the navy is employed in peacetime.

World War-related Strategic Requirements

To a military planner in Moscow, studying how best to counter the strategic threat from Western naval strike units (and to fight a "world war" if necessary), the ordering of geographic priorities is fairly clear. Set at 1,500 nm and centered on Moscow, his dividers take in the Barents and Norwegian Seas and the Eastern Mediterranean. Extended to 2,500 nm, they reach beyond the tip of Greenland, describe an arc through the Eastern Atlantic that cuts the African Coast abreast the Canary Islands and then crosses the Arabian Sea between the Horn of Africa and Bombay. That is the area-defense problem.

A separate but equally important problem is to cover the fleet bases on the East coast of the United States and the transit routes to the operating areas. Such cover is needed for the initial acquisition of Western strike units and to deny the option of withholding them from the initial exchange. This would require a forward base located close to the United States, so as to reduce transit time to the attachment area. The deployment route to the Northeast Atlantic could be covered by reaching out from the extended maritime defense zones. The Central Atlantic is more difficult, but the Soviet Union would gain considerable strategic advantage by establishing a naval capability on NATO's southern perimeter along the Tropic of Cancer. This would expose the maritime lines of communication between the United States and the Mediterranean; it would place Soviet forces in a blocking position should U.S. carriers sail for the South Atlantic at the outbreak of war;[46] and a <u>pointe d'appuie</u> in the eastern half of the area would give access to the northern arm of the Cape Verde Basin. This extends northward as far as the Bay of Biscay, straddling the Mediterranean approaches.

The extending pattern of Soviet naval deployments matches these strategic requirements almost exactly, reflecting a judicious mixture of priorities and possibilities.

The Northeast Atlantic and the Mediterranean

The progressive extension of fleet activity into the Norwegian Sea, which began in 1961, was followed in 1964 by the first substantial deployment into the Mediterranean. Measured in ship days, naval activity in both areas continued to build up until it leveled off in 1971.[47] Operations in the South Norwegian Sea and Northeast Atlantic were supported directly from the Northern Fleet area, but in the Mediterranean it was necessary to establish a forward operating base in the eastern basin. After some difficulties, such facilities were acquired in Egypt and included the provision of extensive shore-based air support.[48] The Soviet naval presence having been successfully established in the Eastern Mediterranean, the mixture of inducements and pressure is perhaps now being applied to the Arab countries at the western end, either of whom could threaten free passage through the straits. This strategic encroachment has been mirrored in the pattern of naval arms supplied to states in the area since the early 1960s, which more often reflect the Soviet Union's strategic priorities than the client state's particular requirements. These military developments have been matched by diplomatic measures, which initially aimed to exclude the Sixth Fleet by having the Mediterranean declared a nuclear-free zone; after 1967 the line was switched to establishing the Soviet claim to have "natural rights" as a Mediterranean state, and the corollary that the navies of non-Mediterranean powers should be excluded from the area.

The U.S. East Coast Bases

A Soviet detachment deployed to the Caribbean in July/August 1969, calling at Havana, Martinique, and Barbados. There was a second deployment in May/June 1970, the main force remaining in Cienfuegos for 19 days; in September a third deployment delivered barges used for supporting nuclear submarines and a submarine tender stayed on for several weeks. Soviet detachments, usually including submarines and/or a tender have continued to visit Cienfuegos regularly.[49] A stern U.S. warning about the use of Cuba for supporting nuclear submarines may have had an effect, but the present arrangements are the next best thing to a forward operating base, which would bring considerable advantages to the Soviet Union. Norfolk, Virginia and Charleston, South Carolina are the main bases for U.S. carriers and Polaris submarines, respectively; they lie some 4,000 miles from the Northern Fleet bases but are only 900 and 600 miles north of Cuba. Submarines operating out of Cuba would spend three to four days in transit to and from the attachment area, compared with about 20 days out of Kola. This represents a 40 percent increase in time-on-task for a 60-day deployment, and a consequential reduction in the number of units required.[50]

The Central Atlantic

The choice of Martinique and Barbados for the two side visits during the 1969 Caribbean deployment may have been related to their physical location in respect to the West Atlantic Basin. On the other side of the Atlantic, Conakry in Guinea is situated on roughly the same latitude and was also visited for the first time in February 1969. Some 18 months previously, in the summer of 1967, a Soviet submarine tender and a missile support ship spent five months at sea in the general area of the Cape Verde Islands (that is, some 600-700 nm northwest of Conakry) in company with a number of submarines. The possible connection becomes more interesting in light of the combatant patrol established off Guinea in the aftermath of the Portuguese-supported seaborne attack on Conakry in November 1970.[51] Although the immediate purpose of the attack was the fall of Sekou Toure's government, the longer-range purpose was to promote a regime that would withdraw support from the Guinea-Bissau liberation movement, the African Party for Independence in Guinea and the Cape Verde Islands (PAIGC) for which Toure's Guinea was a major source of sustenance.[52] The PAIGC was the unchallenged representative of the Bissau anticolonial movement and many of its leaders were themselves from the islands. Hence the Guinea Patrol could be seen as earning favor both with Sekou Toure[53] and with the likely "inheritors" of the Cape Verde Islands. Access to the latter would bring substantial strategic advantage to the Soviet Navy, particularly in terms of fixed acoustic surveillance systems.

One payoff has been the use of Conakry airfield by Soviet Tu-95 Bear aircraft, carrying out reconnaissance over the Central Atlantic.[54] The triangle is now complete. As long ago as April/May 1970, Soviet Bears flew reconnaissance out of Northern Fleet airfields, refueling in Cuba and then returning home. In September 1972, Soviet Bears flew sustained reconnaissance of the Western Atlantic from Cuban fields and returned there. Since 1963, Soviet long-range aircraft flying out of Northern Fleet bases have regularly probed carriers en route to and from the Mediterranean, but the availability of airfields at the three corners of the surveillance area is a considerable advantage and will greatly increase flexibility and time on task.

Before leaving the Atlantic we should note that Soviet naval visits to Guinea's neighbors, Senegal and Sierra Leone, fit the general strategic pattern.[55] Dakar in particular is a major port located near Cape Verde, the westernmost point of Africa, which juts out into the Atlantic and has a very narrow continental shelf. None of these three small countries qualify as an entre to political influence among African states, and the Soviet Union's solicitude would seem to be predominantly strategic.

The Indian Ocean

Meanwhile, in March 1968 a detachment of Pacific Fleet units had entered the Indian Ocean and visited nine ports in eight countries in 3.5 months. A second deployment toward the end of the year was arranged to coincide with the transfer of combatants from the western fleets. An analysis of the pattern of port visits suggests a survey of prospects in the area, and by September 1969 naval attention was clearly focused on Somalia and South Yemen.56 Although Aden is visited regularly, Berbera is now used continuously and extensively, and is being developed as a Soviet base facility.57 This includes the building of a naval communications station, and a tender has been berthed there since 1972. A local airfield is being expanded, using Soviet technicians.

Some kind of continuous presence had been achieved by the end of 1969, but a regular deployment cycle was not established until mid-1970, with ships deploying from Vladivostok and spending about five months on station. Until late 1971, the regular deployment usually comprised a destroyer, a landing ship, a submarine and submarine escort, plus some six to seven auxiliaries. From 1972, the number rose slightly to perhaps five to six warships and eight to nine auxiliaries on regular deployment. Numbers increase during periodic "flag showing" cruises, and specially formed groups are deployed from the Pacific Fleet in times of crisis.

The Northeast Pacific

The Trident SSBN, which carries twenty-four 6,000 nm missiles, will begin to enter service in 1978. All 10 will operate out of Bangor in Washington State, partly in order to outflank Soviet ASW.58 It is not clear when this basing decision became widely known, but it may not be unrelated to the apparently purposeless cruise by a Pacific Fleet detachment of Soviet warships in August/September 1971. They first took an easterly course south of the Aleutians, and after encountering Canadian naval units they turned south toward Hawaii.59 This is not atypical of the "reconnaissances in force" that have presaged Soviet deployments into other areas.

Overview

The purpose of this section has been to outline the more obvious responses that the West's seaborne strike capability would evoke in any moderately competent strategic planner in Moscow; and to show how both the geographical pattern and the sequence of events in the shift to forward deployment match these requirements. Of course the original decisions were taken 10 years ago, since when there have been major developments on both sides. But although time lags and the overlay of ancillary objectives tend to confuse the picture, there is every indication that the Soviet Union still takes this basic strategic requirement very seriously. Their doctrinal writings and public pronouncements certainly support this assessment. So does the greater part of the operational activity in the areas concerned, the long periods spent inactive at strategically located anchorages and the heavy emphasis on hydrographic survey work of the kind required for submarine operations.[60]

The outline of the geographical framework can now be discerned, although it is still not complete in all its parts. The full operational capability to make use of this strategic infrastructure has yet to be developed and deployed. Meanwhile, as is her wont, the Soviet Union is making the best use of the relatively limited means at her disposal.

Peacetime Tasks and Employment

The progressive encroachment into sea areas that have traditionally been dominated by Western maritime power has been done cautiously and on the whole skillfully.* The low profile maintained during the first three years of the Mediterranean deployment suggests genuine doubts about the scale and nature of Western reactions, which resurface with major new initiatives such as the shipping of Moroccan troops to Syria in April/May 1973.[61] In sensitive areas a game of "grandmother's footsteps" has been played in establishing precedents for further Soviet activity, a good example being the port visits to Cuba by G II-class SSB on both the occasions that "grandmother" (that is, the United States) was preoccupied in her concern to reach agreement over SALT.[62]

The employment of Soviet naval forces on forward deployment divides into the two main categories of "general war-related" and "peacetime" tasks. The war-related tasks have already been discussed, and it is only necessary to reiterate that they have the highest priority and take up a substantial proportion of the naval effort invested in forward deployment.

---

*Perhaps the one exception was the initial splurge of visits to the Indian Ocean in March-July 1968, which caused so much alarm in the West. There are indications that the full itinerary was only arranged piecemeal after the success of the initial visits to Madras and Bombay, and their judgment may have been affected by Western reactions to the Soviet naval presence during the Middle East crisis.

The peacetime role divides into three main tasks. Many operations will support more than one of these, but the distinction is still useful: (1) securing state interests; (2) increasing prestige and influence; and (3) countering imperialist aggression.

The Soviets themselves refer to these categories and distinguish between them, just as they distinguish between their war-related and peacetime tasks. They have a tendency to lump all three peacetime tasks under the one heading "securing state interests," but at other times, the same writers will clearly distinguish between "state interests" and "prestige and influence." The scope of each task is a matter of argument, but I would assess them as follows.

Securing State Interests

The Russian verb (obespechivat') translated here as "to secure" has the meaning of "providing for" and in this context could be rendered as "to promote and protect." The most important element of this task is building up, consolidating, and preserving the infrastructure (physical, political, and operational) on which the war-related mission depends, both now and in the future. This task provides the primary motivations for a wide span of operational decisions ranging from preventing the overthrow of a client regime that is important to the strategic plan, to acquiring base rights by barely disguised coercion. The consequential policies should not be seen as being either "for" or "against" the affected third parties, but as being in support of Russia's strategic interests. Thus the decision to berth Soviet naval units in Port Said and Alexandria in July 1967 was not directed against Israel, but toward ensuring Soviet access to Egyptian base facilities.

The lesser element of this task is the classical one of "protecting national interests" in the shape of lives and property as represented by merchant shipping, fishing vessels, and Soviet citizens and property ashore. This task has not been put properly to the test. The one clear-cut example was the deployment of naval units off Ghana in February/March 1969, which achieved the release of two Soviet trawlers, impounded over four months previously by the Ghanaian authorities.[63] But while the events are reasonably clear, the underlying purpose of this somewhat atypical incident[64] is much less certain. It appears that the naval demonstration was arranged at very short notice and the units involved sailed directly from Conakry.[65] That visit, made one month after an abortive army coup, by a group of ships detached from the Mediterranean squadron, was the first to any West African port and could be seen as demonstrating support for Sekou Toure's regime. Perhaps more relevant, Nkrumah had been offered hospitality by Toure after his overthrow as President of Ghana in 1966, and was titular copresident of Guinea.[66] It could well be that during the naval visit, some awkward questions were posed about the value of Soviet patronage if they were not even capable of effecting the release of Soviet vessels from a weak African state hostile to Guinea; hence the vigorous reaction, which involved a sudden shift in the instrumentalities of Soviet policy toward Ghana. This incident may therefore come within the main strategic category of "state interests."

Although the Russians speak of merchant shipping in the context of "state interests," the Soviet Union has established a fairly consistent record of accepting the seizure of property and the expulsion and even loss of personnel, in the interests of longer-term foreign policy objectives. This is not to say that she wouldn't react vigorously to an incident involving her ships on the high seas, but this has yet to be demonstrated.

Increasing Prestige and Influence

This term seems to imply what the West refers to as "the use of naval power as an instrument of foreign policy." Taken literally in that sense, it should obviously include the task of "countering imperialist aggression" but the distinction is worth preserving and reflects the "target" of the policy. Activities that fall within this category would include flag-showing naval visits, port-clearing operations in Bangladesh, naval involvement in the Iraq/Kuwait border dispute, ferrying Moroccan troops to Syria,[67] and certain aspects of the "Okean" naval exercise/demonstration in April 1970. So far, the navy's role in discharging this task has been entirely supportive, and if for the reasons just given we exclude the Ghanaian incident, the coercive use of naval power has not been involved.

Countering Imperialist Aggression

This term implies the use of Soviet naval forces to contest the West's use of the sea for the projection of conventional military power in support of their foreign policy objectives. Primarily aimed at the United States it also covers the activities of other "imperialist" navies when employed in a manner the Soviet Union defines as being inimical to the interests of the world liberation movement.

In certain areas such as the Eastern Mediterranean, the necessary operational reactions are superficially similar to those demanded by the more deadly war-related task in times of crisis. In other areas such as the South China Sea or the Bay of Bengal, this task generates its own operational reactions and deployments.

In the Mediterranean, we have the example of the close shadowing of Sixth Fleet units during the 1967, 1970, and 1973 Middle East crises.

Elsewhere we have the deployment to the Indian Ocean at the time of the British withdrawal from the Persian Gulf, which overlapped the start of the Indo-Pakistani War in 1971/72;[68] to the South China Sea in response to the mining of Haiphong in May 1972;[69] and again to the Indian Ocean during the 1973 Middle East crisis. In all three cases, the Soviet deployments were in response to Western initiatives.

It is difficult to discern the depth of political resolution that underlies this task. In the Mediterranean, such operations are "for real," but this is an area where the "war-related" requirements apply. Elsewhere, one is less certain, although the Soviets are past masters at retrospective claims of influencing events, when the detailed chronology has become blurred. In one sense, this task was a useful answer to Chinese and international critics, who claimed that Russia was in collusion with America and did nothing to prevent the latter from stifling national liberation movements. And there seems little doubt that in 1965 Soviet leaders were genuinely concerned by the trend in U.S. overseas' involvements. But in another sense, such deployments raise expectations about tangible results, which the Soviet Union may not be prepared to fulfill.

Review of the Period

In the four years 1968-71, the international image of the Soviet Navy changed radically. Although the groundwork had been laid during the previous four years, the full impact of the major decision taken in 1961 and of the consequential policies argued out through 1963/64 were not visible until 1968. Before the Arab-Israeli war, the Soviet Navy was stretched to maintain a Mediterranean presence of two to four submarines and three to five destroyers, backed by a gun cruiser and/or submarine tender, and the occasional SSM-armed unit, and this for only nine months of the year. By mid-1969, having gained access to Egyptian port facilities, numbers were up to 7-10 submarines and 9-13 surface combatants, with one or more SSM-armed unit always on station and a substantial measure of shore-based air support. By mid-1969, the Soviet Navy had visited the Caribbean and had made significant contact with Guinea on the eastern side of the Atlantic. In the Indian Ocean, a form of continuous presence had been achieved by using ships transferring from the West to supplement Pacific Fleet units. In 1968/69 there was an upsurge in the number of naval visits, and if one was to judge from the Western press, Soviet ships appeared to be everywhere.

The impression of vigorous growth was reinforced by the new classes of ship that began to enter service during the period, by the twofold increase in nuclear-building capacity and by the appearance of the Moskva helicopter carrier, which suggested new departures.

And with it all came the increasing use of Soviet naval units for specifically political purposes in distant parts of the world.

Their success (or lack of failure) in this last respect was due to caution, discretion, and luck.[70] But these qualities do not sail ships, and operationally, the Soviet Navy was severely overstretched. (This can be deduced by comparing their order of battle with the level of deployments. But specific indicators include the uneven lengths of individual deployments in the Mediterranean;[71] the extensive use of 1,000-ton escort ships on distant deployment and the regular appearance of Z-class submarines in the Mediterranean; the imaginative expedient of deploying submarines on the surface in escorted groups; and the rather large number of submarine breakdowns. Regular schedules were achieved in the Indian Ocean by overly long deployments, with consequent loss of operational efficiency.)

Throughout the 1960s, the demands levied on the navy rose inexorably, with a sharp increase after 1967. But ocean-going new construction was joining the fleet at a relative trickle* and the majority of the new generation of nuclear submarines was being configured as task-specific SSBN.

In the discharge of its primary mission of "protecting the homeland from attack from the ocean expanses," the Soviet Navy was required to carry out continuously in peacetime what were essentially wartime tasks and was usually operating in conditions of maritime inferiority. The great majority of the fleet had been designed in the late 1940s and mid-1950s for tasks that were radically different in concept and geographical scenario. And although considerably improved classes of surface ship were due to enter service in the early seventies, the overall delivery rate would show little improvement. Nor did past Soviet warship construction compare favorably with that of her potential opponents.† Over the 14 years spanning 1958-71, the West had taken delivery of two to three times the number of major combatants as had the Soviet Navy, and if account were taken of relative size and combat capability, the disparity was more like three to four.[72] In the 12 years up to 1969, the West had built more attack carriers than the Soviet Union built missile cruisers and the latter had turned out to be unsuccessful. Until 1968, the West was outbuilding the Russians in numbers of nuclear submarines. Meanwhile, 40-50 percent of the West's nuclear delivery capability was seaborne.

---

*An average of three ships a year of destroyer-size and above joined the fleet. Another 11 underwent major conversion during the 10-year period.

†That is, the NATO powers, Australia, New Zealand, and Japan. The comparison favors Russia by ignoring natural opponents such as Sweden and Spain.

By 1971, the Soviet Navy could be proud of what it had accomplished, particularly in the face of so many difficulties. In 10 years it had managed to reverse completely its traditional deployment patterns and had evolved new operational concepts to meet a substantial maritime threat whose nature has radically changed over the past 15-20 years. It had acquired high visibility and made a distinct political impact. It has however never been put to the test. To judge from the extended length of deployments and the limited facilities for support in the forward areas, the average level of combat effectiveness is unlikely to be high.

1971-1974

As we turn to assess contemporary Soviet naval policy, it is salutary to look back 10 years and consider what we could have foreseen of the present situation with the evidence then available. We were aware that a defense debate was in progress during 1961-64, but because we did not know the underlying assumptions, we missed the main thrust of the discussion and tended to fasten on the details of the argument. We must therefore recognize that we just don't know what decisions were made concerning future naval policy during the debate leading up to and following the 24th Party Congress in March 1971. Furthermore Soviet naval policy is inextricably involved with arguments about detente, strategic-arms limitations, conventional arms control, the likelihood of nuclear war, the competition for world influence, and a whole range of internal domestic problems of which the allocation of scarce resources is probably by far the most important.

In such circumstances, perhaps the most useful approach is to review briefly what we know of past policy decisions and priorities and what capabilities are likely to be available in the next few years and to suggest some of the main points that may have been under discussion.

The War-related Strategic Requirements

The main policy decision was that the navy should develop the means to counter the West's seaborne strategic delivery capability, the aim being to reduce the damage of strikes from this direction and to deny the West the option of withholding these systems from the initial exchange.

The operational concept was to extend the maritime defense zones to cover the sea areas whence such strikes could be launched, and by establishing an increasingly active naval presence in these areas, to first contest and perhaps ultimately deny their use by the West for the deployment of strategic systems. At the same time the capability would be developed to follow the movement of all strategic delivery units when they departed their U.S. bases.

To the extent the navy needed new types of ships to carry out this new policy, the requirement had to be met from their existing allocation of shipbuilding capacity. To achieve economy in numbers, the policy was predicated on the availability of naval facilities in the forward operating areas. This was such a fundamental assumption that in several key areas the requirement for bases became a major determinant of Soviet foreign policy. It is known that bases still remain a problem today,* and in relation to their requirements, the provision of specialized afloat support has been minimal.

Although a plateau in the buildup of deployments has been reached in most areas, the Soviet Navy is still a long way from being able to discharge its war-related tasks fully. Nevertheless operational behavior continues to indicate that these tasks are given the highest priority.

The Naval Capability Required to Discharge These Tasks

In order to meet their requirements for surface ships within existing yard capacity, the navy has had to make a sharp distinction in terms of seakeeping qualities and combat capabilities between their ocean-going units and those intended for operations within range of shore based support systems. In other words, they are building up a new ocean-going surface fleet that at the present rate will comprise more than 100 units of destroyer-size and above by 1985.[73] In the immediate future they have about 110 such ships, only 35 of which do they seem to consider fully adequate for forward deployment. (Other units are either too old or are the pipeline inertial products of canceled programs.) A sorting-out process has been apparent over the last few years, with the Pacific getting more than its share of the less effective (although still impressive-looking) ships, while the Black Sea (Mediterranean) and Northern Fleets have the lion's share in quality and numbers. The configuration of recent new construction and major conversions is heavily biased toward ASW, but

---

*Admiral Sergeev, Chief of Naval Staff, was asked recently what his greatest problem was with the Soviet Navy's shift to forward deployment. He replied without hesitation: "Bases."

the major surface units carry a punch. Future policy toward carriers remains uncertain.

The present indicators are that by 1985 the Soviets will have a force of some 200 nuclear submarines, of which 60 will be SSBN, and more than 200 diesel submarines, of which a large proportion are likely to be high-performance closed-cycle units.74 At present they have about 120 nuclear submarines, of which about 80 are task-specific, and another 200 diesel submarines. They are short of general-purpose nuclear attack submarines. However, they are introducing a new tactical ballistic missile system, which if retrofitted into the E-class would considerably upgrade its versatility and capability.

Developments in naval aviation are harder to perceive because of the close dependence on the Soviet Air Force for aircraft and for certain types of operational support, including some long-range reconnaissance. The missile-armed strike aircraft continues to be linked with the nuclear submarine as the "qualitatively new" capability of the fleet. There is considerable use of long-range aircraft for reconnaissance over the Mediterranean and the broader reaches of the Atlantic, and it looks as if they are developing the capability to operate out of Somalia. They are progressively improving their shore-based fixed-wing ASW capability. Their future intentions concerning carrier aviation are not at all clear, and the decisions in this area will be an important indicator of the navy's future role.

It must be assumed that they are already exploiting the satellite's potential for ocean surveillance and targeting and are trying to develop some capability for detecting submerged submarines by such means. There are fairly clear indications that naval formations are targeted by land-based ballistic missiles,* and in the Mediterranean this may well have been the case for the 10 years 1964-73. And assuming that it can be located, a 16-missile Polaris submarine is a more worthwhile target than several Minutemen silos.

The provision of afloat support ships is relatively slow, and there are no indications that the Soviets are building the type of logistic support capability that would relieve the ships of their dependence on bases in foreign lands.

The SS-N-8 SLBM system allows them to strike at North America from the comparative safety of their home fleet areas, and it seems likely that they will retrofit all their modern SSBN with this system.

---

*The 1968 edition of <u>Military Strategy</u> could be read in this way, see p. 366. More recently there have been specific references to land-based missiles being able to strike at naval formations. In the early days of the Mediterranean deployment, the Sixth Fleet was "marked" but there were frequently no missile units on station.

It is possible that the SSBN force is now considered as an organic component of the strategic rocket forces.[75]

Operational and Deployment Patterns

Measured in ship days, the great majority of Soviet naval operations in distant waters can be directly linked in one way or another to the war-related tasks that prompted the initial shift to forward deployment and that have to be discharged continuously in peace.

There has meanwhile been a steady increase in the employment of these same forces on "peacetime" tasks, which fall into three categories. Each of these appears to have distinct characteristics in terms of relative importance, political commitment, and permitted operational modalities. And although any particular operation can usually be seen as furthering the objectives of more than one of these three tasks, the subclassification serves to clarify the issues involved.

The most important task is "to secure state interests." This task has accounted for the great majority of politically oriented activity since the middle-1960s and aims to establish the strategic infrastructure for the Soviet Navy to discharge its war-related tasks. Because of that aim's importance, this task has been backed by a high level of political commitment and a willingness to accept political costs as long as the main strategic objective was being furthered. Egypt provides a good example of the type of costs this task can incur, and it is significant that denial of access to naval base facilities was the one thing the Soviet Union would not concede when forced to withdraw from Egypt in 1972.

A second task of "countering imperialist aggression" was probably originally articulated in the middle-1960s, but the first deployment that can be clearly ascribed to this task, as distinct from the war-related tasks in the Mediterranean, was the deployment of two detachments of combatants to the Indian Ocean at the end of 1971.

It is hard to tell whether this represented a new departure of policy or reflected the increased numbers of suitable ships in the Pacific Fleet and the existing presence of Soviet naval units and support facilities in the area. The different circumstances of the South China Sea deployment in May 1972 tends one in favor of this being a new departure, and in any case the precedent now seems to have become the norm. At this early stage it is difficult to be certain of the level of political commitment that lies behind this task. How far is the Soviet Union prepared to go toward physical confrontation with the United States and to what extent is it an exercise in international public relations, making use of forces that are already available?

The third task of "increasing prestige and influence," whereby naval forces are used to promote other foreign policy objectives, gradually increased in importance after 1968 with a more active policy of "naval visits." It has acquired new prominence since 1972 with three clear-cut cases of strong naval initiatives, each of them "supportive."

The objectives of these tasks can be in harmony or in conflict. The interposition of Soviet naval forces to prevent U.S. naval support of a Western-oriented coup in Somalia would promote all three. But if regular reactive deployments by Soviet naval units consistently failed to affect the achievement of U.S. objectives, then "countering imperialist aggression" would work against "increasing prestige and influence," as indeed was the case in June 1967.

The objective of securing state interests has been persistently pursued with a high level of commitment since the very beginning and is likely to continue as an important determinant of policy. Trends for the other two tasks are both upward, and this form of activity seems likely to increase.

The Main Areas of Debate

Publication of the series of articles by Admiral Gorshkov under the title "Navies in War and Peace"[76] threw a shaft of light on what appears to have been a much wider debate about the likelihood of war with the West, the opportunities and dangers presented by arms' limitation and detente, and the role of armed forces in Soviet foreign policy. The naval element of this debate is suggested by the basic questions: (1) What is the best way of discharging existing tasks? (2) Are all these tasks still necessary or practical? (3) Are there more advantageous ways of employing naval forces?

Discharging Existing Tasks

There are long-standing arguments about the best mix of naval forces to meet existing requirements and whether certain land-based systems would not be able to discharge traditionally "naval" tasks more cost-effectively. The navy has always found it particularly difficult to justify the essential role of surface ships in a forward-deployment strategy. (Talking to Western officials at the end of 1971, Gorshkov commented on the difficulty of justifying the construction of more surface ships.)

Are The Tasks All Necessary?

This goes to the root of Soviet military doctrine, which is based on a "war-fighting" (and winning) capability, and has consistently refused to rely on the Western concept of "nuclear deterrence." But, given present weapon inventories, is the concept of waging nuclear war realistic? And what is the likelihood of such a war? Perhaps the mutual understanding achieved during SALT convinced some of the Soviet leadership that "deterrence theory" was not just an elaborate subterfuge, but that the West actually believes it and can itself be "deterred." If the possibility of a Western-initiated war is infinitesimal, can the navy's war-related tasks be justified unless they serve some additional purpose? Even allowing that the problems of the counter-Polaris task can be mastered, what are the implications of the Trident system?

More Advantageous Employment

Do strategic parity, the moves toward detente, and the probability that the United States can be deterred all add up to lower risks of escalation and greater opportunities for the use of Soviet military power in support of foreign policy objectives? What does the Nixon Doctrine and the shift to a "blue-water" policy portend in terms of U.S. involvements overseas? Can this best be met by the interposition of countervailing Soviet naval forces? Or is it both cheaper and more effective to negotiate mutual withdrawals and limitations on naval operations?[77]

The Naval Viewpoint

Gorshkov draws on a broad range of mainly valid examples to demonstrate the vital importance of a powerful fleet to the Soviet Union. Four propositions appear central to what is a classical "sea power" argument: (1) military power determines the outcome of international interactions; (2) lack of naval power has in the past allowed Russia's enemies to dictate the outcome of events in areas of primary concern to her; (3) navies have an increased importance in war; and (4) in peacetime navies are the most effective means of projecting military and other forms of state power and influence beyond a country's borders.[78]

Both the structure and the flavor of the underlying argument leave little doubt that the primary purpose of publishing this series was to persuade, both in the sense of advocating and justifying;* in

---

*This was the nearly universal opinion of the participants at the meeting of the Soviet Naval Studies Group in October 1973. McConnell dissented in that he considered that Gorshkov was making

the process Gorshkov also makes a substantial number of debating points.* But while the main thrust of his argument is extremely clear, we do not know the reasons that caused Gorshkov to advance it at this time and in this way.79 Nor do we know whether he himself is fully committed to his own line of argument or whether he sees it as the best way of achieving some intermediate goal. In view of the record prior to 1961, we should be careful when imputing "natural" (often Western) professional/institutional attitudes and ambitions to the Soviet naval leadership; and if we do, we must draw on the full range of such attributes, including a "natural" concern for the vulnerability of far-flung ships, heavily dependent on shore support in foreign lands.

As Weinland points out, in advocating an assertive naval policy in peacetime, Gorshkov went against the military leadership's distaste for an activist role in the international arena, which they see as being adventurist and inimical to Soviet security.80 On the other hand, Gorshkov devotes considerable attention to the problems of war at sea and lists the navy's tasks in general war; he thereby implicitly gives his support to the official military doctrine of defense based on a war-fighting capability. In the same way, Gorshkov goes directly against Brezhnev's proposal for a mutual limitation on the deployment of ships on war-related tasks in the Mediterranean and in the Indian Ocean; but on the other hand he strongly supports the use of warships in peacetime to promote the goals of foreign policy.

The common threat that links all of Gorshkov's arguments about the role of navies in war and peace is that each one of them justifies a stronger navy,† not only in terms of numbers, but in the broadest sense of a properly balanced fleet, with a sufficient diversity of ship types and adequate afloat support. And while we can not be

---

a significant announcement concerning a new "withholding strategy" whereby SSBN would be available for deterrent/political use during the course of the war.

*Seemingly inconsequential parts of the Gorshkov series fall into place when seen as debating points. There are four main categories: (1) reassure those who have doubts but are still uncommitted; (2) rally support by emphasizing doctrinal respectability; (3) rebut earlier attacks on his case and personal record; and (4) attack certain opposing viewpoints by analogy. There are relatively few attacks and there is the impression of avoiding unnecessary provocation and of being careful not to antagonize whole groups or interests.

†Robert Herrick illustrates this point in an unpublished table that reviews various of the key quotations and asks the question "Did Gorshkov indicate that more ships were needed?"

certain what prompted the Gorshkov series, we do know that between 1961 and 1970 the Soviet Navy was required to shift to forward deployment and to develop a whole range of new operational concepts, but that there was no significant reallocation of shipyard facilities to meet these new requirements. And since 1968, the navy has been required to undertake an increasing number of peacetime tasks, over and above the new demands imposed by its extended war-related mission.

In his final section, Gorshkov itemizes the minimum requirements (in terms of operational characteristics) that will permit sustained deployment. This is in the context of the special demands of the "nuclear era," when opposing forces remain in company with each other in peacetime, requiring continuous, instantaneous readiness to fire the "first salvo." Although this is a list of what he needs, rather than what he already has, the context suggests that these particular requirements will be met.[81] But this is not the impression given by his repeated stress on the need for balanced forces able to discharge a wide range of missions, the inherent weakness of task-specific navies built to discharge narrowly defined missions, the inescapable need for a wide range of surface ship types, and how multipurpose surface ships have always been proven unsatisfactory.[82]

The crux of the naval debate lies in the allocation of resources to future warship construction. By 1971, decisions were being made on the scale of new construction for deliveries beginning about 1980 and on the production runs (and configuration mixes) for the remainder of the 1970s. We will have to wait several years for shipyard evidence of what was decided, but official pronouncements made around the time of the 24th Congress suggest that the previous pattern was to be maintained.[83] Although the navy had avoided a cutback in their allocations, it seems probable that they got less than what they considered essential to meet their extending commitments.

Future Prospects

The formulation of Soviet naval policy used to be fairly simple. Given certain doctrinal tenets about war and military strategy, and confronted with Western maritime preponderance, the Soviet leadership was faced by clear-cut imperatives and the problem was one of allocating scarce resources among competing priorities.

Today the options are far more complex. And one out of the six substantive points that Gorshkov selects to ram home in his final conclusions suggests that Soviet naval policy lacks clear-cut objectives and a coherent supporting program. The wording is somewhat obscure (perhaps deliberately), but I understand him to be saying that maritime power is not some all-purpose commodity one buys by the ton, but that its type and quality stem directly from a country's perception of its <u>particular requirements</u> for maritime power, and from the <u>naval policy</u> it decides to adopt.[84] From this, I infer him to be saying that the leadership must make a conscious decision on the role of naval power in Soviet foreign policy, and they must not expect to be able to rely on the by-products of a naval policy tailored to the specialized mission of nuclear-missile war.

It is too early to say what effect, if any, Gorshkov's arguments will have had on policy, although it is worth recalling that at the equivalent period in the 1960s the navy was still arguing its case, apparently to some useful purpose. But the present debate is much wider in scope, and now that it has moved away from the imperatives of national security it raises fundamental questions about the relative utility of naval power compared to other instruments of foreign policy and whether the political costs of its use outweigh the benefits.

There are no indications that the navy's relative standing within the political leadership has improved, and if membership of the Central Committee is any guide, Ground Force domination of the military leadership has increased progressively since 1961.[85] Therefore, one suspects that Gorshkov's strictures will fall on deaf ears and that the political leadership will choose to make increasing use of the Soviet Navy as an instrument of foreign policy, but without increasing its relative share of resources. Within certain limits this can be achieved by adjusting existing assumptions about the dangers of escalation and the probability of Western-initiated nuclear war. More active efforts to increase prestige and influence offer few problems, but countering imperialist aggression is fraught with risks, and one wonders how much political commitment the Soviet Union can afford to give to this task.

Meanwhile, the task of securing state interests by creating the strategic infrastructure to support the navy's war-related tasks is likely to continue as hitherto. Although the coercive overtones of this task tend to incur the highest political costs, the navy's forward deployment is premised on the availability of this support. The incremental and continuous nature of this task and its essential place in the Soviet Union's strategic posture does not allow the experimental approach that is possible with the other two peacetime tasks.

The decision to base strategic plans on the availability of overseas base facilities, as opposed to circumventing this requirement by building large numbers of warships and specialized support facilities provides an interesting insight on the Soviet Union's internal priorities. It also gives some indication of the economic constraints within which Soviet naval policy is forced to operate.

NOTES

1. For an outline of the postwar building programs, see The Soviet Union in Europe and the Near East (London: Royal United Service Institute, 1970), pp. 74-82. For justification of the medium-type submarine programs, see M. MccGwire, "The Turning Points in Soviet Naval Policy," MccGwire, ed., Soviet Naval Developments: Capability and Context (New York: Praeger Publishers, 1973), pp. 184-85.
2. Gorshkov worked with Khrushchev on the Southern Front during World War II and can be seen as the naval member of the "Stalingrad Group." See R. Kolkowicz, The Soviet Military and the Communist Party (Princeton, N.J.: Princeton University Press, 1967), Appendix A.
3. For justification of these decision periods see MccGwire, "Turning Points . . . ," op. cit., pp. 200-204.
4. See ibid., pp. 185-86.
5. The choice of name encouraged the widespread assumption that Polaris (whose construction was authorized in 1958) was intended to close the northern arc of threat against the USSR. This was reinforced by Nautilus's submerged polar transit and subsequent Western comment. See, for example, the report of a statement by NATO's C-in-C, North, in Krasnaya zvezda, 11 November 1958.
6. The construction of 14 Polaris had been authorized progressively during the three years 1958-60. On 29 January 1961, Kennedy authorized the construction of a further 27, of which 15 were to start building within six months. The 1962 edition of V. D. Sokolovskij's Military Strategy (Moscow) makes specific reference to this increase in the rate of production. See Chapter 27 of this volume for a discussion of this point and Table 27.1 for the Minuteman buildup.
7. See Chapters 25 and 26 of this volume for a discussion of the Soviet understanding of deterrence. See P. Vigor in MccGwire, Soviet Naval Developments, op. cit., p. 22 for the Soviet definition of "world war."
8. See Chapter 27 of this volume for an extended discussion of Europe's place in Soviet military strategy.
9. Krasnaya zvezda, 5 February 1963.
10. Morskoi sbornik, May 1966, p. 9 and February 1967, p. 16.
11. Gorshkov gives this as one of the navy's three main tasks in general war; Morskoi sbornik, February 1973, p. 21. For a discussion of this role, see N. M. Kharlamov, ibid., January 1966, pp. 31-36, and the editorial, ibid., July 1966, pp. 3-7.
12. Krasnaya zvezda, 5 February 1963.
13. Ibid., 11 February 1968.
14. T. W. Wolfe, Soviet Strategy at the Crossroads (Harvard University Press 1965), pp. 112-17.

15. Although the SSBN had always had the task of carrying out strikes against enemy land targets, the latter were predominantly ports, shipyards, and the like. In 1957/58, when it was decided to build up a nuclear submarine force, its projected primary mission was to counter the seaborne threat inherent in the growing number of U.S. attack carriers. The derivation of the 1957/58 figures is explained in Chapter 23 of this volume, p. 431. The 1961 figure is based on the final pattern of production (Table 23.1), making allowance for the further adjustment in 1963/64.

16. See Chapter 23 of this volume, section entitled "Submarine Weapon Systems," for a discussion of the drawbacks of long-range SSM and the potential advantages of tactical SLBM systems. Targeting location data could be provided by long-range aircraft, or by N-class SSN. The latter are reported to have trailed U.S. carriers (New American, 15 September 1968).

17. Senator Stuart Symington in Washington Post, 29 May 1966.

18. See Norman Polmar on this point, quoting personal and public Soviet sources. U.S. Congress, Senate Armed Services Committee, Fiscal Year 1974 Authorization Hearings, p. 6047.

19. See Ullman on the details of this debate, Chapter 31 of this volume, pp. 587-589.

20. I am indebted to R. G. Weinland for this point.

21. See David Cox, "Sea Power and Soviet Foreign Policy," United States Naval Institute Proceedings, June 1969, p. 37.

22. See Chapter 13 of this volume, note 59.

23. V. D. Sokolovskij, Voennaya Strategiya, Moscow 1963, p. 398; and V. D. Sokolovskij and M. I. Cherednichenko, "Military Art at a New Stage," Krasnaya zvezda, August 25 and 28, 1964.

24. See Chapter 33 in this volume, pp. 632-33 for a discussion of this concept.

25. Past ship-building practice suggested a minimum "production run" of six to eight units. See MccGwire, Soviet Naval Developments, op. cit., p. 182.

26. See Chapter 23 of this volume, pp. 442-43. T. W. Wolfe, Soviet Strategy at the Crossroads (Cambridge: Harvard University Press, 1965), pp. 112-17.

27. See Table 27.1 in this volume.

28. A total of three Y-class are maintained on station, two in the Atlantic and one in the Pacific. Some 25 units were operational at the time. Washington Star, 20 April 1973.

29. See note 14 above and Chapter 23 of this volume, pp. 431-432.

30. Admiral Gorshkov in Krasnaya zvezda, 11 February 1968. He does not specify the date but says "It was decided to build up an air and submarine force capable of fulfilling its task in a nuclear

world war if the imperialists unleashed one. For this we needed ships with great independence of movement and unlimited range, high striking power and combat stability which were designed to inflict blows on the enemy both at sea on his maritime territories, and to defend targets in our own coast areas."

31. See Chapter 23 of this volume, section entitled "Submarine Weapon Systems."

32. See Harlan Ullman's "The Counter-Polaris Task," Chapter 31 of this volume, from which I have taken these ideas.

33. Air et Cosmos, 24 November 1973.

34. Admiral Rickover's testimony to the U.S. Congress, Joint Committee on Atomic Energy, Naval Nuclear Propulsion Program 1972-73, pp. 128-29.

35. See George Dragnich on this general point in "The Soviet Quest for Naval Facilities in Egypt," Chapter 13 in this volume.

36. T. W. Wolfe, Soviet Power in Europe (Baltimore: Johns Hopkins Press, 1971), p. 266.

37. Admiral Kasatanov, Izvestia, 8 January 1966.

38. See the 4th part of the 1st "Resolution on the World Situation" of the 23d Congress, which discusses foreign policy in terms of Soviet interests and international revolutionary duty. In considering the aggressive forces of imperialism, which are "aggravating international tension and creating hotbeds of war" (the previous paragraph had discussed Vietnam), the resolution states that the CPSU would continue "to reinforce the defense potential of the U.S.S.R. so that the Soviet armed forces would be ever ready to defend the gains of socialism dependably and deal a crushing blow to any imperialist aggression"; 23d Congress of the CPSU (Moscow: Novosti Press, 1966), p. 289. See also Gorshkov's report on the congress, which refers to being ready to "protect the achievements of Socialism and inflict a crushing rebuff to any imperialist aggression": Morskoi sbornik, May 1966, p. 10. Also note 37 above.

39. See Dragnich, Chapter 13 of this volume, p. 256.

40. Ibid., p. 44.

41. At a conference of European Communist Parties, 23-26 April at Karlovy Vary, Czechoslovakia. The purpose was to discuss proposals for all-European security and increased West-East technical cooperation and to agitate against the U.S. presence in Europe. See Wolfe, op. cit., p. 325.

42. See R. E. Athay, The Economics of Soviet Merchant Shipping (Chapel Hill: University of North Carolina Press, 1971).

43. B. Chilikin was delivered in 1970 (USNIP, February 1973) and was an adaptation of the Velikij Oktyabr' class of tanker, which had been building in Leningrad since 1966/67.

44. See table on page 427.

45. See Chapter 23 of this volume, notes 10 and 24.

46. Admiral Kharlamov discusses the likelihood that carriers will be transferred to the "second wave"; Morskoi sbornik, January 1966, p. 33. U.S. Admirals were discussing this option in the early 1960s.

47. See Robert Weinland, "Soviet Naval Operations," Chapter 20 this volume.

48. See MccGwire, Soviet Naval Developments, op. cit., pp. 344-47 and 351-54 for a detailed discussion.

49. James D. Theberge, Russia in the Caribbean (Washington, D.C.: Georgetown University Press, 1973), pp. 97-102.

50. See MccGwire, Soviet Naval Developments, op. cit., pp. 231-35 and 472-81 for a discussion of the advantages of a forward operating base in the Caribbean.

51. See Robert Weinland in MccGwire, Soviet Naval Developments, op. cit., p. 302.

52. Kaye Whiteman, "Guinea in West African Politics," World Today, August 1971, p. 354.

53. Although Toure refused to join the French Communaute in 1958 and had dealings both with Russia and China, he was always fiercely independent. In 1961 he expelled the Soviet ambassador for being involved in a plot against his regime. Ibid., p. 351.

54. Marine Rundschau, February 1974, p. 118 quoting a U.S. Defense Department release dated 7 December. At the time of the Cuban missile crisis Guinea refused landing rights to Soviet jets. Whiteman, op. cit., p. 351, quoting William Attwood, The Reds and Blacks (London: Hutchinson, 1967), p. 109.

55. The first Soviet naval visit to Sierra Leone in May 1971 coincided with a threat to Stevens's presidency (Weinland, op. cit., in MccGwire, Soviet Naval Developments). Guinea had a defense treaty with Sierra Leone and had sent troops to Freetown at Stevens's request, following the abortive coup of 23d March, when the Sierra Leone army nearly fell apart (Whiteman, op. cit., p. 357). Although 15 months apart, there is a certain similarity between the circumstances of the initial Soviet naval visit to each of these neighboring states, and in their supportive, legitimizing aspects. See also note 65 below.

56. See MccGwire, Soviet Naval Developments, op. cit., pp. 429-36 for an analysis of the pattern of visits 1968-71.

57. U.S. Department of Defense release DTG 072217Z April.

58. U.S. Congress, Senate Armed Service Committee, Appropriations Hearings FY 1974, p. 728. One of the implications of moving to the west coast of the United States so as to outflank Soviet ASW is that the Soviet Union is making some progress in developing their ASW capability in the Atlantic. Specific reference is also made

to the advantages of conducting patrols in the Indian Ocean from the Pacific (as opposed to Atlantic) coast.

59. Halifax Chronicle Herald, 6 November 1971; Naval Review, 1972, p. 344.

60. See for example the chart of Soviet naval hydrographic activity in the Caribbean, in Theberge, op. cit., p. 79.

61. See Weinland, "Soviet Naval Operations," Chapter 20 this volume.

62. Ibid., pp. 10, 13. The 1972 deployment either showed poor coordination or suggests the importance the Soviet Navy attaches to establishing the precedent of SSB visiting Cuban ports. In the final stages of SALT, an impasse was reached over the inclusion of the "older" ballistic-missile submarines. It was at this stage (May 4-15) that a G-II visited Nipe in Cuba, an initiative that might be expected to increase the U.S. Navy's concern to see these units included in the final calculations. In the event, the H-class SSBN were counted in and the G-class SSB were left out. The Soviet Navy established the precedent, but it was a risky finesse. (The children's game "grandmother's footsteps" is known elsewhere as "One, two, three—Red Light!")

63. See Robert Weinland in MccGwire, Soviet Naval Developments, op. cit., pp. 301-2.

64. Ibid. The two trawlers were detained at Takoradi on 10 October and formally charged with "operating without a license in Ghanaian waters." There were however suspicions that they were linked to an alleged plot to restore Nkrumah to the presidency. The trawlers were released on 28 February, the two masters and one second mate being retained to appear before the commission investigating the plot allegations. They were allowed to leave in mid-March, and in due course the Amissah commission exonerated the Ghanaian senior officers who had been accused of complicity (Keetings Contemporary Archives 1969-70 23803B). As far as concerns normal fishery offenses, the Soviet Union usually pays up without prejudice to her position on the legal issues. In this case she was not given the option.

65. Weinland (op. cit., in MccGwire, Soviet Naval Developments) refers to a "hastily arranged visit" to Lagos. This also implies that originally, the visit to Conakry was the sole objective of this detachment from the Mediterranean, which increases its significance.

66. Whiteman (op. cit.), p. 358.

67. See Chapters 21, 17, 19, and 20 (respectively) of this volume for more information.

68. See McConnell and Kelly in MccGwire, Soviet Naval Developments, op. cit., pp. 442-55.

69. See Weinland, Chapter 20 of this volume, p. 384.

70. Weinland, "The Changing Mission Structure of the Soviet Navy" in MccGwire, Soviet Naval Developments, op. cit., p. 272. Weinland updates his analysis of the political use of Soviet naval forces in Chapter 20 of this volume.

71. See Table 27.1 in MccGwire, Soviet Naval Developments, op. cit.

72. See Chapter 12, MccGwire, Soviet Naval Developments, op. cit.

73. See Chapter 23 of this volume, p. 444.

74. Ibid.

75. This might be inferred from the fact that in a standard listing of the five arms of service, in which the navy (as usual) is listed last, the SSBN force is neither listed as a co-equal of the SRF (as is usual) nor is it referred to in the naval section. Krasnaya Zvezda, 23 February 1974; S. L. Sokolov.

76. Morskoi sbornik, 1972, Nos. 2-6, 8-12; 1973, No. 2.

77. Brezhnev suggested this in 1971, both through private soundings and in a public speech. See Barry Blechman, "Soviet Interests in Naval Arms Control," Soviet Naval Developments, op. cit., p. 522.

78. These five paragraphs are based on an extended analysis of the Gorshkov series completed in April 1973, which is being published elsewhere.

79. See Robert Weinland, "Analysis of Admiral Gorshkov's 'Navies in War and Peace,'" Chapter 29 of this volume.

80. Ibid., p. 568.

81. Morskoi sbornik, February 1973, p. 22.

82. Ibid., November 1972, p. 32; February 1973, pp. 20-21.

83. Herrick quotes Marshalls Zakharov and Grechko as saying in January and February 1971 (respectively) that time had proven that the right course has been selected for the development of the navy; Sovetskaya Rossiya, 19 January, and Pravda, 23 February. These statements come within the shadow of the 24th Congress and taken together with Grechko's Navy Day article in Morskoi sbornik (July 1971) would seem to imply no change in naval allocations.

84. The statement reads: "Given [an adequate economic capacity], it is the policy, premised on the country's need for maritime power, which then becomes the important factor, determining the type of fleet which is built . . . and it is an indispensable condition for the development of maritime power." Morskoi sbornik, February 1973, p. 24.

85. Ground Force "full members" have risen from 8 to 15; Naval "full members" went from 2 in 1961 to 1 in 1966 and stayed there. See John McDonnell, Chapter 6 of this volume, Figure 6.3, "Membership of the Central Committee."

CHAPTER

29

ANALYSIS OF ADMIRAL GORSHKOV'S "NAVIES IN WAR AND PEACE"

Robert Weinland

In February 1972, <u>Morskoi sbornik</u> (Naval review, the highly respected and widely read professional journal of the Soviet Navy), began publishing a series of articles by the commander in chief of the Soviet Navy, Admiral of the Fleet of the Soviet Union S. G. Gorshkov, under the general title "Navies in War and Peace." The 11th and final installment was published in February of 1973, and the series totaled some 54,000 words—and clearly represented an important statement.[1]

Its importance was attributable in part to its author. Admiral Gorshkov has been commander in chief of the Soviet Navy for some 18 years; he is a deputy minister of Defense; and he is a full member of the CPSU Central Committee. Given his position, he must choose his words carefully; and what he chooses to say must be given careful consideration—since it is inevitably considered to carry the weight of authority.

The importance of his statement was also attributable to anomalies in its appearance. Admiral Gorshkov publishes infrequently. His last major article in <u>Morskoi sbornik</u> had appeared in February 1967. In the five years between its appearance and the initial installment of "Navies in War and Peace," Admiral Gorshkov published only two minor articles in <u>Morskoi sbornik</u>. Furthermore, <u>Morskoi sbornik</u> normally publishes neither works of this length nor serialized

---

The ideas expressed in this chapter are those of the author. They do not necessarily reflect the views of the Center for Naval Analysis, nor of the Department of Defense and the Department of the Navy. The author is indebted to Robert W. Herrick, Michael K. MccGwire, and James M. McConnell for the great help he got from their analyses of Gorshkov's articles.

articles. Thus, the form in which "Navies in War and Peace" appeared in itself lent import to Gorshkov's statement.

In the final analysis, however, the importance of Admiral Gorshkov's statement lay primarily in its content. Taken as a whole, and viewed in context, the articles appeared to advance a system of views that was at variance on critical points with established Soviet foreign and military policy—as these were generally understood in the West.

These apparent discontinuities raised important questions about the antecedents of Admiral Gorshkov's statement, and equally important questions about its implications. In order to understand fully what he was saying, it became imperative to determine whether these apparent discontinuities were real; and, if so, whether Admiral Gorshkov was "advocating" or "announcing" modifications to established Soviet foreign and military policy.

Several intensive analyses of "Navies in War and Peace" have been undertaken in the attempt to make these determinations—without, however, achieving complete unanimity on the answers. This discussion incorporates most of the findings of those analyses, and has three objectives. The first is to describe the publication of "Navies in War and Peace" and summarize the arguments advanced there by Gorshkov. The second is to examine potential links between the publication of his statement and the domestic and international political context in which it appeared. The third is to present some summary judgments on the meaning and import of "Navies in War and Peace."

FACTS OF PUBLICATION

Morskoi sbornik is published under the aegis of the Ministry of Defense and serves as the navy leadership's principal medium of mass communication. Its primary function is the dissemination of "military scientific knowledge" to, and the political and military education of, the officer corps of the navy—its primary audience.[2] It has other audiences, however,* and it probably has other functions—for instance, mobilizing support for the navy's case in debates within the military establishment, and possibly at the national decision-making level as well.

"Navies in War and Peace" was published without fanfare. Neither the preceding issues of Morskoi sbornik nor the annual production plans released at the beginning of the year by the publishing house of the Ministry of Defense gave advance notice of its appearance.

---

*It has been available to subscribers outside the Soviet Union since 1963.

With the exception of a one-paragraph editors' introduction to the first installment, Morskoi sbornik made no reference to it in the course of its publication, nor was it discussed during that period in any of the other major organs of the Soviet military press.

As indicated in Table 29.1, which provides the basic details of its publication, the 11 installments of "Navies in War and Peace" appeared in the 13 issues of Morskoi sbornik published from February 1972 through February 1973. There were two interruptions in the series. While the July 1972 issue did contain an article by Admiral Gorshkov, it was a discussion of the role of the commanding officer of a ship rather than a continuation of the series. The January 1973 issue contained no article by Admiral Gorshkov.

THE STATEMENT

"Navies in War and Peace" was published as a series of separate but related articles. However, it was obviously meant to be considered as a whole—an integrated, 18-chapter, 54,000-word statement (see Table 29.2).

Subject and Objectives

In his introduction, Admiral Gorshkov outlines both the subject and the objectives of this statement. He identifies two principal subjects of discussion:

- the role and place of navies [within the system of component branches of the armed forces]
- in various historical eras, and
- at different stages in the development of military equipment and the military art, [and]
- the dialectical relationship between the development of naval forces and the goals of the state policies they were intended to serve.

He also places a number of explicit caveats on the discussion: it is restricted to questions applicable to the navy; it is not intended to be a history of the "naval art"; and it does not attempt to define "the prospects for the development of naval forces."

He states that his proximate objective in conducting this examination of "the employment of various branches of armed forces in time of war or in peacetime" is to "determine the trends and principles of change" in (1) "the role and place of navies in wars," and (2) navies'

TABLE 29.1

Publication of "Navies in War and Peace" in
Morskoi sbornik

Installment	Issue	Pages	Chapter Headings
1	2-72	20-29	[Introduction] The Distant Past, but Important for Understanding the Role of Navies.
2	3-72	20-32	Russia's Difficult Road to the Sea. The Russians in the Mediterranean.
3	4-72	9-23	Into the Oceans on Behalf of Science. The Russian Fleet during the Industrial Revolution and the Transition from Sailing Vessels to Steam Vessels. Navies at the Beginning of the Era of Imperialism.
4	5-72	12-24	The First World War.
5	6-72	11-21	The Soviet Navy. The Development of the Soviet Navy in the Period from the End of the Civil War to the Outbreak of the Great Patriotic War [1921-1928].
	7-72	*	
6	8-72	14-24	" [1925-1941] "
7	9-72	14-24	The Second World War
8	10-72	13-21	The Soviet Navy in the Great Patriotic War.
9	11-72	24-34	The Basic Missions Executed by Navies in the Second World War.
10	12-72	14-22	Navies as a Weapon of the Aggressive Policy of Imperialist States in Peacetime. [Intermediate Conclusion]
	1-73	*	
11	2-73	13-25	Some Problems in Mastering the World Ocean. The Problems of a Modern Navy. [Conclusion]

*Interruption in publication of the series.

"employment in peacetime as instruments of state policy" (emphasis added). He implies that his ultimate objectives in presenting the discussion are not only "the development of the military art" but "the development of a unity of operational views in the command personnel of the armed forces"—which is fostered by the "command personnel" understanding "the specific features with which each of the branches of the armed forces is imbued."

Central Argument

In essence, Admiral Gorshkov argues that the navy's status within the armed forces should be redefined to reflect the increasing importance of the navy in wartime and in peacetime— although he does not identify the specific policy implications of this argument.

The argument contains five fundamental theses:

1. Given the increasing importance of the oceans as an arena of potential military conflict, and the navy's special military features,* the wartime importance of the navy is increasing—although, since Soviet military doctrine considers concerted action by all branches of the armed forces to be essential to victory, this increased importance does not give the navy a unique position within the armed forces;

2. Despite the introduction of nuclear weapons and the advent of detente, the armed forces have not lost their historic importance as instruments of state policy in either wartime or peacetime (if anything, the political influence of demonstrably superior military potential has increased);

3. Given the increasing economic and hence political importance of the oceans, and the navy's special political features,† the peacetime utility and importance of the navy are increasing, which gives it a unique position, compared to the other branches of the armed forces, as an instrument of foreign policy;

4. The structures of armed forces and the roles and places of their component branches can and do change (such changes are situationally dependent, can occur in peacetime as well as in the course

---

*High maneuverability, capability for covert concentration, and relative invulnerability to the effects of nuclear weapons—when compared with land-based forces.

†It can be employed in peacetime for demonstrating the economic and military power of the state beyond its borders; it is the only branch of the armed forces capable of protecting the state's overseas interests.

of war, but have limits—for example, maritime states must have navies as well as armies; and if they are to achieve and maintain great-power status, their navies must be commensurate with the full range of their interests); and

5. There is a necessary link between the acquisition and maintenance of armed forces and the goals of the state policies they are intended to serve; and, in order to insure the achievement of those goals, <u>command echelons must have a shared understanding of the relative capabilities and optimal modes of employment of each branch of the armed forces</u>.

These fundamental theses are outlined at the outset, and the bulk of the subsequent discussion is devoted to elaborating and supporting them.

Historical Discussion

In introducing this discussion, Admiral Gorshkov states that he intends "only to express a few thoughts" on the "historical" and "problem," or contemporary, aspects of the subjects under discussion. Contemporary military questions are rarely addressed directly in the Soviet open literature, unless in the context of a formal doctrinal pronouncement. Generally, military questions are approached indirectly in their "historical aspect" or, when in their "problem aspect," then less in terms of the Soviet than of Western armed forces. "Navies in War and Peace" contains few exceptions to this pattern.

As Table 29.2 illustrates, less than 20 percent of the discussion is devoted to the contemporary or "problem" aspect of the questions at hand. The bulk of the discussion is devoted to history: more than a third to the pre-Soviet era; more than half to the period before World War II; more than three-quarters to the prenuclear era.

In this historical discussion, Gorshkov makes no attempt to present a balanced, comprehensive description of the "role and place" of seapower in Russian and Soviet, not to mention Western, history. Rather, he employs history forensically to support the "system of views" he is advancing. His discussion of the past develops five major themes.

1. Exploitation of the sea, and seapower—in all of its forms—are necessary to achieve and maintain great-power status and consequently have always been and will always be important to maritime states in general, and Russia in particular.

2. Large and modern* naval forces are the sine qua non of effective seapower.

---

*Modernity—which is measured in terms of the quality of ships, the level of training of their crews, and the level of development of

3. Seapower can be used in peacetime as well as in wartime to implement state policy.

4. These facts were often overlooked by the Tsarist leadership. When they did not recognize the importance of, and hence neglected, the navy, Russia lost wars, or lost the gains of those wars it won, and was often unable to implement its policies in peacetime. On the other hand, when they did appreciate and support the navy, both the wartime and the peacetime goals of state policy were achieved.

5. The Soviet leadership—in contrast to their predecessors—have consistently recognized the importance of seapower. But, because of the economic and technological constraints that prevailed during the early years of the Soviet era, and the necessary concentration on land warfare that characterized Soviet military experience in the Civil War and the Great Patriotic War/World War II,* it was not until relatively recently that the Soviet Union acquired the effective seapower it now possesses.

Each major theme is illustrated with specific examples, the majority of which focus on the consequences of prevailing policies: losses sustained, benefits derived. In general, since the Tsarist leadership did not understand seapower, and until recently the Soviet could not afford it, this historical discussion is largely a chronicle of losses.

One chapter—"The Russians in the Mediterranean"—stands out as an exception. It concentrates on the benefits derived from the possession and use of naval forces and explicitly links the past and the present—that is, it treats the subject in both its "historical" and its contemporary or "problem" aspects. Historically, Russia's intermittent deployments of naval forces to the Mediterranean have been undertaken primarily to insure its security from the Southwest. Militarily, Russian naval forces in the Mediterranean acted as a "forward defense," providing strategic support to the army. Politically, they acted as a powerful, at times the most powerful, weapon of Russian foreign policy, contributing to change in the political situation in Europe. At present, Soviet naval forces are deployed in the Mediterranean primarily as a defensive response to the presence of opposing naval forces that threaten direct attack on the Soviet

---

the tactics they employ—is a critical factor. Naval victory is a function of numerical superiority, given equivalent modernity of opposing forces; however, given parity in numbers, victory becomes a function of modernity—because the more modern and hence more capable force is able to seize the initiative.

*To which the Soviet navy nevertheless made vital contributions (the description of which absorbs a large share of the discussion).

TABLE 29.2

Distribution of Attention in "Navies in War and Peace"

Chapter	Approx. No. Words/ Chapter	Coverage/ Historical Period*
[Introduction]	1,520	
The Distant Past, but Important for Understanding the Role of Navies	2,490	
Russia's Difficult Road to the Sea	3,050	
The Russians in the Mediterranean	2,440	
Into the Oceans on Behalf of Science	860	28 percent Pre-World War I
The Russian Fleet during the Industrial Revolution and the Transition from Sailing Vessels to Steam Vessels	1,685	
Navies at the Beginning of the Era of Imperialism	3,550	
The First World War	5,045	9 percent World War I
The Soviet Navy	3,770	18 percent Interwar Period
The Development of the Soviet Navy in the Period from the End of the Civil War to the Outbreak of the Great Patriotic War [1921-1928; 1928-1941]	5,335	
The Second World War	4,565	26 percent World War II/ Great Patriotic War
The Soviet Navy in the Great Patriotic War	3,700	
The Basic Missions Executed by Navies in the the Second World War	4,840	
Navies as a Weapon of the Aggressive Policy of the Imperialist States in Peacetime	3,555	19 percent Present Era
[Intermediate Conclusion]	935	
Some Problems in Mastering the World Ocean	2,655	
The Problems of a Modern Navy	3,480	
[Conclusion]	520	
Total	53,995	

*Calculated after subtraction of introductory and concluding chapters from total.

homeland. They are also there in the "active defense of peace" to deter intervention in littoral affairs by the U.S. Sixth Fleet and to deter aggression by littoral powers supported by the Sixth Fleet. The presence of Soviet naval forces in the Mediterranean thus achieves three objectives: the strategic defense of the Soviet Union, increasing Soviet "international authority" (i.e., its coercive or negative political influence), and evoking "international sympathy for the Soviet Union" (that is, its positive political influence).

Discussion of the Present

His discussion of the contemporary or "problem" aspects of the subject continues two of the main themes developed in his historical discussion. The first of these is the importance of the sea and all forms of seapower to a maritime state such as the Soviet Union. This is treated in terms of observations on the current movement to redefine questions of maritime boundaries and the ownership of oceanic resources. The second theme, which is closely linked with the first, is the utility of seapower as an instrument of state policy in peacetime. This is treated largely in terms of the Western experience. In addition, in the chapter on "The Problems of a Modern Navy" and the two concluding sections, he deals directly with three current military questions: (1) the threat posed to the Soviet Union and its interests by imperialist naval forces, (2) Soviet policy regarding the response to those threats, and (3) the material and other requirements of implementing that policy. In keeping with standard practice, much of this portion of the discussion focuses on Western armed forces, although his treatment of the contemporary Soviet naval mission and force structures is unusually explicit.

According to Admiral Gorshkov, the two principal features of the contemporary era are the continuing aggressiveness of imperialism and the growth of the economic power and defense capability of the Soviet Union. The latter provide for the security of the entire socialist community and—given the new balance of forces in the international arena—are altering the structure and content of international relations in favor of the forces of "peace and progress." The growth of the Soviet Navy, and its emergence onto the high seas, have made a major contribution to these changes.

The imperialists (primarily the United States, in which the navy is the "pet" instrument of foreign policy) have actively exploited the special political features of navies—their ability to demonstrate economic and military power, and the capability of protecting overseas interests through coercion, largely without resort to force. They have used their navies in a variety of ways—including general support for their diplomacy—but their primary employment has been (1) by demonstrations of force, to put pressure on the Soviet Union and the other countries of the socialist community; and (2) by the threatened or actual use of force, to retain or restore supremacy over former colonies and other victims of economic oppression. Their specific objectives in undertaking such demonstrations have included showing their own resolve, deterrence of the intended actions of opponents, and provision of support to friendly states. The imperialists are also employing their naval forces in support of their efforts to achieve domination over the resources of the ocean.

Admiral Gorshkov states that the Soviet Union has decisively opposed all of these imperialist actions. In contrast to the imperialist modus operandi, the Soviet Navy, while fulfilling its mission of defending the Soviet Union against attacks from the sea, has by its presence acted as a diplomatic force of deterrence and containment of aggression. It has thus been employed as an important political weapon in its own right, as well as providing significant support to other instruments of foreign policy—especially through the increasing number of official visits and business calls by Soviet warships to foreign ports, which improve interstate relations and strengthen the "international authority" of the Soviet Union.

Given the imperialists' actions, the Soviet Union needs a powerful navy not only to defend its state interests on the seas and oceans but to defend itself against attack from the sea. The magnitude of the latter problem has grown since the end of World War II; and the threat of sea-based nuclear missile attack against the Soviet Union has elicited, in response, the construction of a new, ocean-going Soviet Navy—which now poses to a potential aggressor the same threat posed to the Soviet Union. This new Soviet Navy is nevertheless unique. Its composition has been determined by the threat, the material-technical base upon which it is constructed, and its assigned missions. It has acquired nuclear weapons and ballistic missiles (which gave it strategic capability and a strategic role), cruise missiles (for use against surface targets), SAMs (the main means of AAW at sea), AAA guns, electronics, and nuclear propulsion for its submarines (which gave it ASW capability). As a result, its combat capabilities have been greatly increased.

Submarines—especially nuclear-powered submarines—and aircraft have become its primary strike forces, although there is a continuing need for various types of surface ships—to "give combat stability to" (that is, protect) submarines, and to carry out a wide variety of tasks in both peacetime and wartime. The diversity of those tasks requires the construction of numerous types of surface ships, with different armament for each type.

The acquisition of these capabilities is a reflection of the changing role of the Soviet Navy, which now has three components:

1. strategic defense—"participating in crushing an enemy's military-economic potential;"

2. strategic deterrence—"becoming one of the most important factors in deterring an enemy's nuclear attack" (SSBNs are more survivable than land-based launchers and consequently represent a more effective deterrent); and

3. peacetime political influence—"visibly demonstrating in peacetime to the peoples of friendly and hostile countries" both the extent of Soviet capabilities and Soviet readiness to use those capabilities in defense of its state interests and for the security of the socialist commonwealth.

The navy has acquired these capabilities because only a force capable of blocking aggression can deter it; and—together with the strategic missile forces—the navy now represents such a force.

THE QUESTION OF CONTEXT

On first inspection, "Navies in War and Peace" appears to be a relatively straightforward presentation of a coherent system of views—presumably those of Admiral Gorshkov himself—on a complex and obviously important subject: the changing wartime role and place of navies and their employment in peacetime as an instrument of state policy. That is what the introduction leads the reader to expect, and it is easy to find just that in Gorshkov's discussion.

It is just as easy—and very tempting—to read more than that into it. While the reader is told at the outset what the discussion is about, and what benefits such a discussion can produce—for example, an understanding on the part of "command personnel" of the "special features" of the navy, and hence a "unity of operational views" on its role, place, and employment—nowhere in the discussion itself is it made clear why Admiral Gorshkov felt it necessary that it be written, let alone published.

Were such a publication routine, this would not present a significant problem. However, the discussion itself, and its manner of publication, are anomalous. The reader is therefore left—no, forced—to speculate on (and search the text for clues to) the occasion for its appearance; and the rationale postulated for the appearance of the discussion cannot fail to influence the interpretation given its content. The opportunity for circular reasoning is obvious. To minimize that danger and improve the chances of determining not only what the discussion says, but what it means, an analysis must go beyond contents to context and attempt to establish the antecedents of the statement.

We do not know when or under what circumstances the statement was written:* But we do know where and when it was published and, therefore, something about the surrounding domestic and international circumstances, which may provide some clues to its significance. Also, Admiral Gorshkov and other members of the Soviet leadership have written elsewhere on the subject, and attention to the continuities and discontinuities between "Navies in War and Peace" and the other statements should also prove rewarding.

Anomalies in Publication

Morskoi sbornik appears monthly. Like all Soviet publications, it must be cleared by the censors before printing, and the dates that it is "signed to typesetting" and "signed to press" are duly noted in each issue.† Examination of these dates for the issues containing installments of "Navies in War and Peace" reveals significant deviations from established practice. These dates are listed in Table 29.3. Figures 29.1 and 29.2 depict the same information in terms of days ahead of (or behind) month of issue for 1967-73. The issues of primary interest (February 1972 through February 1973) are represented on those graphs by dashed lines. Inspection of the data reveals that (1) the April 1972 issue was "signed to typesetting" a month late; (2) the July 1972 issue—the initial interruption in the publication of the series—was "signed to press" a month early; (3) the August 1972 issue—which resumed the series—and all subsequent issues—including the January 1973 issue, which was the second interruption in its publication—were late being "signed to press"; and (4) regular publications practices were not restored until after the conclusion of the series. Given the relative stability of publication practices during the preceding five years, these deviations are obviously unusual; and they deserve to be explained, if possible.

One approach is to attempt linking the contents of the installments with external events at the time each was being prepared for publication. Table 29.1 above identifies the contents of each installment. Table 29.4 below identifies potentially relevant contemporary events.

---

*Except for the eighth installment—"The Soviet Navy in the Great Patriotic War"—which is a near word-for-word duplicate of an article by Admiral Gorshkov entitled "The Navy in the Great Patriotic War" that appeared in the May 1970 Morskoi sbornik.

†It has not been possible to establish the actual publication dates of individual issues.

TABLE 29.3

Dates Morskoi sbornik
"Signed to Typesetting" and "Signed to Press"

Issue	Signed to Typesetting	Signed to Press
2-72	21 Dec 71	28 Jan 72
3-72	20 Jan 72	25 Feb 72
4-72	19 Mar 72	31 Mar 72
5-72	22 Mar 72	29 Apr 72
6-72	19 Apr 72	29 May 72
7-72	19 May 72	04 Jun 72[a]
8-72	22 Jun 72	11 Aug 72
9-72	21 Jul 72	07 Sep 72
10-72	22 Aug 72	06 Oct 72
11-72	21 Sep 72	02 Nov 72
12-72	21 Oct 72	08 Dec 72
1-73	21 Nov 72	10 Jan 73[b]
2-73	20 Dec 72	05 Feb 73

[a]First interruption in the series.
[b]Second interruption in the series.

The delay in typesetting the April 1972 issue of Morskoi sbornik might only reflect apprehension over the effect of the initial installment. Then again, this third installment contains exceptionally forceful denunciations of the Tsarist lack of appreciation of seapower—for example, references to "dull" figures and emigre reactionaries in the naval leadership, who "dismantled" the fleet. These cannot have been welcome words to the Soviet leadership in February and March 1972, when they were preparing for the opening of the final SALT negotiating session, in which limitations on naval systems were to figure prominently.

The second major anomaly in the publication of the series occurred with the July 1972 issue of Morskoi sbornik, which was not only "signed to press" a month early but did not contain an installment of "Navies in War and Peace"—although it did have an article by Admiral Gorshkov on the role of the commanding officer of a ship. It is difficult to escape the impression that this latter article was a "filler," inserted perhaps to mitigate embarrassment at the interruption in publication of "Navies in War and Peace." If so, then it is

FIGURE 29.1

Date Morskoi sbornik Signed to Typesetting

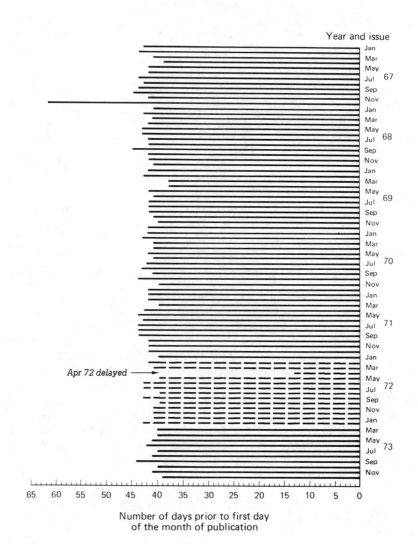

FIGURE 29.2

Date <u>Morskoi sbornik</u> Signed to Press

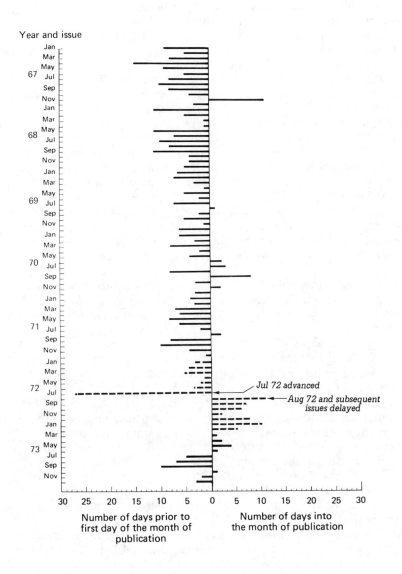

likely that there were objections to publication of the sixth installment—at least at that time, since it eventually appeared in the August 1972 issue. There was no loss of continuity between the fifth and sixth installments, and it appears reasonable to conclude that the material that appeared in the August 1972 issue of Morskoi sbornik could as well have appeared in the July 1972 issue.

The sixth installment, which covers the 1928-41 period, contains disparaging references to both the naval arms limitations conferences of the interwar period—"the war of the diplomats for supremacy at sea"—and various "minuses" in the prewar construction and training of the Soviet fleet, most stemming from underestimation of the combat capabilities and potential strategic contributions of naval forces. The July 1972 issue was "signed to press" five days after the signature of the SALT agreement placed rigid constraints on the further growth of the Soviet SLBM force, at a point where the Soviet leadership would have been sensitive to the appearance of critical words from the navy on naval arms limitations and the failure to appreciate the potential of naval forces.

The alternative explanation—that the July 1972 article on the role of the commanding officer is itself an important statement, perhaps more important than the interruption in the publication of "Navies in War and Peace"—must also be considered. Party control of the Soviet military is not contested directly. However, it is contested indirectly, through emphasis on military professionalism; and Admiral Gorshkov's discussion of the role of the naval officer epitomizes that professionalist argument. Perhaps then it was a deliberate, though veiled, warning.

The third major anomaly—the second interruption in the series—occurred with the January 1973 issue of Morskoi sbornik. This interruption is more difficult to explain. The final installment, in the February 1973 issue, contains the only portion of the text that, judged on the grounds of content, style, and continuity with the general thrust of the discussion, appears "out of place" in the statement: an excursion into questions of the Law of the Sea and the ownership and exploitation of oceanic resources. This segment may well be an afterthought, and the delay in publication of the final installment due to technical difficulties encountered in including the material.

However, the concluding installment also contains a vigorous exposition of the navy's role in the "active defense of peace"—including an explicit claim that the navy's presence on the high seas is a vivid demonstration of the willingness and capability of the Soviet Union to defend not only itself and its own interests, but the security of socialist countries as well. But, when the January 1973 issue was being prepared for publication, the Soviet Union was not "actively defending" the security of one socialist country, the Democratic

Republic of Vietnam (DRVN), which was under heavy attack by the U.S. forces in the Linebacker II operation. It may have been felt that the conspicuous absence of Soviet naval forces in the Gulf of Tonkin would raise embarrassing questions of Soviet credibility and that a month's delay in the appearance of the conclusion of "Navies in War and Peace" would be "the better part of valor."

It is also worth noting that beginning with the resumption of publication of the series after the July 1972 interruption Morskoi sbornik was consistently late being "signed to press" and that the publication schedule that had obtained over at least the previous five years was not reestablished until after the series had been concluded.* This implies the establishment of some sort of additional review and screening procedures for "Navies in War and Peace," probably coincident with the May 1972 decision to interrupt publication. It also suggests—as do the other, more dramatic anomalies in its publication— that Admiral Gorshkov's military or political superiors (or both) were not entirely happy with "Navies in War and Peace," at least as initially proposed by Admiral Gorshkov.

Other—and far less dramatic—publication anomalies also support this suggestion. Morskoi sbornik normally contains some 30 individual articles, averaging 5,000 words each, organized under the following general headings: Lead Articles—often containing one or more special subdivisions of articles focused on a current event such as "The Decisions of the 24th CPSU Party Congress" or "The 50th Anniversary of the Formation of the USSR"; The Naval Art; Combat and Political Training; The Pages of History; Armament and Technology; Phenomena of Nature; Foreign Navies; and Critique and Bibliography.

Given the position Admiral Gorshkov holds, the relationship between that office and Morskoi sbornik, and the obvious importance of what he was saying, one might expect the installments of "Navies in War and Peace" not only to have been published without interruption but to have been given a prominent place in each issue of Morskoi sbornik. While all 11 installments appeared in the "lead articles" section, none appeared as the "lead article"—although it should be noted that his July 1972 article on the role of the commanding officer of a ship did.

The placement of the installments of "Navies in War and Peace" within the "lead articles" section varied widely:

---

*Only 11 percent of the 61 issues published prior to the appearance of "Navies in War and Peace" were late being "signed to press"; on the other hand, 54 percent of the 13 issues spanning the publication of the series were late.

Issue	Placement	Issue	Placement
2-72	5	8-72	3
3-72	4	9-72	3
4-72	3	10-72	8
5-72	3	11-72	10
6-72	3	12-72	7
		2-73	3

With one exception, where special subdivisions of the "lead articles" section appeared, they were given precedence over the installments of "Navies in War and Peace." The February and March 1972 issues contained no such special subdivisions; all subsequent issues from April 1972 through February 1973 did. The subordinate placement of the first two installments is therefore not readily explicable—at least on these grounds. The exception to the "rule" was the May 1972 issue, in which the fourth installment of "Navies in War and Peace" appeared <u>ahead</u> of a special subdivision devoted to "The 50th Anniversary of the Formation of the USSR." The apparent "burial" of the 8th through 10th installments in the October through December 1972 issues resulted from the natural expansion of that "50th Anniversary" subdivision coincident with the anniversary itself and can be safely ignored. The subordinate placement of the first two installments is probably a further reflection of apprehension over the reaction that the appearance of the series would elicit and as such can also be safely ignored.

The exception to the "rule" on the other hand, cannot be ignored. The placement of the fourth installment ahead of the special subdivision was an obvious attempt to give it prominence. It appeared in the May 1972 issue, and covered the period of World War I—an ostensibly innocuous period in Russian and Soviet naval history. The contents of that installment were, however, anything but innocuous. It (1) explicitly criticized the prevailing technique for determining the required size and composition of the Soviet armed forces;* (2) implicitly claimed an expanded role for and increased effectiveness of naval strategic strike forces—especially in achieving the political objectives of a war; (3) advocated the establishment of naval

---

*The "retrospective method," which—in conformity with Marxist-Leninist methodology—emphasizes the discovery and generalization of the "lessons" of experience (that is, the last war); and according to Admiral Gorshkov must be augmented by "scientific prediction" of the future development of science and technology and the probable conditions of future combat if misleading results are to be avoided.

supremacy as much, if not more, for its peacetime political impact as for its wartime military utility; and (4) explicitly attacked naval arms limitations.

It was, of course, in May 1972 that SALT moved into its final stages—and produced an agreement. The appearance of such criticism at this time may have played a major role in the decision to interrupt the publication of "Navies in War and Peace" and to introduce the implied review process that disrupted Morskoi sbornik's publication cycle. Given these circumstances, the placement of the July 1972 article on the role of the commanding officer as the "lead article" lends credence to the hypothesis that it was intended to be more than just a "filler" and—however implicitly—conveyed a message.

Context of Publication

It is obvious that the SALT negotiations had considerable influence on the publication of "Navies in War and Peace." However, other events must also have affected the series. (See Table 29.4.)

The 24th CPSU Party Congress and the Ninth Five-Year Plan (1971-75), which implemented the programs promulgated by the Congress, were significant events that occurred prior to its publication that also had an effect on the series. The Soviet military establishment cannot have been overly pleased with either. The Congress signaled quicker movement toward detente with the West—specifically endorsing efforts to achieve an arms limitation agreement and to implement other conflict-dampening measures. It also signaled increased attention to Soviet domestic needs; and this latter emphasis was reflected in the plan. In essence, the Congress gave the military this message: The Soviet Union's ability to influence international events—including its ability to deter an attack on itself—was a function of its economic as well as its military power; strengthening the economic and technological capabilities of the Soviet Union was the most effective way to strengthen its defense capability; consequently there would be more emphasis on the industrial base supporting, but no major shift in resources toward, the armed forces per se; so that, in order to achieve the increase in combat capability it desired, the military would have to make better use of the resources they already had.[3]

In endorsing detente with the West, however, the Congress was by no means endorsing Soviet "isolationism"—quite the contrary! The relaxation of tension in direct Soviet-Western relationships was but one of four component parts of the "Peace Program" promulgated by the Congress, and those other three components called for the Soviet Union and the Soviet armed forces to play a more active role

TABLE 29.4

Context of Publication of "Navies in War and Peace"

Typeset		To Press		Events	
				1971	
				Mar.	24th CPSU Party Congress
				May 20	U.S.-Soviet agreement to resolve SALT problems within one year
				Nov.	Approval of Ninth Five-Year Plan (1971-75)
				Nov. 15	SALT VI negotiating session initiated
				Dec.	Indo-Pakistani War; U.S. and Soviet Deployment to Indian Ocean
I	Dec. 21				
				1972	
II	Jan. 20				
		I	Jan. 28		
				Feb. 4	SALT VI negotiating session concluded
		II	Feb. 25		
				Mar. 9	Soviet offer to clear ports accepted by Bangladesh
III	Mar. 19				
IV	Mar. 22				
				Mar. 28	SALT VII negotiating session initiated
				Mar. 30	DRVN "Easter offensive" initiated
		III	Mar. 31		
V	Apr. 19				
				Apr. 20-24	Kissinger visit to Soviet Union (discussions on SALT, summit meeting, and Vietnam)
		IV	April. 29	Apr. 29- May 6	Soviet ballistic-missile submarine (SSB) visit to Cuba
				May 8	Operation Linebacker initiated
				May 15-16	Soviet air force and naval firepower demonstrations in Egypt
				May 19	CPSU Central Committee plenum/Politburo membership changes
		V	May 29	May 24-30	Summit meeting, SALT and detente-related U.S.-Soviet agreements
VI	Jun. 22				
				Jul. 17	Soviet forces expelled from Egypt
VII	Jul. 21				
		VI	Aug. 11		
VIII	Aug. 22				
		VII	Sep. 7		
				Sep. 10-13	Kissinger visit to Soviet Union
IX	Sep. 21				
		VIII	Oct. 6		
X	Oct. 21				
				Oct. 31	Open break in U.S.-DRVN negotiations
		IX	Nov. 2		
		X	Dec. 8		
				Dec. 18	Operation Linebacker II initiated
XI	Dec. 20				
				1973	
				Jan. 13	U.S.-DRVN agreement
		XI	Feb. 5		

in the international arena. These other components were increased cooperation with the member states of the socialist commonwealth; an intensification of efforts to achieve international agreements that would alter the form and structure of international relations to minimize opportunities for conflict; and the "active defense of peace"— which encompassed not only deterrence of Western attempts to coerce the members of the socialist commonwealth, the newly independent states, and national liberation forces, but active support of them in case they should be attacked. In effect, cooperation in direct Soviet-Western relations was to be complemented by continued—if not increased—competition with the West in the "third world."[4]

"Navies in War and Peace" explicitly incorporated—and exploited—this line. It argued that by demonstrating its military—and hence its economic and technological—capability in the international arena the Soviet Union acquired influence over events in that arena, and that the navy was the branch of the armed forces best suited to making such demonstrations in peacetime—that the navy through its "forward deployments" was, in fact, effectively engaged in the "active defense of peace."

A third event that occurred prior to the publication of "Navies in War and Peace"—the December 1971 Indo-Pakistani war—provided the navy with two occasions to conduct such demonstrations. One demonstration was the deployment of countervailing Soviet surface combatant and submarine forces into the Indian Ocean in response to the "threat" posed to India by British and U.S. carrier task forces.[5] The other was the deployment of Soviet minesweeping and salvage units to clear the ports in Bangladesh—which coincided with the appearance of the initial installments of the series and may well have been undertaken to make the same points to the Soviet leadership in action language that Admiral Gorshkov was then beginning to elaborate in Morskoi sbornik.[6]

Three additional demonstrations were undertaken in the course of the publication of "Navies in War and Peace." The first was the movement of a ballistic missile submarine to Cuba in April 1972 (probably intended to have an influence on the final SALT rounds, and possibly also undertaken as a further illustration of Admiral Gorshkov's message). The second was a firepower demonstration off the Egyptian coast in May 1972 (probably part of a Soviet attempt to forestall the imminent expulsion of Soviet forces from Egypt). The third was the deployment of countervailing Soviet surface combatant and submarine forces into the South China Sea in May 1972 in response to the concentration there of U.S. carrier strike forces involved in Operation Linebacker (a response that defies ready explanation, since the Soviet forces simply deployed, anchored for a while, and then sailed home—having undertaken and accomplished nothing). In addition, there was an opportunity for a naval demonstration in response to Operation Linebacker II in December 1972, and no action was taken.

Comparison of "Navies in War and Peace" with Other Statements

There are significant differences between "Navies in War and Peace" and earlier statements by Admiral Gorshkov—in particular his last major article, "The Development of Soviet Naval Art," in the February 1967 issue of Morskoi sbornik. It also focused on the role, place, and employment of the Soviet Navy but dealt almost exclusively with the military aspect of these questions: the evolution of the navy's general roles, specific missions, and actual uses in wartime; the relationship between those roles, missions, and uses and the evolving naval force structure; and the course of future development of the navy's combat capabilities. Like "Navies in War and Peace" it also drew heavily on the lessons of history, but its focus was restricted to the Soviet era, which meant that it dealt almost exclusively with "military history": World War I, the Civil War, the Intervention, the Great Patriotic War/World War II.

"Navies in War and Peace," on the other hand, focuses less on the military than on the foreign political aspect of the navy's role, place, and employment and consequently employs a far broader historical base as a source of "lessons" in support of its argument. It could not be otherwise. Marxist-Leninist methodology requires that Admiral Gorshkov's argument be based on the lessons of experience; but the only relevant Russian experience he can cite is that of the Tsarist era. Until quite recently, the Soviets have not really used their fleet, and in any event political sensitivities obviously preclude his reference to those more recent uses.

There is a second reason for Admiral Gorshkov's taking "Navies in War and Peace" far beyond Soviet military history. When discussing that history for military purposes, he is bound by the dictates of Soviet military doctrine. However, his principal topic—the employment of naval forces in peacetime and the political purposes—lies outside the purview of military doctrine, and to the extent that he can keep the discussion beyond these boundaries, he retains considerable freedom of expression.

Furthermore, were "Navies in War and Peace" a doctrinal discussion like "The Development of Soviet Naval Art," it would cover not only the role, place, and employment of the fleet but its future development as well. In 1967 Admiral Gorshkov stated that discussion of the latter was a "natural" concomitant of a discussion of the former;

and one of the principal functions of Soviet military doctrine is the control of future developments. In 1972-73, however, he explicitly eschewed discussion of the future course of naval construction—and deviated from that guideline on very few occasions.

"Navies in War and Peace" is unique in other respects as well. There are significant differences between what Admiral Gorshkov said there and what his military superiors were saying at the time and have said since—although a recent statement by Marshal Grechko provides some evidence that the Soviet military leadership may be moving toward endorsement, or may have already endorsed, the fundamentally activist Soviet military role in the international arena that Admiral Gorshkov was advocating. The political leadership had already endorsed the expansion of the "internationalist functions" of the Soviet armed forces at least as early as the 24th Party Congress; but—except for Admiral Gorshkov—the military leadership in general, and Marshal Grechko in particular, appeared reluctant to even discuss, let alone embrace, that mission. Until May of 1974, Grechko, in describing the missions of the Soviet armed forces, had routinely limited them to the defense of the Soviet Union per se, and the defense—in concert with the armed forces of the other member states of the Warsaw Pact—of the entire "Socialist Commonwealth." On rare occasions, he had stated that the mission might also include the defense of the state interests of the Soviet Union, but without indicating what or where those interests might be. (At times he even appeared to equate the defense of Soviet state interests with the defense of the Soviet Union itself.) In the May 1974 issue of <u>Voprosy istorii KPSS</u> (Questions of CPSU history), however, he explicitly stated that the armed forces' "internationalist functions" had been given "new content" and clearly implied that this "new content" was the protection and promotion of the overseas interests of the Soviet Union—which is precisely what Admiral Gorshkov was discussing in "Navies in War and Peace."

SUMMARY JUDGMENTS

Perhaps the best way to present summary judgments on "Navies in War and Peace" is to focus on a limited number of fundamental questions:

- Why does it exist at all? What was the occasion for its writing? What was the occasion for its publication?
- Why does it exist as it does—a question of both form and content? It is a book-length piece. Why was it published in serial form? It is clearly directed to an audience that includes but goes

beyond the navy. Why was it published in a naval forum, rather than an armed forces-wide, or national, medium? Why is it cast in terms of an extended discussion of historical rather than contemporary actors and actions?
- What does it say?
- What does it mean (if something other than what it says)?
- What does its publication imply—first for the Soviet Navy, and second for the Soviet Navy's potential opponents?

Few of these questions can be answered in full here, and some not at all.

First, while we know quite a bit about its publication, we know relatively little about its origins, and should really reserve judgment on that issue—although it is reasonable to conclude that the bulk of the statement was written prior to December 1971, when the initial installment was "signed to typesetting." Given the context of its publication, it seems reasonable to conclude that the primary stimulus of its appearance was a review of Soviet state policy, occasioned by the impending conclusion of a SALT agreement, conducted at the central decision-making level and covering Soviet foreign policy, its attendant military requirements, and the resources necessary to meet those requirements.

Second, if such a review did take place, it seems reasonable to conclude that the navy's top leaders and their supporters at the central decision-making level feared that its outcome would prove unfavorable to the hard-won momentum for the development and employment of Soviet naval capabilities. They attempted to mobilize the additional support necessary at least to continue that momentum. Given the sensitive nature of the subject, the highly political atmosphere in which such a review would have been taking place, and the fluid state of affairs in the international arena at the time, incremental publications of the statement provided an opportunity not only to minimize unfavorable reactions but to take advantage of developments that would strengthen the argument it presented. Its appearance in a navy publication is probably attributable to high-level opposition to the policy it advocated—opposition sufficient to block access to broader forums but insufficient to prevent its appearance in a navy-controlled journal. It is an extended historical discussion for two reasons: Because it is only in terms of oblique discussion that current or anticipated Soviet state policy can be openly criticized by the military; and because of the domestic and international sensitivity of its argument, recourse to pre-Soviet naval history was the only available oblique way to advocate a more active anti-imperialist policy—and the navy as the principal instrument of that policy.

Third, as just noted, it says that navies are important, effective, and utilitarian instruments of state policy in both war and peace: more important and effective in wartime than has heretofore been recognized in Soviet military doctrine; more effective and utilitarian in peacetime than any of the other branches of the armed forces— not only as deterrents, but also in the accomplishment of positive ends. This effectiveness of navies is based not only on their combat capabilities but on the absolute import of the medium in which they operate and derives largely from their potential for the exercise of political influence. That potential cannot be exploited, however, unless its existence is recognized and adequate steps are taken to meet its necessary conditions—that is, provision must be made for the acquisition of a large and modern navy; and, once acquired, both its numerical and its technological strengths must be maintained.

Fourth, given its context and its content, it appears to contain the following general message, addressed to the Soviet central decision-makers. Despite domestic and international pressures from the other branches of the Soviet armed forces (which do not understand what the navy can do and covet the resources now being allocated to it), from the Soviet military leadership (who are drawn from and favor the army, are in general insufficiently "progressive," and in any event tend to equate "internationalism" with "adventurism"), and from detente-oriented factions in the Soviet political leadership (who underestimate the threat posed to both the Soviet Union itself and its overseas interests by Western naval forces), as well as from the U.S. (for obvious reasons) the Soviet Union cannot afford to limit expansion of the capabilities and the scope of employment of its navy. If it does, it will be unable adequately to protect and promote its overseas interests in peacetime, deter attacks on itself, protect itself if attacked, or achieve the ultimate policy objectives of any war it might have to fight with the imperialist powers. On the other hand, if it maintains a large and modern navy commensurate with its interests as a great power and exploits the political influence potential provided by such a navy, it will be able to implement its policies more effectively both in peacetime (as is well known) and in wartime (a fact not generally appreciated).

Fifth, the appearance of "Navies in War and Peace" implied that at the time serious consideration was being given in the Soviet Union to placing limitations on the growth and employment of the Soviet Navy—at least some of which were subsequently incorporated in the May 1972 SALT and other agreements. At the same time, however, its appearance implied that a significant element of the Soviet leadership appreciated the potential benefits to be derived from the acquisition, maintenance, and employment of a large and modern navy—an appreciation that not only kept those limitations from degrading the capabilities of the navy but resulted in a subsequent increase in its utilization as an active instrument of Soviet state policy. In neither case did its publication imply that Western navies could "rest on their oars"!

NOTES

1. The complete text of "Navies in War and Peace" is published in translation in the January through November 1974 issues of the U.S. Naval Institute Proceedings.
2. (Anonymous), "About the Journal Morskoi sbornik," Morskoi sbornik, February 1971, pp. 33-34; Fleet Admiral N. Sergeyev, "Friend and Advisor of Naval Officers," Morskoi sbornik (March 1973), pp. 17-22.
3. Capt. 1st Rank A. N. Kramar' and Capt. 1st Rank S. M. Yefimov, "The New Five Year Plan and Questions of Strengthening the Defense Capability of the Country," Morskoi sbornik, August 1971, pp. 3-9.
4. V. V. Zhurkin and Y. M. Primakov, eds., Mezhdunarodnyye Konflicty (International conflicts) (Moscow: "International Relations Publishing House, signed to press: 15 August 1972), translated in Joint Publications Research Service (JPRS) 58443, 12 March 1973; V. M. Kulish, ed., Voyennaya Sila I Mezhdunarodnyye Otnosheniya (Military force and international relations) (Moscow: "International Relations" Publishing House, signed to press 14 September 1972) translated in JPRS 58947, 8 May 1973.
5. James M. McConnell and Anne M. Kelly, "Superpower Naval Diplomacy in the Indo-Pakistani Crisis," Arlington, Va.: Center for Naval Analyses, Professional Paper no. 108, February 1973; reprinted in Michael MccGwire, ed., Soviet Naval Developments: Capability and Context (New York: Praeger Publishers, 1973), pp. 325-43.
6. Charles C. Petersen, Chapter 17 of this volume.

CHAPTER

30

THE SOVIET NAVAL
GENERAL PURPOSE FORCES:
ROLES AND MISSIONS
IN WARTIME
Bradford Dismukes

Technology and diplomacy have combined in the 1970s to make the world ocean the main stage of U.S.-Soviet military interaction for the extended future. This interaction now occurs at two levels:

The first is when both sides use naval general purpose forces (GPF)* in their "political presence" missions targeted against each other and against third-world states. Interactions at this level (analyzed elsewhere in this volume) probably offer the most likely contingencies in which combat between the superpowers may erupt in the immediate future.

Secondly, the naval GPF balance almost certainly will have a significant effect on high-level strategic relationships between the two countries. At that level in the past, the question has often been asked, What threat do Soviet naval forces pose to the security of the U.S. SSBN force? The official answer appears to be little or none. For example, throughout the debate on the Trident system and its predecessor, ULMS (under water long-range missile system) U.S. naval spokesmen expressed the view that, while no specific ASW threat to Polaris/Poseidon now exists, Trident is a hedge against the possibility of such a threat and (more recently) a "counter" to the Soviets' D-class SSBN.[1]

This paper deals with the relationship between GPF force levels and capabilities and the SSBN security issue from the Soviet point of

---

The ideas expressed in this chapter are those of the author. They do not necessarily reflect the views of the Center for Naval Analysis, nor of the Department of Defense and the Department of the Navy.

*General purpose force is a term of Western origin referring to all naval forces except ballistic missile launching submarines.

view. Its purpose is to urge students of Soviet naval policy to reexamine the body of available evidence by asking two questions that have thus far been given scant attention: What kind of threat to their own SSBNs do the Soviets perceive in Western ASW systems? What has been their reaction, if any, to this potential threat?

One should say at the outset that these questions do not arise because of explicit U.S. policy. The United States has generally (and in the author's opinion, properly) avoided policy statements that imply that its ASW capabilities are targeted against Soviet ballistic missile submarines. At the national level, spokesmen like Admiral Elmo Zumwalt do not list ASW defense against SSBNs as one of the four "capabilities" of the U.S. Navy. (These capabilities are strategic offense, sea control, projection of force ashore, and political "presence.")[2] However, the U.S. has not yet gone as far as specifically excluding defense against SSBNs as a naval task; the utilization of U.S. ASW capabilities in a damage-limiting role to blunt a possible Soviet SLBM attack is a logical policy option. Indeed during the early years of the McNamara incumbency in the Department of Defense, naval GPF systems were described as a "defense against submarine launched ballistic missiles." This characterization disappeared by February 1965, at the same time that assured destruction became the dominant objective of U.S. strategic programs.[3] At the national level, U.S. policy, based on the requirements of mutually assured destruction, now reflects little desire to threaten the security of Soviet sea-based strategic systems. However, ASW defense against SSNs can be brought to bear against SSBNs. For example, a recent public affairs brochure describing the U.S. Pacific Fleet states, in connection with the missions of Commander Third Fleet (ASW forces in the Pacific), "Protection of the United States against the threat of submarine-launched ballistic missiles is the newest and most complex of ASW responsibilities."[4]

U.S. BEHAVIOR

Although statements of this kind may be a source of concern to the Soviets, Soviet naval planners probably give equal or greater weight to three other factors in assessing the current and future security of their SSBN force: (1) the possibility of a breakthrough in ASW sensor technology that would permit ASW forces to maintain continuous fire control solutions on SSBNs; (2) patterns in U.S. resource allocation between various types of ASW (area search, convoy defense, and so on) and between ASW and other areas of naval warfare; and (3) observed trends in U.S. naval operations, exercises, and tactical development and evaluation.

What little information is available about Soviet views on these subjects will be described later; it will be useful at this point to review the evidence publicly available in the West on U.S. ASW activities in order to conjecture about its possible impact on the Soviets. The basic question is, Have those activities been of a scale and character likely to have produced concern in the USSR with SSBN security?

It is the hypothesis of this paper that, whatever the U.S. intentions, the answer is yes. Unofficial arms control advocates in the United States have made SSBN security a central theme of recent investigations. Scientists like Herbert Scoville and Richard L. Garwin have proposed measures that would isolate "conventional" ASW for the protection of surface ships from "strategic" ASW, which threatens ballistic missile submarines.[5] While these technical specialists generally agree that "there is no danger to the invulnerability of the sea-based deterrent either immediately foreseeable or on the horizon," the fact of raising the issue to the level of public discourse could encourage Soviet planners to reexamine their own assessments closely.[6]

The evidence from U.S. resource allocation patterns probably further impels the Soviets toward concern with SSBN security. Soviet writings indicate they estimate that a third of the U.S. Navy's procurement budget goes to antisubmarine warfare. To offset the retirement of specialized, antisubmarine carriers, U.S. attack aircraft carriers (CVAs) have recently been programed to become CVs, embarking ASW aircraft in addition to fighter and tactical strike aircraft. The newly designated CVs have been assigned a "sea control" mission; however, the objective for which sea control is sought is left somewhat open-ended.

In general, U.S. ASW programs are justified in U.S. writings by the requirement to defend U.S. naval forces wherever they may operate, especially when they protect military and economic sealift to NATO in the context of a European land war. This latter justification the Soviets may find puzzling if a short-duration general war scenario can properly be imputed to them. In brief, although the wartime missions of the U.S. GPF navy may be well understood by U.S. officials and analysts, Soviet understanding of those missions may be substantially less clear.

In addition to the development of the CV, "Sea Control" and other ASW ships of more radical design are planned. The United States has invested in large-scale undersea surveillance systems, which in some cases may appear to the Soviets to be targeted against the operating areas of their strategic submarines. The United States is currently acquiring advanced land- and sea-based ASW aircraft (the P3C and the S3A), whose combined total appropriations had reached nearly $4 billion by FY 1975.[7] The budget justification for the U.S. nuclear attack submarine program has long characterized

SSNs as effective ASW platforms. The mission description of the latest, SSN-688-class, continues that justification. The 688 is being acquired to "destroy enemy ships, primarily submarines, in order to prohibit the employment of such forces in attacks and destruction of U.S. or allied targets."[8] Which targets are being threatened by which enemy "submarines" is not specifically stated; the ambiguity of the mission description at least offers the possibility that the SSN-688 and U.S. ASW systems in general may have Soviet SSBNs as targets.

SOVIET VIEWS

Senior Soviet spokesmen have not publicly addressed the broad question of the wartime missions of all U.S. ASW forces. However, at the technical level they have discussed the role of Western nuclear attack submarines. These they apparently regard as having primary pro- and anti-SSBN missions. In 1971, a review of Western ASW in the Soviet Naval Digest referred to the primary mission of U.S. nuclear submarines as "combatting other submarines and protecting their own guided missile submarines."[9] The following year a second naval spokesman observed in the same forum that Western SSN's were intended "to track and destroy nuclear powered guided missile submarines and attack submarines."[10] This discussion of submarine missions is particularly significant, because the Soviets have made quite clear that they regard the submarine itself as the most effective antisubmarine platform.[11]

Little information is available on the critical question of Soviet assessments of the quality of Western ASW systems. They have publicly articulated an interest in U.S. ASW detection and attack systems, ascribing ranges of tens of kilometers to U.S. sonars.[12] They appear to be attentive students of Western ASW developments, writing in detail (always citing "specialists in the West" as a source) on U.S. fixed underwater surveillance systems, ASW mine developments like Captor, and the organizational structure of U.S. ASW forces. In general, however, the available evidence only permits one to conclude that the Soviets are aware of the range of U.S. efforts in ASW and, at the technical level at least, perceive these capabilities can be targeted against their SSBNs.

There is a final and important reason to suspect that the Soviets may in fact be more concerned with the SSBN defense problem than they have publicly revealed thus far. This has to do with the effects of uncertainties about an adversary's capabilities that are always inherent in defense planning and the common technique of assessing a threat via the mode of pure capabilities analysis. It seems possible that if the Soviets apply some version of pure capabilities analysis to the body of evidence on U.S. behavior described above, they may already have concluded that the U.S. Navy has adopted, but not announced (or will soon adopt) a basic mission of ASW defense against Soviet SSBNs.

While we have no direct evidence of the role it plays in Soviet planning, pure capabilities analysis is a familiar tool of defense analysts. In the pure capabilities perspective one of an opponent's principal capabilities—in this case, U.S. ASW systems—is matched against a very high value and potentially vulnerable stake—the Soviet SSBN force—in a scenario involving several worst-case assumptions. If the Soviets employ this mode of analysis, their conclusions do not hinge so much on overt U.S. verbal and operational behavior as on potential U.S. actions, which prudent Soviet planners must prepare for.

The Soviet SSBN force is almost certainly the highest-value entity that the Soviet Union must maintain in the world ocean; further, its importance is likely to increase as long as mutually assured destruction remains the basic theme of de facto U.S.-Soviet strategic relations. Because of large investments and substantial experience over an extended time period and perhaps because of the organizational emphasis of the U.S. Navy on ASW, the United States may appear to possess a serious capability to threaten that force. It thus appears that Soviet maritime strategists in 1974 have substantial reason to accord significant attention to SSBN defense, particularly in light of the Soviets' historical concern with defensive missions.

THE RANGE OF POSSIBLE SOVIET RESPONSES

If one surveys the record of Soviet naval behavior, is it possible to detect evidence of concern with a Western ASW threat to Soviet SSBNs? The answer to this question is quite uncertain; the evidence is mixed. In theory, at least two responses* employed independently or in combination appear open to the Soviets:

• They can attempt to dilute the effectiveness of any potential U.S. threat by increasing the range of their SLBMs, thus enlarging the area in which their SSBNs can operate while remaining on station;

---

*Other responses are theoretically possible, including mounting additional self-defense systems on their SSBNs, an alternative that might tend to nullify the SSBN's ability to remain concealed if prelaunch employment of defensive weapons systems were necessary.

- They can adopt what might be termed a sea control approach with the object of (1) threatening U.S. SSBNs, or (2) defending their own SSBNs on station.

The Soviets have clearly adopted the first approach. The SS-N-8 missile mounted on the new D-class SSBN has a range publicly announced in the West on the order of 4,000 nm. This represents about a threefold increase in range over the SS-N-6 missile currently carried by the Y-class and permits the D-class to be on station while in or near home waters. This increase in range affords the opportunity to disperse deterrent platforms more widely. It also presents the possibility of concentrating D-class submarines (or other platforms that potentially might carry the SS-N-8) in order to defend them with naval and land-based forces in areas near to the USSR. The Soviets have not publicly described such a possible tactic for their own forces, although it would be quite congruent with the defense-of-the-homeland tasks that make up the primary missions of the Soviet naval GPF in wartime. It is quite noteworthy however that in a recent article Soviet nonnaval political analysts have suggested that the United States might concentrate its Trident submarines for defensive purposes. Active defense of a unilaterally declared safe area for Soviet SSBNs in the Norwegian Sea or off the Kamchatka coast would appear to be attractive alternatives in a time of severe tension, after SS-N-8 inventories permit. (Declaration of such an area in the Norwegian Sea before the longer-range missiles of the U.S. Trident system are operational in significant numbers might, of course, produce a severe dilemma for U.S. planners if the Atlantic Polaris/Poseidon force operated in the same area.)

In the near term, however, until the SS-N-8 is operational in sufficient numbers, the Soviets may be seriously concerned with the second approach—sea control as it relates to the strategic platforms of both sides. A considerable portion of Soviet ASW effort has reportedly focused on the U.S. Polaris/Poseidon force in the last five years. One could expect explicit anti-Polaris/Poseidon operations to intensify as SALT II negotiations proceed. But according to the public statements of the U.S. leadership the Soviets currently lack any significant capability against SSBNs. Indeed in the Gorshkov series in <u>Morskoi sbornik</u> (February 1972-February 1973) the possibilities of ASW defense against ballistic missile submarines are not discussed favorably. (The evidence on this point tends to be of a negative character. This interpretation is based primarily on an inference comparing Gorshkov's treatment of the anti-SSBN ASW with that of Sokolovskiy.)[13] On the other hand, it's probably fair to say that the Soviets already possess significant capabilities to contend with at least the surface component of any ASW forces the U.S. might seek to bring to bear against Soviet SSBNs in transit to or operating on their current launch stations. This reality would appear strongly to encourage the Soviets to place high priority on countering such forces.*

The Soviets can significantly improve their ability to operate in the high-threat environment implied by this mission if they employ a force of large air-capable ships such as those currently under construction at Nikolayev on the Black Sea. It is the conjecture of this paper that the Nikolayev ship is being developed principally for the purpose of enhancing Soviet GPF sea control capabilities in the pro- and anti-SSBN mission area. However, alternative missions are clearly possible and a review of the evidence regarding other roles for the ship is required.

ROLE OF THE AIR-CAPABLE SHIP

Although intelligence apparently remains sketchy, the air-capable ship is publicly described in the West as displacing on the order of 40,000 tons and being capable of operating both helicopters and fixed-wing V/STOL aircraft.[14] Two major missions appear possible for this ship: projection of power ashore on the U.S. attack carrier or amphibious helicopter carrier models; and sea control.

Considerable overlap exists among the capabilities required to perform these missions, but, on balance, the evidence seems to support the conjecture that the new air-capable ship will be designed primarily for a sea control mission.

First, sea control is a general prerequisite to projection of power ashore. The ship is not large enough to combine major sea control systems—ASW and AAW—with significant capabilities to project power ashore—strike aircraft or amphibious helicopter lift. If it is to be exclusively an amphibious helicopter ship (the Soviets described the use of helicopters in the amphibious phase of Maneuver Okean),[15] it would appear excessively large for the intercoastal mission on the Eurasian continent traditionally assigned to Soviet amphibious forces.

It is clearly too early to foreclose on the possibility that the ship is a commando carrier (which also exist in the navies of France,

---

*Advocates of mutual deterrence in the United States may have difficulty in imagining the possible contingencies in which the United States might choose deliberately to threaten the Soviet SSBN force; however it is doubtful that decision-makers responsible for the development of the D-class (or the Trident) would be similarly troubled.

the United Kingdom, and Australia), designed for essentially political missions in the third world. If so, it would presumably rely on other Soviet units for AAW and ASW support. However, operation of a commando carrier would clearly be the most radical single step the Soviet Navy could take and would strongly intensify and accelerate the trend toward political intervention in the operations of the Soviet Navy that has emerged since 1967. In this regard, the ship, whatever its primary mission, will no doubt engage in distant deployments and almost certainly will be used for political effect. The ship will be a major symbol of both the global capability of the Soviet Navy and its claim to equality with the Western naval power.*

In the final analysis, however, the strategic relationship with the United States has been the governing concern of Soviet naval planners. The Moskva-class antisubmarine cruiser's ASW capabilities were widely interpreted by naval analysts in this perspective (that is, Moskva was estimated to be primarily a counter-SSBN platform).[16] It seems therefore likely that the Nikolayev ships will be in this sense a second-generation Moskva, which (presumably) bring fixed-wing, sea-based air to contend with the strategic defensive problems that the Soviets perceive.

If this line of argument is valid, two aspects of the Moskva-class CHG are particularly important. First, by providing organic sea-based air in the form of helicopters, the CHG significantly extended the range of sensors in the field of ASW. The Nikolayev ships will extend sensor range in the fields of AAW and antisurface ship warfare. The latter is particularly important because of the sensor (and to a lesser degree) the attack range advantage that a CVA task force appears to enjoy over Soviet surface units. Moskva also represented a quantum improvement in ASW and AAW self-defense capabilities; the Nikolayev ship will almost certainly improve on Moskva, especially if air defense fighter aircraft are carried.

A second, and perhaps less obvious aspect of Moskva-class operations has been the CHGs' possible role in task force coordination.[17] It is an occasionally controversial tenet of Soviet naval planning, according to Soviet doctrinal writings, that success in combat requires, where possible, the coordinated employment of surface, subsurface, and air units. Soviet descriptions of their exercise activity frequently reflect this concept. In addition, the Soviets have been particularly forthright in expressing the view that the two major

---

*And equality with other navies as well. For example, the Soviets probably find the lack of sea-based air capability embarrassing when they deploy to the Indian Ocean, where the Indian Navy operates an attack carrier.

obstacles to be overcome in ASW are submarine detection and coordination of Soviet attack systems.[18] Because of her size and implied capacity to carry major communications and signal processing systems, and because of her potential to communicate reliably with submarine, surface, and air units, the new air-capable ship seems likely to have major responsibilities for coordination of Soviet operations designed to achieve tactical superiority in an ocean area of operations.[19] If this is the case, one might expect to see increasing Soviet attention to tactical data links and communications relay aircraft in addition to the helicopters and early warning and electronic warfare aircraft that that task implies.

However, it cannot be specifically determined at this time whether the Nikolayev ships are to be employed to protect Soviet SSBNs against a possible Western threat or whether they will be used as the nucleus of a force designed to attack the sea-based strategic systems of the West. Because of the great similarity of shipboard sensors and weapons systems necessary to accomplish either task, it may be that the question will remain unanswered for some time to come.

IMPLICATIONS OF CURRENT SOVIET SSBN OPERATING POLICY

The crucial evidence regarding possible Soviet concern with SSBN security necessarily comes from current SSBN operating policy itself. In 1973 the Soviets apparently maintained only two to three Yankee-class SSBNs within range of the United States.[20] Assuming they continue this practice, it would appear to have two obvious implications: (1) the Soviets do not expect a sudden surprise attack by the West that would destroy these forces; and (2) they are apparently confident that they can move the remainder of the SSBN force into launch position when they desire to do so.

It is the latter point that is of direct concern here. There appear to be four possible sources of Soviet confidence: (1) They simply do not believe that existing or planned Western naval forces have the capacity to impede their SSBN deployments. For the reasons argued above, this is unlikely to be the case. (2) They will deploy their SSBN force during the period of increasing tension preceding hostilities and depend on the protection of the peacetime situation for its security. Although Western planners have apparently rejected such a policy for their SSBNs on the grounds of risk and the felt need for immediate retaliation, we do not know enough about Soviet views to reject this alternative as entirely implausible.* (3) The Soviets can withhold a significant portion of their SSBNs from the initial nuclear exchange by deploying it in presumably safe territorial or near-shore waters. It has been argued elsewhere in this volume that Admiral Gorshkov advocates such a "withholding strategy" on the grounds that naval forces in being are a vital necessity for intrawar or postwar bargaining. This strategy implies that the force could then be deployed at will, if one assumes that some significant portion of Western capabilities to oppose it would have been eliminated. (4) Finally, the Soviets could employ their GPF naval systems (and other elements of their armed forces) to protect a surge deployment of the force as the war begins. Whatever other benefits existing Soviet SSBN operating policy affords, it probably makes a larger portion of the force available for surge deployment when compared to a policy of maintaining peacetime deployments continuously at a relatively high level.†

The latter two cases imply a range of wartime tasks for the GPF sector of the Soviet Navy associated with a pro-SSBN mission. In the case of the "withholding strategy" GPF requirements would vary with the timing of SSBN employment and the size of the remaining Western force. In the last case—in which the force might fight its way to launch position—GPF requirements might be very significant indeed. This is particularly so because the tactical situation would be exactly the reverse of the defense-of-the-homeland role long associated with the Soviet GPF Navy. The Soviets would not adopt blocking positions for a defense in depth against an expected Western penetration; instead, the Soviets might be required to penetrate a Western defense in-depth.

CONCLUSIONS

It has been argued here that the character and scale of U.S. (and other Western) ASW activities has been such that one could plausibly expect the Soviets to be concerned with the security of their SSBN force; that current technical realities appear to encourage them

---

*No evidence is available from Soviet or Western sources regarding Soviet SSBN deployment policies during recent international crises.

†Assuming that their SSBN force faces no physical constraints on its deployability—crew availability, system reliability, and so on—current Soviet SSBN deployment practice appears to offer the advantages of lower operating costs and reduced risk against accident.

to increase their capabilities to contend with a potential threat to that force from Western ASW systems; that the air-capable ship is likely to be assigned both pro- and anti-SSBN missions; and that the pro-SSBN task may lie closer to possible achievement.

The evidence from Soviet verbal and operational behavior—particularly the current peacetime SSBN deployment policy—is difficult to interpret until the question of the wartime role of the SSBN force is resolved. The possible existence of a "withholding strategy" may in itself reflect Soviet concern with the SSBN security problem. The evidence does not permit a conclusive judgment. At this time it would appear useful for analysts to add the pro-SSBN mission to the list of possible wartime tasks of the Soviet navy.

NOTES

1. See Admiral Elmo Zumwalt's remarks in U.S. Congress, Senate Committee on Appropriations, Hearings on Department of Defense Appropriations, Fiscal Year 1973, 92d Cong., 2d sess., February 23, 1972, p. 101.

2. U.S. Congress, Senate Committee on Appropriations, Hearings, Department of Defense Appropriations Fiscal Year 1973, 92d Cong., 2d sess., February 23, 1972, p. 68.

3. See U.S. Congress, Senate Committee on Armed Service, Hearings on Military Procurement Fiscal Year 1964, 88th Cong., 1st sess., February 19, 1963, p. 71; Senate Committee on Appropriations, Hearings on Appropriations for the Department of Defense, Fiscal Year 1965, 88th Cong., 2d sess., February 5, 1964; p. 109; and Senate Committee on Armed Services, Hearings on Military Procurement Authorization, Fiscal Year 1966, 89th Cong., 1st sess., February 8, 1965, p. 66.

4. Public Affairs Office, Commander in Chief, U.S. Pacific Fleet, U.S. Pacific Fleet, Fleet Post Office, San Francisco, 90610, no date, p. 12.

5. Herbert Scoville, "Missile Submarines and National Security," Scientific American 226 (June 1972), pp. 15-27; Richard L. Garwin, "Anti-Submarine Warfare and National Security," Scientific American 227 (July 1972), pp. 14-25.

6. Kosta Tsipis, Anne H. Cahn, and Bernard T. Feld, eds., The Future of the Sea-based Deterrent (Cambridge: MIT Press, 1973), p. ix.

7. U.S. Congress, Senate Committee on Armed Services, Hearings on Department of Defense Authorizations, 93d Cong., 2d sess., March 4, 1974, Part 3, pp. 1149 and 1151. Procurement costs through FY-75 for the P3C and the S3A amounted to $2.0752 billion and $1.8192 billion respectively.

8. U.S. Congress, Senate Committee on Appropriations, Subcommittee on Defense Appropriations, Hearings, 93d Cong., 1st sess., Part 3, p. 1153.

9. Captain First Rank D. P. Sokha, "The Past and Present of Submarine Forces," Morskoi sbornik, no. 9, 1971.

10. Captain First Rank N. Aleshkin, "Some Trends in the Development of Naval Forces," Morskoi sbornik, no. 1, 1972. In addition Vice-Admiral A. Sorokin and Captain First Rank V. Krasnov, "Anti-Submarine Defense Reviewed," Nauka i Zhizn', no. 1 (January 1972), signed to press 7 December 1971, pp. 48-55. Joint Publications Research Service (JPRS) 55386, Translations in USSR Military Affairs no. 788, 8 March 1972.

11. Among other works, see N. I. Suzdalev, Podvodniye Lodki Protiv Podvodnikh Lodok (Submarines vs. submarines) (Moscow: Military Publishing House, November 1970).

12. G. Svyatov and A. Kokoshin, "Sea Power in the Plans of the American Strategists, "Mezhdunarodnaya Zhizn', no. 3, 1973 signed to press 21 February 1973, pp. 77-86, JPRS 58538, Translations on USSR Political and Sociological Affairs, no. 350, 22 March 1973.

13. See James McConnell, "Admiral Gorshkov on the Soviet Navy in War and Peace," Center for Naval Analyses Working Paper, mimeo, 20 July 1973.

14. New York Times, 17 October and 27 February 1973.

15. N. I. Shablikov, "Ocean" Maneuvers of the USSR Navy Conducted in April-May, 1970, JPRS translation, 19 April 1971.

16. See, inter alia, Michael MccGwire, "Soviet Naval Capabilities and Intentions," The Soviet Union in Europe and the Near East (London: Royal United Service Institution, 1970), pp. 33-51.

17. Moskva's massive and complex array of visible electronic and communications equipment has been widely commented on. See for example Sigfried Breyer, Guide to the Soviet Navy (Annapolis, Md.: U.S. Naval Institute, 1970), p. 77.

18. See particularly Captain First Rank V. G. Yefremenko, "Development and Perfection of ASW Forces and their Tactics," Morskoi sbornik, no. 10, 1970, who intensifies in a narrower context the basic principles expressed in Suzdalev, op. cit., chapter 3.

19. As Yefremenko (op. cit.) reports, "modern surface ships as a rule carry helicopters and aircraft and have everything necessary to organize the control of various types of ASW forces and means."

20. Washington Star-News, 20 April 1973.

CHAPTER
31

THE COUNTER-POLARIS TASK
Harlan Ullman

Western interpretations of Soviet naval policy generally adopt one of two viewpoints. There are those who see the primary motivation that underlies current policy as a traditional reaction to a rival power's maritime superiority, as just one aspect of the Soviet Union's challenge to the United States. The other viewpoint sees this motivation as a more sharply focused response to the direct threat posed to the Soviet Union by the West's maritime capability. Whereas the first group perceives a clear break in Soviet naval motivations since Cuba, the second group considers that the primary motivation remains the same but that the drastic increase in the seaborne nuclear threat has engendered radical changes in Soviet operational concepts and consequential policies. For the sake of simplicity, the naval policies implicit in these two viewpoints will be referred to as "traditional" and "counterdeterrent."[1]

Against the second viewpoint, it is sometimes contended that (1) it is no longer the primary mission of U.S. aircraft carriers to launch nuclear strikes at Russia, and (2) the problems of countering Polaris are so great as to deter the Soviets from the attempt. Neither argument is very satisfactory. The first overlooks the fact that as long as U.S. carrier-borne aircraft have the capability to strike at Russia, the potential threat remains. The second disregards the Soviet Union's readiness to devote scarce resources toward providing a partial counter to some Western threat, and toward increasing the cost to the West of launching a successful attack.

This chapter is based on one part of the argument in the author's doctoral dissertation, "From Despair to Euphoria" (Boston, Tufts University, 1973).

Nonetheless these arguments cannot be ignored. For the "counter-deterrent" viewpoint to be convincing, it must be demonstrated that the Soviet Union takes the threat from seaborne strike units seriously, that it attaches high priority to countering this Western capability, and that on the basis of its own criteria, it has some chance of success. The matter of whose criteria are used to evaluate success is an important one.

Countering Polaris is obviously the nub of the problem. The purpose of this chapter is to test the hypothesis that counter-Polaris has been assigned to the Soviet Navy as a priority task, and that pending the development of a comprehensive solution, there are interim measures which (from the Soviet viewpoint) would have worthwhile results. Since we are concerned to identify the causes that underlay the shift in naval policy during the early 1960s, the analysis concentrates mainly on the internal Soviet debate during the first half of the decade. But when considering the feasibility of counterdeterrence, it looks at force levels from the middle 1960s on, since it would be these that would figure in their interim plans.

OFFICIAL SOVIET PRONOUNCEMENTS

Analysis of Soviet naval pronouncements since 1959 indicates that the primary requirement of Soviet naval doctrine has been to meet strategic nuclear criteria. Soviet writers have therefore considered the aircraft carrier, and later, Polaris as significant naval threats to the Soviet Union. The Soviet naval press, almost without exception (Gorshkov's recent series of articles included) has never attempted to challenge or deny the most important mission in "ocean theaters of war"—namely the destruction of the "missiles of nuclear missile-firing submarines." Since 1967, two adjuncts to the counter-deterrent requirement have received increasing attention—the peacetime role of naval forces in securing prestige and influence and protecting state interests; and the wartime requirements and capabilities, which are necessary in conditions of less than general nuclear war. A third adjunct became explicit by 1970, the specific naval mission of deterrence. But despite these additions, there has been no indication in the press, doctrine or in fleet exercises, that counterdeterrence is out of favor or that it is now regarded as a junior partner to these additional requirements.

Official Soviet military statements and doctrine have almost uniformly addressed the problems raised by the possibility of deterrence failing. Sokolovskij's Military Strategy was quite specific in its "war-fighting" direction and explicit notion that "mutual deterrence" was insufficient. And, since 1955, the military has been increasingly vocal in its plea for military-technical superiority. On the naval level, close examination of official statements and doctrine reaffirms the explicit meaning of Soviet war-fighting strategies, should deterrence fail.

In 1959 Khrushchev resurrected the minimum deterrence argument in an attempt to maximize security while minimizing expenditures. The strategic rocket forces were established in 1959 and the impact of nuclear-missile war was judged as sufficient reason for replacing large numbers of "obsolescing" ground forces. The army stood the most to lose under the new scheme and thus became a major adversary in the debate between the "traditionalists" who opposed Khrushchev's strategic new look and the "radicals" who supported it.

The navy, with an oar in both camps, chose a path compatible with Khrushchev's strategic designs. For several years naval strategy had been directed against the aircraft carrier and (since 1958), the specter of Polaris,[2] and it could therefore continue to build around a growing (but still small) counterdeterrent capability. The counter-Polaris task became the navy's unquestioned raison d'etre, if only because it provided a justification for increased capabilities. Therefore, in open literature, the navy never questioned the primacy of counterdeterrence, although Admiral Platonov, perhaps in an unguarded moment, let it be known that "not all the problems" related to hunting Polaris had been solved.

This unanimity on naval missions did not however extend to the composition of naval forces, and from 1960 until late 1961, a debate took place over the kind of navy required for counter-deterrence. The traditional "balanced fleet" school of thought, propounded by Kasatonov, believed that a judicious mix of submarines, aircraft, and surface ships was required:

> The essence of the problem is to create effective means for the distant destruction of submarines from the air which will make it possible to employ for their destruction the most effective modern means of destruction— missiles with nuclear charges launched from submarines, aircraft and [surface] ships and possibly also from shore launching mounts.
>
> The use of a single weapon [the submarines of PLO defense] in the struggle against such a formidable enemy as missile-carrying submarines is an emergency measure caused by the status and capabilities of the ASW forces and weapons at the given moment.[3]

A second more radical school of thought argued that the mission could be effected by submarines and aircraft with substantially less surface ship support. This position, which was probably only preferred by a minority of the navy, had the merit of conforming with Khrushchev's views and would allow the navy to operate on a reduced share of the defense budget. The "proponent" of this view was Admiral Platonov.[4]

In line with Khrushchev's strategic new look, elements of the strategic rocket forces were included in the counter-Polaris mission, at least in a secondary role, as was indicated by Gorshkov when he wrote "a frequent assertion at the time [1959-61] was that single missiles from land-based launchers would be sufficient for destroying strike dispositions of surface warships and even submarines."[5]

The vital issue of the naval "debate" was the role and the existence of the surface ship because, unless the navy could justify these ships under strategic nuclear criteria, then it was likely that surface-ship-building programs would be in jeopardy.[6]

By the time of the 22d Congress of the CPSU (October 1961), the "traditionalists" had won a few concessions from the "radicals." On the naval level, this meant that for the time being, "conventionally powered and armed ships hold an honored position and are assigned a great role in naval combat operations."[7] The navy had argued that the role of the surface ship would have several aspects in nuclear war. First, surface vessels in coordination with other branches of the naval and air forces support ASW operations, lend defense in depth, and thereby increase the effectiveness of the anti-Polaris forces. Second, surface ships would screen the deployment and operation of Soviet submarines engaged in anti-Polaris missions. Third, surface ships were necessary for antimining operations. Last, ships would support the ground force operations on land. And it should be noted that all of this occurred well before the Cuban Missile Crisis of October 1962.

But the best public statement of naval position came in the wake of the first edition of Sokolovskij's Military Strategy, released in the fall of 1962. Sokolovskij recognized that the navy must not be tied to the land theaters of operations and that the main aim was to defeat the enemy fleet and sever his maritime communications. "One of the most important tasks from the first minutes of war" would be to destroy enemy strike carrier forces. But combat against missile-carrying nuclear submarines was only rated "an important task."[8]

The naval reaction was vehement. Leading the attack was Admiral V. A. Alafuzov, who stated his purpose as being to "concentrate on those aspects of naval strategy which the book Military Strategy dealt with inadequately."[9] Alafuzov's criticism covered three broad areas: the threat from Polaris, the viability of the surface ship in modern warfare, and the growing importance of the Soviet Navy. Alafuzov stressed the imperative need to counter the

Polaris threat. American "missile-carrying nuclear submarines" were assigned a "very essential mission . . . side by side" with the aircraft carrier in making "nuclear strikes deep into enemy (USSR) territory." Repeating the need for surface ships four separate times, Alafuzov specifically noted tHat "The usefulness of surface ships against the missile-carrying submarine is acknowledged here in contrast to earlier statements sending the surface navy into discard."[10]

Last, Alafuzov accused the editors of restricting the operations of nuclear war to the strategic rocket forces and to the ground forces. In engaging a "maritime enemy . . . one whose territory can be occupied only after negotiating a sea obstacle. . . . one cannot get along without a navy." Alafuzov concluded by saying, "We think that success in modern war is achieved by the actions of all components of the armed forces, while the role of each one of them will be different in different phases."[11]

For the next two years, the naval "campaign" continued along the lines set by Alafuzov.[12] And, although he may not have been the main target, in August 1964 Sokolovskij came to the realization that "the foremost task of military operations in ocean theaters will be the destruction of the nuclear missile-carrying submarines."[13]

Intra- and extramural debate continued after Khrushchev's retirement in October 1964, involving the same institutional participants. In summarizing the statements and doctrine and the latter Khrushchev years, one can say that the political leadership's view was oriented toward strategic nuclear and not general-purpose forces. The army was relegated almost by definition to a conventional argument for military strategy favoring or, at least, not downgrading general purpose forces. The navy had no such constraints and, even if the naval staff favored a more traditional mission, Gorshkov realized that naval forces could be most certainly guaranteed under strategic nuclear criteria. Thus, although the Soviet Navy declared counterdeterrence as its primary mission, the requirement to equip a fleet that could operate in a multithreat environment brought with it significant capabilities having wider application. Whether or not the naval leadership was sufficiently far-sighted to reach this appreciation and, then, adopt Byzantine-like methods of achieving that end cannot be known. More likely (and simply) the general penchant that admirals have for ships, combined with the inherent flexibility of warships, gave rise to capabilities that exceeded the specific intentions and requirements of the political leadership.

The assumption of power by the Brezhnev-Kosygin leadership brought to a head two crucial and interdependent issues that came to dominate naval planning. The first focused on forward deployment and the requirement for overseas basing (including air rights) in order that a permanent presence could be established against Polaris and the carrier. The second and more complex issue derived from the attainment of near strategic parity with the United States and the dynamic impact this changed "balance" had on Soviet perceptions of power arrangements.

Much of the Soviet naval response to these issues, particularly in the press, seemed directed toward the conventional capabilities of navies in other than general war situations. This condition was fostered by the Soviet trend toward military "flexibility" and by the entry into service of new construction designated for counterdeterrence but that might be needed in confrontations arising from the shift to forward deployment. Implicit in these statements may also have been a campaign for larger naval appropriations.

Mid-1968 ushered in the formal era of strategic arms negotiations and the impact these arrangements would have across the entire Soviet spectrum of party, government, military, and industrial bureaucracies. The military was determined to guarantee some form of military superiority and, despite occasional arguments, which viewed nuclear war as potential "suicide for both sides,"[14] Soviet military doctrine and strategy conformed to the wartime requirements set forth in the third edition of Military Strategy (1968). The raison d'etre of the armed forces was to prevent a nuclear attack against the homeland; in the event deterrence failed, the mission became "war-fighting." However, as nuclear inferiority approached nuclear "parity," the exact demarcation of "wartime" and "peacetime" requirements appeared, at least in the press, slightly more blurred.

The Navy Day 1968 statements continued to emphasize the main forces of the navy: the nuclear submarine, naval aviation, and the surface ship.[15] Indeed, Gorshkov labeled the surface ship as "the pride of our navy." Meanwhile, V. A. Kasatonov noted, "The main means of deterring and, if need be, of decisively defeating an aggressor are now our strategic rocket forces and the missile-carrying nuclear submarine."[16]

With the entry into service of the Y-class SSBN, references to Soviet naval capability included both counterdeterrence and deterrence. However, the counterdeterrent element was not downgraded. A. Chabanenko, then the navy's representative on the General Staff, concluded that ". . . nuclear submarines [are] . . . the most effective means for the struggle against enemy missile submarines,"[17] and Moscow Radio added, "They [Polaris] are not invulnerable. . . . They can be spotted in time by ASW submarines, by planes and surface ships. . . ."[18]

But, within the navy, it was noted that "the problems of warfare against the newest submarines is far from solved."[19] It may be that, failing deterrence, the navy saw "damage-imposition" as the most efficient means of countering Polaris. On a parallel tack, the war-fighting naval mission received more specific definition from Gorshkov at least in terms of nuclear versus conventional weaponry.[20] Clearly, in wartime, any Soviet inferiority in the number of ships was to be offset by nuclear missilery.

However, by 1970 while continuing to show a very strong mixture of counterdeterrence and deterrence, these same statements contained a subtle and perhaps subjective inclination toward the peacetime application of Soviet strategic might as a means of constraining imperialist aggression.[21] Placed in the perspective of the political debate and modified by other statements by the military, these Soviet naval pronouncements suggested a crystallization of Soviet decisions concerning the strategic balance. First, mutual deterrence, which had not proven wholly disadvantageous to the Soviet Union during the period of nuclear inferiority, was appearing more credible and useful now that the strategic balance had become more "equal." The minimum deterrence formula of the early 1960s had spawned much greater Soviet force levels, levels that were then operational. Second, as the political attractiveness of mutual deterrence increased, the need to emphasize counterdeterrence as a measure to offset strategic inferiority was reduced; this was also because serious negotiations were in process with the United States, whose attention the Soviets wished to concentrate on defensive systems. Third, the Brezhnev-Kosygin trend toward a Soviet military equivalent of flexible response implied that the political utility of military force was increasing. The United States, meanwhile, was replacing flexible response with the Nixon doctrine. On the naval level, this "blue-water" strategy had particular general-purpose force application. The Soviet Navy had been developed for general nuclear war and was less well equipped to wage conventional war. But the threat to Soviet state interests implicit in such a "blue-water" strategy could under conditions of nuclear parity be countered by a naval peacetime presence.

This Soviet naval presence remains more or less coincident with likely Polaris operating areas. But these naval concentrations in the sensitive eastern Mediterranean and the Arabian Sea are at the same time providing a buffer against possible maritime encroachment by the United States and are therefore protecting the Soviet Union's state interests. Whether or not these peacetime missions with political objectives will provide a new rationale for forward deployment has yet to be determined.

Meanwhile, a review of Soviet exercises such as Sever 1968, Okean 1970, and Yug 1971 tends to confirm the proposition that Soviet doctrine is primarily concerned with satisfying conditions of general nuclear war after deterrence has failed, when war-fighting strategies will be required. The counter-Polaris role has been freely reported in the press, but in none of the exercises was it the sole consideration.

The current trend in naval pronouncements suggests that no one solution (or solutions) to the problems posed by global nuclear war has radically changed Soviet strategic thought. Statements continue to stress the new trinity (deterrence, war-fighting, and peacetime missions) and, although tempered by the recent diplomatic initiatives and SALT, they show no evidence of deviating from the holy writ.

THE FEASIBILITY OF COUNTERDETERRENCE

A counterforce strategy[22] is directed at denying the adversary the wartime use of his strategic nuclear weapons. It basically consists of offensive and defensive components. The offensive component attempts to destroy the adversary's strategic capabilities by an attack normally before the adversary's weapons can be launched. This is first strike. The defensive component attempts to destroy the adversary's strategic weapons after the other has been launched and enemy weapons are en route to strike the homeland. This is the damage-limitation aspect. At sea, counterforce means the attempt to destroy the strategic systems of an adversary, while protecting one's own strategic forces from attack.

The central problem in countering Polaris is detection and location. Ideally, the Soviet Union would like to know the exact location of each Polaris submarine all the time, which entails some form of continuous tracking. But even with the extensive underwater surveillance systems of the United States and the continuing research and development effort, detection is still a relatively short-ranged proposition. There are three basic means of detection: acoustic, electromagnetic, and infrared. Thus far, the most efficient means has been sonar, with electromagnetic and infrared detection limited to extremely short ranges against submerged submarines.[23] The problems are so severe that, excepting an ASW breakthrough and enormous expenditures by the Soviet Union, a reliable and effective very long-range tracking capability is fairly unlikely. To the extent that detection is possible, it would normally rely on a combination of ASW submarines, aircraft, surface ships, satellite, and underwater surveillance systems, supported perhaps by certain units of the extensive Russian trawler fleet.

Continuous tracking is the ideal solution because it offers the possibility of killing the submarine before it fires. But a Polaris submarine carries 16 missiles, and the damage-limitation aspect of counterforce will have been served if one or more of those missiles are prevented from reaching their target. Once the submarine launches its first missile, it is detectable by "backtracking" the missile path through infrared or laser means. Much would depend on the speed of reaction, but this may be increased by using statistical techniques to focus on areas that have a high probability for submarine operations. One such approach might be to develop reverse targeting data— that is, knowing likely targets and plotting backward to the possible launch areas. Another method would be to use data obtained from Polaris operations over the last 12 years, combined with information supplied by one's own ballistic missile submarine force and applied to the potential adversary in order to hypothesize his form of operations. The United States has in fact, asserted that none of its ballistic missile submarines when on station has ever been tracked or even detected by the Soviets.[24] The latter did however speculate on the location of the George Washington during her very first Polaris patrol undertaken in November 1960,[25] and articles in their press appear from time to time on the Soviet Navy's proven ability to track "enemy" ballistic submarines during exercises and training missions.[26] What information is available, tends to support the U.S. claim; however, it is altogether probable that the detection problem is not completely impossible and perhaps substantially less so than United States' planners overtly maintain.

AN OPERATIONAL CONCEPT BASED ON DAMAGE LIMITATION

Given the basic background to counterdeterrence, a relatively simple operational concept will be hypothesized, drawing where possible on implicit Soviet references.[27] The underlying assumption is that large nuclear warheads used by prepositioned surface, submarine, and aviation units, coordinated with area sanitization (or barrage bombing) and the "radical" detection and localization techniques described above, can produce some margin of effectiveness against the Polaris submarine. This measure of effectiveness would also require the use of more conventional ASW methods.

The use of large-megaton weapons raises the question of Polaris vulnerability. How much peak overpressure can the hull withstand; at what overpressures are the missiles and guidance systems rendered inoperative; and what overpressures can prevent the missile doors from opening? None of these questions can be readily answered but the power of these large weapons can be seen in the table below.[28] The slant range would not apply under water, but it gives a general impression.

Peak Overpressure (pounds per square inch)	Warhead Size (megatons)	Slant Range (statute miles)
150	1	20
	5	34

Peak Overpressure (pounds per square inch)	Warhead Size (megatons)	Slant Range (statute miles)
	8	40
	27	60
300	1	12
	5	20
	8	24
	27	36
500	1	7.5
	5	12.5
	8	15
	27	22.5
700	1	5
	5	8.5
	8	10
	27	15

Disregarding for the moment the cruise missiles of air, surface, and submarine units (because of time of flight delays over long distances), the area coverage provided by the SSB/SSBN force alone (using suborbital trajectories) is fairly extensive. The G-II and H-II SS-N-5 missile has a maximum range of about 650 nautical miles, which gives a theoretical area coverage of 1.3 million miles per submarine. The G-I with its SS-N-4 missile of 350 nautical miles range provides a theoretical area coverage of 390,000 square miles. The Soviet inventory includes 10 H-II SSBN, 12 G-II, and 10 G-I SSB, each with three missiles. These submarines are stationed with a total of 21 in the Northern Fleet and II in the Pacific Fleet. Assuming that under increasing tension the Soviets could deploy roughly half (55 percent) of their inventory, the mix would be as follows:

	Northern Fleet on Station	Pacific Fleet on Station
G I	3	2
G II	5	2
H II	4	2
	33 390,000	2 × 390,000
	9 × 1,300,000	4 × 1,300,000
Total theoretical area coverage	23,000,000 sq. mi.	6,000,000 sq. mi.
Total missiles on station	36	18

Additionally, the Y-class SSBN may also have a specific anti-Polaris mission. The point is that, at least theoretically, the Soviets possess sufficient naval assets to give broad weapons coverage over all the possible Polaris operating areas.

Besides prepositioning missile submarines for long-range destruction, the Soviets might assign chunks of ocean to the strategic rocket forces and coastal missile forces for area sanitization. These areas might include high Polaris probability areas in locations where it would be difficult to maintain Soviet units on station. There would be no guarantee of success, and a counterargument suggests the missiles risk being wasted. However, with the advent of MIRV, a nuclear "surplus" might make this option less wasteful. Also by electing to defend specific areas (for example, European Russia west of the Urals), likely operating areas could be reduced still further.

In time of war, destruction of U.S. Polaris bases would occur automatically. This point is continually emphasized by Soviet naval spokesmen as well as by Military Strategy. The importance of these targets suggests dual coverage by the strategic rocket forces and the soviet missile-firing submarines with possible support from long-range aviation (and perhaps surface ship cruise missiles) against Rota and Holy Loch.

Destruction, disruption, and incapacitation of Polaris submarines during transit and before firing is probably the responsibility of ASW submarines and naval aviation with surface ships most likely restricted to close-in and intermediate ranges, perhaps no further distant than the Poseidon outer firing ranges. The requirement of "marking" may be left to trawlers and a few ASW submarines on the theory that division of forces outside Polaris firing range is both unproductive and not without risk. Additionally, the strategic rocket forces may have some responsibilities in long-range area sterilization.

Shorter-range detection and destruction would rest with the conventional ASW forces. Coordination between the areas to be sanitized and Soviet naval forces would be necessary, and the close-in forces would have to coordinate closely with the prepositioned missile submarines. The missile submarines could be brought in as contact was made either through detection of the enemy submarine or of its missile once fired.

Despite the great difficulties, the theoretical possibility of rendering a portion, even though small, of the Polaris force ineffective has not been disproven. Add to this the "fog of war" and the unknown effects of multiple nuclear explosions in the ocean, and it is impossible to deny the possibility of Soviet "disruption, frustration, and destruction" of Western missile-firing nuclear submarines. And subjectively, there is a strong body of evidence that demonstrates the Soviets have not foregone any particular strategy because it seemed ineffective at the time.[29]

Obviously, the problem of command and control and weapons delay time would be critical. Assuming that detection of a missile could be transmitted to a prepositioned submarine, would the weapon's time of flight allow the Polaris submarine's destruction before it had launched its full salvo of missiles? Could communications between prepositioned units even be maintained, let alone guaranteed? And what would be the effects of thermonuclear weapons on Soviet command and control systems?

Finally, do the Soviets possess a sufficient number of large warheads for them to be used against Polaris? A simple count of potential Soviet launchers of all types suggests that enough would be available for this task. And this without benefit of the Soviet vantage point or knowledge of their particular sensitivities.

CONCLUSIONS

This chapter set out to demonstrate that the Soviet Union takes the threat from seaborne strike units seriously, that it attaches high priority to countering this Western capability, and that on the basis of the Soviet's own criteria, it has some chance of success. In particular it has sought to test the hypothesis that counter-Polaris has been assigned to the Soviet Navy as a priority task and that, pending development of a comprehensive solution to the problem, there are interim measures that (from the Soviet viewpoint) would have worthwhile results.

On the basis of official pronouncements there can be little doubt that the Soviet Union has always given high priority to what has been termed a "counterdeterrent" strategy and that the naval share of this task has been of increasing importance as the United States shifted their strategic emphasis from land-based to sea-based systems. It is quite clear that by 1961 the Soviet Navy considered counter-Polaris as the primary element of its share of this task and that this had been generally accepted within the military leadership by 1964 at the latest; to the extent there was argument, it was over means and not ends. There is no evidence implying that the importance of the counter-deterrent task has been downgraded, although it may now have to coexist with a more active peacetime role.

The pattern of operations and the output of the building programs are fully consistent with this hypothesis, and although some of the evidence is inherently ambivalent, other parts can only be read one way.

The importance of counterdeterrence in general and of the counter-Polaris task in particular stems from the planning assumption that mutual deterrence may fail and that the Soviet Union must be prepared to fight (and win) a nuclear war. Given such assumptions, damage limitation ceases to be an abstraction related to the credibility of mutual deterrence and becomes a vital part of the strategy covering the "initial stage of war," which figures so largely in Soviet military thought. In such circumstances, "some" limitation is better than "no" limitation and relatively low chances of success may be worth pursuing. It is noteworthy that when speaking of the Soviet Navy's counterdeterrence tasks, Admiral Gorshkov talks of "weakening" the enemy's nuclear strikes and not of preventing them.[30]

Pursuit of a counterdeterrent strategy is exceedingly difficult, but <u>providing</u> the expectation of effectiveness is well below 50 percent, and perhaps as low as 10 or 20 percent, it is still feasible * To the rational Western analyst, schooled in the abstractions of deterrence theory, such a return may seem unacceptable. But to the Soviet war planner, working out his strategy for waging nuclear war should deterrence fail, 10 or 20 percent may be sufficient to justify the costs, on the grounds that it could make the difference between victory (that is, survival) and defeat (obliteration). There can be no clear-cut conclusion as to the likely effectiveness of such measures, but for the advocates of a counterdeterrent strategy, the anti-Polaris task is perhaps not as <u>impossible</u> as conventional Western thinking tends to assume.

NOTES

1. Counterdeterrent is used in its literal sense of "against a deterrent" as in F.C. Ikle "Nuclear Deterrence," <u>Foreign Affairs</u>, January 1971, p. 271.
2. Admiral L. Vladimirskij, "Missile Weapons and the Conduct of Naval Combat Operations," <u>Sovetskij Flot</u>, September 21, 1956.
3. V. A. Kasatonov, "On the Problem of the Navy and Methods for Resolving Them," October 1961.
4. V. Platonov, "The Tasks of the Navy and the Methods for Carrying Them Out," January 1961.
5. S. Gorshkov, "The Development of Soviet Naval Science," <u>Morskoi sbornik</u>, February 1967, p. 17.

---

*A 50 percent effectiveness against a 50 percent Polaris/ Poseidon deployment level will leave as "survivors" approximately 100 deliverable megatons and 1,400 warheads on station—that is, 16 Poseidon and 5 Polaris on station, of which 8 Poseidon and 2 Polaris destroyed, leaves $8 \times 16 \times 10 \times 50$ KT plus $3 \times 16 \times 2 \times 200$ KT.

6. See Khrushchev's remarks in "A Cruiser Is Struck from the Lists," Leningradskaya Pravda, March 23, 1960.
7. Editorial, "Combat Watch of Soviet Sailors," Krasnaya Zvezda, January 9, 1962. See also N. V. Isachenkov, "New Weapons of Warships," Krasnaya Zvezda, November 18, 1961. This also fits MccGwire's estimate that the 1957/1958 decisions produced fewer surface ships than required.
8. V. Sokolovskij, Soviet Military Strategy, Rand Corp. translation (Englewood Cliffs, Note: Prentice-Hall, 1963), pp. 420 and 421.
9. V. A. Alafuzov, "On the Appearance of the Work Military Strategy," Morskoi sbornik, January 1963, pp. 88-96. Alafuzov was a wartime Deputy c in c and held the title of professor.
10. Ibid., p. 91.
11. Ibid., p. 95.
12. See S. G. Gorshkov, "Concern of the Party for the Fleet," Morskoi sbornik, no. 7 1963, pp. 9-18; V. P. Rogov, "U.S. Imperialists form a 'Polaris' High Command," Morskoi sbornik, May 1963, pp. 77-85; S. G. Gorshkov, "The Navy of our Fatherland," Kommunist Vooryzhenikh Sil, no. 13 July 1963, pp. 18-25.
13. V. D. Sokolovskij and M. I. Cherednichenko, "Military Art at a New Stage," Krasnaya Zvezda, August 25, 28 1964. The article also stated that "It is noted in the American press that submarines are very sensitive to underwater nuclear explosions." The one missing item from this happy naval picture was the surface ship to which Sokolovskij never attached much importance.
14. A. I. Krylov, "October and the Strategy of Peace," Voprosi Filosofii, no. 3 March 1968.
15. S. G. Gorshkov, Pravda, July 28, 1968.
16. V. A. Kasatonov, Krasnaya Zvezda, July 28, 1968.
17. A. Chabenenko, Literaturnaya Rossiya, July 25, 1969.
18. Moscow Radio (English) 0100GMT November 2, 1969. This broadcast, of course, may have been more concerned with the propaganda effects and the SALT negotiations which had just begun.
19. Kvitnitskij, "Aircraft Against Submarines."
20. S. G. Gorshkov, Pravda, July 27, 1969. Gorshkov also noted, "The Soviet fleet has become a world factor and its operations give support to Soviet foreign policy . . . the modern fleet is capable not only of frustrating the attacks of any aggressor from the sea but also of striking crushing blows at his fleet in the most remote areas."
21. See, for example, Gorshkov, Izvestia, February 27, 1970.
22. For a Russian discussion of counterforce see Yu. N. Listvinov, First Strike: Some Trends in the Development of American Conceptions (Moscow: International Publishing House, 1971), p. 252.
23. The Soviets have gone to great lengths to record in the unclassified press, Western ASW efforts. See, for example, I. K. Firsov,

"Reconnaissance Forces and Equipment in Foreign Navies," Morskoi sbornik, March 1970, pp. 85-93; D. P. Sokha, "The Past and Present Submarine Forces," Morskoi sbornik, September 1971, pp. 20-29, and "Foreign Naval Chronicle," Morskoi sbornik, August 1972, pp. 103-4; A. Prostakov, "Sonar Communications: What Are Its Special Features?" Morskoi sbornik, December 1972, pp. 67-69; V. P. Zhukov, "ASW Aviation," Morskoi sbornik, February 1971, pp. 117-18; V. A. Kovalenko and M. N. Ostroumov, Handbook on the Foreign Fleets (Moscow: Voenizdat, 1971). So far as exotic Soviet ASW research and development efforts on lasers, see A. Prokhorov, "The Light of Lasers," Izvestia, March 10, 1970, p. 3, in particular: ". . . one can obtain a powerful infrared laser . . . for underwater vision systems." Moscow Radio (English) 0710GMT October 20, 1970, in FBIS, October 20, 1970, VIII, vol. 204, p. 1: "The use of lasers for the . . . ocean from orbital space stations opens up immense prospects [for science]." And V. Etkin, "The Space Robots' Sense Organs," Pravda, March 13, 1971, p. 2 cites laser sensors in Cosmo's satellites. And "Lasers," Morskoi sbornik, October 1972, pp. 90-91.

24. Admiral Thomas Moorer, now chairman of the U.S. Joint Chiefs of Staff, in an interview appearing in Ordnance, January/February 1970, pp. 23-24, responded, when asked if there was any possibility that Polaris submarines will lose that immunity currently enjoyed: "One would be less than prudent if he didn't watch all the time for this development you have mentioned. I can only say that right now I am not aware of such a development. We are watching every day to see if such an event will occur. So far we have seen no such indication." When asked if the Soviet ASW cruiser Moskva had been able to detect or track Polaris, Moorer answered, "I don't think so. That is just a professional opinion on my part."

25. Rear Admiral O. Zhukovskii, "Combat against Enemy Missile Submarines," Morskoi sbornik, October 1961. Zhukovskii noted that since November 1960, one or two American ballistic missile submarines were on patrol near Lofoten Island. He was referring to George Washington and Patrick Henry by name.

26. See N. Kharlamov, Radio Moscow, July 24, 1968 and his discussion of Operation "Sever." Also Yu. Dmitriev, "Ocean Patrol," Trud, August 18, 1970, p. 3 in which the writer implies that a submarine detected by Soviet naval air units was a missile firer. In reference to a British military commentator noting the "most difficult problem" being posed by missile submarines' invulnerability, Dmitriev noted, "but there never has been, nor is there, any absolutely undetectable weapon . . ."

27. The author must stress the implicit nature of these references and admit the real possibility of attempting to draw from these sources, more than was intended. See, for example Okean,

op. cit., p. 75. "For several hours the crew kept the enemy submarine under surveillance. Contact was steady . . . the command headquarters of the maneuvers did not give orders to destroy this 'Southern' missile-firing submarine. It had its own reasons for this." And also see Kvitnitskij, "Aircraft against Submarines," op. cit.; A. P. Anokhin, "Aviation in Combat against Naval Strike Forces," Morskoi sbornik, June 1970, pp. 33-36; V. G. Yefremenko, "The Development and Perfection of ASW Forces and Their Tactics," Morskoi sbornik, October 1970, pp. 16-23; V. P. Zhukov, "ASW Aviation," Morskoi sbornik, February 1971, pp. 117-18; V. G. Yefremenko, "The Evolution of ASW," Morskoi sbornik, November 1971, pp. 18-25; N. Semenov, "The Use of Aviation in Naval Warfare," Morskoi sbornik, December 1972, pp. 22-36; and I. M. Sotnikov and N. A. Bruznetsov, ASW Aviation (Moscow: Voenizdat, 1970), 207 pp.

28. All figures from Glasstone, Effects of Nuclear Weapons (Washington, D.C.: Government Printing Office, 1964), Table 6-73, p. 303; based on 1 KT overpressures using cube root scaling.

29. This is a point made by both MacKintosh and MccGwire who observe that the Soviets have oftentimes allocated resources for defense programs that the West perceived as "cost-ineffective."

30. Morskoi sbornik, February 1973, p. 21.

CHAPTER

32

SOVIET UNDERSTANDING OF "COMMAND OF THE SEA"
Peter H. Vigor

The famous series of articles by Admiral Gorshkov that appeared in Morskoi sbornik during 1972 and 1973 contain a number of instances of the use of the Russian expression gospodstvo na more, the English translation for which is "command of the sea." But just as the meaning that has been read into "command of the sea" by successive generations of Englishmen and Americans has varied considerably from epoch to epoch and even from person to person, so too the significance attached by Russians to "gospodstvo na more" has undergone important fluctuations. The purpose of this background paper is therefore simply to set down a kind of outline history of these various fluctuations, and to see what, if anything, can be learned from them about Soviet naval doctrine.

This is perhaps a suitable point at which to explain that the Western expression "command of the sea" or "mastery of the sea" was often translated into Russian as vladenie morem in the late 19th and early 20th centuries, as well as by "gospodstvo na more"; and in the works of some of the writers that appeared in the 1920s and 1930s both phrases occur with seeming indifference. Subsequently, "vladenie morem" became progressively rarer, and "gospodstvo na more" progressively more common; though the latest edition of the Great Soviet Encyclopedia contains at least one instance of the former. Consequently, although one is tempted to think that there must be a difference of meaning between the phrases, I can detect none; and accordingly throughout this paper I have treated them as synonymous; but of course it is perfectly possible that I have been wrong to do this.

THE MEANING ATTACHED TO THE TERM

To return to the question of the meaning of these expressions, I should begin by saying that, so far as I can trace it, their first appearance in the Russian language was due to the translation into that tongue of the classic Western writers on naval matters who flourished in the 19th century.

Colomb and Mahan

Of such men, Mahan and Colomb are the ones most closely studied in the Soviet Union today, though in the 1950s at any rate Corbett was studied too; and since the West also regards the works of these three men as naval classics, it will suit the requirements of this paper with regard to concision as well as accuracy if we base our discussion on Mahan's, Colomb's, and Corbett's writings, when we need to consider the traditional Western concept of "command of the sea."

At this juncture it might be as well to make the point that the USSR considers (or used to consider until recently) Colomb, not Mahan, to be the "onlie begotter" of this theory.

Thus, both the first and the second editions of the Great Soviet Encyclopedia unite in declaring that "gospodstvo na more" was Colomb's concept. Mahan, indeed, is not even given an entry in the first edition; while though the second edition accords him one, it gives singularly little space to it and merely says that Mahan was the author of a "theory of sea-power" (teoriya morskoi sily).[1] In neither edition of the encyclopedia does the phrase "gospodstvo na more" occur in connection with Mahan.

Despite this, however, it seems to be ascribed to Mahan in the fourth of Gorshkov's recent series of articles, though the actual words used, mekhenovskaya teoriya 'gospodstva na more',[2] may mean no more than that the theory is conformable to Mahan's ideas, the sort of theory that Mahan might well have propagated, rather than one he actually originated. The article on Colomb in the third (1973) edition of the encyclopedia appears to confirm this; for there Colomb is described as originating a "so-called theory of 'vladenie morem'" (not, it will be noted, "gospodstvo na more") to which Mahan's "theory of sea power" is said to be analogous.[3]

Colomb's book Naval Warfare, which was first published in 1891, was almost at once translated into the major European languages, the first Russian edition appearing in 1894 under the title Morskaya Voina, and subsequent ones being published as late as 1940. It soon came to be regarded as essential reading for every Russian naval officer who took his profession seriously.

It is therefore appropriate that at this point we should stop and set down Colomb's understanding of the phrase "command of the sea," because it is obviously from this that "gospodstvo na more" derived.4 Interestingly enough, in Colomb's view it is something that was wholly incapable of attainment until the middle of the 16th century; because it was only then that ships were built that could stay at sea for any significant period. Once such ships were invented, however, it became open to their possessors to use them to try and win the mastery of the seas; though naturally enough they would make the necessary effort only in those sea areas that were of real importance to them, and only then if they were shrewd enough to appreciate the advantages it would bring.

According to Colomb, not every nation possessed this necessary shrewdness: The Spaniards, for instance, merely regarded the sea as a means of transit, not as something the command of which was of enormous value in its own right; and, according to Colomb, the downfall of the Spanish Empire was due to the failure of her rulers to perceive this vital distinction. A similar failure is ascribed by Colomb to Napoleon, whose celebrated wish to command the Channel for only a matter of an hour or two is taken by him as evidence that Bonaparte's view of the sea was that of the Spaniards; he saw it merely as a means of transit for his armies, whose ultimate defeat he thereby rendered certain.

But Colomb is emphatic that, even if a nation has a true understanding of the value of commanding the sea, the actual areas it will wish to command may well be very small. Thus, in the Dutch wars of the 17th century it was really only the English Channel that either the Dutch or the English sought to command; while in the wars between Sweden and Russia in the 17th and 18th centuries it was solely the eastern Baltic that either sought to dominate (though this is not, incidentally, an example that Colomb adduces, as indignant Soviet critics are quick to point out).

"Command of sea," as Colomb propounded it, consequently by no means implied command of all the seas the wide world over, or even of the majority. Obviously, there would be countries whose interests, being extensive, demanded that more than just home waters should be attempted to be commanded; and in the 19th century, at the time that Colomb was writing, Britain was clearly a prominent example of this. Consequently Colomb, a forvid British patriot, argued strongly for a big navy as a means of attaining that command over the world's great oceans that he believed to be necessary for Britain. But this was Colomb the publicist, not Colomb the naval strategist; and the two elements in his writing must be clearly distinguished by anyone trying to assess the value of his theories of naval warfare. A similar distinction should be made in the case of Mahan.

On one point, however, Colomb was unyieldingly insistent; and this was that, whatever the size of the sea to be commanded, it could only be done by destroying the enemy's fleet, or at least by rendering it inoperative by bottling it up in its ports.

The Russian Usage

If we now turn from Colomb's own exposition of his theory to consider the Russian understanding of it, we may take as an early example the writings of Admiral S. O. Makarov, whose "Rassuzhdeniya po Voprosam Morskoi Taktiki" was published in Morskoi sbornik in 1897. According to him, "gospodstvo na more" in Tsarist Russia meant the winning, and subsequent retention, of complete command of the sea either by destroying the enemy's naval forces in a big battle or else by blockading them in harbor, and so rendering them equally useless by another method. The resultant situation meant that the sea, in Makarov's words, was the "absolute property" of the victor; though Makarov went on to express his doubts about whether what had been proved to be true about naval warfare in the age of sail would necessarily be true for the age of steam; and he supported his doubts by reference to the strategy of the Sino-Japanese War of 1894-95.[5]

By the early 20th century, however, Tsarist naval officers, including some of Makarov's closest friends and associates, had ceased to believe that the change from sail to steam had made any difference. Vladimir Semenov, for instance, in his Flot i Morskoe Vedomstvo do Tsusimy i Posle, published in 1911, states categorically (p. 25) that the decisive role in winning "gospodstvo na more" belongs to battleships.

The early Soviet period seems to have understood "gospodstvo na more" in the same sense as Semenov had done. A representative figure of this period was B. B. Gervais, part of whose lectures to Bolshevik naval officers at Leningrad during the period 1919 to 1921 have recently been republished. They show clearly that "gospodstvo na more" meant, so far as he was concerned, the attainment of complete command of the sea by means of a general engagement, or else by means of blockading the enemy's naval forces "within their operational base."[6]

Gervais, of course was a former Tsarist naval officer, and his understanding of strategic and tactical concepts had naturally continued basically unaltered from the days of Nicholas II; but such an interpretation of the concept of "gospadstvo na more" was by no means confined to former members of the old Imperial Navy. Frunze, a soldier, a proletarian, a senior communist, and People's Commissar for War in the middle 1920s, had essentially the same understanding of the expression as Gervais did, though Frunze's interpretation was confined to the advantages to be had by obtaining "gospodstvo," especially for a maritime nation, and ignored the means whereby it was to be acquired.[7]

In the late 1930s, "gospodstvo na more" was defined as meaning

> the attainment of a superiority of force over the enemy in the principal direction and the pinning down of his forces in the secondary directions throughout an operation . . . that is to say, the creation of a situation where the enemy is paralyzed or gravely hampered in the conduct of his own operations, or else weakened and thereby hindered from preventing us from carrying out a given operation or completing an operational task.[8]

The author of this definition, the celebrated V. A. Belli, elsewhere stated that such "gospodstvo" might still be attained by a general engagement, even under the conditions of the 1930s; but that it was more likely that submarines and aircraft would have as big a part to play in the winning of it as conventional surface vessels.[9] He seems to have had no doubt, however, that "gospodstvo" could still be won.

In 1969, an authoritative Soviet publication defined "gospodstvo na more" a good deal more succinctly. It is said that this was based upon the reactionary theory of seapower of Mahan and Colomb and meant "the complete removal of the enemy's fleet from the naval theaters of war." The two Western writers are then cited as believing that such "gospodstvo" could be achieved by ships of the line acting on their own and that they could either achieve it by a general engagement or else by a blockade.[10]

The actual meaning attached by the Russians to the expression "gospodstvo na more," in the sense in which they derived it (and derive it) from the works of Colomb and Mahan, has therefore varied little from the time it was first originated to the present day. Admiral Makarov may have had reservations about whether it would prove to be fully applicable in the age of steam, and V. A. Belli, writing in the Stalin era, may have felt it behoved him to be guardedly circumlocutional; but the general sense ascribed to Colomb's expression did not, and has not, changed. It should be observed, furthermore, that it is wholly and completely conformable to the sense that Colomb ascribed to it himself (Admiral Makarov's reservation excepted).

CHANGE IN ATTITUDE TOWARD THE CONCEPT

Tsarist and Postrevolutionary

What has changed enormously has been the attitude adapted toward it. In Tsarist times, the Imperial Russian Navy accepted and thoroughly approved of it. In nautical matters, as in so much else, Western concepts, Western technology, and Western operational experience were widely admired in Tsarist Russia, and their validity was almost unquestioned. It is therefore no accident that, after the destruction of the old navy in the Russo-Japanese War, the replacement navy ordered by Nicholas II was deliberately modeled on the pattern of the contemporary Western navies, as Gorshkov bitterly complains.[11]

The point of so doing was that the new Russian Navy was intended to fight the same sort of war as that which the Western navies were intended to fight; and we all know that the latter were built to attain, and subsequently exercise, what Colomb called in English "command of the sea."

At this point it is important to make clear to the reader that the Russian Navy at the time of World War II had no hope of attaining "gospodstvo na more," still less of exercising it. This desirable attribute could only be acquired, it will be remembered, as a result of victory in a general engagement or of the maintenance of a more or less total blockade of the enemy's fleet in his ports. The Tsarist Navy of 1914 was quite incapable of winning a general engagement against a first-class naval power and equally incapable of blockading its ships in harbor.

The Russians claim that the imperial ships a-building were technically much superior to their Western counterparts; but they admit that, when World War I broke out, there were far too few in commission.[12]

Comments like these were based on the tacit assumption that, if only the outbreak of World War I could have been postponed for three or four years, the Tsarist Government would have possessed a navy that would have been fully equal to other navies in respect of both quality and quantity; and that, if this had happened, it would have made use of its improved performance to make a serious effort to win "gospodstvo na more," of those sea areas that were most obviously vital to it. If the Imperial Navy had had "gospodstvo" over the Black Sea, for instance, the course of World War I might well have been different.

In other words, in the minds of the Tsarist admirals there was no doubt that "gospodstvo na more" was the correct objective to strive for; and both their own writings and Gorshkov's article agree that this is so.[13] It does not, however, follow that the "gospodstvo" for which they were aiming was that command of the bulk of the world's oceans that was the objective of the Royal Navy and that, according to Colomb the publicist, was properly its objective.

In the years immediately following the revolution (that is to say, until somewhere in the early 1920s) the doctrine of "gospodstvo na more" was still accepted as correct in Soviet Russia, and men like Gervais and Petrov were still asserting that the only way to win it was by a general engagement or else by blockading the enemy's ships in their ports. It is essential to bear in mind, however, that at this period the Soviet Navy was far less capable even than the Tsarist of any attempt to win it, being fully occupied with the business of sheer survival (Lenin, it will be remembered, ordered the scuttling of the Black Sea Fleet in the summer of 1918, which meant that one whole fleet area was without a navy at all).

As a consequence, the Communist Party leaders, the rulers of the country, showed scant concern for the truth or falsehood of theories of "command of the sea" when Soviet Russia, in their opinion, was not likely to be concerned with naval warfare, except of the most minor kind, for the foreseeable future. Either the "imperialists" would seek to invade their country, in which case the naval war would be a matter of coastal defense, or else events would have moved so much to the communist advantage that they would be well placed to launch "revolutionary war"—but that would be by land.[14]

The Interwar Years

With the rebirth of the Soviet Navy, however, things began to alter. Gorshkov dates that rebirth as beginning in 1922;[15] and Ludri, as quoted by Herrick, appears to agree with him.[16] Certainly by the end of the 1920s Stalin's government was beginning to take a vigorous interest in Soviet naval doctrine.

And the now doctrine stemmed from the new circumstances. The industrial base envisaged by the five-year plans would allow Russia to build a respectable navy, if only time were given her. In the early stages however (that is to say, during the first plan and the first half of the second), the industrial base was weak and therefore inevitably was only capable of engendering a weak navy. For such a navy it was obviously impossible to think of winning "gospodstvo na more" by means of general engagements or total blockades.

The result of this situation was that the Soviet Government built the ships it was technically capable of building and then pronounced in favor of a naval strategy that would suit the ships it had built. The strategy in question is termed in Russian malaya voina, which is usually rendered into English as "small war" and which consists essentially of not attempting to win "gospodstvo na more" except in the limited, local context of the coastal waters of Russia and even then by methods other than the classic general engagement or total blockade.

The merits of the "small war" strategy had been much debated in Soviet naval circles for a number of years prior to its official adoption; and it is not to be supposed that Stalin himself invented it. Nor does it seem to be true that, when adopted, it was intended to be the permanent naval doctrine of the Soviet Union. As has been mentioned above, the "small war" theory stemmed from the circumstances of the time; and when those altered (and every loyal Russian of the 1930s convinced they were going to alter), the theory would alter too. Marxism-Leninism propounds as a basic principle the close interaction between thesis, antithesis, and synthesis, (or, in the concrete language of defense studies, between the weapons systems and the theory of how to use them); and this would make it particularly easy to win general acceptance for an expectation of a coming change in the theory.[17]

Indeed, the notion that command of the sea was desirable per se, and that a country that could afford battleships might well be advised to build them is to be found propounded even by the younger, more progressive naval officers. Thus, Yakimichev, a prominent member of the "Young School" of Soviet naval strategy, declared in 1928 that the theory of "small war" had arisen simply because of circumstances according to which the task of the Soviet Navy was to give the maximum support to the land forces; and to defend the Soviet coasts. For this purpose battleships could not really be thought essential; and furthermore at that time Soviet Russia could not afford to build them.

But, said Yakimychev (and it is a most significant admission), the "Young School" did not reject in principle the notion of a fleet of battleships; they merely knew that they would have to wait for them.[18]

Some Western scholars today are doubtful about this proposition. Thus Herrick, in his paper "The Gorshkov Interpretation of Russian Naval History," makes it plain that, in his view, Soviet writings appearing today in the 1970s that put forward the thesis that I have outlined in the preceding paragraph are distorting the past to suit the needs of the present. Because Gorshkov today is arguing powerfully in favor of a "big-war," openly offensive strategy, it would help him greatly, says Herrick, if he could show that what he wants is not a new departure for the Soviet Navy but something that has been innate in it from the moment of its first inception. Consequently, argues Herrick, what we have got in Gorshkov's quotations from the past are not the whole, unvarnished truth, but the truth "slanted" to favor Gorshkov's arguments.[19]

That Soviet admirals, like Western admirals, can indulge in special pleading is not to be doubted; but to my mind the disproof of Herrick's theory is to be found in the Soviet writings of the 1920s and 1930s contained in the "Voprosy. . . ." In particular, Belli's and Yakimychev's articles can only, I think, be understood as meaning that, so soon as Russia was in a position to build a large conventional surface fleet, she would obviously do so.

It is, of course, just possible that the text of the Belli and Yakimychev articles has been tampered with and that what appears in the "Voprosy . . ." is not the same as what those gentlemen wrote. Unfortunately, I have not been able to see copies of the early Morskoi sborniki in which they were originally published; so I have been unable to check this supposition. I must say that it seems to me unlikely that in a book published as late as 1965 (the date of the publication of the "Voprosy . . .") any such clumsy falsification was indulged in; but one cannot say for certain until one has checked.

Assuming that there was no falsification, it is surely implicit in Belli's and Yakimychev's words that a big conventional surface fleet was intended. This assumption is reinforced by a whole string of books on naval matters published in Moscow and Leningrad throughout the 1930s. These speak of battleships as the backbone of the fleet, not as obsolescent monsters. They are admittedly portrayed to their readers as being vulnerable to submarines and aircraft; but the general impression created is that no navy worth its salt could possibly be without them.[20] Moreover, this assumption is still further reinforced by the actual shipbuilding program of the middle and late 1930s. From which it follows that the strategy that would inevitably be pursued by such a fleet when built would be the attempt to attain some form or other of "gospodstvo na more" over certain vital sea areas.

Consequently we may reasonably assert that, up to the outbreak of World War II, the Soviet Union considered "gospodstvo na more" to be a correct piece of naval doctrine, in the sense that it was something intrinsically desirable; that the USSR would much have liked to have exercised it; and that she had no doubt that the capitalist navies (and in particular the British Navy) did actually do so. This was accepted by the Soviet writers of the time as something that was naturally regrettable but that was also a fact of life, in much the same way as a volcanic eruption is a fact of life, and one that similarly cannot be ignored. It was not merely former Tsarist naval officers who thought and wrote in this view, but impeccable "proletarians" (Frunze, for instance) also did so.[21]

To me, however, it seems probable that if at that time the Soviet Navy had been a bit stronger than it actually was in surface vessels, and so had been in a rather better position to offer at least some challenge to the capitalists' "gospodstvo na more,"* it might well have done so, in which case it is quite likely that the Soviet Union's cool, detached attitude toward the concept of "gospodstvo na more" would have disappeared, to be replaced by something more partisan.

THE POSTWAR DENIGRATION OF THE "MAHAN" CONCEPT

What in any case is certain is that World War II produced a tremendous upsurge of Russian national patriotism, and that this was accepted, and indeed encouraged, by the communist authorities. Furthermore, they continued to encourage and foster it after the war was over; though they sought to disguise its ideological impurity by calling it "Soviet patriotism," and ascribing to Soviet patriotism a range of virtues that, so they averred, could not be matched by any bourgeois patriotism.[22] Furthermore by the early 1950s it had become official doctrine in the USSR that World War II had been won by the Soviet Union almost single-handed and that her Western allies had made at best no more than a marginal contribution.

But if this were so, it had to be proclaimed that the course of the war, the issue of defeat or victory had been decided essentially on land; because as everyone knows, the Soviet contribution to the war at sea cannot even be described as marginal. Consequently, any theory purporting to explain that the outcome of wars can be, and has been, decided at sea was at once considered suspect; and it is that moment that marks the beginning of the Soviet denigration of Mahan, Colomb, and Corbett, who had believed otherwise.

Thus, the article on Colomb contained in the second edition of the Great Soviet Encyclopedia, which was published in 1953, is in markedly different vein from that of the first edition, published in 1938. Whereas the latter, the earlier, edition merely states that his Naval Warfare is a "monumental work," and that its basic idea is that "gospodstvo na more" is an essential precondition for the success of any large-scale naval operation, the second edition declares that Colomb was a military ideologue of the English bourgeoisie; and that his book Naval Warfare, like Mahan's The Influence of Seapower on History, was an attempt to justify an adventuristic strategy by the

---

*I am here using "challenge," not to mean an actual recourse to war (because at the time of which I am speaking it was Soviet policy to avoid war), but a "challenge" in the sense of "having a capability to exercise "gospodstvo na more" in certain waters, and of being seen to have that capability."

theory of "gospodstvo na more." Further objectionable features of Colomb's ideas are then listed, all of which are said to have been completely refuted by the events of the two world wars.

The attacks on the work of these three men was therefore principally directed against what the Russians claim to be Mahan's theory of seapower and Colomb's of "gospodstvo na more." I say deliberately "against what the Russians claim" to be these two strategists' theories, rather than "against their theories," because if one actually reads Mahan's The Influence of Seapower on History or Colomb's Naval Warfare and then compares them with most of the Soviet summaries of them, one is tempted to wonder whether the Soviet summarists had ever read the originals.[23]

Leaving that point to one side, however, the next thing to be noted is that, in the way in which it was generally presented to the Soviet reading public, the work of Mahan, Colomb, and Corbett was said, as has already been mentioned, to contain a number of extremely objectionable features.

The first was that Colomb's theory of "gospodstvo na more" and Mahan's "theory of seapower" served as a basis for the expansionist policies of Britain and America, respectively.[24] In other words, Britain's acquisition of a far-flung colonial empire and America's acquisition of worldwide economic influence both depended upon the possession by them of tremendous strength at sea. Since overseas empires and worldwide economic influence are, when in the hands of an imperialist nation, damnable in Soviet eyes, the naval theory said to have underpinned them became damned by association too.

The second objectionable feature is that, according to the Soviet critics, Mahan, Colomb, and Corbett were agreed in saying that this "seapower," this "gospodstvo na more" could only be acquired by the near total destruction of the enemy's fleets or else by their eviction from the major theaters of war. Furthermore, these two aims were themselves only attainable (so Mahan, Colomb, and Corbett are reported by the Russians as saying) either by means of a general engagement between fleets of battleships or else by a blockade. It then becomes a simple matter for the Soviet critics to "disprove" the two Western strategists by saying that in neither the First nor the Second World War was there any general engagement between fleets of battleships; so consequently "gospodstvo na more" a la Colomb was a purely mythical concept (the Soviet critics, interestingly enough, say nothing about blockade in this connection).

It is clear, I think, that what the Russians really object to in "gospodstvo na more," as formulated by Colomb, Mahan, and Corbett, is that it seems to them to assume as an eternal axiom that it is Western nations that are destined by History to exercise this command of the sea, and not the Soviet Union. Furthermore the doctrine itself, at least in the form in which the Russians choose to present it, is a very simplistic, very full-blooded one, which asserts that all the seas and all the oceans are subject to the nation exercising it. Such a claim, which might perhaps have been made by the Royal Navy at the height of British naval supremacy, and which even then would have been something of a caricature, is presented to the Soviet reading public as the claim that is made continually even today by the major imperialist maritime countries, especially America.

Such a doctrine is absolutely intolerable to the USSR. For if the oceans are indeed commanded by the navies of the imperialists, it follows that Soviet ships could only cross those oceans by imperialist permission, whether tacit or overt. In the new mood of communist/nationalist chauvinism, which is strongly marked today in Gorshkov's writings, such a thing is simply not to be borne. Readers of Gorshkov's second article, for instance, will note his indignation at the mere idea that there could be any waters, such as the Mediterranean, where Soviet ships were only allowed on sufferance.[25] Apart from purely territorial waters, from which he concedes they can properly be excluded, all other seas are "high seas" and as such are as freely navigable by Russians as by Britons or Americans. A similar attitude is to be found in the article entitled "Voenno-Morskoe Iskusstvo" in the third edition of the Great Soviet Encyclopedia.

But if this version of the doctrine of "gospodstvo na more" is anathema to the Soviet Union, it is worth remarking that the essence of the concept itself, taken in isolation, has only very seldom been attacked in Soviet writing. Indeed, so far as my own knowledge goes (admittedly defective), I am aware of only three instances of it, and even there it might be objected that in one case the concept is not being treated in complete isolation, but rather that what is the object of the criticism is a particular way of applying "gospodstvo na more."[26]

It is obviously extremely lucky for the Soviet admirals that this is so. For underneath these interesting polemics lies the incontrovertible fact that, just as it is not possible nowadays to undertake any important naval operation without local air superiority (gospodstvo na vozdukhe), so no such operation can be undertaken without local command of the sea. The word "local" is vital. For although it may be nice to be ruler of the waves in all the four corners of the world, what is essential is naval superiority in those waters through which it will be supplied.

Up to sometime in the 1960s, the Russians could ignore this contretemps, because until then the only operations they expected to wage at sea, or were capable of waging, were in waters adjacent to the Soviet Union; and there they had that essential "local command." There was therefore no need to talk about it; they could treat it as an eternal axiom whose existence need not be discussed. But once they had built an ocean-going fleet, which was intended to wage, and has frequently practiced waging, large-scale operations on the high seas, they would need in wartime to gain command of the seas in those areas where they intended to conduct their operations, despite the fact that the waters concerned might be hundreds or thousands of miles away from the Soviet Union's coasts.

Furthermore, a general attack on the intrinsic meaning of "gospodstvo na more" per se would be bound to cause great difficulties for the Soviet admirals. For if war at sea is indeed a secondary matter, the service designed to wage it can only be secondary too. Not only that, but Soviet official doctrine, as expounded in the Soviet textbooks, declares that, in the case of a "continental" war (that is, in the case of a war against a predominantly land power such as Russia) the effect of war at sea is not even secondary; it is so small as to be scarcely worth the mentioning; and Gorshkov in his ninth article dutifully concurs with this opinion.[27]

But if that is so, it becomes extremely difficult for the Soviet admirals to justify the expense of enormous sums of money on the maintenance, improvement, and enlargement of that one of the fighting services that by definition has a nugatory part to play in any conflict in which their country is involved (the ballistic missile submarines, of course, are a case apart).

Nor is this the only difficulty that would be created by wholesale denigration of the concept of "gospodstvo na more." For the Soviet Union, as we all know, has been busily building an "oceanic fleet." But the sole purpose of an "oceanic fleet" (again omitting the missile-carrying submarines) must be to exercise seapower. In which case the crews of the ships must be trained in the meaning of seapower, and in particular the senior officers must learn how best to wield it.

It is, however, obviously extremely difficult for the Soviet c in c to urge his men to study a strategic concept, and to spend time in meditating how they can best apply it in modern conditions, if the very concept were condemned by party doctrine and declared to be "bourgeois" rubbish.

"COMMAND OF THE SEA" IN A NEW CONTEXT

In this context, therefore, it is interesting to note that a new approach to "gospodstvo na more" began in the middle 1950s. Whereas, prior to then, it was only mentioned in connection with the reactionary Colomb, after that date it was sometimes presented as a concept that could, though obviously did not always, exist in its own right. The article on the Battle of Trafalgar in the second edition of the Encyclopedia, for instance, was published in 1956. After giving details of the battle, it simply says that the British were the victors at Trafalgar, and that as a result they acquired "gospodstvo na more." It does not label it a "reactionary" gospodstvo; it does not term it a "bourgeois" gospodstvo; it merely presents it as something acquired as a result of the Battle of Trafalgar.28

In 1958 the Soviet Ministry of Defense published a history of World War II called Vtoraya Mirovaya Voina, which was designed to be read by the officers of all three services up to the rank of colonel or the equivalent. For this purpose it was issued as one of the titles of a series called Biblioteka Ofitsera (The officers library), which serves as a course of recommended reading for junior and medium-grade officers.

In this volume the notion of "gospodstvo na more" is expounded as a self-evident fact; and no pejorative adjectives are used to describe it. The British, for instance, are shown as having acquired "gospodstvo" in the Western and Eastern Mediterranean in the middle of 1941 as a result of the successful attacks they had launched on the Italians29 and later that year to have lost it to the Germans,30 while in 1941 the Japanese Navy is said to have acquired "gospodstvo" in the Western Pacific.

In 1961, the first of the six volumes of the well-known Istoriya Velikoi Otechestvennoi Voiny Sovietskogo Soyuza, 1941-1945 was published, the final volume of which appeared in 1965. The phrase "gospodstvo na more" occurs on several occasions in this monumental history of the Great Fatherland War, each time without any denigratory comment attached to it. Thus, in the first volume the Italians are said to have aspired to it in the Mediterranean;31 in the second it was the Japanese who aspired to it in the Pacific.32 Other instances can be found in other volumes.

If it be objected that all the countries mentioned in these examples are in fact "imperialist," so that the "gospodstvo" they exercised was a damnable, imperialist form of it, I can only answer that the tone of the writing, especially when read in context, gives no ground for such an assertion; and that I have no reason to assume that the Soviet Navy would not be shown as exercising "gospodstvo," were it not for the fact that, as we all know, it never actually did so.

I am confirmed in this latter hypothesis as a result of noting that, whenever the party historians feel they can claim that the Soviet Air Force won "command of the air" ("gospodstvo na vozdukhe"), they are very happy to do so; and this same fact disposes of another supposition to the effect that, the basic meaning of the word "gospodstvo" being "lordship" in the old feudal meaning of the expression (see also the related words gospod' and gospodin), it would of itself be very repugnant to party susceptibilities.

In 1965, another important work on Soviet strategy rolled off the presses of the publishing house of the Soviet Ministry of Defense. This was the "Voprosy Strategii i Operativnogo Iskusstva v Sovietskikh Voennykh Trudakh 1917-1940." As its name implies, the work was a selection taken from Soviet military writings of the period 1917-40.

The section devoted to naval strategy has less than 90 pages, so that what is presented is obviously only a small sample of the vast choice available. From which it follows that either what is presented has been specially selected so as to provide a representative sample of the whole range of Soviet naval writing in the period under discussion, or else it has been equally specially selected to inculcate a particular point of view. I personally have no doubt that the latter alternative is correct; so it must be borne in mind that in what follows I am basing myself on that assumption.

Once one does this, once one bases oneself on that assumption, the extract from Gervais that begins the section at once assumes a particular significance in connection with our present discussion; for he makes it clear that, in his opinion, it is the duty of every navy that is conducting an offensive strategy to do its utmost to secure "gospodstvo na more"; and he then goes on to explain the means by which this should be done (Gorshkov, it should be remembered, is explicitly in favor of the Soviet Navy's adopting an offensive strategy).

It could of course be objected that Gervais is a relic of the Tsarist past and that his views are presented in this symposium only to be knocked down. The tone of the editorial comment, however, makes this very unlikely. The extract from his writings is prefaced with the note that "despite a number of serious mistakes, mostly of a methodological character," Gervais's work played an important part in the formation of Soviet naval doctrine;[33] and it appears clear from this passage and from the "Introduction" to the naval section that his "methodological mistakes" consisted of advocating the wrong method (namely, a general engagement between squadrons of battleships) of attaining a correct aim ("gospodstvo na more"). The aim itself is nowhere criticized in this volume; on the contrary, it is given general support.

It should be added that, in further support for this view, Gervais is given an entry in the latest edition of the Great Soviet Encyclopedia, where his work is described as "fundamental" and as "exercising a significant influence on the development of Soviet theoretical military thought." The date of this volume is 1971, so he is clearly in favor at present.

In 1967 the Soviet publishing house Nauka produced an important study of some aspects of naval warfare under the title Blokada i Kontriblokada. One of the contributors to this symposium was the distinguished naval strategist V. A. Belli.

On pp. 22-23 of the work in question Belli agrees that the advent of large-scale submarine warfare had made the old concept of "gospodstvo na more" outdated; but he argues powerfully in favor of the view that the concept itself is perfectly valid and that what is needed is merely a reformulation to take account of the modern weapons. Since Gorshkov quotes this book of Belli's on a number of occasions in his famous series of articles, it is clear that Belli's views on the matter enjoy influential support.

Also in 1965 the Soviet Ministry of Defense published Strokov's famous volume on the art of war in the capitalist era. He devotes a whole section of it to a discussion of "gospodstvo na more," in the course of which he displays a thorough knowledge of the writings of Colomb and Mahan. He disapproves of their notion of "gospodstvo"; because, he says, it served as a basis for American and British expansionism; because their theory of seapower served as the means by which both these nations might exercise their "natural right" to rule over other peoples and enslave them.[34] However, says Strokov, the expression "gospodstvo na more" must not be used solely in connection with the names of Colomb and Mahan, as "bourgeois" military writers are accustomed to do. When divorced from them, it is often he says, a valuable and wholly realistic concept.[35]

In 1969, the Soviet Ministry of Defense published that Istoriya Voenno-Morskogo Iskusstva that has been the object of so much study by Western sovietologists and naval experts. There are three passages in it where "gospodstvo na more" is mentioned.

The first occurs in the sixth chapter, written by I. A. Kozlov. There the concept is pilloried by implication (but it would appear on close inspection, only by implication) because it is said to be based on the "reactionary" and "imperialist" theory of seapower of Mahan and Colomb, the purpose of which (so Kozlov says) was to justify the "aggressive policies of American imperialism."* But all that is said about "gospodstvo na more" itself is that its essence consisted of the total removal of the enemy's fleet from the theater of naval operations, and that its authors, Mahan and Colomb, were so misguided as to imagine that this could be done by battleships alone.[36] This is an attack on a particular variety of "gospodstvo," not on "gospodstvo" itself.

The second instance of "gospodstvo na more" in the "Istoriya . . ." is on p. 524. Here the author is Admiral Stalbo; and he states uncompromisingly that the theory of "gospodstvo na more" is

---

*Though why Colomb, an English admiral of the old school (he had taken part in the Crimean War) should wish to justify American imperialism is beyond my comprehension.

"worthless" (niesamostoyatel' nyi). He says that the theory postulates, as a necessary condition for attaining one's objective in a naval operation, that the enemy fleet must either be sunk or else driven out of the area where the operation is to be conducted.

The evidence he adduces to demonstrate that the theory is worthless is derived from World War II, and is itself worthless; for it is based on the German ability to mount their invasion of Norway in 1940. This, says Stalbo, was done despite the fact that the British Navy exercised its "so-called 'gospodstvo'."

Now there are three things to be said about this extraordinary remark. One is that, to the best of my recollection (and I speak as one who took part in the operations that followed upon that invasion), the Royal Navy in 1940 did not claim to exercise "gospodstvo" in the waters through which the German invasion fleet passed. In particular, they never made any pretensions to exercise command of the sea in the Skagerrack.

The second is that, supposing my memory to be faulty, supposing the Royal Navy did do as Stalbo says, and did claim to have mastery of the sea in the areas adjacent to Norway, the facts of the history of that period show that their claim was false and that they simply had not got it. In which case the success of the German invasion does nothing whatever to discredit "gospodstvo na more"; because you cannot discredit a theory where it does not exist.

Thirdly, that and the other example quoted by Admiral Stalbo strongly indicate that he had in his mind the old, full-blooded doctrine of "gospodstvo na more" that derived from nineteenth-century British naval supremacy and that claimed "gospodstvo" over all the oceans of the world.[37] If this is true, then here again what is being attacked is not the basic doctrine of "gospodstvo na more" but a particular variant of it. I am bound to admit, however, that it comes closer to an all-out attack on the fundamental essence of the doctrine than any other recent example known to me.

GORSHKOV AND "COMMAND OF THE SEA"

The year 1970 saw the start of the publication of the new edition of the Great Soviet Encyclopedia; and the fifth and twelfth volumes, published in 1971 and 1973 respectively, each carry articles on naval warfare. In both of them Mahan and Colomb are mentioned; and they are there said to have tried to work out a theory of "gospodstvo na more" by means of building up an overwhelming superiority of ships of the line and smashing the enemy in a general engagement. The encyclopedia avers that the notion of winning "gospodstvo" by one single general engagement is an integral part of Mahan's and Colomb's theory and declares that the theory was thoroughly refuted by the experience of the two world wars.

Here again, therefore, it is a particular variant of the "gospodstvo" concept that seems to have incurred displeasure; though it is also possible to argue that, since the articles were published under Soviet conditions, what has really happened is that it is the basic theory that has in fact incurred the displeasure, and that the denigration of Mahan and Colomb is merely a means of attacking it. In other words, according to this latter view, Mahan and Colomb are really irrelevant to the whole business and what is under attack is the theory itself.

We come now to 1972, in which year were published the three articles in Gorshkov's series that mention "gospodstvo na more."[38] In all, they contain five separate mentions of the doctrine. Three of them are concerned with World Wars I and II; and there Gorshkov alludes to the concept in the detached, analytical way that, as we have seen, was characteristic of the Soviet approach to the subject prior to 1941. Thus, the British Navy, he says, aimed at acquiring "gospodstvo na more" in World War I; and he thinks that its tactics at Jutland were reasonably suited to this purpose.[39] In World War II, he observes that Japan's successes of 1941-42 won her "gospodstvo na more" in the Pacific; while later in the war the brilliant achievements of the American aircraft and shipbuilding industries in providing the U.S. Navy with such enormous quantities of both allowed the Americans to acquire "gospodstvo" in those areas where they were conducting operations.[40]

As already noted, Gorshkov attaches no pejorative adjectives to the doctrine in these instances; he merely remarks them as part of the facts of life. In 1941, he says, Japanese naval victories ensured the acquisition of "gospodstvo"; by 1944, American naval and air superiority had caused the Japanese to lose it. The change in circumstances is described with a detachment worthy of Thucydides; though I hasten to add that the general tone of Admiral Gorshkov's writing is far from Thucydidean.

When one comes to the fourth instance, however, pejorative adjectives appear. This occurs in the fourth of Gorshkov's articles, where he describes the theory of "gospodstvo na more" as "Mahanist."[41] Here, however, he is concerned with the way that the navies of the World were developing in the years preceding the outbreak of World War I; because their exaggerated estimate of the value of battleships and their underestimation of the value of submarines and aircraft had led to their fleets being composed of what in Gorshkov's view was an excessive number of the former and an insufficient number of the latter. This, he says, led to the navies being unbalanced in their composition; and everyone knows how much Gorshkov abhors the notion of an unbalanced fleet.

In this context, therefore, "gospodstvo na more" led to a result that Gorshkov considered unfortunate; and it is perhaps for this reason that he attaches to it the pejorative adjective "Mahanist." I am quite unable to decide, however, whether he truly believes that it was Mahan's thinking that led to the creation of unbalanced navies, or whether he uses it as a purely conventional epithet, in the way in which Homer so frequently uses ϵἴλιπους when he wishes to talk about oxen.

The final mention of "gospodstvo na more" in Gorshkov's famous articles is the most important of the lot; and it occurs on p. 21 of the August issue of Morskoi sbornik. It is here that he makes the distinction between the Soviet and Western notions of the expression; and it is this comment and the way in which he elaborates it that seem to me to provide a valuable clue to some likely paths of future development of Soviet naval strategy.

For what he says there is that those who think that the Soviet Navy is capable of carrying out operations beyond the limits of its coastal waters (and Gorshkov himself today is certainly one of them) will naturally support the theory of "gospodstvo na more"; and that this means that it must aim to acquire superiority of force over its enemy in the principal theater of operations and be strong enough in the secondary areas to prevent him from interfering. It also means that the Soviet Navy must be able to stop the enemy's navy from carrying out its own operations. Finally Gorshkov emphasizes by means of his quotation from Belli, that this wholly admirable Soviet "gospodstvo na more" does not imply a claim to command every sea and ocean in the world (unlike the horrible British and U.S. navies and the reactionary Mahan, and Colomb) but that the area over which "gospodstvo" is to be exercised is restricted to the actual theater of operations or even to part of that theater.[42] In saying this, he is merely repeating what Frunze said in 1922.

So with Gorshkov's series of articles we are back at a presentation of the doctrine that Colomb would have had no difficulty in recognizing as his, though Gorshkov omits any mention of the means by which the doctrine was to be implemented. As Colomb conceived it, this was to be done by the total removal of the threat to one's command of the sea that was posed by the mere existence of an enemy fleet. Consequently, that fleet had either to be destroyed or else completely neutralized (for example, by being blockaded in its ports); Colomb was insistent that this must always be done.

So why does Gorshkov say nothing at all about it? And why do those who seem less than enamored of the doctrine concentrate their fire upon the means of implementing it and make little adverse criticism of the doctrine itself? Why do these people allege that Colomb declared that "command of the sea" can only be won by a single general engagement when Colomb said nothing of the sort? Why do they talk of the surface fleet as being, in Colomb's opinion, the essential instrument for attaining "command of the sea"? Admittedly, Colomb and Mahan asserted repeatedly that this was so; but this merely means that they underestimated the capabilities of the aeroplane and submarine; and their theories therefore need to be updated, as Belli agreed and as he has in fact updated them. But if the whole concept has in fact been rendered obsolete by modern technology, why on earth does not the party unequivocally say so? If the concept itself in its strictly Colombian limited sense remains valid, why is Gorshkov silent on the means by which "gospodstvo" is to be won? As Sir Thomas Browne said, these are puzzling questions, though the things that puzzled that 17th-century philosopher were what song the Sirens sang, and what name Achilles assumed when he hid himself among women. The problems posed by the Soviet Navy might have puzzled him even more.

NOTES

1. Bol'shaya Sovietskaya Entsiklopediya (hereafter referred to as BSE), 2d ed., vol. 28, p. 637. Inverted commas in the original.
2. Morskoi sbornik (hereafter referred to as MSb), May 1972, p. 13.
3. BSE, 3d ed., vol. 12, p. 441.
4. The following summary is from Chaps. 1-4 of Colomb's Naval Warfare: Its Ruling Principles and Practice Historically Treated (London: W. H. Allen, 1891).
5. The text of Makarov's articles can also be found in Russkaya Voenno-Teoreticheskaya Mysl' (Moscow: Voenizdat, 1960).
6. Voprosy Strategii i Operativnogo Iskusstva v Sovietskikh Voonnykh Trudakh, 1917-1940 (Moscow: (Voonizdat, 1965), hereafter referred to as Voprosy.
7. M. N. Frunze, Izbranniye Proizvedeniya, vol. 2, p. 12, (Moscow: Voenizdat, 1957).
8. Quoted by Gorshkov in MSb, August 1972, p. 21.
9. See his article in Voprosy.
10. Istoriya Voenno-Morskogo Iskusstva (Moscow: Voenizdat, 1969), p. 80.
11. MSb, April 1972, p. 22.
12. For Gorshkov on this point see MSb, May 1972. p. 14.
13. For Gorshkov, see ibid. To a Marxist-Leninist, Tsarist Russia would naturally be one of the "glavnymi imperialisticheskimi gosudarstvami"; and Russia's subsequent ship-building program agrees with his diagnosis.

14. For the validity of the first two of these propositions, see Yakimychev on pp. 700-701 of Voprosy. There is no one reference by which to support the validity of the second proposition, but readers are referred to my forthcoming "War, Peace and Neutrality: The Soviet View" (London: Routledge, 1975).

15. MSb, June 1972, p. 20.

16. Robert Waring Herrick, Soviet Naval Strategy (Annapolis, Md.: U.S. Naval Institute, 1963), p. 15.

17. For Gorshkov on the interaction between weapons systems and strategy, see MSb, February 1972, p. 21.

18. Voprosy, pp. 698-703.

19. Herricks' paper is contained in "Soviet Naval Developments; Context and Capability" (Michael MccGwire, ed., Center for Foreign Policy Studies, Dalhousie University, Halifax, Canada), pp. 275-89.

20. See, for instance, Voonno-Morskoe Delo (Moscow: Voenizdat, 1937); Sovremennie Boevie Sredstva Morskogo Flota, written by the staff of the Voroshilov Naval Academy, and published by Voenizdat (Moscow) in 1938; and Popov's and Gordon's Lineinie Korabli (Redaktsiy Sudostroital' noi Literatury, 1938).

21. For Frunze's view, see his Voenno-Politicheskoye Vospitaniye Krasnoi Armii (Moscow, 1922).

22. For an exposition of the characteristics of "Soviet patriotism" see the entry "Sovietskii patriotizm" in the second edition of BSE.

23. There are a very few honorable exceptions, such as A. A. Strokov; but, even so, their scholarly accuracy finds no echo even in a work with such academic pretensions as BSE.

24. See, for instance, the article on Colomb in the third edition of BSE.

25. MSb, March 1972, p. 31.

26. The instances in question are Admiral Stalbo's formulation of it (which he then proceeds to condemn), which are to be found on pp. 524 and 530 of the Istoriya Veonno-Morskogo Iskusstva. The third instance can be found in "Voprosy" and dates from the middle 1930s.

27. MSb, November, 1972, p. 30.

28. The first edition goes even further and says that Trafalgar gave the British "gospodstvo na more" for the rest of the 19th century; but that article was published before Stalin's reign and before the Great Fatherland War.

29. Vtoraya Mirovaya Voina (Moscow: Voenizdat, 1958), p. 342.

30. Ibid., p. 344.

31. Istoriya Velikoi Otechestvennoi Voiny Sovietskogo Soyuza, 1941-1945 (Moscow: Voenizdat, 1961), vol. 1, p. 294.

32. Ibid., vol. 2, p. 590.
33. op. cit., p. 684.
34. A. A. Strokov, "Istoriya Voennogo Iskusstva," pp. 597-604 (Moscow: Voenizdat, 1965).
35. Ibid., p. 604.
36. Istoriya Voenno-Morskogo Iskusstva, op. cit., p. 80.
37. Stalbo's other use of it occurs on p. 530 of the Istoriya Voenno-Morskogo Iskusstva.
38. They are contained in the May, August, and September issues of Morskoi sbornik for that year.
39. See pp. 15-16 of the May issue.
40. See pp. 20 and 22 of the September issue.
41. See p. 13. of the May issue.
42. MSb, August 1972, p. 21.

CHAPTER
33
COMMAND OF THE SEA IN SOVIET NAVAL STRATEGY
Michael MccGwire

When drawing inferences from the Soviet use of an expression such as "command of the sea" (gospodstvo na more), both the textual and the strategic contexts must be analyzed.[1] The purpose of this chapter is to review the use of this term in relation to Soviet naval strategy[2] so as to assess its significance as a contemporary operational concept.

DIFFERING USAGE

At the outset, we must distinguish between two very different ways in which the term "command of the sea" is employed, both by the Soviets and ourselves. On the one hand it is used to describe a situation, either existing or desired. And on the other, it is used as a label for a theory or concept.

In Soviet usage there are two different theories, one right and the other wrong. There is the "Mahanist" concept of "extensive" command of the sea, which, until the middle 1920s, was generally accepted in Russia as a valid theory, even if neither the Imperial nor the Red Fleets were in a position to apply it. However, by the late 1920s, its obvious irrelevance to the Soviet situation had led to a thorough reevaluation of the doctrine and, ultimately, to its total rejection.[3] Since that time, the "Mahanist" theory of command of the sea has been considered fundamentally defective. This assessment was reconfirmed by analysis of naval operations in World War II[4] and the condemnation is repeated as late as 1973.[5] As we shall see below, it has recently been claimed that in the 1930s there emerged in place of Mahan's "extensive" doctrine, a theory of "limited" command of the sea, which we will call the "Belli" concept, after one of its exponents.[6] This concept is unexceptionable, but it is not really a theory of command of the sea. It is an operational concept for conducting naval operations in the face of superior forces.

In its descriptive sense, for Russia as for the rest of the world, command of the sea is a self-evident good. Maritime strategy is about the use of the sea, and if one can achieve "command," then one can use the sea for one's own purposes and prevent the enemy from using it for his. To recognize this implies nothing about how important that use of the sea is, and therefore, when employed in this sense, the term is ideologically neutral. More particularly, the term does not impinge directly on the Soviet doctrine that the outcome of war is determined by the battle on land, and the claim that victory in World War II was decided on the Eastern Front. The Soviets are quite willing to acknowledge that the Western contribution depended on their having command of the sea in certain areas, but they assert that this contribution was in no way central to the outcome of the war. The Mediterranean was a sideshow and the Normandy landings did not take place until after the Soviet advance had drawn the great majority of Axis forces to the Eastern Front. In the Pacific the Japanese did not surrender until Soviet forces advanced against the main body of the still intact Kwantung Army in Manchuria and Korea.[7]

Since the descriptive use of command of the sea is ideologically neutral, nothing can be inferred from Gorshkov's factual remark that the Japanese gained command of the sea at the outbreak of the war in the Pacific and that the United States gained command throughout their operating areas by 1945.[8] Similarly, nothing can be inferred from the nonpejorative comment published in 1954 that the British acquired command of the sea after Trafalgar,[9] except that the postwar veto on the descriptive use of the term had been lifted. There are no grounds for inferring that Mahanist doctrine was creeping back into favor at this period, since the one-line comment does not claim that this British command of the sea changed the course of the Napoleonic wars; nor should it. British naval historians do not see it as a turning point, nor even as a major event in the war;[10] it took eight more years before Napoleon was defeated at the decisive land battle near Leipzig in 1813, and this only after his disastrous retreat from Moscow. In any case, 1954 was the year that saw the massive cutback in the postwar naval building programs, and the adoption of an operational concept that was predicated on operating within range of shore-based air cover. Neither this nor the submarine-based concept adopted after 1957/58 were remotely Mahanist; indeed, very much the reverse.

THE IMPERIAL RUSSIAN NAVY

It would seem that the Russians only entertained truly "Mahanist" ideas about command of the sea for a fairly short period. On the other hand, they have always been aware of the advantages of gaining command of their main fleet areas, although this does not mean that it was always (or even usually) given high priority, relative to other demands. In their successive wars with the Ottoman Empire they achieved such command several times in the Black Sea, but they could not retain it if Britain or France intervened against them. Russia had tended to bow to force majeure in this respect, but after she was denied the fruits of victory at the Treaty of Berlin (1878), largely as the result of British naval pressures, she set out to build up a fleet that would be strong enough to challenge such incursions into the Baltic and the Black Sea.[11] Meanwhile, over the next 15 years there was a major shift in international alignments and comparative naval strengths.[12] In 1893 she joined with France in the "Dual Alliance,"[13] and at about this period Russian building programs turned to constructing a fleet of sea-going battleships,[14] and one might say that between 1895 and 1905, Russian naval strategy was generally conceived in "Mahanist" terms.

(At this period, Russia's "colonial" interest was in China. Expansionist naval policies were very much in vogue. The 1880s were the vital formative phase of the "new imperial expansion" by European powers into Africa, Asia, and the Pacific. The revolution in naval architecture that started around 1830 came to an end in about 1880, and its fruits were ripe for application. Circumstances could not have been more propitious for the appearance in 1890 of Mahan's theory of seapower, which dramatically publicized the arguments the Colomb brothers had been advancing since 1867.)

It took a long time to sort out the aftermath of the Russo-Japanese War (1904-5) and the destruction of the Russian fleet at Tsushima, and the new construction program that slowly got underway in 1910 reflected Russia's limited industrial capacity and underlined the perennial "four fleets" dilemma. The Pacific was to be sacrificed and the Black Sea was to be built up to 1.5 times the relatively weak strength of the other coastal states. The main effort was to be concentrated in the Baltic where the German Fleet had long been seen as a serious threat.[15] If the program had run its full course the Russians would have had a powerful fleet by 1930,[16] but this must be seen in the context of the very large foreign navies that already existed in 1910, and naval arms race then in progress. Given the resources, no doubt the Russians would have liked to have planned for achieving command of both the Baltic and Black Seas. But things being as they were, one suspects that their more immediate concern

concern was to be able to deny command to the foreign fleets that could concentrate their forces in one or other sea; to protect and support the army's flank; and, in the Baltic, to help defend St. Petersberg. Thus, at the outbreak of war in 1914, the Baltic Fleet was subordinated to the Commander of the Seventh Army, who was responsible for the defense of the capital.[17]

THE INTERWAR YEARS

The Rejection of Mahan

If Mahan's ideas had little practical application for Russia before the war, they had even less relevance during the 1920s. By 1930 a more pragmatic approach to establishing naval requirements had emerged, but it would be wrong to conclude that Mahan's ideas were rejected, wholly, or even mainly, because they were bourgeois or because the Soviet Union lacked the means to implement them. The reasons were more fundamental and are generally valid. Mahan stated that the object of his book was to "estimate the effect of seapower upon the course of history"; but having argued that seapower was essential to the growth of national strength and prosperity he went on to draw up a set of principles by which seapower could be achieved or exercised. Since he was saying what people wanted to hear in 1890, there was little critical analysis of his arguments although, as Graham points out, Mahan's exposition occurred at the very time when the new instruments and economics of the Industrial Revolution were beginning to erode principles and theories on which his doctrine was based.[18] But more important than whether or not Mahan's principles were well-founded, was the emergence of the tacit assumption that his "theory" embodied a set of universal rules for success at sea. Of course military strategy is not a science and does not have universal rules, and even its principles need to be applied with selective care. But on the basis of the forceful analyses carried out by the Colomb brothers and Mahan, there grew up the idea that there was such a thing as "correct" naval strategy, and by implication any other strategy was ill-founded and wrong; command of the sea became an end in itself.

The Small War Theorists

In fact the "correctness" of a strategy must depend on the particular situation and circumstances of the time, and it was on this elementary truth that A. P. Aleksandrov based his monograph entitled "A critical analysis of the theory of Command of the Sea," which was published in <u>Morskoi sbornik</u> in 1929-30.[19] His arguments provided the basis for Soviet policy in the first half of the 1930s, which gave up all ideas of universality in naval doctrine (as exemplified by the Mahan/Colomb theories) and concentrated on the practical conditions in the Soviet Union. From this emerged the "small war" theory of naval strategy. Aleksandrov's analysis was wide-ranging, and many of his basic conclusions can be seen as underlying post-1945 naval policy, through to the present day. But of immediate interest are the main reasons that led him to reject the prevailing doctrine, with its emphasis on command of the sea, achieved through major fleet engagements. He considered that the theory tended to generate its own strategic aims and problems, many of which had little relevance to, and diverted resources from, the primary military objectives on land. On the other hand, it engendered passivist attitudes in those navies that were not in a position to achieve such command. Both criticisms derive from the tacit appreciation that the primary objective of naval strategy is "to enable the sea to be used for one's own purposes." The first was reinforced by Lenin's principle of the unity of land and sea operations, which remains operative today.

The naval policy that emerged from the theoretical debate in the early 1930s was based on a strategy that avoided fleet engagement and relied on successive "combined attacks." These would be launched by aircraft, submarines, and torpedo craft, with no one weapon of primary importance and tactics characterized by aggressiveness and persistance. Heavy emphasis was placed on the offensive and defensive use of mines, and there would be strikes against enemy bases and airfields. This policy did not deny the role of larger surface units: They would be slotted into the all-arms concept of operations, as and when they became available. But the strategy was essentially defensive and saw no requirement to control a greater area than the narrow strip of sea needed to support army operations on land.

Belli's Concept of Limited Command

Command of the sea continued to be universally condemned as an end in itself. But the "small war" school of thought also rejected it as a means and argued for the continuing applicability of their defensive theories, even when forces became available for more extensive operations. Writing in 1967, Gorshkov gave the impression that this approach prevailed through into the early 1950s. However, Gorshkov has since alleged (in 1972) that there also existed an opposing faction that favored the offensive employment of naval forces and that held to a "theory of command of the sea," although it was rather different from the Western concept.[20] Quoting Belli, Gorshkov defines the Soviet concept as follows:

> To attain superiority in forces over the enemy in the principal direction and to fetter his forces in the secondary directions during the period of an operation, that means to achieve <u>command of the sea</u> over a theatre or some part of a theatre; that is to say, to create such a situation whereby the enemy will be paralysed or constrained in his operations, or weakened and thereby hindered from interfering with our execution of a given operation or with carrying out our operational task.[21]

Gorshkov goes on to comment that "it was precisely this interpretation of command of the sea which underlay the employment of the navy's forces in naval warfare,"[22] and leaves the impression that he is referring to World War II.

Assuming for the moment that Gorshkov's second version is the correct one, the reintroduction of the term in the 1930s did not imply any rehabilitation of Mahan's doctrine, which continued to be rejected as totally ill-founded.[23] A glance at the fleet strengths of the major naval powers in the latter 1930s* is sufficient to dispel as absurd the notion that Stalin entertained Mahanist aspirations;[24] indeed Gorshkov has bewailed the navy's defensive orientation in the prewar period. Morale-boosting claims about the future ocean-going capability of the navy must be read in the context of the predominantly coastal capability that existed up to this time; for "ocean-going" one should read "out of sight of land." The limited industrial capacity that was available for building large ships, and the projected order of battle for 1943,[25] dictated a naval policy that concentrated on the four fleet areas, and the wider ocean expanses had little relevance to Russia's strategic situation.

---

*For example, in 1939 Germany had 4 battleships, 11 cruisers, 37 destroyers, and 57 submarines in service with another 2 battleships, 2 aircraft carriers, 4 cruisers, and 8 submarines under construction. The Japanese had 10 battleships, 10 aircraft carriers, 35 cruisers, 106 destroyers, and 58 submarines, with another 1 carrier, 2 cruisers, 8 destroyers, and 8 submarines under construction.

THE POSTWAR PERIOD

Increased Denigration of the "Mahanist" Concept

Given the Soviet belief that information should be tailored to support current policies, there were sound reasons for the newly acerbic tone of the references to Colomb and Mahan that appeared after the war. On the one hand, the Soviet Union could argue with some justification that the war had confirmed the primacy of land powers and armies. It was Hitler's armies that had conquered Europe; it was Russia's armies that first checked and then threw them back in the "meat grinder" of the eastern front. Victory stemmed directly from land operations, even though at a very heavy cost, and maritime operations were secondary. On the other hand, the one country that had emerged from the war with her strength immeasurably increased was the United States, now a maritime power <u>par excellence</u>. Mahan's geopolitical and distinctly deterministic theories were particularly offensive to a sorely weakened Soviet Union, which had its own conception of the future course of history and was seriously concerned about the "inevitable" attack by the capitalist camp, now comprising the "traditional maritime powers."[26]

In the second place, the ideas of Colomb and Mahan were ideologically unacceptable. They believed that sea power was essential to the growth of national strength and prosperity and advocated its accretion. Mahan gave the components of sea power as production, shipping, and colonies or markets; he also spoke of sea power as one of three interlocking rings, the other two being colonies and commerce. The fact that imperial expansion took place both before and after the Colomb/Mahan theses were articulated, does not invalidate the assertion that these theories provided the justification for British and American expansionist policies,[27] although it probably gives them more credit than they are due. Both Colomb and Mahan were openly concerned to promote the growth of their countries' strength, and saw sea power (and its corollary, colonies) as the enabling instrument. Mahan states categorically (and repeatedly) that "naval strategy has for its end to found, support and increase, as well in peace as in war, the sea power of a country."[28] Their theories were an invitation to naval arms racing, and Western historians have suggested that they contributed to the causes of World War I.[29]

Finally, the Colomb/Mahan theories on command of the sea, achieved through major fleet engagements, did not stand up well to the test of maritime operations in World War II. If we take all these reasons and add them to those already advanced by Aleksandrov in 1929, there would seem to be a strong case for arguing from the

Soviet point of view that the "Mahanist theory" of command of the sea is "worthless," and most of those reasons remain valid today.

The Descriptive Use

The descriptive use of "command of the sea" also ceased in the immediate postwar years. Partly perhaps because it would be confused with the Mahanist theory, but more probably because its relevance to the situation that faced the Soviet Navy in 1945 was purely negative. With the Anglo-American fleets as the obvious opponents, the problem was how to prevent the enemy from gaining command of the Baltic or the Black Sea, as had occurred four times in the past hundred years. As an interim measure, some variant of the "small war" strategy was the only option. However, by 1952, the postwar shipbuilding programs were well under way, and new construction was beginning to join the fleet in appreciable numbers. The Soviet Naval Air Force was being equipped with jet fighters and torpedo bombers and posed an increasingly serious threat to any surface intruders who dared venture into the confined waters of the Baltic and Black Seas. Gaining command of the fleet areas became a practical option, and the term was restored to normal use.

The Shift to Forward Deployment

In 1961, the Soviet Navy once again faced the requirement to conduct naval operations against much superior forces, only this time the "small war" strategy was not available as an option. Their forces had to move forward in strategic defense of Russia and conduct sustained operations in a potentially hostile environment at great distances from her shores. Early in 1968 Gorshkov told the Soviet armed forces that although the requirement had now been met, it had not been easy, and the necessary organic restructuring of the fleet did not go smoothly: "Besides the need to solve the technical problems involved, it required . . . the redeployment and retraining of key personnel, and the working out of the basic principles that would underlie the operational concepts and tactics required to meet these new demands."[30] As the pattern of forward naval operations built up since 1964, we have gradually obtained some idea of the consequential policies and operational requirements, although the strategic picture has been obscured by the overlay of peacetime tasks.

Introduction of Belli's Concept

In 1972, discussing the interwar years in the course of his 11-article series, Gorshkov described the "Belli" concept of limited command of the sea that was defined earlier in this chapter.[31] Interesting in itself, it assumes greater significance when this article is compared to the one published exactly five years previously, on much the same subject and likewise signed by Gorshkov.[32] The greater part of the 1967 article was devoted to criticizing the way Soviet naval strategy developed between 1930 and 1955, at which date Gorshkov took over the navy. There are substantial differences in tone and content between the articles, two points being of immediate interest to this analysis. (1) In 1967 there is no reference to an offensive school of thought, to a "command of the sea" theory, nor to Belli's definition of limited command. Discussion concentrates on the defensive viewpoint stemming from "small war" theories. (2) In 1967 he clearly states that the defensive view held sway through into the 1950s, but in 1972 he implies that the offensive school gained ascendancy by the end of the 1930s.

We have no means of knowing what underlies this substantial divergence, and it is hard to interpret its significance. It is unlikely that Gorshkov was introducing the Belli definition as a new departure since it is a classical strategic concept. It does indeed have relevance to current operations in the face of superior forces, but it would seem unnecessary to reach back into the 1930s to define such a common-sense concept. One suggestion that derives from a wider analysis of the complete "Gorshkov series" is that in 1972 Gorshkov was rebutting accusations that he was harboring Mahanist aspirations and at the same time seeking to emphasize the historical roots of contemporary policy.*

THE RELEVANCE TO CONTEMPORARY STRATEGY

What does this analysis suggest in terms of Soviet naval strategy? Starting with the negative inference, the Soviet Union has valid reasons for rejecting the universal relevance of Mahan's theory of sea power. Their own ideology is antagonistic to geopolitical theories, but if they came to espouse one, Mackinder's would be more appropriate.

---

*I found that many of Gorshkov's comments only made sense if it were assumed that among other things, he was concerned to rebut various charges that were being levied against him by political opponents.

This does not imply that they "don't understand sea power," and over the last 250 years, they have repeatedly demonstrated their awareness of the strategic utility of naval forces. But it does mean that different priorities will apply in assessing competing demands on national resources. There are of course those who argue the advantages of naval strength in classical terms, and we would expect the admirals to be among them. But the evidence suggests that Soviet leaders recognize that the significance of sea power differs with national circumstances, and they have yet to be convinced that in Russia's case it has the relative importance implied by Mahan's theories.

The Soviet Navy does not accept the Colomb/Mahan thesis that command of the sea is essential for the successful prosecution of naval operations nor the emphasis placed on gaining command by bringing the enemy's forces to battle. The original reasons for advocating this were sound; only by destroying (or bottling up) the enemy's fleet can one be certain that he will not be able to contrive local superiority (and hence victory) in some distant area. It was a good theory for a nation both willing and capable of building up the forces to achieve such victory. For others it was a recipe for inaction, and from the earliest days the Soviet theoretical debate has been concerned with the problem of how to conduct naval operations without command of the sea.

This does not mean that the Soviet Navy ignores the advantages of achieving control over a sea area, where this can be obtained at reasonable cost. It has always been easier to establish command close to one's shores, and developments in shore- and ship-based weapon and sensor systems now permit the de facto extension of national frontiers to include the sea areas that are directly adjacent to the Soviet Union. Latent command of these areas exists in peacetime and could be imposed in war. In the Baltic and Black Seas, this would include the seizure of the exits, effecting a form of "command by exclusion." But the underlying concept is really one of area defense.

The Soviets are not persuaded that the more extensive Colomb-Mahan concept of command of the sea has any practical relevance to modern warfare. Considering the proliferation of different types of weapon vehicle, the vastly increased range of the weapons themselves, and the intermixing of land, sea, and submerged systems in the maritime battle, "command" is no longer a realistic concept, outside certain geographical areas that are amenable to area defense. And it is hard to see what military actions would be particular to achieving command, as distinct from the more general processes of war at sea. But neither are the Soviets prepared to concede command of strategically significant areas to their opponents by default. Modern weapon systems allow a middle zone of "no command," and the Soviet

Navy aims to make this as extensive as possible. Such a policy of "command denial" is particularly appropriate in countering the West's seaborne strategic delivery capability.

What then of Belli's concept of "limited" command? We do not know what prompted Gorshkov to spell it out in his review of the interwar years, but it has obvious relevance to Soviet naval units on forward deployment, operating in the face of Western naval preponderance. Bearing in mind the timing (and the absence of any reference to this concept in the 1967 article), perhaps its most immediate application is to naval operations in support of "state interests." The requirement to "create such a situation whereby the enemy will be paralysed or constrained in his operations, or weakened and thereby hindered from interfering with our execution of a given operation" is suggestive. The "given operation" could perhaps be the concern of some other instrument of government, with the Soviet Navy's task being to prevent interference by Western naval power.

We can only conclude that the Western concept of "command of the sea" (to the extent it is still alive) is not a major factor in Soviet naval strategy, except in the limited context of defense of the home fleet areas. This does not mean that the Soviet Navy will eschew the classical paths to command, of battle and/or blockade. But command of the sea is not an end in itself, and their concentration on submarines demonstrates their intention to bypass the requirement. Where this has not been possible, their primary concern has been to enable the conduct of operations in a hostile maritime environment and to provide the means of discharging their three main tasks in the event of nuclear war. Belli's concept is the classic recipe (on land as at sea) for achieving local superiority in the face of stronger forces, and it is particularly apposite for a navy that places such heavy emphasis on mines. But despite Gorshkov's claim, it does not constitute a theory of "command of the sea" except in the narrowest sense.

SUMMARY AND CONCLUSIONS

In Soviet usage, three different meanings can be attached to the term "command of the sea." It is used (1) To describe a situation past, present, or future. (2) As a label for the Colomb/Mahan "theory" of extensive command, achieved through fleet action or blockade. Mahan's theory of seapower is often subsumed under the term. (3) As a label for Belli's concept of limited and temporary command through contrived local superiority.

The term is used relatively infrequently and almost wholly either in its prejorative (Mahanist) sense, or else descriptively. The

Belli definition has only been noticed once, in a discussion of the interwar period, but it may have contemporary relevance.

In Soviet usage, "command of the sea" does not carry the emotive baggage that it has in the English language. In circumstances where Western navies might think in terms of "command of the sea," the Soviet Navy is more likely to speak of "defending the country's maritime frontiers," "concentration of force," or "combat stability."*

Since its earliest days, the Soviet Navy has addressed a great deal of thought to the problems of conducting naval operations without having command of the sea and usually in the face of superior enemy forces. Here, the concept of "combat stability," or the ability to successfully discharge one's task under adverse conditions, has particular relevance. This does not mean that they ignore the advantages that command of a given area can bring. Close to their own shores they apply area-defense concepts, which results in a form of command by exclusion; in strategically significant distant areas they adopt a policy of command denial; and in a specific area they may seek to achieve temporary command through contriving a local superiority of force.

But as a general rule they do not see the "extensive" concept of command of the sea as the proper objective of navies in war. Besides being unrealizable, except as the outcome of the all-arms battle, its pursuit as an objective tends to divert the navy from its proper mission and tasks. The Soviet Union is not challenging the United States for command of the sea. But it is challenging the command that the Western navies used formerly to exercise, through default of any opposition. And she is posing this challenge partly by means of the forward deployment of her own forces and partly by providing third-world countries with navies whereby to defend themselves.

While the Soviet Union has valid reasons for rejecting "Mahanist" theory as a basis for national policy, this does not imply that her leaders lack an awareness of the advantages that sea power can bring or of the underlying strategic principles. It is therefore appropriate to end this chapter with an extract from Mahan. In concluding the introduction to his first book, he quotes a French author as saying that naval strategy "differs from military strategy in that it is as necessary in peace as in war. Indeed, in peace it may gain its most decisive victories by occupying in a country, either by purchase or treaty, excellent positions which perhaps would hardly be got by war. It learns to profit by all opportunities of settling on some chosen point

---

*Boevaya ustojchivost'. A concept that is widely used, particularly in the navy and covers the ability to give, take, and recover from damage, often in adverse conditions.

of a coast, and to render definitive an occupation which at first was only transient."[33]

This dictum seems very relevant to contemporary Soviet naval strategy, which relys on the use of overseas bases to discharge their war-related tasks in peacetime.[34]

NOTES

1. Peter Vigor's masterly review of the usage of "gospodstvo na more" (Chapter 32 of this volume) prompted this paper, and I have drawn heavily on his research.

2. For examples of the difference in the interpretation of a given pronouncement that this leads to, see the author's "The Turning Points in Soviet Naval Policy," Chapter 16 in M. MccGwire, ed., Soviet Naval Developments: Capability and Context (New York: Praeger Publishers, 1973), especially pp. 187-94.

3. D. Fedotoff White, "Soviet Naval Doctrine," Royal United Services Institute Journal, August 1935, pp. 607-15. White drew his material from then current issues of Morskoi sbornik (MSb). See also S. G. Gorshkov, MSb., February 1967, pp. 9-12.

4. S. E. Zakharov, ed., Istoriya Voenno-Morskogo Iskusstva (IVMI), (Moscow, 1969).

5. Bol'shaya Sovetskaya Entsiklopediya (BSE), 3d ed., quoted by Vigor, op. cit.

6. V. A. Belli, who was a commander (Capt. 2) in 1938. MSb, August 1972, p. 21.

7. MSb., September 1972, p. 32.

8. MSb., July 1972, pp. 20, 22.

9. BSE, 2d ed., vol. 43 (1956), p. 172. This draws on the 1954 edition of IVMI, vol. 2, p. 25.

10. G. S. Graham, The Politics of Naval Supermacy, (London: Cambridge University Press, 1965), p. 7; Brian Tunstall, The Realities of Naval History (London: Allen and Unwin, 1936), p. 173. Tunstall was a history professor at the Royal Naval College, Greenwich, in the 1930s.

11. D. W. Mitchell, A History of Russian and Soviet Sea Power (London: Macmillan, 1974), p. 190.

12. In addition to increases in French, Russian, and American navies, the Italian and Japanese navies had recently emerged on the scene. In 1898 Germany indicated her intention to build a large navy. Tunstall, op. cit., p. 204.

13. Graham, op. cit., p. 123. In 1896 the combined strengths of the French and Russian fleets inhibited the despatch of British reinforcements to the North American and West Indies Squadrons during the Venezuelan crisis of that year.

14. Mitchell, op. cit., p. 197.
15. Mitchell, op. cit., p. 294.
16. W. Hadeler in M. G. Saunders, ed., The Soviet Navy (London: Weidenfeld and Nicolson 1958), p. 141.
17. Mitchell, op. cit., p. 288.
18. Graham, op. cit., p. 124. The naval historian Sir Julian Corbett was however a discriminating critic of Mahan's strategic theories. And Mackinder advanced his "Heartland" thesis as a rebuttal to Mahan's geopolitical theory.
19. Fedotoff White, op. cit., p. 608. I have not read these articles myself and have drawn on White's summary and analysis.
20. MSb., August 1972, p. 21.
21. Ibid., quoted by Gorshkov who states that this definition was being taught at the Naval Academy in the 1930s.
22. Ibid.
23. Admiral N. G. Kuznetsov, "Pered Voinoj," Oktyabr' no. 8, 1965, p. 170. Quoted by G. E. Hudson in "Soviet Naval Doctrine under Lenin and Stalin" prepared for the Central Slavic Conference, November 1972.
24. See R. W. Herrick on this point. Soviet Naval Strategy (Annapolis, Md.: U.S. Naval Institute, 1968), pp. 33-40.
25. M. V. Zakharov, "Nakanune Vtoroj Mirovoj Vojny," Novaya i novejshchaya istoriya, September-October 1970, p. 11. This gives the planned fleet strength at 1 January 1939 as 3 battleships, 7 cruisers, 55 destroyers, and 221 submarines; of these, all 3 battleships, 5 cruisers, 24 destroyers, and 18 submarines were World War I vintage. The plan for 1 January 1943 would have added 16 battleships, 13 cruisers, 110 destroyers, and 120 submarines. In the case of the surface ships, this was wildly optimistic and relied on the Soviets' being able to buy the majority of the battleships abroad. The Soviet Union in fact laid down two battleships and three battlecruisers, but none had been launched in June 1941. By then, she had a total of six new cruisers partly complete or in commission, a seventh having been destroyed on the building ways.
26. MSb., February 1967, p. 16.
27. Quoted by Peter Vigor, Chapter 32 of this volume, note 24.
28. A. T. Mahan, The Influence of Sea Power Upon History, 1660-1783 (London: Methuen, 1965), pp. 23, 89.
29. Graham, op. cit., p. 5, who also quotes Sir Charles Webster.
30. Krasnaya Zvezda, 11 February 1968.
31. MSb., August 1972, p. 22.
32. MSb., February 1967.
33. Mahan, op. cit., p. 22.
34. See chapter 28 in this book, pp. 522-29.

CHAPTER 34

THE TACTICAL USES OF NAVAL ARMS CONTROL

Franklyn Griffiths

A vigorous naval establishment does not usually have a vigorous interest in naval arms control. The Soviet naval establishment is evidently a vigorous one. This suggests that if there are to be formal and tacit agreements that limit the development and deployment of Soviet naval forces, they will be imposed upon a reluctant Soviet Navy and its supporters by winning coalitions that have other military-strategic, foreign political, and economic priorities in mind. The recent series of articles by the commander in chief of the Soviet Navy, Admiral S. G. Gorshkov, may similarly be interpreted to suggest the navy is siding with those in Moscow who are opposed to arms control. It is the purpose of this chapter to challenge some of these judgments.

ARMS CONTROL AS AN AID TO ARMS RACING

The ideology and practice of what we in the West call arms control have proven to be quite elastic. For many of its advocates in East and West alike, arms control or "partial measures of disarmament" represent the only workable solution to the twin problems of reducing the likelihood of nuclear war and limiting the arms race. Comprehensive disarmament is at present impractical. Unilateral restraint is vulnerable to internal political opposition, and to unpleasant surprise from abroad. Faute de mieux, step-by-step arms control has offered the best prospect for stabilizing and eventually transforming the international military environment. Indeed, tacit and formal arms control does seem to have enhanced a sense of mutual security between East and West, though it has not thus far had a major dampening effect upon the East-West arms race.

At the same time, arms control is also an important component in continuing national military efforts. Arms agreements have served to stabilize, redirect, and reduce the costs of military preparations that will presumably continue so long as there are nation-states and technological innovation. Arms agreements have been purchased at the price of concessions to domestic opponents in the form of enhanced r and d and procurement programs in weapons systems not barred by treaty. Arms negotiations have similarly been used to justify the continued development and acquisition of weapons on the ground that this improves the nation's bargaining position. Viewed from this perspective, arms control does not stand in opposition to energetic military programs. It supports them. There remains the hope that the cumulative effect of a series of arms control measures consisting essentially of short-term tactical moves for national advantage may eventually produce a largely unintended demilitarization of East-West relations. But the hope is slender indeed.

In any event, it is fair to say that arms control is not necessarily incompatible with vigorous arms programs and may in fact be of assistance to fast if not exhausting competition in an arms race. Soviet arms race behavior seems to reflect an awareness of this possibility.

In retrospect, the limited nuclear test ban and the accompanying measures of 1963-64 were employed by Moscow as a means of reinforcing existing constraints on U.S. arms programs while the USSR proceeded to carry out an extraordinary military buildup during a transitional period of unacceptable strategic inferiority. Arms control was of use in generating psychopolitical reassurance effects, and in reducing U.S. readiness to exploit revealed Soviet vulnerability by means of diplomatic demands. Nevertheless, the lead times required for the strategic and conventional forces that began to appear in impressive numbers in the mid-1960s indicate that the decisions to press ahead with the acquisition of new weapons had already been taken by 1963-64. This when Khrushchev was speaking publicly of reasonable compromises, "the policy of mutual example," the need for arms agreements, and the desirability of channeling Soviet resources away from defense. Similarly, the numerical advantages and also the opportunity to focus resources on qualitative improvement that Moscow obtained in the strategic arms limitation agreements of May 1972, seem thus far to have been exploited primarily if not exclusively for tactical purposes. Against a background of continued professions of interest in agreement to bring the nuclear arms race under control, the strategic arms limitation talks (SALT) appear to be serving the Soviet system as a means of disorienting the U.S. military effort, while allowing improved conditions for the Soviet Union to overcome the qualitative advantages enjoyed by the United States in offensive weapons.

But this is not to say that the Soviets have been practicing a systematic policy of deception through arms control with the United States. Some Soviet leaders could have had this in mind. But Soviet policy-making on issues that effect the entire structure of foreign relations and domestic appropriations is not likely to be the product of a single deliberate intention alone. Rather it would seem that the major conflicting tendencies in Soviet foreign and military policy vis-a-vis the United States have summed up into a resultant force that makes for a tactical approach to arms control. The tendencies that favor a fast arms race, on the one hand, and that militate for a stabilization of political-military relations, on the other, have succeeded in canceling one another out, thus producing outcomes that probably suit neither the militants nor the advocates of relatively moderate responses. These outcomes have limited the scope and pace of the Soviet arms effort in ways that must have proved exasperating to those who were especially concerned to increase Soviet strategic power even more rapidly than has been the case. And they have permitted costly and destabilizing investments in alternate military programs that doubtless chagrined those whose primary responsibilities lay in the civilian economy. The net effect has been an arms policy equivalent to that which might have been decreed by a Soviet hierarchy fully united on the utility of a tactical approach to arms control.

The propensity to unhappy compromise is not limited to the Soviet approach to arms control. A similar interpretation might be made of U.S. conduct on the issue of naval arms control following the Washington Disarmament Conference of 1921-22. Whatever the intentions of the principal players in the making of U.S. naval policy, the effect of their actions was to employ arms control negotiations and agreements not as ways of stabilizing the military balance but as means of securing British acceptance of parity with the United States in all classes of ship by 1930. For domestic economic and political reasons and for purposes of international stability, Presidents Harding, Coolidge, and Hoover sought to limit naval development at the Washington, Geneva, and London conferences. The big navy advocates in the United States, aided by the naval moves of Great Britain and Japan, sought ambitious building programs in areas not covered by treaty. These were justified by reference to the need to compel the British to negotiate, and by interpretations of the naval balance that suggested the United States had either been duped or war falling behind the others. Naval arms control negotiations that the White House and State Department sought for purposes of limitation were employed by the opposition as cause for sustained naval building. Negotiations made it possible to focus attention on naval rivalry, to undermine confidence in existing and projected U.S. capabilities, and to argue for expansion in replacement programs and categories not excluded by agreement. At the same time, naval negotiations and agreements in the 1920s and 1930s definitely reduced the extent and rate of U.S. naval construction. On the face of it, the record of gradual increments in U.S. naval power, accompanied by efforts to achieve arms control with the British between 1921 and 1930, could be interpreted to suggest that the U.S. leadership was ultimately guided by a commitment to arms control deception as a means of arms-racing. In fact, U.S. policy-makers were caught up in processes with a dynamic of their own, as the conflicting tendencies to militance and moderation in naval matters combined to produce a tactical commitment to arms control not unlike that which seems to characterize recent Soviet behavior.

But there are significant differences in the Soviet case. For one thing, it could well be that on military questions within the Soviet elite there are no consistent liberals, but instead militants and conservatives who divide over policy priorities and the assessment of the situation, with some advocating dovish positions on specific issues in response to situational pressures. In this case it would be objective circumstance rather more than subjective preferences that served to push the Soviet system in the direction of collaboration with the United States. Moreover, while the conflicting tendencies to secure greater military power and stabilize relations with Washington may involuntarily be resolved in favor of the tactical use of agreements, readiness to pay close attention to tactical matters is a hallmark of a political leadership steeped in the Leninist mode of thought. It is no doubt acknowledged at least by some within the Soviet elite that arms control does have tactical utility in the search for unilateral military-strategic and foreign political advantage when resources are scare.[1]

If Soviet officialdom is to some extent persuaded of the tactical uses of arms control, it follows that coalition politics in the formation of Soviet arms control moves may be considerably more complex than we have usually been willing to admit. Though it is customary to think of "the military" as a force in opposition to arms agreements and cooperation with the United States, elements of the Soviet military establishment may not in fact be opposed to all forms of arms control at all times. The same could well hold true for Soviet "ideologues" or party ideological secretaries, who are frequently depicted as having a vested interest in international tension alone.[2] Nor should we dismiss the possibility that the Soviet Navy may have a positive attitude toward arms control measures that limit the missions and budget claims of other branches of the armed forces, and even a positive attitude toward certain forms of naval arms control as well. So long as arms control served to promote the immediate objectives of the navy in dealing with its Soviet opponents and the United States alike, so long as it also helped the navy to realize its peacetime foreign policy missions and hence its claims for supporting appropriations, tactical agreements could reasonably be endorsed by the naval establishment and its supporters.

In principle, naval spokesmen may have cause to take a dovish position on certain arms control issues. This would not be dovishness in the liberal sense of a preference for agreement and unilateral restraint as means of reducing conflict. But if we define a "dove" as someone who for whatever reason favors negotiation or agreements with an adversary on a specific issue, the Soviet Navy may turn out on occasion to support dovish positions on arms control even as it pursues a vigorous overall arms race strategy.

GORSHKOV AS A DOVE

The best statement of the Soviet Navy's views on Soviet naval policy is to be found in the series of articles that appeared under the name of Admiral Gorshkov in the journal Morskoi sbornik between February 1972 and a year later. There is dwindling debate as to whether Gorshkov was advocating or announcing policy in this presentation. Increasingly it is believed that Gorshkov was engaged in advocacy. There is also a divergence of opinion on the implications of Gorshkov's remarks for future Soviet capability requirements and their utilization in war and peacetime. Some believe that the basic decisions for Soviet capabilities in the 1980s may already have been taken and that Gorshkov is explaining the new situation. Others suggest that while the direction of Soviet shipbuilding programs may already have been determined, the rapidity and extent of construction remains subject to decision, as does the manner in which existing and projected forces are to be operationally employed. Opting for these latter assumptions, let us consider the Gorshkov series for what it may tell us about Soviet naval approaches to arms control.

The first thing to be said about Admiral Gorshkov is that he presents himself as a great Russian nationalist dressed up in Marxist-Leninist garb. He sees himself as the bearer of a centuries-old naval tradition and bitterly laments the failure of the Tsarist autocracy to implement naval development policies that would have allowed it consistently to prevail over the other imperialist powers. What he has to say about Imperial Russia applies equally to the current situation. In the first place, he deplores the narrow thinking of Tsarist officials who opposed expenditures on naval construction in the belief that Russia was "not a maritime power, but rather a continental one and therefore [did] not need a fleet."[3] Not only was this a lamentable failure of the Russian imagination, which continued into the Soviet period,[4] but it amounted to an acceptance of traditional foreign imperialist opposition to a large Russian fleet.[5] Accordingly, Gorshkov rejects Nixon's 1970 view that the USSR is a land power and argues that this assertion flies in the face of Russian state interests past and present.[6] In part, Gorshkov seems here to be presenting naval preferences to the Soviet General Staff. The latter is dominated by ground force officers who reportedly in the navy's estimation do not know the bow from the stern when it comes to naval affairs. However, Gorshkov's references to "high officials" and to appropriations matters in Tsarist times indicate that he is also concerned with today's decision-makers who have a truncated conception of the Soviet identity as a land power and who are reluctant to make large investments in naval capability. Evidently the navy cannot have everything it wants. Instead it must seek to persuade the political and military leadership that its preferred projects deserve to be funded even if they require increased appropriations.

Gorshkov's response to this situation is to indicate that from the navy's viewpoint, nuclear ballistic missile submarine (SSBN) programs are already adequate to perform the priority task of strategic defense, and that there are considerable advantages to be had from a substantial increase in appropriations for high-speed, long-range, task-specific surface combatants.[7] Gorshkov does not appear to be calling for an increase in the development and construction of SSBNs, either because existing forces are adequate or because decisions on new SSBN capability are acceptable, and he evidently believes the United States is already deterred from launching nuclear strikes at the Soviet Union. The strategic missile forces and the navy have in his estimation succeeded in promoting a "sobering effect on the aggressively minded circles of the imperialist camp," compelling the United States to modify its objectives and to act with a degree of restraint in the pursuit of its expansionist objectives.[8] Add to this the implied existence of less aggressive or "sober-moderate circles" in the United States, and the principal adversary does not pose a dramatic strategic threat so long as there is no sudden change in the access of the right wing to the White House. In this situation, strategic defense remains the priority role of the Soviet Navy, but there is no pressing need for additional nuclear missile submarine construction.

On the other hand, Soviet great-power status and the pursuit of a global and activist foreign policy during war and in peacetime require much more in the way of surface combatants. A good deal of Gorshkov's presentation is devoted to the political advantages that can be derived from a powerful surface fleet. Drawing on the experience of history, he identifies a wide range of peacetime political tasks, some of which are nominally attributed to imperialist naval policy alone, that an effective surface presence can perform. These are essentially as follows:

- to demonstrate the power of the USSR beyond its borders for purposes of prestige and influence;[9]
- to increase Soviet control over the oceans and their food, industrial raw material, and energy resources, and to protect Soviet trading operations;[10]
- to perfect Soviet naval capabilities so as to surprise adversaries with Soviet power, lower their morale, intimidate them, and to secure local political objectives on shore without fighting;[11]
- to deny the credibility of American naval power, and hence American success in constraining centrifugal tendencies in NATO, exercising pressure, and supporting local reactionary groups on shore;[12]
- and ultimately to apply direct pressure on the United States, including threats and nuclear blackmail, as the United States has done to the USSR.[13]

The achievement of these political goals, as well as the related improvement in Soviet war-fighting capability, requires impressive new naval building programs. It also requires the permanent deployment of balanced task forces in various regions of the world. As matters stand, the navy's surface fleet is now stretched fairly thin merely in the performance of strategic defense against U.S. and allied SSBNs and carrier-based strategic strike forces. Somehow, appropriations have to be found for substantial additional surface fleet construction and deployment. It is here that the tactical use of arms control enters the picture.

At first glance, Gorshkov appears to be opposed to arms control as such. He commends the party and government for having seen that "the way out" of the situation created by the growing threat to Soviet security lies in the development of strategic defensive counterforces.[14] He does not say that "the real escape" is to be had by "other, peaceful means."[15] Nor does he cite in this connection Brezhnev's statement of June 1971, which declared a willingness to negotiate limitations on distant area deployment of naval forces in the Mediterranean and Indian Oceans.[16] Moreover, Gorshkov's comments on the interwar naval negotiations are disparaging and seem to suggest a lack of interest in arms control. He describes the negotiating efforts of the period as an attempt of the powers to achieve by the "diplomatic route" an improved armaments relationship.[17] And he writes off the various interwar conferences and bilateral discussions as ultimately a failure in limiting the naval construction programs of the largest states.[18] However, none of this is incompatible with a tactical interest in arms control. The fact that negotiations and agreements do not offer "the way out" presents no problem, for this is not the objective in tactical arms control, which is merely a means of supporting unilateral naval development efforts. Similarly, to observe that arms control does not really limit military programs but instead represents a "diplomatic route" to an improved power position, is to state precisely what a tactical approach to arms control is all about.

At the same time, Gorshkov or the panel that wrote his presentation has some favorable things to say about arms control. Though he is presumably under no obligation to do so, he makes mention of the need for "a cessation of the ruinous and wasteful arms race."[19] He notes that in 1936, the Soviet Union entered into negotiations with Great Britain "in order to check the naval arms race to some degree."[20] He also observes that the interwar negotiations performed "only a delaying function in the naval construction of the largest states."[21] In addition, he comments late in 1972 that, "arms control is still only being extended to strategic missiles, including those belonging to the navies," and adds that it is "interesting" that similar attempts have not been made to limit "other branches of the armed forces."[22] Further, he makes two quite positive references to the Seabed Treaty of February 1971, which prohibits the emplacement of weapons of mass destruction on or under the seabed.[23] What Gorshkov seems to be saying in all of this is that arms control is of use (1) in inhibiting the naval development of an adversary when one is behind and striving to catch up despite economic constraints; (2) in permitting a redirection of development and building programs to reduce "waste;" (3) in sealing off areas of military competition in which Moscow is reluctant to become involved, such as the seabed; and (4) in inhibiting the growth of some branches of the armed forces as opposed to others.

If we put the pieces together, we find Gorshkov making the following general statement to his readers. The General Staff and the political leadership are resisting our claims for resources for additional surface warship construction. We cannot obtain the capability required without trimming existing programs and without procuring additional funds either from the state budget or from cutbacks in the programs of the other armed services. The United States is basically deterred, and we are in a position to use SALT to obtain permanent limitations on offensive strategic forces. Funds may thereby be liberated not merely from our own budget, but also from the strategic missile forces and the air force as well.* At the

---

*The technological, industrial, and manpower requirements of surface fleet and SSBN construction, or surface fleet and air force

same time we favor mutual force reductions in Europe (MBFR) as a means of limiting the ground forces and providing slack in budget demands. Arms control agreements with the United States and its allies will not of course end the arms race. They will however create a situation in which we are better able to obtain the added appropriations for surface warship construction.

If the foregoing bears any resemblance to reality, it suggests that behind Gorshkov's bluster and the increasingly assertive Soviet naval presence in various regions of the world there lies a tactical interest in arms control as a means of redirecting Soviet naval development for arms race purposes. Obviously the navy cannot take on all the other armed services single-handed and will have to act with prudence in seeking favorable SALT outcomes. Nevertheless, the navy's arguments may prove particularly compelling, for they should appeal simultaneously to those in the Soviet leadership who wish for security and economic reasons to see SALT prove successful, and also to those who desire global predominance for the USSR and a continued political offensive against "imperialism." In effect, Gorshkov and his colleagues are trying to manipulate the innate ambiguity of arms control to the Navy's advantage. By making the Navy the beneficiary of a coincidence of interests among actors and institutions who normally would be opposed to one another, Gorshkov stands to improve his chances of gaining the political support necessary to weaken the claims of the other armed services and to move into a new era of naval competition through the use of arms control.

Should Gorshkov have his way and should a SALT II agreement on a "holiday" for offensive weapons acquisition lead to an increase in the scope and pace of Soviet surface combatant construction, we could expect to see a continued naval interest in the tactical use of arms control. Gorshkov's references to U.S. naval policy in the interwar years are revealing on this point. He notes that United States shipbuilding and disarmament negotiating efforts saw it win "international recognition of the 'parity' of its naval forces with the British."[24] Furthermore, he observes that the Americans subsequently displaced Britain as "Mistress of the Seas" and achieved naval superiority. "In this connection," he adds, "the Americans succeeded, without a war with her, in achieving what Germany could not achieve in two world wars."[25] Gorshkov evidently has some respect for the U.S. achievement. Whatever the results of SALT II, the Soviet Navy seems likely to display a persistent interest in the

---

programs, are obviously not fully competitive. However, all defense programs share a common requirement for investment funds, which are in short supply.

"diplomatic route" to a favorable naval relationship with the United States. As it moves ahead with new naval construction and enlarges the pattern of naval deployment, the navy may find it expedient to make arms control proposals, to enter into negotiations, and to strike agreements in order to limit U.S. construction programs one by one and to secure U.S. acceptance of Soviet naval power. From the navy's perspective, this would be a tactical use of arms control as resources were repeatedly recommitted to bring Soviet forces up to parity, if not beyond. It would be arms control in the service of arms racing, but arms control nevertheless.

TAKING THE "DIPLOMATIC ROUTE"

As of April 1974, the actual outlook for the Soviet Navy seems rather bleak. Broadly speaking, the Soviet Union stands now in relation to the United States roughly where the latter stood in relation to Britain at the conclusion of the Washington Disarmament Conference: short of actual parity in the principal strategic weapons, but with strategic parity being recognized in principle by the superior power. The Soviet Navy's perception of its task, as was previously the case with the American, is presumably to obtain authentic strategic parity and the recognition of its right to parity in all classes of ship from an adversary of superior power in a setting of continued economic constraints on rapid shipbuilding. This is obviously going to take time, so much so that the prospect of naval parity, much less superiority, seems quite remote. Again, in the Washington Treaty of 1922, the superior power was willing to grant actual parity only by 1942, and then only in strategic weapons to a state with which it had many affinities. Moreover, U.S. economic potential in relation to Britain was considerably greater than is the case with Soviet economic and technological potential today. Present American advantages in multiple independently targeted warhead capability, missile accuracy, on-station time for SSBNs, submarine silencing, and forward-based systems aboard carriers confront the Soviet Navy with the prospect of factual strategic inferiority for years to come. At lower levels of naval power, including the ability to wage general or theater war at sea and to land intervention forces in the face of opposition, the United States is also clearly superior. Furthermore, in the year or so since the Gorshkov series of articles was completed, Soviet surface warship construction seems barely to have kept pace with replacement requirements. Thus far, naval building decisions do not seem to support Gorshkov and his colleagues.

In the event of failure to reach a permanent agreement on the limitation of offensive strategic weapons at SALT II, the prospects for surface fleet construction may continue to be unfavorable, particularly when Brezhnev is proposing to invest 45 billion roubles in the Siberian agriculture in the 10th Five-Year-Plan.[26] On the other hand, if SALT II opens the way to more rapid and extensive surface warship construction, the navy will still be faced with a choice of emphasis between arming up or compelling the United States to negotiate limitations that would improve the Soviet position by reducing the U.S. capacity to wage war at sea. A SALT agreement that stabilized the Soviet-American strategic relationship would moreover emphasize the importance of this choice. For the limited ability of the Soviet Navy to act at "war at sea" levels in the expression of naval power is largely dependent upon the threat of escalation to strategic nuclear war.[27] If the latter is made less likely by further strategic arms control, existing U.S. superiority at "war at sea" levels will be enhanced, thereby reducing the political effectiveness of the Soviet Navy and increasing the urgency of offsetting Soviet shipbuilding and/or arms control moves. In this situation, economic and internal political constraints may again favor a consideration of the "diplomatic route."

Apart from arms race interests in arms control, the Soviet Navy may in the future be guided by political interests in proposals, negotiations, and even agreements on specific measures. A need could be perceived to minimize reactions of alarm in third-world countries as well as the United States, as new Soviet surface vessels were acquired and deployed in various naval theaters. Also, if the navy is increasingly to be the instrument of an activist foreign policy, the projection of naval power would probably be associated with an exploitation of the peace issue to promote on-shore sympathy for the USSR. Calls for nonnuclear zones, detente at sea, naval disarmament, and so on could be employed to expose the predatory nature of "imperialism," to cast the Soviet Union in an exemplary role, and to gain support for specific objectives in the third world. In addition, considerations of prestige and recognition could help to make agreements acceptable to the navy once it had achieved sufficient progress in solving specific problems of construction and deployment to permit reorientations of further naval development by treaty.

A variety of factors may be expected to influence the extent to which the navy endorses tacit and formal agreements, as opposed to nonnegotiable proposals and nonproductive negotiations, on foreseeable issues of naval arms control. Among these factors are the state of Soviet-U.S. relations, the level of Soviet naval appropriations, the development and operations of the U.S. and Western navies, the opportunities for an activist emphasis in Soviet foreign policy, and the degree to which new naval tasks require agreement in themselves or in prior areas of competition that have become less productive. Evidently there is little room for prediction here. It would seem nevertheless that with a further elaboration of Soviet-U.S. cooperation, corresponding constraints on the activist tendency in Soviet behavior, and only modest investment in surface combatant construction, the navy may be brought to an increased interest in the tactical use of agreements to restrain U.S. naval development, to reduce the risk of war at sea, and to permit a focused expression of available Soviet forces for political purposes. Alternatively with greater Soviet-U.S. rivalry, increased Soviet activism, and moderate naval appropriations, the navy's interests would be largely confined to the propaganda and atmospheric effects of proposals and negotiations designed to enlarge Soviet influence in the third world and tantalize Americans with possible Soviet reciprocation for U.S. unilateral restraint and concessions.

The variations in a tactical approach to arms control may become clear on considering possible future naval preferences regarding deployment limitations, confidence-building measures, force limitations, and ad hoc United Nations maritime peacekeeping forces.

DEPLOYMENT LIMITATIONS

If Soviet naval power is to be projected outward in the form of permanent task forces in various regions of the world, agreements to limit deployment will presumably be resisted by the navy either until it has built up adequate regional presences or until it has made sufficient progress to permit trading for spheres of maritime influence with the United States (that is, the Caribbean for the Eastern Mediterranean).[28] Propagandistic proposals and protracted negotiations on surface deployment limitations might however become attractive to the navy well before interests in agreement emerge.

The concern of third-world littoral states such as India to avert Soviet-American competition in offshore waters will make it increasingly profitable to project Soviet naval power under the cover of a concern for peace. At present the navy is evidently not inclined to endorse proposals for denuclearized zones and mutual withdrawals, as is indicated by Gorshkov's failure to cite previous Soviet offers to this effect. To lend support to such proposals at this time might be to risk premature constraints on the political uses of naval power in third-world situations, and hence on the ability of the navy to argue for increased appropriations. But once it had begun to maintain permanent and substantial task forces in the Indian Ocean, South Atlantic, and other areas, proposals for deployment limitations could become more acceptable. Proposals to this effect, and possibly negotiations as well, could include not merely mutual withdrawals but also regional limitations on the number of combatants, time on station, and the use of base facilities on islands and littoral states.[29]

The political effects of such moves would be to enhance Soviet prestige, to further political objectives on shore, and to provide ammunition for Americans seeking to limit naval arms-racing and the risk of war at sea. Relatively unworkable proposals and extended negotiations on deployment limitations might also serve to ease the navy through the delicate stage at which it begins to match United States and allied naval presences without provoking offsetting Western moves.

On the other hand, the navy may already have an interest in discussing limitations on the deployment of U.S. carrier-based strategic strike forces. At SALT II the Soviets have insisted on across-the-board negotiations that include land-based and submarine-launched missiles, bombers, and NATO forward-based systems (FBS) in a single package. It is conceivable however that the FBS issue may be transferred from SALT to MBFR. The United States shows signs of favoring such a move,[30] the result of which would be to write off a solution for years to come, if the widespread pessimism concerning the complexity and political prospects of the Vienna talks proves justified. The Soviet Navy may also have a preference for discussing FBS at MBFR in order both to lower the risks of an agreed withdrawal from the Mediterranean and to inject itself as an active participant in the formation of Soviet policies for Vienna. As Gorshkov suggested, it would be "interesting" from the navy's viewpoint to see Soviet ground force capabilities become the subject of serious East-West negotiations. If naval deployment issues were to be included on the MBFR agenda, the navy would stand to gain increased leverage in dealing with the budget claims of the other armed services either by allying with some others in support of ground force reductions, or by trading cooperation with the army for the latter's support on other arms decisions affecting naval interests. Moreover, if NATO forward-based systems were traded primarily against Soviet forces in Europe, the navy could expect to remain in the Mediterranean in force, and its capabilities devoted to a counter-FBS role would be liberated for other purposes.

In the event Soviet military and foreign policies failed to develop in a direction favorable to the vigorous political use of a rapidly growing Soviet surface fleet, the navy's interests in deployment limitation agreements might mature relatively rapidly. With Soviet foreign policy activism increasingly subordinated to the maintenance of Soviet-American cooperation, the prospects for naval development would become increasingly dim, and deployment agreements would seem more attractive as a means both of denying the U.S. capacity to make political use of naval power and of reducing the risks of confrontation with a superior power at sea. If reflected in Soviet policy, the navy's interests in this case could be signaled by unilateral restraint, leading possibly to tacit deployment arrangements

.ght be made the topic of formal negotiations. Presumably
 /y would prefer tacit understandings, for they preserve freedom
 ion and allow for changes in the pattern of naval deployment in
 ,onse to developing situations.

CONFIDENCE-BUILDING MEASURES

As the inferior power at sea in important respects, the navy should have an interest in confidence-building measures (CBMs), in the same way as the United States as the inferior power in Europe has an interest in extracting CBMs from the Warsaw Pact. Such measures could include the exchange of naval observers, advance notification of maneuvers and force rotations, the prohibition of maneuvers in certain areas, regional limitations on fleet size, and agreements on "basic principles" or "rules of the game" in Soviet and U.S. naval operations. Agreements to this effect could complicate the task of the Americans in coping with situations on shore, assist the development of local conflict situations according to their own dynamic, and make it easier for the Soviet Navy to respond to U.S. naval operations. As was the case with deployment limitations, CBMs would also sow doubt concerning the reliability of U.S. commitments to its European allies and to client governments in the third world. And if the initiatives were Soviet ones, they would engender support for Soviet diplomacy as well as an acceptance of the growth of Soviet naval power. CBM negotiations and agreements would thus favor the pursuit of arms racing and offensive foreign policy goals by the Soviet Navy, while reducing the risk of war if Soviet-American competition in the third world and the European theater intensified.

However, the Soviet Navy also has an interest in maximum freedom of action in projecting its power for political purposes. It seems likely at present to resist CBMs insofar as it seeks a free hand to be where the trouble is and in a manner that intimidates and denies confidence to the adversary. Though naval power may be projected for offensive political purposes within a framework of CBMs, limitations on Soviet naval operations would necessitate relatively subtle and hence long-term forms of influence more suitable to a setting of high U.S.-Soviet cooperation. CBMs would also attenuate the climate of Soviet-American conflict necessary for substantial Soviet naval spending. Political and funding considerations of this kind may in part explain Gorshkov's reference to the Incidents at Sea Treaty of May 1972, which is curt and noncommittal by comparison with his endorsement of the seabed agreement.[31]

Until the navy has succeeded in extending and diversifying its surface capabilities, or until this possibility and the opportunity to support an activist foreign policy have been effectively reduced, Gorshkov and his successors are likely to oppose agreement on CBMs that in themselves connote acceptance of an adverse balance of power and lend assurance to the superior force. But as the Soviet capacity to deal with war-at-sea situations grows in various regions, proposals and negotiations to enhance confidence at sea may become tactically useful in reinforcing and ratifying an increasingly favorable situation.

At some point, therefore, the navy will have to chose between an emphasis on contention and coordination of expectations with the United States in its approach to confidence-building at sea. This problem may be illustrated by reference to the question of codifying "rules of the game" in U.S.-Soviet naval operations. U.S. observers, for example, have pointed to the emergence of tacit rules in the expression of naval power in support of client states. According to this assessment, the evolving practice of both sides points to implicit rules of behavior according to which the Soviet Union and the United States are free to assist their clients against the clients of the other so long as the latter are on the defensive strategically and not seeking an immediate change in the status quo.[32] Agreement on a rule of this kind would be to the advantage of Soviet Navy, for it would be accorded equal rights even though it was the inferior naval power.[33] But it would also circumscribe possible offensive political missions of the navy in lending limited support to clients who desire to alter the status quo in short order and in landing intervention forces to transform the situation in areas contested with the United States. To eliminate such missions before the requisite capabilities were obtained, the navy might conclude, would be to dilute the activist rationale for high levels of naval spending. Better to take advantage of U.S. interests in "rules of the game" by lending ambiguous support to the idea and leaving proposals and agreement on basic principles of naval operations for the day when shifts in the balance of naval power give the USSR an edge in defining rules of assistance to client states and movements.

FORCE LIMITATIONS

Thus far it has been suggested that the navy is interested in matching U.S. capabilities through the full spectrum of available naval expression, and that economic constraints together with the need to minimize offsetting U.S. construction will cause it to consider the tactical use of arms control in support of its own building efforts. The navy may however have an additional choice. On the one hand, it could try arming up and then negotiating in order to match the U.S. ability to act at "war at sea" levels. Alternatively, it could attempt

the United States to refrain from the replacement and new
...tion necessary to maintain its "war at sea" capabilities in
... of age and technological obsolesence. If the navy seeks to
... capabilities to some extent, its approach to the issue of force
...tions would emphasize increasingly conciliatory propaganda
... rotracted negotiations that might eventually be expected to yield
...ements fixing Soviet rights to parity in major surface combatants.
...he other hand, if it endeavored to shape U.S. naval power to its
... image, an interest in agreements might be displayed more rapidly.
...this case it would be necessary to build only so far as necessary
... approach parity at relatively low force levels compared to those
...ntailed in an effort to equal the present and future ability of the United
States to engage in war at sea. The resources saved might in turn be
reallocated for further Soviet building at lower levels of naval expression that offer relatively high political payoffs for comparatively
low investment. An effort to negotiate a change in the profile of naval
power would also be in keeping with the Soviet Navy's pride in not
emulating Western experience in the development and structuring of
its forces.[34] It would however require overcoming the traditional
U.S. preference for a navy that is capable of virtually any mission
at the drop of a hat.

Whether the navy seeks to arm up to imitate the U.S. fleet or
to make the Americans conform to Soviet practice, it would approach
force limitations as a means of altering the balance of forces at "war
at sea" levels to Soviet advantage. The development of its interests
from propaganda through negotiations to agreements should depend
substantially upon its shipbuilding progress. As it moved ahead to
counter U.S. superiority in sea-war situations, it might first be willing
to entertain proposals and negotiations to limit new and replacement
building. Proposals to this effect would make Soviet construction
programs ambiguous and less threatening. They would strengthen
sympathy for the Soviet Union among the usual publics in the third
world and Western Europe. Negotiations on naval building limitations
might subsequently become appropriate. These would no doubt be
lengthy and complex, given the differences in structure and armaments
between Soviet and U.S. fleets,[35] to say nothing of other naval powers
that might have to be brought in. Negotiations could also be employed
to mask Soviet building, to justify continued construction programs
in debate within the Soviet Union, and to increase U.S. desires to halt
the growth of the Soviet Navy by unilateral restraint or agreement.

Eventually, however, a tactical interest in agreements might
materialize. By then the choice of emphasis would have been made
between matching the U.S. force structure and seeking to make the
latter conform to the Soviet practice. In principle the strategy of
emulation, which would push agreements into the distant future,

might best suit the navy as a means of assured growth. But it would presuppose both a conversion of the political leadership and the General Staff to the navy's view of the Soviet Union as a great sea power, and a degree of unregulated U.S.-Soviet competition that may not readily be achieved despite inevitable fluctuations in the level of superpower tension. On the other hand, if appropriations for rapid shipbuilding were not available, and if the option of attempting to secure U.S. conformance to the profile of Soviet forces were selected, an interest in building limitation agreements could develop quite quickly. For a lengthy assymetry in the power to wage war at sea would be costly to the navy's ability to carry out foreign policy missions, particularly if additional strategic arms control gave the United States greater confidence that nuclear war would not result from the vigorous expression of its naval power at subnuclear levels. Simultaneously the United States might itself come to a growing realization that the maintenance of a full array of capabilities to wage war at sea was uneconomical in the light of growing Soviet-American cooperation in other areas. Circumstances could thus arise in which the navy might prefer agreements to limit new and replacement construction of major surface combatants in order to emphasize procurement at lower levels of naval expression offering substantial returns to a political leadership still unreconciled to the power of the United States.

MARITIME PEACEKEEPING FORCES

In the final article of his series, Gorshkov draws attention to the problem of sharpening international competition for the ocean's resources. He asserts that capitalist states are engaged in a struggle to divide the ocean into spheres of influence for economic as well as military purposes and adds that proposals are being made in the West to take over areas of the seabed and prohibit freedom of navigation in waters over undersea work.[36] Navies may come into play, he notes, and with them the possibility of crisis situations.[37] The Soviet Union, it should be added, is reported to have been considering deep seabed exploration in the Atlantic, Indian, and Pacific oceans.[38] Gorshkov evidently believes the scramble for seabed resources entails the possibility of "a new arms race" with the West that would impose additional requirements on Soviet surface forces, while also risking the exclusion of Soviet vessels from potentially large areas of the ocean.[39] At the same time he acknowledges the existence of differences with third-world states that seek far-reaching revisions in the law of the sea that would limit freedom of navigation and establish a United Nations seabed agency to control exploitation of the bottom's resources. This latter arrangement he rejects as a vehicle for seabed control by multinational corporations.[40]

The problem is obviously a complex one that cannot be dealt with by Soviet naval forces alone. Competition for control over the sea's resources could accelerate more rapidly than the navy's capacity to cope, thus confronting it with the emergence of de facto areas of Western maritime control or international arrangements that discriminate against Soviet interests and prop up the capitalist system by giving it ready access to new resources. The navy may be expected to seek additional capabilities if it is called upon to undertake new missions required by seabed developments. However, Gorshkov's presentation also emphasizes the multilateral aspects of the problem, and calls for the maintenance of the seabed as a "sphere of peaceful international competition."[41] Somehow a response will have to be devised that looks after Soviet naval and economic interests, appeals to third-world states, and limits Western freedom of action in establishing sea control and conducting seabed operations unilaterally or under the umbrella of a United Nations agency.

In addition to United Nations licensing activities and other measures that might be negotiated at the forthcoming Law of the Sea Conference, one possible response of the navy might be to support a proposal for the establishment of lightly armed maritime peacekeeping forces on a regional basis under the Security Council. A proposal of this nature would help to counter Western efforts to establish spheres of influence at sea, reduce the need for littoral states to extend their territorial waters outward, lend security to Soviet seabed operations, and appeal to third-world governments that desire a regulatory role for the United Nations. Small maritime peacekeeping forces could be drawn from littoral states and the major powers, and their subordination to the Security Council would allow the Soviet Union to influence their operations in the event the navy was used in local conflict situations on shore, or in disputes over offshore seabed rights when the Soviet Union wished to tilt in favor of one side. Alternatively, peacekeeping operations could be confined to deep sea areas, thus avoiding a United Nations maritime presence in local conflict situations and allowing the Soviet Union and littoral countries a role in the regulation of Western seabed activities. (A proposal of this nature, which might be presented within a disarmament context, would echo certain provisions of the 1928 Soviet draft convention on immediate and complete disarmament. Chapter III of this plan provided for a division of the world's oceans into 16 zones extending from the Baltic to the Southeastern Pacific. Each zone was to be supervised by a "naval militia" comprised of national combatants drawn from regional groups of states.)[42]

Whether or not the navy would be interested in the establishment of such forces would depend upon the outcome of the Law of the Sea Conference and the relative rate of development of seabed operations and Soviet naval capabilities. Should the latter be retarded, an interest in ad hoc operations could materialize. Given the highly restrictive attitude of the Soviet Union toward United Nations peacekeeping on land, this seems unlikely. But in some respects the position of the Soviet Union at sea is the reverse of its situation on land.[43] Where processes of change on land can readily be exploited for offensive political purposes by unilateral action and minimal United Nations intervention, the context at sea favors the Western states with their superior naval and technological power. In this situation, the use of ad hoc United Nations maritime peacekeeping forces, particularly if there were third-world support, would assist the navy both in making up for its lack of capability and in rendering the ocean environment more suitable for the performance of activist foreign policy roles in peacetime.

THE "DIPLOMATIC ROUTE" REASSESSED

In sum, the Soviet Navy may in the future have reason to display a continuing interest in the "diplomatic route" to an improved armaments, deployment, and operational situation. The tension between its ambitions for a large surface fleet and the official persuasion that the Soviet Union is essentially a land power may make it necessary to exploit naval arms control as a means of doing more with less. The tactical use of arms control could reduce the strength of U.S. shipbuilding reactions, assist in the performance of foreign political missions, and promote favorable conditions for a gradual but sustained development in the pattern of deployment and operations. For the time being the navy's interests in arms control are likely to be confined to the achievement of propaganda and atmospheric effects through proposals and later negotiations that stop short of agreements. But as it began to approach its initial building and deployment objectives, tacit and formal understandings could become a valid means both of securing recognition of Soviet progress and of redirecting naval development into more profitable areas of competition. While others in the Soviet Union and the West might approach naval arms control as a step-by-step process of enhancing security and reducing defense spending, for the navy, arms control would be used to promote incremental growth in Soviet naval power.

THE DUALITY OF ARMS CONTROL

Insofar as the navy attempts to turn future naval arms control issues to tactical advantage and does not oppose arms control as such, it will contribute to the larger flow of influence within the USSR that makes for a commitment to formal and tacit arms agreements. During the past two decades the propensity of the regime to go beyond the atmospheric effects of detente and to collaborate with its Western adversaries has increased, together with increases in Soviet military power and political influence. Thus far, Soviet participation in East-West arms control arrangements appears to be primarily tactical and subordinate to arms race and offensive political priorities. But should coming decades witness a continuation of the Soviet trend to collaboration with the United States and its allies, the scope of East-West arms control could prove to be quite substantial, as could Soviet gains in military strength and foreign influence. Indeed, there is a possibility that collaboration among adversaries could become less tactical in nature. Acquiring a momentum of its own, it could increasingly express shared as opposed to converging unilateral interests, even as the East-West arms race continued. In time, arms control arrangements could accumulate to a point where they effectively began to choke the arms race and reduce the oportunity costs of high levels of defense preparedness. On the other hand, if the relationship among the opposing tendencies in Soviet foreign and military policies remained essentially as it is today, the overall Soviet interest in arms control would of course be considerably more favorable to the navy. Protracted detente and tactical agreements, relieved by bouts of renewed tension, would further the pursuit of offensive arms racing and foreign political objectives even at relatively high levels of East-West agreement.

Arms control is inherently dualistic. No one can say for sure which way it will go. It offers advantages and risks both for the exponents of enhanced military power and external influence and for the advocates of civilian priorities and international security.

The armed service that enjoys substantial backing from the political and military leadership of the Soviet Union can be expected to oppose military collaboration with an adversary, especially when its own capabilities are at stake. It would be reluctant to see arms control proposals and concessions being made, for an array of other actors and institutions might later come together to use negotiations to block its capability requests and get on with other things in domestic and foreign affairs. Moreover, the well-placed armed service could be leery of the notion of tactical arms control, because of its susceptibility to exploitation by proponents of limited military spending who could argue that collaboration was merely tactical and did not represent an acceptance of the status quo.[44] The armed service in a good internal position would accordingly attack the proposals, negotiations, and agreements advocated by others in policy debate, throwing doubt on Soviet military power and pressing for increased capabilities necessary to create a situation of strength. If this were the case with the Soviet Navy, the tactical approach to naval arms control that has been considered here would be the preference primarily of political officials seeking to reconcile the conflicting naval and economic-political requirements of the Soviet state.

But as Gorshkov indicates, the navy does not possess a degree of support within the General Staff and the political leadership that is commensurate with its ambitions to increase Soviet seapower. It is difficult to see how the navy can readily improve its situation against a superior adversary and in the absence of effective internal backing without relying on the "diplomatic route" to sustained building and deployment. A readiness to make tactical use of naval arms control would give it needed leverage on U.S. and Western behavior. More important perhaps, it would resonate with the tendencies in the Soviet leadership that favor an activist policy of peace directed against "imperialism" and an effort to collaborate with the adversary to reduce the opportunity costs of defense. The risks of premature and undesirable agreement associated with tactical uses of arms control would have to be accepted. But then militant opposition to naval arms control might stimulate opposition within the leadership and thus produce neglect where higher levels of support for naval development were required. Moreover, in running the risks of the "diplomatic route" the navy might expect to count on Western naval activities to help justify its arguments for further building and to generate support for Soviet naval perseverance from the activist side of the house.

Key issues in internal Soviet discussions of naval arms control may therefore concern not so much the merits of arms control as such but the timing and scope of proposals, negotiations, and agreements. While the navy's aim would be to see that arms control remained the prisoner of arms-racing, the appropriate Western response would lie not merely in reducing the Soviet naval buildup but in shifting the balance of Soviet preferences from tactical to what may be termed opportunity cost arms control. This would entail denying the Soviet Navy its construction or deployment breakthroughs, and rejecting Soviet naval arms control proposals that were unmitigated tactical ploys. It would also require a patient effort to employ negotiations to influence the play of forces in Soviet policy-making so as to promote opportunity cost arms control for the West as well as the USSR. By utilizing internal Soviet differences over the timing of moves along the "diplomatic route," it might be possible to advance Soviet acceptance dates, thereby prompting a deceleration of naval development and a redirection of attention to more pressing needs.

NOTES

1. Direct evidence is lacking on this point. Inferences can however be drawn from some remarks attributed to Khrushchev by Academician Andrei Sakharov. At a meeting in the summer of 1961, Sakharov suggested to the Soviet leader that a unilateral Soviet resumption of atmospheric nuclear testing after a three-year moratorium would undermine the test ban talks and lead to a new round in the arms race. Khrushchev's reply was, "Sakharov is a good scientist. But leave it to us, who are specialists in this tricky business, to make foreign policy. Only force—only the disorientation of the enemy [works]. We cannot say aloud that we are carrying out our policy from a position of strength, but that is the way it must be. I would be a slob, and not Chairman of the Council of Ministers, if I listened to the likes of Sakharov." New York Times, March 5, 1974. If the rupture of an arms control measure could be used to disorient the adversary, it follows that Soviet entry into arms control arrangements could also be intended to serve tactical and offensive purposes. Nevertheless it would be unwise to conclude that Khrushchev, even in 1961, viewed arms control exclusively in a tactical light.

2. For an account of the daily work of the regional ideological secretaries that draws attention to their heavy involvement in "health, education and welfare" matters, see Jerry F. Hough, "The Party Apparatchiki," in H. Gordon Skilling and Franklyn Griffiths, eds., Interest Groups in Soviet Politics (Princeton, N.J.: Princeton University Press, 1971), pp. 72-79.

3. Morskoi sbornik, no. 3, 1972, p. 20.

4. Ibid., no. 8, 1972, pp. 22-23. On the failure of Tsarist naval thought and practice, see also ibid., no. 3, 1972, p. 21; no. 4, 1972, pp. 14 and 18-23; and no. 6, 1972, p. 20.

5. Ibid., no. 3, 1972, p. 20.

6. Ibid.

7. He states, "Therefore we, in giving priority to the development of submarine forces, believe we have a need not only for submarines but also for various types of surface ships." Ibid., no. 2, 1973, p. 21. On the characteristics of new surface warships, see ibid., pp. 21 and 22.

8. Ibid., no. 12, 1972, p. 15.

9. Ibid., p. 16 and no. 2, 1972, p. 23.

10. Ibid., no. 2, 1972, pp. 23-25.

11. Ibid., no. 12, 1972, p. 16, and no. 2, 1972, p. 20.
12. Ibid., no. 12, 1972, pp. 19-20.
13. Ibid., p. 20.
14. Ibid.
15. This is the assertion of B. L. Teplinskii, "Mirovoi okean i voennaya strategiya SShA," SShA, no. 10, 1972, p. 24.
16. Ibid.
17. Morskoi sbornik, no. 8, 1972, p. 14.
18. Ibid., no. 12, 1972, p. 18.
19. Ibid., no. 2, 1973, p. 18.
20. Ibid., no. 8, 1972, p. 24.
21. Ibid., no. 12, 1972, p. 18.
22. Ibid.
23. Ibid., no. 2, 1973, pp. 15 and 18.
24. Morskoi sbornik, no. 8, 1972, p. 14.
25. Ibid., no. 12, 1972, p. 18.
26. Pravda, March 6, 1974. Again, naval development and agriculture are noncompetitive in many respects but do compete for investment funds. It is notable that the Soviet Defense minister accompanied Brezhnev on the occasion of his speech announcing the new agricultural program.
27. Steven F. Kime, "The Nuclear Age and the Navies of the Superpowers," unpublished paper presented to the Seminar on Soviet Naval Developments, Halifax, 14-17 October 1973.
28. Michael MccGwire, "Soviet Naval Interests and Intentions in the Carribean," in MccGwire, ed., Soviet Naval Developments: Capability and Context (New York: Praeger Publishers, 1973), p. 482.
29. Barry M. Blechman, "Soviet Interests in Naval Arms Control: Prospects for Disengagement in the Mediterranean," in MccGwire, Soviet Naval Developments, op. cit., p. 525.
30. New York Times, March 3, 1974.
31. Morskoi sbornik, no. 12, 1972, p. 20.
32. J. McConnell and A. Kelly "Superpower Naval Diplomacy in the Indo-Pakistani Crisis" in MccGwire, Soviet Naval Developments, op. cit., pp. 449-52.
33. See "Summary of Discussion in Part II," p. 370 of this volume.
34. On this point, see Gorshkov in Morskoi sbornik, no. 2, 1973, p. 19.
35. Gorshkov observes that comparison in terms of tonnage or hulls will not do and that the relevant criteria are "strength of combat might calculated by a method of mathematical analysis solving multi-criterial problems for different variations of the situation and different combinations of heterogeneous forces and means." Ibid., p. 21. This seems to be a recipe for protracted negotiation at best.
36. Ibid., pp. 13-14.

37. Ibid.

38. United Nations, "Report of the Secretary General to the Committee on the Peaceful Uses of the Seabed and the Ocean Floor Beyond the Limits of National Jurisdiction," GA Report A/AC. 138/73, 12 May 1972, paragraph 9.

39. For the reference to arms racing, see Morskoi sbornik, no. 2, 1973, p. 15.

40. Ibid., p. 16.

41. Ibid., p. 15.

42. See 50 let borby SSSR za razoruzhenie: sbornik dokumentov (Moscow: "Nauka," 1967), pp. 82-90.

43. Compare Kime, op. cit.

44. In another context, this could have been Brezhnev's ploy in reportedly informing East European leaders that rapprochement with the West was a temporary tactic designed to allow the Soviet Union to achieve military supremacy. William R. Frye, Toronto Star, September 23, 1973. If Brezhnev was making this point to Eastern Europe, he was certainly doing the same within the Soviet Union. Alternatively, it is possible that Brezhnev meant what he said. Who knows? Possibly not even Marshal Grechko.

CONTRIBUTORS TO THIS VOLUME

The views expressed in the various chapters are those of the individual authors; they should not be taken as necessarily reflecting the opinions of the institutions to which they are affiliated.

ROBERT P. BERMAN is a defense consultant of the Center for Defense Information, Washington, D.C., which publishes The Defense Monitor.

KEN BOOTH is Assistant Professor in the Department of International Politics, University College of Wales, Aberystwyth, and author of The Military Instrument in Soviet Foreign Policy, 1917-1972 (1974).

NIGEL BRODEUR is a naval Captain in the Canadian Armed Forces and until recently was Commandant of the Maritime Warfare School, Halifax.

BRADFORD DISMUKES is a Professional Staff Member of the Center for Naval Analyses, University of Rochester.

GEORGE S. DRAGNICH is a Middle East/Mediterranean area specialist, currently enrolled in graduate history studies at Georgetown University.

JOHN ERICKSON is Director of Defense Studies, University of Edinburgh, and author of The Soviet High Command (1962) and Soviet Military Power (1971).

ROBERT O. FREEDMAN is Associate Professor of Political Science at Marquette University, and author of Economic Warfare in the Communist Bloc (1970) and a number of articles on Soviet policy and the Middle East.

ROBERT L. FRIEDHEIM is Director of the Law of the Sea Project of the Center for Naval Analyses, University of Rochester, and has written extensively on ocean policy.

MATTHEW P. GALLAGHER is a specialist on Soviet affairs, currently associated with the Foreign Broadcast Information Service, Washington, D.C., and author of The Soviet History of World War II (1963) and Soviet Decision-Making for Defense (1972).

FRANKLYN GRIFFITHS is an Associate Professor in the Department of Political Economy and a member of the Centre for Russian and East European Studies, University of Toronto. He is coeditor of Interest Groups in Soviet Politics (1971).

PHILIP HANSON is Assistant Professor of Economics, Centre for Russian and East European Studies, The University of Birmingham.

JOHN P. HARDT is Senior Specialist in Soviet Economics, Congressional Reference Service of the Library of Congress, Washington, D.C., and editor of Soviet Economic Prospects for the Seventies (1973).

MARY E. JEHN is a staff member of the Center for Naval Analyses, University of Rochester, and a Ph.D. candidate in Soviet Studies at the University of Chicago.

GEOFFREY JUKES is Senior Fellow of the Department of International Relations, and the Research School of Pacific Affairs, Australian National University, and author of The Indian Ocean in Soviet Naval Policy (1972) and The Soviet Union in Asia (1973).

ANNE M. KELLY is a Professional Staff Member of the Center for Naval Analyses, University of Rochester, and a contributor to Soviet Naval Developments (1973).

MALCOLM MACKINTOSH is a Consultant on Soviet affairs with the Institute of Strategic Studies, London. Author of Strategy and Tactics of Soviet Foreign Policy (1964) and Juggernaut: A History of the Soviet Armed Forces (1967).

JOHN McDONNELL is a Research Associate of the Centre for Foreign Policy Studies, Dalhousie University, and is preparing a Ph.D. dissertation on the Soviet "military-industrial complex."

MICHAEL McCGWIRE is Professor of Maritime and Strategic Studies, Dalhousie University, and editor of Soviet Naval Developments (1973).

CHARLES C. PETERSEN is a Technical Staff Member of the Center for Naval Analyses, University of Rochester.

URI RA'ANAN is Professor of International Politics and Chairman of the International Security Studies Committee, Fletcher School of Law and Diplomacy at Tufts University, and concurrently Associate of the Russian Research Center, Harvard University. He is the author of The USSR Arms the Third World (1969).

ALVIN Z. RUBINSTEIN is Professor of Political Science at the University of Pennsylvania and author of Yugoslavia and the Nonaligned World (1970) and The Foreign Policy of the Soviet Union (3d edition, 1972).

MARSHALL D. SHULMAN is Director of the Russian Institute, Columbia University, and author of Stalin's Foreign Policy Reappraised (1963) and Beyond the Cold War, (1966).

OLES M. SMOLANSKY is Professor of International Politics at Lehigh University, and author of The USSR and the Arab East under Khrushchev (1974).

HARLAN ULLMAN is a Lieutenant Commander, U.S.N., and recently completed a doctorate at the Fletcher School of Law and Diplomacy, Tufts University.

PETER H. VIGOR is Director of Soviet Studies at the Royal Military Academy, Sandhurst, and author of A Guide to Marxism and

Its Effects on Soviet Development (1966) and "The Soviet View of War, Peace and Neutrality" (forthcoming).

EDWARD L. WARNER, III, is a Major, USAF, currently attending the Armed Forces Staff College in Norfolk, Virginia. Coeditor of Comparative Defense Policy (1974), and author of several articles on Soviet military affairs.

ROBERT G. WEINLAND is a Professional Staff Member of the Center for Naval Analyses, University of Rochester, and a contributor to Soviet Naval Developments (1973).

RELATED TITLES
Published by
Praeger Special Studies

ARMED FORCES OF THE WORLD: A Reference Handbook
(fourth edition)
    edited by Robert C. Sellers

CURRENT ISSUES IN U.S. DEFENSE POLICY
    Center for Defense Information
    edited by David T. Johnson and
    Barry R. Schneider

FROM THE COLD WAR TO DETENTE
    edited by Peter J. Potichnyj and
    Jane P. Shapiro

SOVIET-ASIAN RELATIONS IN THE 1970s AND BEYOND:
An Interperceptional Study
    Bhabani Sen Gupta

SOVIET NAVAL DEVELOPMENTS: Capability and Context
    edited by Michael MccGwire

SOVIET NAVAL INFLUENCE: Domestic and Foreign
Dimensions
    edited by Michael MccGwire
    and John McDonnell